2. wissenschaftliche Konferenz der Gesellschaft
Deutscher Naturforscher und Ärzte
Schloß Reinhardsbrunn bei Friedrichroda 1964

*Funktionelle und morphologische
Organisation der Zelle*

Sekretion und Exkretion

Springer-Verlag Berlin Heidelberg GmbH 1965

ISBN 978-3-540-03398-1   ISBN 978-3-642-92908-3 (eBook)
DOI 10.1007/978-3-642-92908-3

Vorwort des Herausgebers

Nach dem Erfolg der 1. wissenschaftlichen Konferenz über die *„Funktionelle und morphologische Organisation der Zelle"* (Hrsg. P. KARLSON, Springer-Verlag, Heidelberg 1963) entschloß sich die *Gesellschaft Deutscher Naturforscher und Ärzte* zu einem weiteren Symposium dieser Art unter Beibehaltung des Rahmen-Themas. Diese 2. Veranstaltung, über die das vorliegende Buch berichtet, fand im Anschluß an die Hauptversammlung unserer Gesellschaft in Weimar Mitte Oktober 1964 in Schloß Reinhardsbrunn bei Friedrichroda statt und vereinigte in der Abgeschiedenheit des Thüringer Waldes 4 Tage lang 44 Anatomen, Biochemiker, Biophysiker, Botaniker, Cytologen, Pathologen, Physiologische Chemiker, Physiologen und Zoologen.

Das Thema *„Sekretion und Exkretion"*, bezogen auf die Organisation der Zelle, ermöglichte einen ausgedehnten Gedankenaustausch zwischen den Vertretern dieser sehr verschiedenen biologischen Fachrichtungen, vor allem zwischen Physiologen einerseits und Morphologen andererseits. Eine vollständige stenographische Aufzeichnung des gedanken- und faktenreichen Meinungsaustausches erleichterte die Publikation der lebhaften Gespräche, die sich an die Referate anschlossen. Naturgemäß war eine starke Straffung bzw. Kürzung vor der Drucklegung wünschenswert und notwendig.

Zweck der Veröffentlichung dieses Berichtes ist es, das in den Referaten enthaltene Wissensgut und die in den Debatten aufgetauchten Gedanken einem breiten Kreis von Lesern verschiedener Disziplinen zugänglich zu machen, um damit zu einer fortschreitenden *Zusammenarbeit von Physiologen und Morphologen* auf dem Gebiet der Zellforschung beizutragen.

Organisation und Dokumentation wurden sehr erleichtert durch die verständnisvolle und großzügige finanzielle Unterstützung durch den Vorstand unserer Gesellschaft. Für besondere Hilfe bei der Vorbereitung danke ich dem Generalsekretär, Herrn Prof. H. J. ANTWEILER (Bonn) und für die ausgezeichnete örtliche Organisation und Gastfreundschaft Herrn Prof. G. BRUNS (Jena) und seinen Mitarbeitern Dr. W. ZSCHIESCHE und Dr. S. FRITSCH.

Dr. H. KOMNICK (Bonn) übernahm eine schnelle und sachverständige redaktionelle Überarbeitung der Diskussion, so daß dank der guten Zusammenarbeit mit dem *Springer-Verlag* dieses Buch schon wenige Monate nach der Konferenz erscheinen kann.

Bonn, im Dezember 1964   K. E. WOHLFARTH-BOTTERMANN

Inhaltsverzeichnis

Teilnehmerverzeichnis

Referenten

Doz. Dr. *H. F. Helander*, Dept. of Medicine, University of Oklahoma Medical Centre, 800 NE 13th Street, Oklahoma City, Oklahoma, USA.

Prof. Dr. *L. E. Hokin*, Department of Physiological Chemistry, University of Wisconsin, Madison 6, USA.

Prof. Dr. *H. Holter*, Carlsberg Laboratory, Physiological Department, Copenhagen-Valby, Gl. Carlsbergvej 10, Denmark.

Prof. Dr. *L. C. U. Junqueira*, Faculdade De Medicina Da Universidade De Sao Paulo, Caixa Postal, 2921, Sao Paulo, Brasil.

Dr. *H. Komnick*, Zentral-Laboratorium für angew. Übermikroskopie der Universität Bonn, 53 Bonn, Poppelsdorfer Schloß.

Doz. Dr. *G. Kümmel*, I. Zoologisches Institut der Freien Universität Berlin, 1 Berlin 33, Königin-Luise-Str. 1—3.

Doz. Dr. *W. Schmidt*, Institut für Histologie und Exp. Biologie der Universität München, 8 München 15, Pettenkoferstr. 11.

Prof. Dr. *K. Schmidt-Nielsen*, Department of Zoology, Duke University, Durham/North Carolina, USA.

Dr. *L. Schneider*, Zentral-Laboratorium für angew. Übermikroskopie der Universität Bonn, 53 Bonn, Poppelsdorfer Schloß.

Doz. Dr. *E. Schnepf*, Pflanzenphysiologisches Institut der Universität Göttingen, 34 Göttingen, Untere Karspüle 2.

Dr. *A. Sievers*, Botanisches Institut der Universität Bonn, 53 Bonn, Meckenheimer Allee 170.

Doz. Dr. *H. Sitte*, Elektronenmikroskopische Abteilung der Med. Fak. der Universität des Saarlandes, 665 Homburg/Saar.

Prof. Dr. *A. K. Solomon*, Harvard Medical School, Biophysical Laboratory, Boston 15, Mass., USA.

Prof. Dr. *J. Staubesand*, Anatomisches Institut der Universität Freiburg, 78 Freiburg i. Br., Albertstr. 17.

Doz. Dr. *W. Thoenes*, Pathologisches Institut der Universität Würzburg, 87 Würzburg, Luitpoldkrankenhaus.

Prof. Dr. *K. J. Ullrich*, Physiologisches Institut der Freien Universität Berlin, 1 Berlin 33 (Dahlem), Arnimallee 22.

Doz. Dr. *A. Wessing*, Zoologisches Institut der Universität Bonn, 53 Bonn, Poppelsdorfer Schloß.

Diskussionsteilnehmer

Dr. *D. K. Bhowmick*, Zentral-Laboratorium für angew. Übermikroskopie der Universität Bonn, 53 Bonn, Poppelsdorfer Schloß.

Prof. Dr. *G. Bruns*, Institut f. Mikrobiologie und experimentelle Therapie der DAW, Jena, Beuthenbergstr. 11.

Prof. Dr. Dr. h. c. *Th. Bücher*, Physiologisch-Chemisches Institut der Universität München, 8 München 15, Goethestr. 33.

Doz. Dr. *H. David*, Pathologisches Institut der Humboldt-Universität Berlin, Berlin N 4, Schumannstr. 20/21.

Dr. *J. M. Diamond*, Harvard Medical School, Biophysical Laboratory, Boston 15, Mass., USA.

Prof. Dr. *F. Duspiva*, Zoologisches Institut der Universität Heidelberg, 69 Heidelberg, Berliner Str.

Dr. *S. Fritsch*, Institut für Mikrobiologie und experimentelle Therapie der DAW, Jena, Beuthenbergstr. 11.

Doz. Dr. *H. Füller*, Zoologisches Institut der Universität Jena, Jena, Erbertstr. 1.

Prof. Dr. *M. Gersch*, Zoologisches Institut der Universität Jena, Jena, Erbertstr. 1.

Prof. Dr. *F. Hartmann*, Med. Poliklinik der Universität Marburg, 355 Marburg/Lahn, Robert-Koch-Str. 7a.

Prof. Dr. *G. C. Hirsch*, 34 Göttingen, v. Ossietzkystr. 24.

Dr. *M. R. Hokin*, Department of Physiological Chemistry, University of Wisconsin, Madison 6, USA.

Doz. Dr. *M. Klingenberg*, Physiologisch-Chemisches Institut der Universität Marburg, 355 Marburg/Lahn, Deutschhausstr. 1—2.

Prof. Dr. *Ch. M. A. Kuyper*, Dept. of Chemical Cytology, University of Nijmegen, Nijmegen, Driehuizerweg 200, The Netherlands.

Doz. Dr. *H. Luppa*, Zoologisches Institut der Universität Leipzig, Leipzig C 1, Talstr. 33.

Dipl.-Biol. *L. Meißner*, Institut für Biochemie der Pflanzen der DAW, Halle/Saale, Weinberg-Weg.

Prof. Dr. Dr. h. c. Dr. h. c. *K. Mothes*, Institut für Biochemie der Pflanzen der DAW, Halle/Saale, Weinberg-Weg.

Dr. *E. Müller*, Institut für Biochemie der Pflanzen der DAW, Halle/Saale, Weinberg-Weg.

Doz. Dr. Dr. *W. Niesel*, Physiologisches Institut der Universität Kiel, 23 Kiel, Neue Universität, Rudolf-Höber-Haus.

Dr. *B. Parthier*, Institut für allgem. Botanik der Univ. Halle, Halle/Saale, Am Kirchtor 1.

Dipl. Biol. *C. Schulze*, Institut für Biochemie der Pflanzen der DAW, Halle/Saale, Weinberg-Weg.

Prof. Dr. *P. Sitte*, Botanisches Institut der Universität Heidelberg, Abt. Elektronenmikroskopie, 69 Heidelberg, Hofmeisterweg 4.

Dr. *K. F. Springer*, 69 Heidelberg, Neuenheimer Landstraße 28—30.

Prof. Dr. *D. C. Tosteson*, Dept. of Physiology and Pharmacology, Duke University, Medical Center, Durham/North Carolina, USA.

Prof. Dr. *K. E. Wohlfarth-Bottermann*, Zentral-Laboratorium für angew. Übermikroskopie der Universität Bonn, 53 Bonn, Poppelsdorfer Schloß.

Dr. *R. Wollgiehn*, Institut f. Biochemie der Pflanzen der DAW, Halle/Saale, Weinberg-Weg.

Doz. Dr. *W. Zschiesche*, Institut f. Mikrobiologie und experimentelle Therapie der DAW, Jena, Beuthenbergstr. 11.

Begrüßung

K. Mothes

Meine Damen und Herren! Ich habe die besondere Freude, in meiner Eigenschaft als Vorsitzender der Gesellschaft Deutscher Naturforscher und Ärzte Sie in Schloß Reinhardsbrunn herzlich willkommen heißen zu können. Ich danke Ihnen, daß Sie unserer Einladung Folge geleistet haben.

Diese wissenschaftlichen Konferenzen unserer Gesellschaft sind eine junge Einrichtung. Vor zwei Jahren hatten wir zum ersten Mal eine Konferenz zu dem Thema „Funktionelle und morphologische Organisation der Zelle". Diese Veranstaltung war möglich geworden durch eine großherzige Stiftung unseres früheren Schatzmeisters Prof. HABERLAND, dessen wir hier in Verehrung gedenken. Und diese Stiftung macht es uns möglich, die Konferenzen fortzusetzen.

Nun umfaßt natürlich der Rahmen einer Gesellschaft Deutscher Naturforscher und Ärzte ein riesig weites Gebiet der Naturwissenschaft als Betätigungsfeld. Wir haben aber diese Bereiche unserer Gesellschaft der Naturforscher und Ärzte immer so verstanden, daß nicht ein beliebiges Vielerlei von angewandter Wissenschaft mit der eigentlichen Naturforschung auf unseren Tagungen in Berührung kommen sollte — dann müßten wir sagen: deutsche Naturforscher, Ärzte, Techniker, Landwirte usw. —, sondern es ist die Berührung zwischen Naturforschung und medizinischer Wissenschaft in unserer Gesellschaft ein sehr altes, aus ihrer Entstehung heraus verständliches Anliegen, wobei ebenso eine wissenschaftliche Durchdringung der medizinischen Berufe im Auge gelassen wird wie auch andererseits eine Anregung, die durch das Beobachtungsfeld der Ärzte der Naturwissenschaft zukommt. Wir haben also insbesondere die biologischen Themen im Auge, wie eben in diesem Symposium „Funktionelle und morphologische Organisation der Zelle", und kämen heute zum Beginn des zweiten Symposiums dieser Art: „Sekretion und Exkretion".

Die Gesellschaft hat Herrn Prof. WOHLFARTH-BOTTERMANN beauftragt, diese Konferenz vorzubereiten und zu leiten, und ich möchte Herrn WOHLFARTH-BOTTERMANN sehr herzlich für die sehr intensive Vorarbeit danken. Die Vorträge sind Ihnen ja in Fotokopien zugegangen, so daß eine gewisse Vorbereitung auf das Programm und auf die Diskussion bereits gegeben ist. Ich glaube, wir werden von dieser Mühe, die sich Herr WOHLFARTH-BOTTERMANN gemacht hat, den größten Vorteil hier im Laufe unserer Tagung spüren. Ich möchte nur hoffen, daß die ausländischen Gäste sich in diesem schön gelegenen Schloß wohlfühlen.

Dann wollen wir gleich in medias res gehen, und ich darf die Konferenz für eröffnet erklären.

I. Sekretion

Einleitung

G. C. Hirsch

Meine Damen und Herren! Ich glaube, ich bin der Älteste unter Ihnen, und das ist ein schweres Los, denn dann fallen einem alle Ehrenaufgaben zu, und je älter man wird, um so bescheidener wird man. Aber ich kann mich dem nicht entziehen. Wenn nun der Organisator, Herr WOHLFARTH-BOTTERMANN, mich gebeten hat, ein paar einleitende Worte über Sekretion und Exkretion zu sagen, so will ich das tun, so gut ich es kann.

Ich möchte versuchen, zwischen Sekretion und Exkretion dadurch zu unterscheiden, daß ich beide einander gegenüberstelle. Beide zerfallen in drei Phasen: in die Ingestion, in die intracelluläre Verarbeitung und in die Extrusion. Aber in einem Punkt können wir — glaube ich — auch heute noch festhalten an der alten Definition, die ich vor 50 Jahren aufzustellen versucht habe, daß die Sekretion in der Mitte zwischen Intrusion und Extrusion einen neuen Stoff aufbaut. Wir werden uns immer am einfachsten verständigen, wenn wir Beispiele aus unserem täglichen Leben nehmen. Dann wäre also die Sekretion eine Fabrik, welche Rohmaterialien, Energie und viele andere Dinge aufnimmt und dann zu typischen, nur dieser Fabrik zugeordneten Produkten verarbeitet. Ich glaube, daß das heute noch durchaus gilt. Das wäre vielleicht die geeignete Definition über Sekretion. Wir werden ja in diesen Tagen hören, wie außerordentlich verschieden das verlaufen kann.

Und nun zur Exkretion. Ursprünglich habe ich geglaubt, man sollte eine Exkretion dadurch kennzeichnen, daß man sagt: Es werden Stoffe ausgeschieden, die nicht in den betreffenden Zellen bereitet worden sind, so daß also gerade dieser synthetische Mittelteil wegfällt. Das gilt bis zu einem gewissen Grade auch noch von unserer Niere, die ja — wenn ich richtig unterrichtet bin — nur Hippursäure in ihren Zellen selber aufbaut, alles übrige empfängt, z. T. sammelt, ausscheidet, rückresorbiert usw. Aber die Schwierigkeit bei der Exkretion liegt — glaube ich — heute darin, daß auch bei der Sekretion eine große Fülle von Stoffen nicht mehr ausschließlich in der betreffenden Zelle aufgebaut und ausgeschieden wird, sondern daß gleichzeitig Nebenprodukte entstehen, welche die Zelle nur passieren.

Ich bin hierhergekommen, um von Ihnen zu lernen, nicht, um zu lehren. Infolgedessen kann ich jetzt ruhig aufhören und darf Herrn Dr. HELANDER bitten, seinen Vortrag zu beginnen.

Morphology of Animal Secretory Gland Cells

by

HERBERT F. HELANDER, Gothenburg

With 10 figures

The ability of gland cells to produce a secretion is associated with certain characteristic features in their morphology. In this paper, confined to mammalian gland cells, only those structures will be discussed, which are believed to be more or less directly involved in the secretory processes.

The process of secretion includes three major steps, viz. the *ingestion* of raw material, the *synthesis* of the secretory substance, and the *extrusion* of the secretory product.

A. Ingestion

The ingestion of raw material into the cytoplasm of the gland cells is most often inaccessible for morphological studies. The minute particles of ingested substances are not visible in the electron microscope and the relatively simple chemical composition of this raw material does not result in any typical staining reactions, should histochemical methods be used.

The distance between the basal cell surface and the nearest blood capillary is usually in the range of 0.1 μ (Fig. 1). This space contains two basement membranes, each a few hundred Å thick. It must be assumed that a very large quantity of a number of different substances moves across this space from the capillaries to the gland cells, but also in the opposite direction. Yet, in the endothelial cells the only morphological indications of this transport are the pinocytotic vesicles [49], and the significance of these vesicles is still largely unknown.

Though little is known about the morphology and biochemistry of the basement membrane, it must be assumed that it acts as some sort of filter and thus obstructs or even prevents the passage of certain substances (cf. [23]).

Some of the secretory gland cells noted for their water transport, exhibit invaginations of the basal plasma membrane (i. e. cell membrane), e. g. the gastric parietal cell [28], the ependymal cells of the chorioid plexus [41], the secretory cells of the sweat glands [12], and of the salivary and lacrimal glands [66]. This arrangement probably facilitates the ingestion of substances by increasing the resorptive surface area.

Although the intercellular space between neighbouring gland cells is usually very narrow (in the order of 100 Å), one cannot exclude the possibility that ingestion of raw material also occurs on the lateral cell surface. Sometimes invaginations or microvilli are found in these regions too [58], and rows of what appear to be pinocytotic vesicles in adjacent parts of the cytoplasm [28] are suggestive of an ingestion. On the whole, however, such vesicles are observed rather seldom in the basal and lateral parts of the cytoplasm of secretory cells.

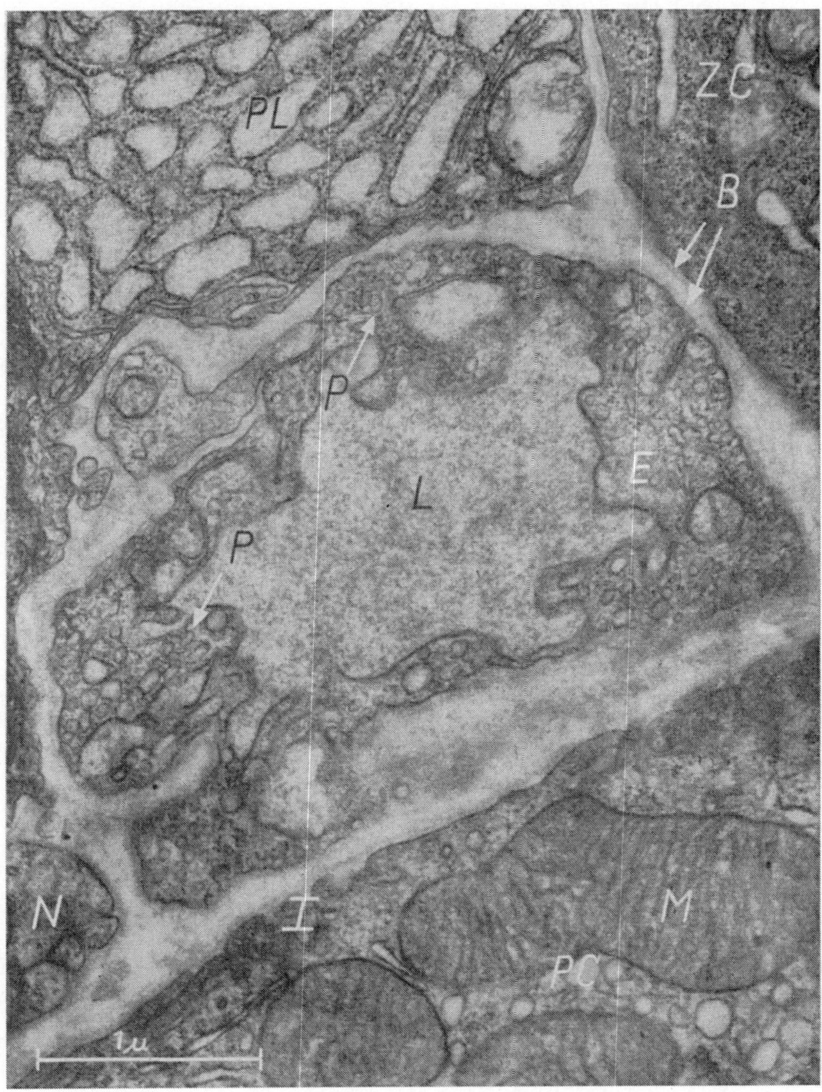

Fig.1.Capillary in the gastric mucosa. The endothelial cell (*E*) which surrounds the capillary lumen (*L*), is rather thin at some places. Numerous pinocytotic vesicles (*P*) are seen in the endothelial cell cytoplasm. The interstitial space between the endothelial cell and the gland cells contains two basement membranes (*B*). The basal cell surface of the parietal cell (*PC*) exhibits minor irregularities (*I*). *M*, mitochondrion; *N*, unmyelinated nerve; *PL*, plasma cell; *ZC*, zymogen cell. Magnification: 30,000 ×

1*

The diameter of the pinocytotic vesicles is in the order of 400 Å. Thus, theoretically, particles of a few hundred Å in diameter could be ingested into the gland cells. How large the ingested particles really may be is not known.

It must be assumed though, that a great proportion of the ingested substances do not pass into the cells enclosed in pinocytotic vesicles. These substances, then, enter the gland cells amorphously across an apparently uninterupted, complete plasma membrane.

The principal structure of the plasma membrane was predicted by Danielli and Davson nearly thirty years ago [15]. These authors, who based their theories on permeability properties, suggested that the plasma membrane was composed of a lipid film lying between two protein films.

The study of various cell membranes in the electron microscope, and the information from X-ray diffraction studies on the myelin sheath, later on has resulted in the unit membrane hypothesis [57]. According to this, the plasma membrane and the various cytoplasmic membranes have the same structure: a 75 Å thick, symmetrical, triple-layered, lipo-protein membrane. The occasional observations of α-cytomembranes (cf. p. 5) being continuous with the plasma membrane [21] support this theory. By means of these connections and at times, also by the incorporation of the surface membrane of pinocytotic vesicles and secretory granules (cf. p. 4), it is believed that the plasma membrane and the various cytoplasmic membranes are constantly exchanging membranous material. This process is called *membrane flow* [4].

However, the results of several recent electron microscopical studies at high resolution, do not conform with the unit membrane theory. The different membranes are not of the same thickness or substructure, and the plasma membrane is not symmetrical [18, 20, 63, 64]. It has been suggested that the two protein layers of the plasma membrane may be of different structure and composition [64]. Still, the unit membrane theory can be valid, if it is assumed that the membrane flow also involves a slight remodelling of the membranes.

The structure of the plasma membrane varies between different types of cells and also between different parts of the circumference of the same cell (cf. p. 15). Different ingestive properties of secretory gland cells may probably be correlated to differences in plasma membrane structure.

The accumulation of ATP-ase and other enzymes splitting high energy compounds at the basal plasma membrane indicates that the transport across this membrane is at least partly an active process [1]. Similarly, the high ATP-ase activity localized to the pinocytotic vesicles of the capillaries, indicates an active transport across the endothelial cells [40].

B. Synthesis

The secretory gland cells may be divided into two different groups with respect to one morphological quality: those with and those without secretory granules.

I. Gland cells containing secretory granules

1. Gland cells mainly synthesizing proteins

Most secretory gland cells belong to this group. Morphological studies on these cells have been very fruitful. In particular the observations on the pancreatic acinar cells have been rewarding. The correlation of functional and morphological data, has here resulted in several comprehensive theories on the synthesis of protein secretory products (for a summary see [55]). Some of these theories are supported by very convincing evidence, but thorough investigation of all steps in the synthetic process is as yet, far from complete.

Several of these steps have been studied in the light microscope, and much information has been obtained, particularly from histochemical and autoradiographic investigations. During the last decade the electron microscope has become a very important instrument for the investigation of the process of synthesis in gland cells. Recently, it has become possible to combine electron microscopy with autoradiography, and the results from studies utilizing this technique on the pancreas, have shed light on the mechanism of synthesis in this gland. The most valuable information has come from biochemical studies of electron microscopically identified subcellular fractions (cf. e. g. [54]). However, these will only be briefly discussed in this paper.

Since the most thorough investigations have been performed on the pancreatic acinar cells, some of the relevant morphological findings in this type of cell will serve as a basis for the following discussion (see text-fig. 1). The figures, however, are taken mostly from gastric zymogen cells, which are very similar in appearance to the pancreatic acinar cells.

The basal part of these cells (Figs. 2, 3) is occupied by a large number of long, parallel membranes, which are referred to as *α-cytomembranes* [62] or rough-surfaced membranes of the endoplasmic reticulum [51]. These are disposed in pairs, and at their ends each membrane is fused with the nearest membrane of the adjacent membrane pair, thereby enclosing flat cytoplasmic regions or *cisternae*. The internal surface of the membranes, facing the cisternae, is smooth, whereas the external surface carries a very large number of dense particles with a diameter of about 150 Å. These particles contain a large proportion of RNA [53] in the form of a protein (RNP) and are therefore called *ribosomes* or RNP particles. This high concentration of RNA causes the basophilia observed by light microscopists in the basal part of protein-synthesizing gland cells (cf. [6, 10]). Not all the ribosomes are attached to the α-cytomembranes: some of them are found free in the cytoplasm, often in clusters of 5 or more. These clusters are called *polysomes* and are believed to operate as functional units [67].

Continuity is often found between the α-cytomembranes of a pair, and by this means thin-necked communications are created between the different cisternae. As pointed out earlier the α-cytomembranes are occasionally also connected with the plasma membrane (cf. p. 4). Similar connections are frequently observed between the α-cytomembranes

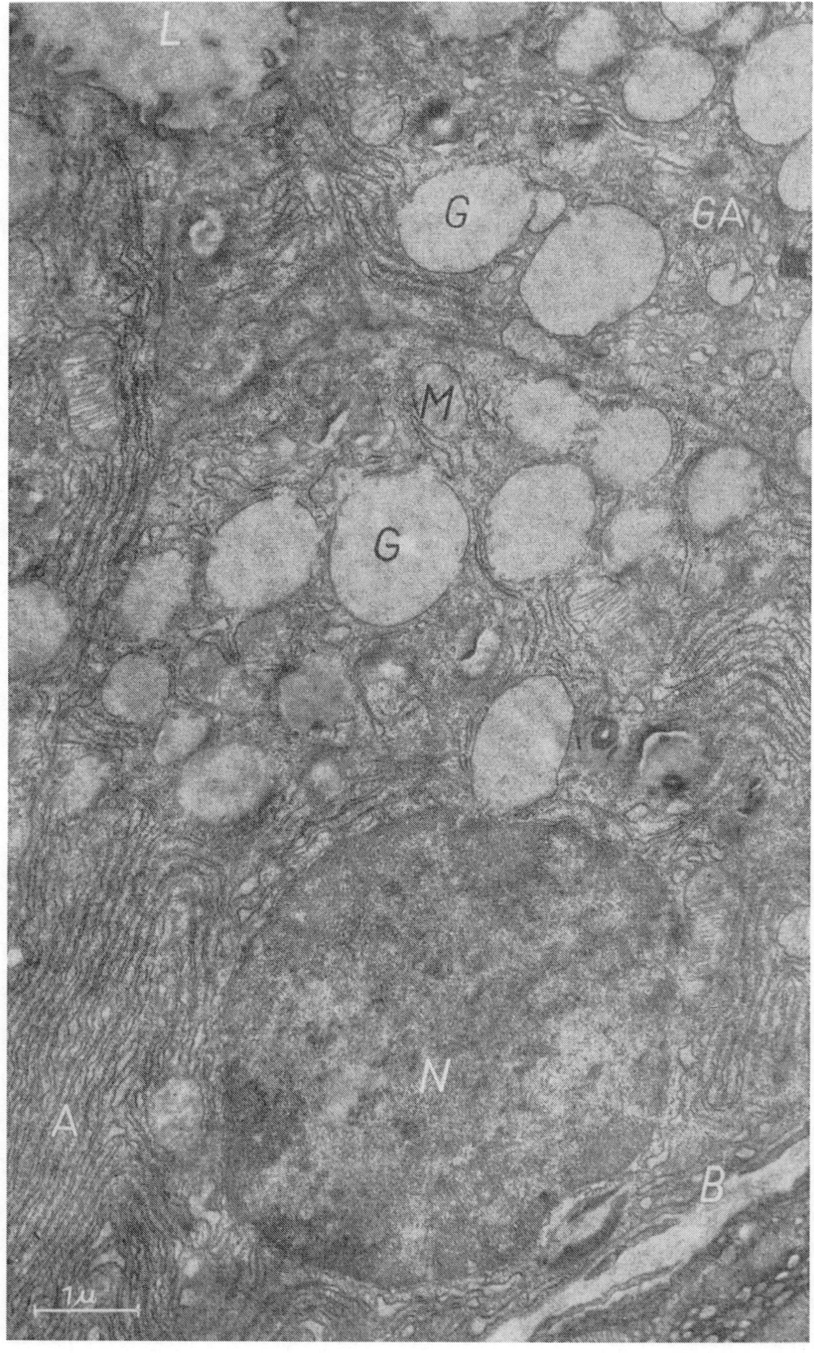

Fig. 2. Survey of gastric zymogen cell. Numerous α-cytomembranes (*A*) are observed in the basal part of the cell. Most of the zymogen granules (*G*) are found in the apical part of the cell. *B*, basal plasma membrane with basement membrane; *GA*, Golgi apparatus; *L*, glandular lumen; *M*, mitochondrion; *N*, nucleus. Magnification: 13,000 ×

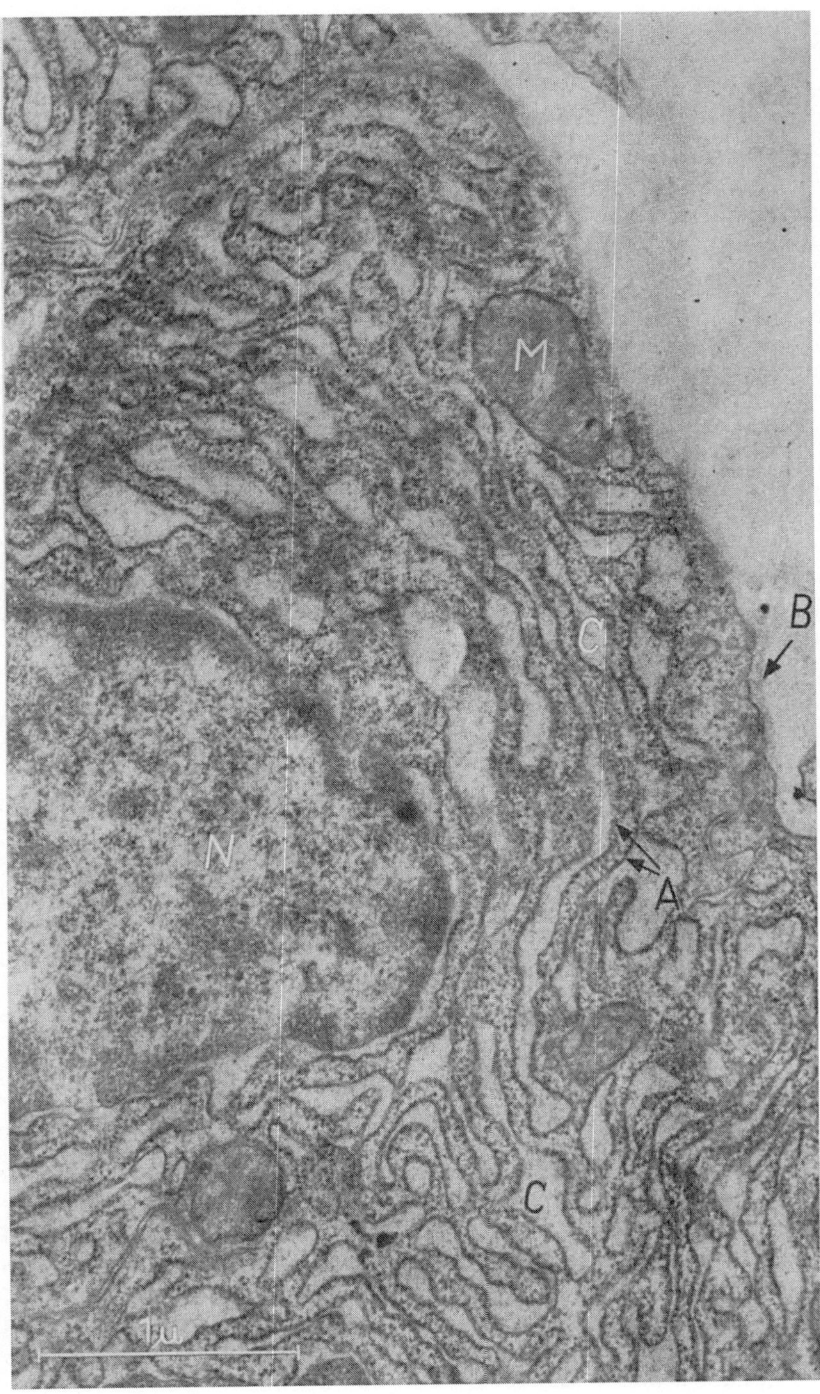

Fig. 3. Basal part of gastric zymogen cell from a recently fed mouse. This part of the cytoplasm is occupied by a large number of α-cytomembranes disposed in pairs (A). Numerous
minute dense particles — ribosomes — are observed attached to the membranes and between the membrane pairs. Irregularly formed cisternae (C) are enclosed by the α-cytomembranes. In stimulated cells the cisternae are often more dilated than in un-stimulated
ones. The nucleus (N) is also surrounded by an α-cytomembrane. B, basement membrane;
M, mitochondrion. Magnification: 33,000 ×

and the nuclear envelope [69]. In the same way α-cytomembranes are believed to be continuous with the smooth-surfaced membranes of the cytoplasm, at least temporarily (cf. p. 9). Thus, the α-cytomembranes and the smooth-surfaced cytoplasmic membranes constitute a single unit system: the *endoplasmic reticulum* [51].

The basal part of the cytoplasm is the locus of the first step in the synthesis. Here, radioactive amino acids accumulate within a few minutes after injection, as has been demonstrated in autoradiographic studies [9, 68]. In vitro experiments have shown that the ribosomes are the organelles where polypeptide chains or proteins are produced from the amino acids (cf. [11]).

In the next step of the elaboration of secretory material, the newly synthesized products are transferred from the ribosomes to the cisternae of the endoplasmic reticulum. Using the electron microscope, small granules without any surface membrane can be observed within the cisternae in a few types of gland cells (cf. e. g. [50]). These intra-cisternal granules, which have been studied most extensively in the pancreatic acinar cells of the guinea pig, were previously believed to contain secretory products [50, 54, 60, 61]. However, this view has recently been abandoned, as these granules showed no activity at any time in autoradiographic investigations designed to follow the protein synthesis [9]. The true nature of the pancreatic intra-cisternal granules remains obscure.

Still, the autoradiographic studies seem to indicate that the secretory products accumulate in the cisternae, although in an amorphous condition [9]. The presence of dilated cisternae with a moderately dense contents in some other protein-synthesizing gland cells is consistent with this observation (Fig. 3) [8, 28, 29, 70].

The *Golgi apparatus* (Figs. 2, 4) consists of a varying number of smooth-surfaced membranes, each enclosing a flat, irregularly formed vacuole with a content of varying density. Peripherally to these vacuoles, numerous vesicles are observed. Small secretory granules are often found in the neighbourhood of the Golgi apparatus. The partially filled Golgi vacuoles or "condensing vacuoles" [9] may be indistinguishable from small secretory granules.

It is believed that the Golgi apparatus is engaged in the third step of the synthesis: the concentration of the secretory substances to secretory granules.

In autoradiographic studies on exocrine pancreas, most of the activity has been traced over the Golgi apparatus 20 to 30 minutes after an injection of a radioactive amino acid. After four hours, the activity has been recorded predominantly over the secretory granules in the apical part of the cytoplasm [9, 68]. In similar studies on gastro-intestinal goblet cells the incorporation of inorganic sulphate in the mucous secretion has been followed. In these cells the activity appeared to accumulate in the Golgi region after one to three hours. After four to six hours the activity was concentrated to most apical part of the cytoplasm which is loaded with secretory granules [34].

Yet the transfer path of secretory substances from the cisternae of the endoplasmic reticulum to the Golgi apparatus, and their fate at this place, remains an enigma. In a few instances continuity has been observed between the α-cytomembranes and the membranes of the Golgi apparatus ([51], Fig. 4). In these cases a direct filling of the Golgi vacuoles from

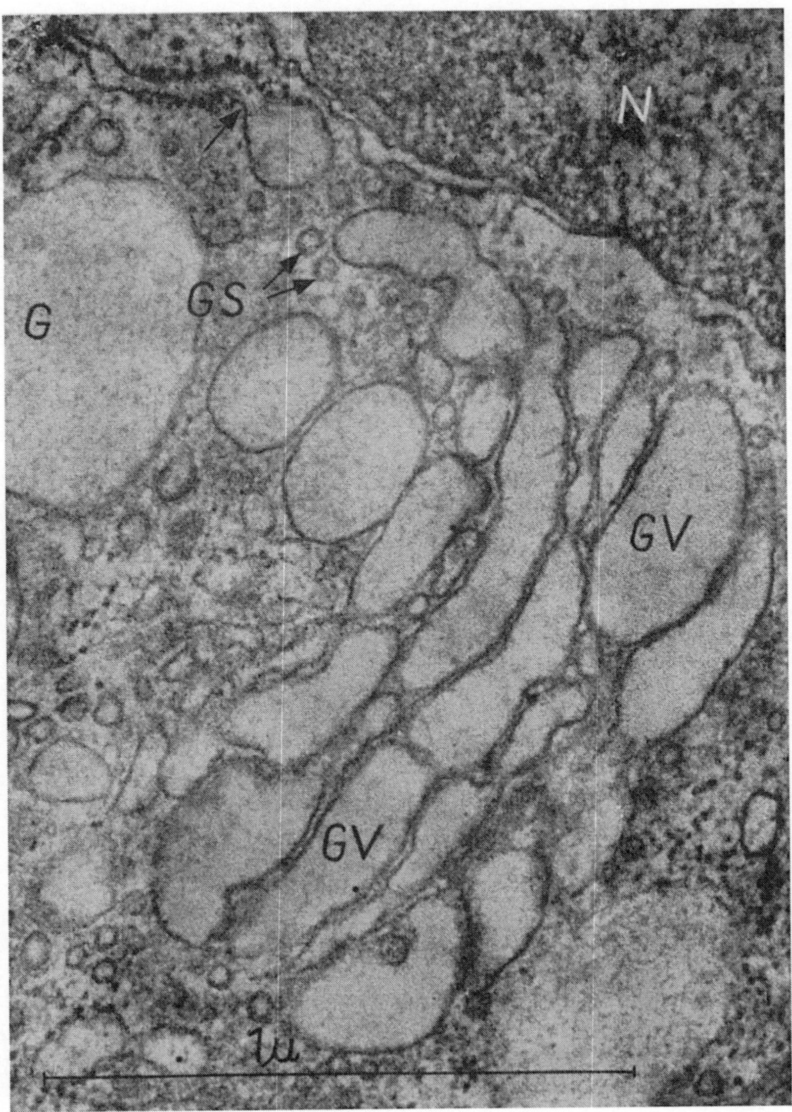

Fig. 4. Golgi apparatus of gastric mucoid cell. Some dilated Golgi vacuoles (*GV*) are seen, and peripherally to these numerous small Golgi vesicles (*GS*) are found. At the arrow it appears that a smooth-surfaced vacuole is being pinched off from a rough-surfaced cisterna. *G*, secretory granule; *N*, nucleus. Magnification: 77,000 ×

the cisternae seems probable. But in the great majority of gland cells there are no such obvious morphological signs of filling.

It has been postulated that small vesicles are pinched off from the cisternae of the endoplasmic reticulum. These vesicles, which contain newly synthesized protein, are then transferred to the Golgi region,

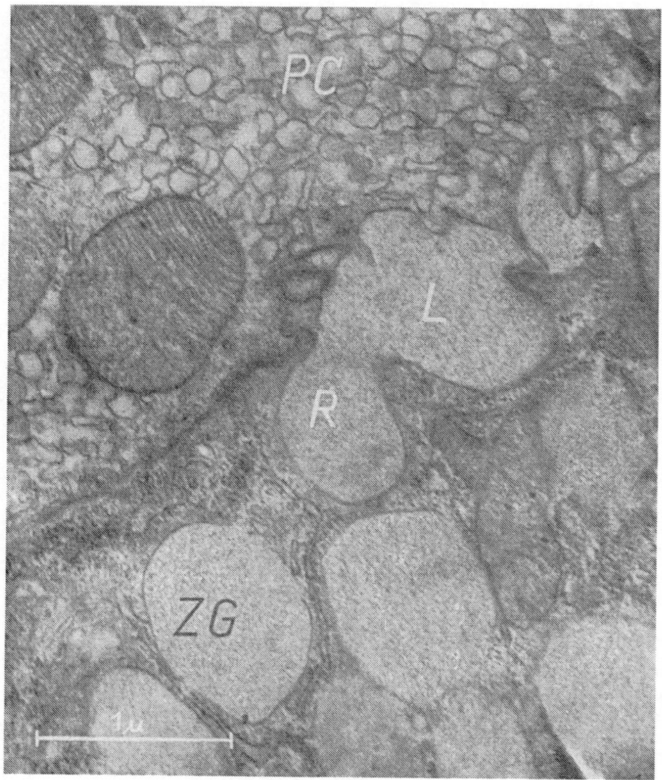

Fig. 5. Apical part of gastric zymogen cell. Some zymogen (= secretory) granules (*ZG*) are present. *R* probably represents a zymogen granule releasing its content into the glandular lumen (*L*). *PC*, parietal cell. Magnification: 26,000 ×

where they are called Golgi vesicles. The vesicles may then empty their contents into the Golgi vacuoles, which ultimately become secretory granules [*9, 30, 72*].

After stimulation of the secretion, the extension of the Golgi apparatus increases in some types of cells [*28, 29, 71*]. Usually the Golgi vacuoles become larger and more numerous. The Golgi vesicles also increase in number.

The last step in the manufacture of protein secretions is the storage, which is in the form of secretory granules. The majority of these granules which measure from 0.3—1.5 μ are found in the apical part of the cytoplasm (Fig. 2). In gastric zymogen cells their mean size is reduced after stimulation [29]. It seems very probable that the largest granules are those which are extruded from the cells primarily, and that the newly formed granules are smaller in size.

Secretory granules within the same cell may display different densities in the electron microscope. This may be taken as an indication of a concentration or "maturation" process of their contents [9, 28, 44, 72].

The conversion of mitochondria into secretory granules has been postulated by several authors (cf. [37]). However, electron microscopical studies on adequately preserved tissues have failed to confirm this theory.

By counting the number of silver grains per unit area in autoradiograms, WARSHAWSKY et al. [68] could measure the radioactivity in different parts of pancreatic acinar cells at several different intervals, after an injection of radioactive amino acids. On the basis of these investigations it was estimated, that the mean life span of newly synthesized "exportable" protein is about 50 minutes in this type of cell. This agrees well with the results of JUNQUEIRA et al. [35] who measured the radioactivity in pancreatic juice after an injection of a radioactive amino acid. No appreciable amount of radioactivity was observed until 50 minutes after the injection.

2. Gland cells mainly producing organic non-protein substances

The large-scale production of *lipids* which occurs in the mammary gland has been studied electron microscopically [2]. Lipid droplets first appear in the basal part of the cytoplasm. As they move towards the apical cell surface they increase in size and are sometimes larger than the nucleus, when they are extruded from the cell. Little is known about how these lipids are produced. It has been proposed that the ribosomes are the sites of fatty acid synthesis, and the mitochondria is the place where the fatty acids combine with alcohols to become fats [26]. It appears that the Golgi apparatus does not take part in the formation on lipid granules [2].

Also, some endocrine cells with rather different synthesizing capacities belong to this group. *Biogenic amines* are produced by enterochromaffine cells. Many of these are granulated, such as the chromaffine cells of the adrenal medulla [17], the pancreatic islet cells [38], and the argyrophil cells of the gastrointestinal tract [28] (Fig. 9). In all these cells the synthesis of the secretory products appears to proceed in the same way as in the protein-synthesizing gland cells.

II. Gland cells devoid of secretory granules

This group comprises a number of cells which are believed mainly to *secrete water and electrolytes*. Examples of these cells in mammals are the non-pigmented cells of the ciliary body, the ependymal cells of the choroid plexus, and the gastric parietal cells. Within this group of secretory cells the gastric parietal cells have been studied most extensively (cf. [28, 29, 33]). Therefore, this cell will supply some of the morphological and biochemical data for the following discussion.

These cells do not exhibit the histological staining properties of RNA-containing cells, and in the electron microscope the number of ribosomes is rather low. The Golgi apparatus is very small, and no secretory granules are present. *Mitochondria* are seen in great numbers, and the cytoplasm often contains a large number of *vacuoles* (Fig. 6).

The *apical cell surface* of many of the cells in this group exhibits gross irregularities, mainly in the form of numerous microvilli. In this way the surface area in contact with the glandular lumen becomes very large. In the parietal cells of the gastric mucosa, the area in functional contact with the lumen is made still much greater by means of intracellular canaliculi which open into the glandular lumen.

Obviously the morphology of the cells in this group points to a different mechanism of synthesis than that in the cells which produce proteins.

The presence of a very large number of mitochondria indicates a high rate of oxidative metabolism, and in the parietal cells, the histochemical demonstration of great activities of succinic dehydrogenase [48] and pyridine nucleotides [7] supports this view.

The role of the vacuoles is not clear. It has been suggested, that they serve as temporary receptacles for secretory substances [36], but this has not been proved. The small Golgi apparatus provides no clues about how the secretory products are made.

Which cytoplasmic structure then, is responsible for the production of secretory material in these cells? In the case of the parietal cells, several authors in the past have suggested, that, the hydrochloric acid is formed at, or very near, the plasma membrane of the intracellular canaliculi (cf. [32]). The production of the secretory substances would then be immediately associated with the extrusion of these products. In this case it is evident that great activities of certain enzymes would be expected at, or very near, the canalicular plasma membrane.

Unfortunately no studies have yet been performed on the localization of enzymes in the parietal cells. However, in other glands such as the liver, the kidney, and the intestinal mucosa, great activities of phosphatases (including ATP-ase) have been demonstrated at the apical cell surface (cf. [13]). It seems very probable that a similar concentration of certain enzymes exists also at the apical plasma membrane of e. g. the parietal cells. The intracellular canaliculi and the microvilli could be a means for these cells to increase their "secretory surface", so that a sufficient quantity of enzymes could be lodged.

Endocrine *steroid-producing cells* of the adrenal cortex and the gonads also belong to this group. Ultrastructurally these cells are distinguished by their mitochondria, whose inner membranes often are arranged as tubules or saccules, instead of parallel membrane pairs as in other gland cells [3, 73]. Some of these cells are heavily loaded with lipid droplets. A multitude of vacuoles constitute another characteristic cytoplasmic component. The Golgi apparatus is comparatively small.

It has been proposed that the mitochondria of the adrenal cortex are more or less directly involved in the synthesis of steroids [27, 43]. It is possible that the lipid droplets contain raw material.

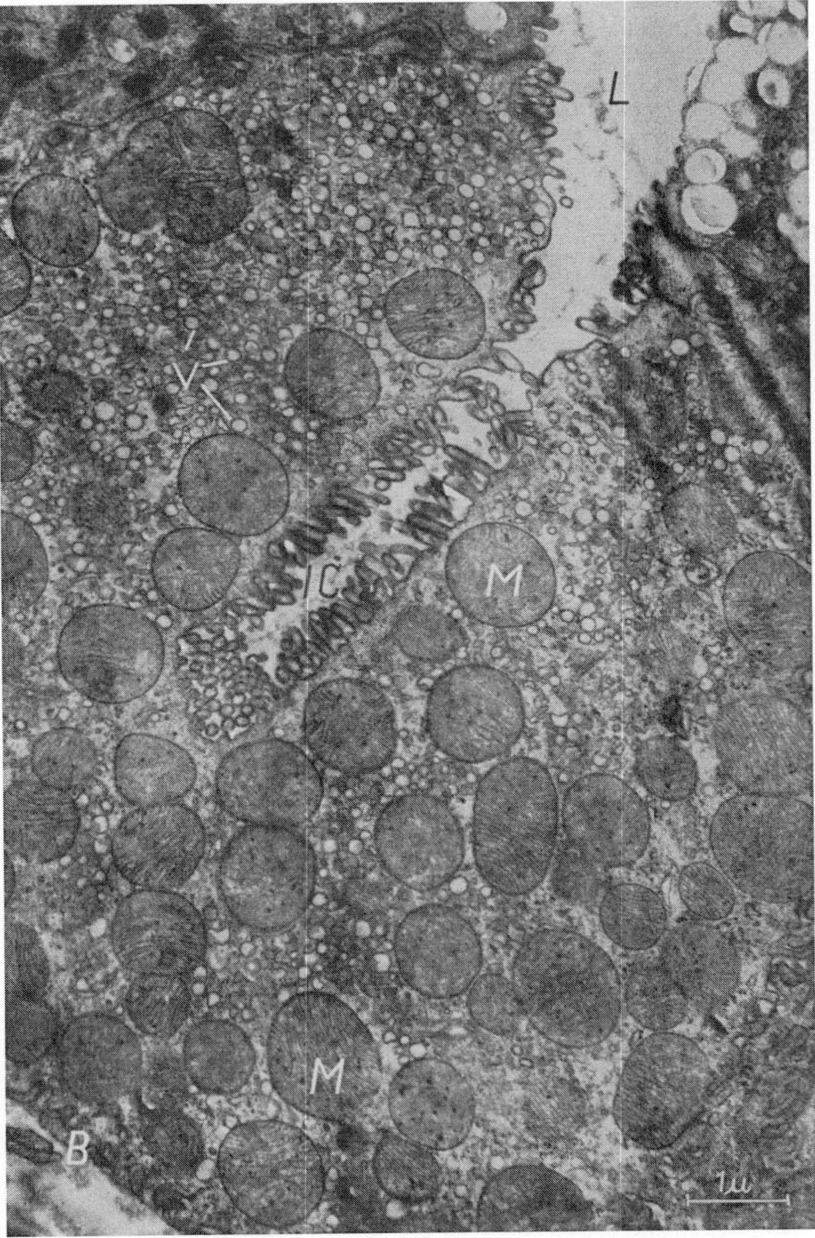

Fig. 6. Gastric parietal cell with an intracellular canaliculus (*IC*) which opens into the glandular lumen (*L*). Numerous microvilli project into the canaliculus. The cytoplasm contains a large number of vacuoles (*V*) and mitochondria (*M*). No secretory granules are observed in this type of cell. The nucleus is not included in the section. *B*, basal plasma membrane. Magnification: 13,000 ×

C. Extrusion

I. Exocrine cells

For a long time exocrine gland cells have been divided into three different groups, with respect to how the secretory products are discharged. The cells of the first group are called *holocrine cells*. In these the secretory products accumulate within the cytoplasm, and later leave the gland as the cells are desquamated.

In the *merocrine cells*, no part of the cytoplasm is lost during the discharge of secretory products.

The secretory substances of the *apocrine cells*, finally, are said to be extruded together with some parts of the cytoplasm.

1. Holocrine secretion

The holocrine type of secretion can be observed in a few types of glands, the classical example being the sebaceous glands of the skin. In other places the holocrine type of secretion may be less evident. However, by means of autoradiography a high rate of cell turn-over can be demonstrated in several glands. The mucoid cells of the gastric mucosa and the epithelial cells of the seminal vesicles belong to this group [42]. Initially these cells are probably merocrine, but after a comparatively short time they are desquamated and thus behave as holocrine gland cells.

2. Merocrine secretion

Most gland cells are believed to belong to this group [14], and many of them contain secretory granules. In these cells, part of the surface membrane of the secretory granules merges with the apical plasma membrane. The joint membrane disappears, and the content of the granules may flow through the orifice into the glandular lumen (Fig. 5). The remaining membranous material is probably incorporated into the apical plasma membrane and thus contributes to the "membrane flow" (cf. p. 4). This mechanism has been observed e. g. in the pancreatic acinar cells [52], in the gastric zymogen cells [29], and the oviduct epithelium [5]. This type of merocrine secretion may be called *granulocrine*.

However, in several types of granulated secretory cells and in all the non-granulated gland cells, there are no such obvious indications of discharge of secretory products. Consequently another release mechanism may be postulated for these cells. It is believed that the secretory substances, passively or actively, pass through the apical plasma membrane in these cases. This type of merocrine secretion may be called *eccrine* [59].

The mechanism of the eccrine extrusion is poorly understood. It appears that the conditions are similar to those of ingestion (cf. p. 2), and again the attention must be focused on the plasma membrane. The accumulation of ATP-ase and/or other enzymes at the apical plasma membrane (cf. p. 12) may be regarded as an indication of a "secretory pump" which transports ions and/or other small particles across the plasma membrane.

After stimulation of the secretory activity in certain gland cells (e. g. pancreas) the turnover of phospholipids is greatly increased. It has been suggested that these phospholipids, which are normal constituents of the cytoplasmic membranes and the plasma membrane, participate

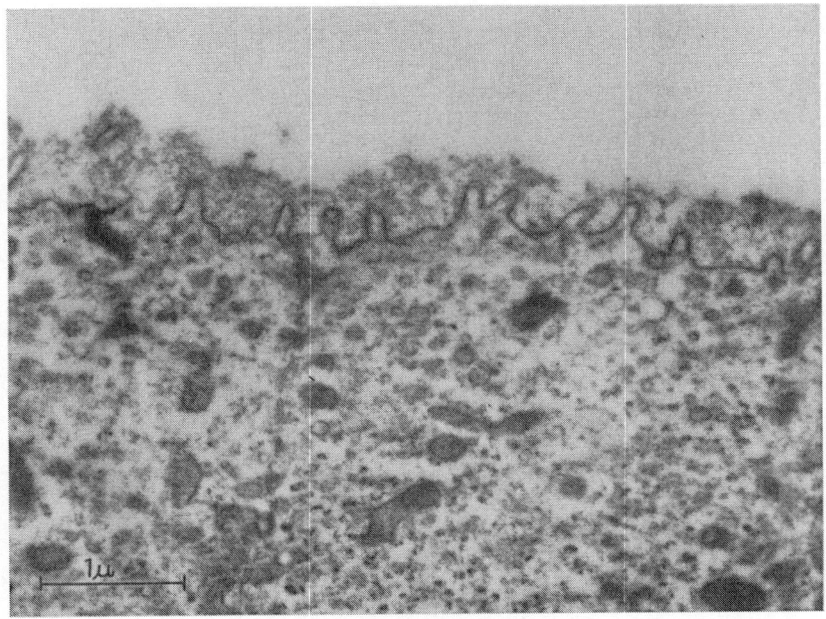

Fig. 7. Apical part of uterine surface epithelial cells. A number of microvilli project from the apical cell surface. These microvilli are covered by an indistinctly defined "furry coat." Magnification: 19,000 × (Courtesy of Dr. T. ZELANDER)

in the transport of certain enzymes across the α-cytomembrane in the second step of the synthesis (cf. p. 8). Similarly there is evidence indicating that phospholipids in some way are engaged in the transport of certain electrolytes across the plasma membrane [31].

The plasma membrane of the apical cell surface is usually thicker than that of the other cell surfaces [28, 74], although the reverse is sometimes the case [19]. The electron microscope, which may achieve a resolution of about 10 Å, has not revealed any pores in the plasma membrane. Neither have any definite alterations been observed in the ultrastructure of the plasma membrane between stimulated and unstimulated cells.

All cells in contact with a gland lumen possess more or less typical microvilli which project from the apical cell surface. Sometimes the luminal surface of the apical plasma membrane, appears to be coated with an indistinctly defined substance forming a layer about 100—300 Å thick (Figs. 7, 8). This has been demonstrated on e. g. the uterine surface epithelium [47], the oviduct secretory cells [24], the pancreatic acinar cells [64], and the zymogen and mucoid cells of the gastric mucosa [28, 33]. As a rule this coating has a filamentous or furry appearance. These strands may represent secretory material immediately before its definite release into the glandular lumen [56].

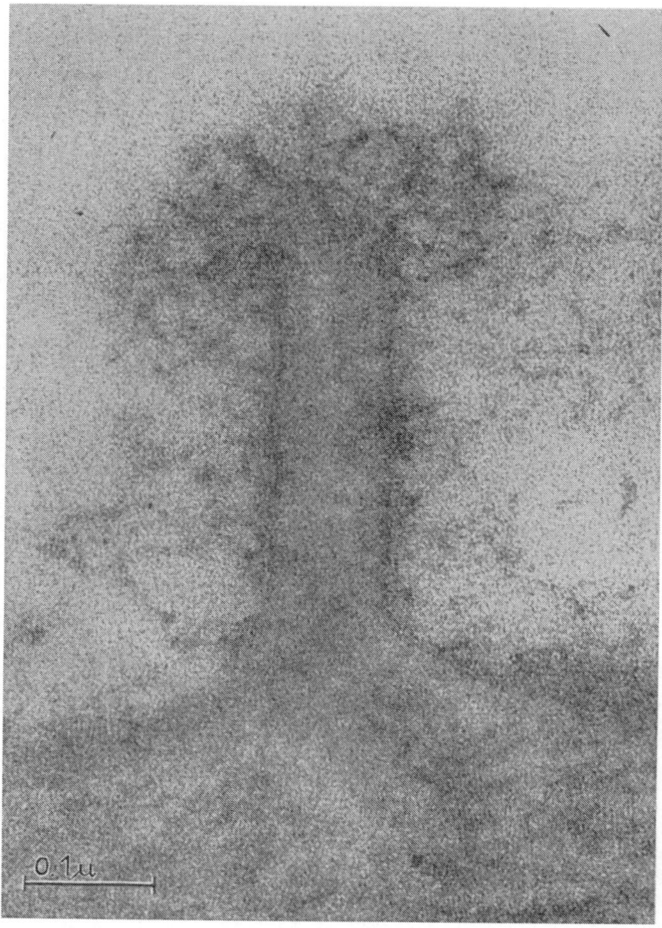

Fig. 8. Microvillus from the uterine surface epithelium. Most of the "furry coat" is concentrated to the tip of the microvillus. Magnification: 170,000 × (Courtesy of Dr. T. Zelander)

It is interesting to note that in a mixed gland, one type of cell may lack the "furry coat". This is the case in the gastric fundus glands, where the parietal cells are devoid of the coating which the other exocrine cells possess. Similarly in the pancreas, the centroacinar cells lack the coating which can be observed on the acinar cells. It is evident, however, that the coating may be lost during the fixation or embedding of the tissue.

3. Apocrine secretion

Apocrine cells are characterized by an excretion, whereby part of the cytoplasm is pinched off together with the secretory substances.

The definition of apocrine secretion was formulated nearly 50 years ago [59] and was based on light microscopical studies. In some cases these seemed to be confirmed by early electron microscopical findings. However, in these investigations, inadequate preservation of the cells often resulted in swelling and bulging of the apical part of the cytoplasm which gave a false impression of an apocrine secretion.

The use of better fixation and embedding methods in more recent electron microscopical studies, have revealed a picture of the gland cells which is probably more genuine. According to these, many of the gland cells classified as apocrine rather seem to be merocrine (cf. also [55]).

In the mammary gland, lipid droplets are released coated with a narrow cytoplasmic brim [2]. Similarly, the goblet cell mucus, seems to be extruded together with small amounts of cytoplasmic contents [25]. It has been debated whether this extrusion of cytoplasm contributes significantly to the secretion and thus may be regarded as an apocrine secretion [2, 65]. On the whole, however, it appears from morphological data available at present, that there does not exist any real apocrine type of secretion [2].

In this connection it must be emphasized, that there may exist more than one type of excretion in a single gland cell. A typical example of this are the mucoid cells of the gastric mucosa [28]: Ordinarily these cells release their secretory products by means of a merocrine secretion. However, from autoradiographic studies, it is known that these cells are shed within a few days after their formation. As the cells which are shed contain ordinary secretory granules they must also be classified as holocrine.

II. Endocrine cells

Endocrine cells are morphologically very similar to exocrine cells. The ingestion and the synthesis of secretory substances appear to proceed along the same principal pathways. (Concerning the production of steroids, however, see p. 12). Quite often, endocrine cells contain dense secretory granules surrounded by a surface membrane. In the pituitary gland and in the adrenal medulla, it has been observed that these surface membranes may fuse with the basal plasma membrane and subsequently rupture. The contents of the granules are then liberated into the interstitial space [16, 22] and transported to an adjacent capillary. This process which has been called *emiocytosis* [39], is rather similar to the granulocrine type of merocrine secretion described above (cf. p. 14). In most cases no remnants of the secretory granules can be observed outside the gland cell, and it has been assumed that the granules disintegrate as they leave the cells.

However, in the parathyroid gland, the secretory granules do not disintegrate when they are extruded from the cells. The granules may

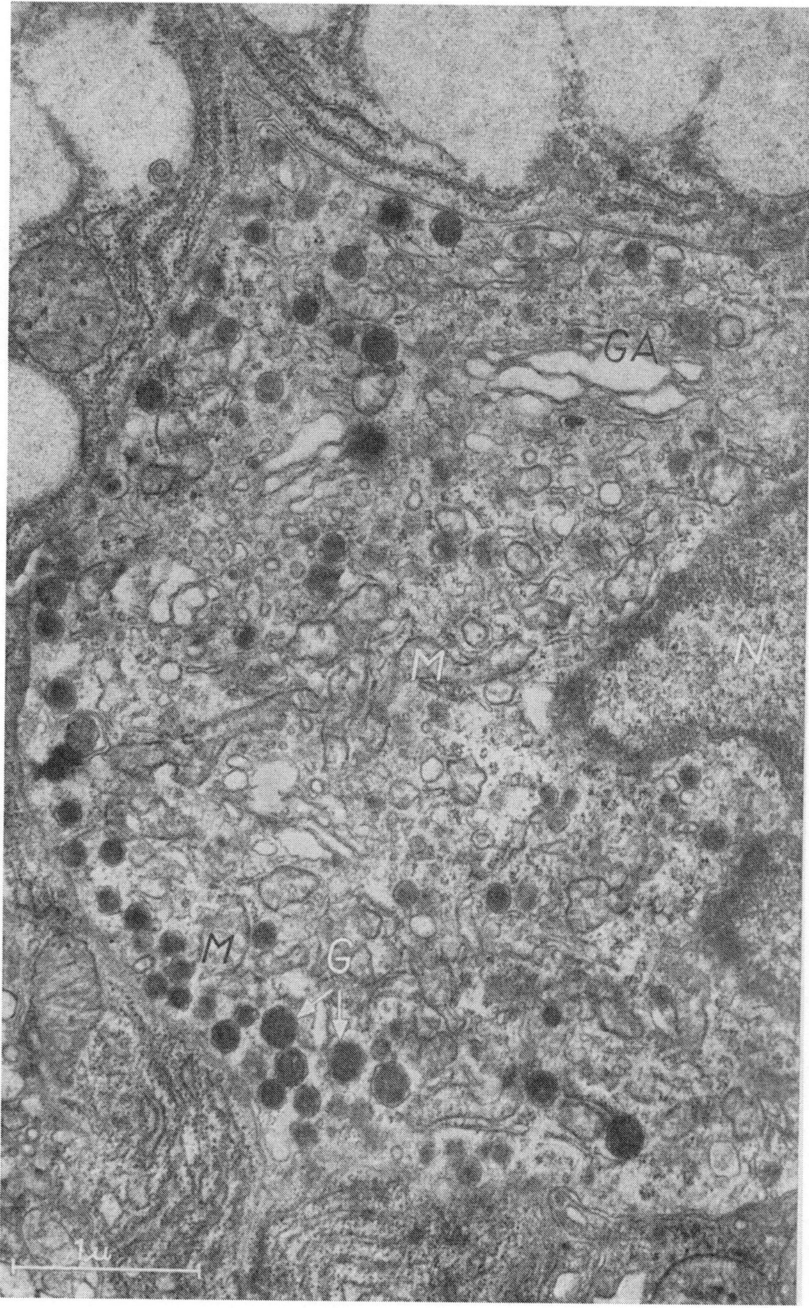

Fig. 9. Part of a gastric argyrophil cell. This type of cell most probably has an endocrine function. Typical for these cells are the small, opaque secretory granules (*G*), surrounded by a loosely fitting membrane. Comparatively few mitochondria (*M*) and α-cytomembranes are visible. *GA*, Golgi apparatus; *N*, nucleus. Magnification: 24,000 ×

be found in the interstitial space and in the cytoplasm of the endothelial cells of adjacent capillaries [46].

The administration of Synthalin produces a rapid degranulation of the α-cells of the pancreatic islets. During this process it has been observed, that the secretory granules break up into appreciably smaller secretory particles. Such particles are also encountered outside the α-cells in the intercellular and interstitial space as well as in the capillary endothelial cells [45].

In other endocrine cells e. g. the cells of the adrenal cortex and the gonads, there are no morphological signs of the release of secretory products. An eccrine type of extrusion across the basal and/or lateral cell surface may be postulated in these cases.

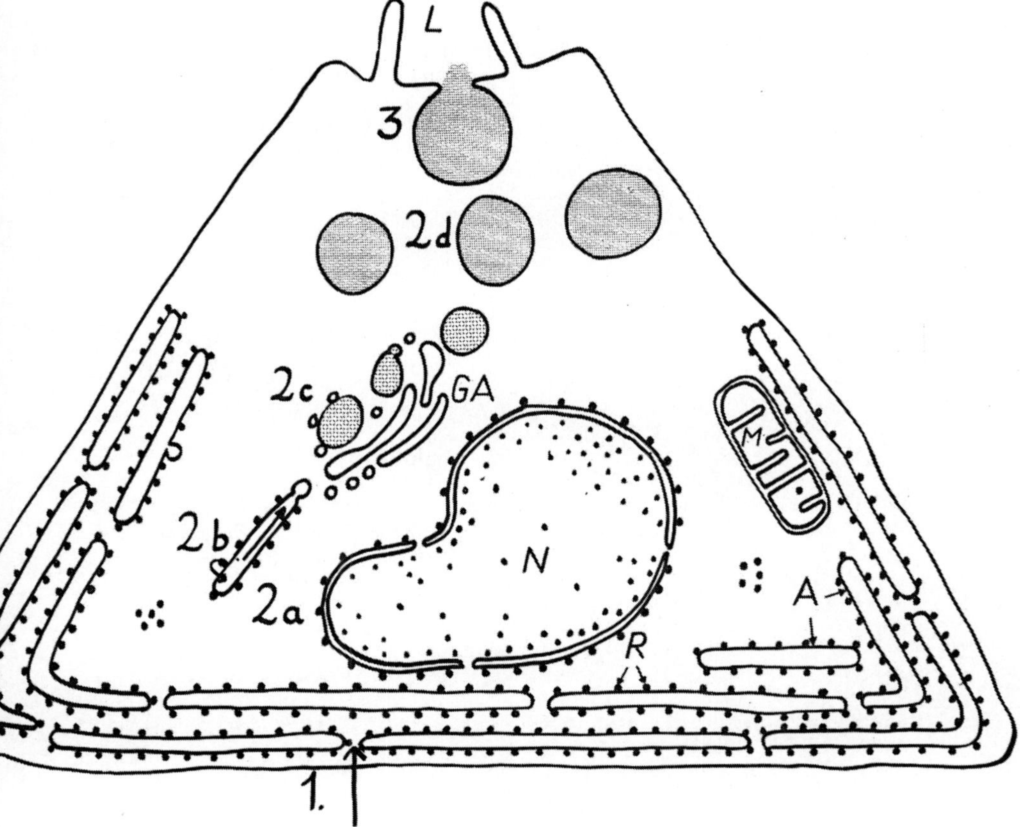

Text-fig. 1. The three hypothecial steps of the secretion in a protein-producing cell, such as the pancreatic acinar cell. 1: amino acids enter the cell across the basal plasma membrane. 2a: At the ribosomes the amino acids are combined to become proteins. 2b: The newly synthesized proteins enter the cisternae. Small vesicles containing these proteins are pinched off from the cisternae. 2c: The vesicles empty their contents into the Golgi vacuoles, which gradually increase in size and finally become secretory granules. 2d: The secretory granules are stored. By a maturation process their contents become more dense. 3: The secretory granules empty their contents into the glandular lumen. Legend: *A*, α-cytomembranes; *GA*, Golgi apparatus; *L*, glandular lumen; *M*, mitochondrion; *N*, nucleus; *R*, ribosomes

References

[1] ASHWORTH, C. T., F. J. LUIBEL, and S. C. STEWART: J. cell Biol. 17, 1 (1963).

[2] BARGMANN, W., K. FLEISCHHAUER, and A. KNOOP: Z. Zellforsch. 53, 545 (1961).

[3] BELT, W. D., and D. C. PEASE: J. biophys. biochem. Cytol. 2, Suppl., 369 (1956).

[4] BENNETT, H. S.: J. biophys. biochem. Cytol. 2, Suppl., 99 (1956).

[5] BJÖRKMAN, N., and B. FREDRICSSON: Z. Zellforsch. 55, 500 (1961).

[6] BRACHET, J.: Biochemical Cytology. New York: Academic Press 1957.

[7] BRADFORD, N. M., R. E. DAVIES, E. ELLIS, and D. E. HUGHES: Biochem. J. 42, lviii (1948).

[8] BRANDES, D., and A. PORTELA: J. biophys. biochem. Cytol. 7, 505 (1960).

[9] CARO, L. G., and G. E. PALADE: J. cell Biol. 20, 473 (1964).

[10] CASPERSSON, T. O.: Cell Growth and Cell Function. New York: W. W. Norton & Co. 1950.

[11] CHANTRENNE, H.: The Biosynthesis of Proteins. Oxford: Pergamon Press 1961.

[12] CHARLES, A.: J. Anat. (Lond.) 93, 226 (1959).

[13] CLARK, S.: Amer. J. Anat. 109, 57 (1961).

[14] COPENHAVER, W. M., and D. D. JOHNSON: Textbook of Histology, 14 ed. Baltimore: Williams & Wilkins 1958.

[15] DANIELLI, J. F., and H. DAVSON: J. cell. comp. Physiol 5, 495 (1935).

[16] DEROBERTIS, E. D. P., and D. D. SABATINI: Fed. Proc. 19, Suppl. 5, 70 (1960).

[17] —, and A. VAZ FERREIRA: Exp. Cell Res. 12, 568 (1957).

[18] EKHOLM, R., H. F. HELANDER, and T. ZELANDER: Z. Zellforsch. 59, 467 (1963).

[19] —, and F. S. SJÖSTRAND: J. Ultrastruct. Res. 1, 178 (1957).

[20] ELFVIN, L.-G.: J. Ultrastruct. Res. 5, 388 (1961).

[21] EPSTEIN, M. A.: J. biophys. biochem. Cytol. 3, 851 (1957).

[22] FARQUHAR, M. G.: Trans. N. Y. Acad. Sci., Ser. II, 23, 346 (1961).

[23] —, and G. E. PALADE: J. cell Biol. 13, 55 (1962).

[24] FREDRICSSON, B., and N. BJÖRKMAN: Z. Zellforsch. 58, 387 (1962).

[25] FREEMAN, J. A.: Anat. Rec. 144, 341 (1962).

[26] GREENE, D. E.: Sci. Amer. 202, 46 (1960).

[27] HAYANO, M., N. SABA, R. I. DORFMAN, and O. HECHTER: Recent Progr. Hormone Res. 12, 79 (1956).

[28] HELANDER, H. F.: J. Ultrastruct. Res. ,Suppl. 4 (1962).

[29] — J. Ultrastruct. Res. 10, 160 (1964).

[30] HIRSCH, G. C.: In: Biological Structure and Function, vol. I, ed. T. W. GOODWIN and O. LINDBERG, p. 195. London: Academic Press Inc. 1961.

[31] HOKIN, L. E., and M. R. HOKIN: In Ciba Foundation Symposium on the Exocrine Pancreas, p. 186. London: Churchill Ltd. 1962.

[32] HOLLANDER, F.: Gastroenterology 1, 401 (1943).

[33] ITO, S., and R. J. WINCHESTER: J. Cell Biol. 16, 541 (1963).

[34] JENNINGS, M. A., and H. W. FLOREY: Quart. J. exp. Physiol. 41, 131 (1956).

[35] JUNQUEIRA, L. C. U., G. C. HIRSCH, and H. A. ROTHSCHILD: Biochem. J. 61, 275 (1955).

[36] KUROSUMI, K., S. SHIBASAKI, G. UCHIDA, and Y. TANAKA: Arch. hist. jap. 15, 587 (1958).

[37] —, M. YAMAGISHI, and M. SEKINE: Z. Zellforsch. 55, 297 (1961).

[38] LACY, P. E.: Anat. Rec. 128, 255 (1957).

[39] — Amer. J. Med. 31, 851 (1961).

[40] MARCHESI, V. T., and R. J. BARRNETT: J. cell Biol. 17, 547 (1963).

[41] MAXWELL, D. S., and D. C. PEASE: J. biophys. biochem. Cytol. 2, 467 (1956).

[42] MESSIER, B., and C. P. LEBLOND: Amer. J. Anat. 106, 247 (1960).

[43] MILLER, R. A.: Amer. J. Anat. 92, 329 (1953).

[44] MUNGER, B. L.: Amer. J. Anat. 103, 1 (1958).

[45] — Lab. Invest. 11, 885 (1962).

[46] —, and S. I. ROTH: J. cell Biol. 16, 379 (1963).

[47] NILSSON, O.: J. Ultrastruct. Res. 1, 375 (1958).

[48] PADYKULA, H. A.: Amer. J. Anat. 91, 107 (1952).

[49] PALADE, G. E.: J. appl. Physiol. 24, 1424 (1953).
[50] — J. biophys. biochem. Cytol. 2, 417 (1956).
[51] — J. biophys. biochem. Cytol. 2, Suppl., 85 (1956).
[52] — In: Subcellular Particles, ed. T. HAYASHI, p. 64. New York: Ronald Press Co. 1959.
[53] — and P. SIEKEVITZ: J. biophys. biochem. Cytol. 2, 671 (1956).
[54] — —, and L. CARO: In: Ciba Foundation Symposium on the Exocrine Pancreas, p. 23. London: Churchill Ltd. 1962.
[55] PALAY, S. L.: In: Frontiers of Cytology, ed. S. L. PALAY, p. 305. New Haven: Yale Univ. Press 1958.
[56] PEACHEY, L. D., and H. RASMUSSEN: J. biophys. biochem. Cytol. 10, 529 (1961).
[57] ROBERTSON, J. D.: Biochem. Soc. Symposia (Cambridge, Eng.) 16, 3 (1959).
[58] RUTBERG, U.: Acta odont. scand. 19, Suppl. 30 (1961).
[59] SCHIEFFERDECKER, P.: Biol. Zbl. 37, 534 (1917).
[60] SIEKEVITZ, P., and G. E. PALADE: J. biophys. biochem. Cytol. 4, 309 (1958).
[61] — — J. biophys. biochem. Cytol. 7, 619 (1960).
[62] SJÖSTRAND, F. S.: Int. Rev. Cytol. 5, 455 (1956).
[63] — J. Ultrastruct. Res. 9, 561 (1963).
[64] —, and L.-G. ELFVIN: J. Ultrastruct. Res. 7, 504 (1962).
[65] STOCKINGER, L., and J. ZARZICKI: Z. Zellforsch. 57, 106 (1962).
[66] TANDLER, B.: Amer. J. Anat. 111, 297 (1962).
[67] WARNER, J. R., P. M. KNOPF, and A. RICH: Proc. nat. Acad. Sci. (Wash.) 42, 122 (1963).
[68] WARSHAWSKY, H., C. P. LEBLOND, and B. DROZ: J. cell Biol. 16, 1 (1963).
[69] WATSON, M. L.: J. biophys. biochem. Cytol. 1, 257 (1955).
[70] WISSIG, S. L.: J. biophys. biochem. Cytol. 7, 419 (1960).
[71] — J. cell Biol. 16, 93 (1963).
[72] ZEIGEL, R. F., and A. J. DALTON: J. cell Biol. 15, 45 (1962).
[73] ZELANDER, T.: J. Ultrastruct. Res. Suppl. 2 (1959).
[74] ZETTERQVIST, H.: The Ultrastructural Organization of the Columnar Absorbing Cells of the Mouse Jejunum. Thesis. Stockholm 1956.

Discussion

Bücher: My questions are very special ones, and I am afraid that they do not go into the centre of your representation. I should like to comment on the role of the mitochondria in these different cells. In your additional figures (for figures see T. ZELANDER, J. Ultrastruct. Res. Suppl. 2, 1959) I saw that the mitochondria of the adrenal cortex are extremely small, and very rich in matrix.

Helander: I should say they are about 0.3 microns in diameter, approximately. I do not know very much about these mitochondria. It is Dr. ZELANDER, who is working with these mitochondria. He has looked at their morphology so far, but he has not isolated mitochondria from the adrenal cortex, and he is presently investigating what capacities these mitochondria have. He thinks that they are able to synthesize or participitate in the synthesis of steroids.

Bücher: And the mitochondria of the parietal cell are the other extreme. In your fig. 6 we saw that they have laminal cristae, one beside the other. They would be the energy producing type, and the other which is rich in matrix is the biosynthetic type?

Helander: Yes, it might be that way. That is quite possible.

Bücher: Is there also some functional aspect? You have parietal cells with different functional states, do you see some changes within the mitochondria?

Helander: I have measured the frequency of inner membranes. It is about 40 per micron in fasted and in stimulated mice. So there is no difference, neither is there any difference in the diameter. I do not know if there is any difference in their frequency. It does not appear so to me.

Hirsch: Ich glaube, daß es recht gut war, daß wir mit den Mitochondrien begonnen haben, denn wir haben keinen besonderen Punkt in unserem Programm, wo wir das behandeln könnten. Deshalb würde es mir sehr lieb sein, Herr Bücher, wenn Sie noch einiges ergänzend dazu sagen würden.

Bücher: Ich glaube, daß die Bilder, die Sie gezeigt haben, mehr die Anschauung fördern, daß die Mitochondrien nichts Direktes mit dem spezifischen Prozeß der Salzsäureproduktion zu tun haben.

Helander: No, I do not think so either. The mitochondria are in the basal part of the cell, and the hydrochlorid acid comes out of the apical part of the cell, so I think they are the energy producers.

Holter: Ich möchte nur eine Frage an Herrn Bücher stellen. Sie haben den Ausdruck „biosynthetische Mitochondrien" benutzt. Den Eindruck bekommt man auch nach dem Bild, das hier gezeigt wurde. Das ist aber eine Unterscheidung, die mir neu ist. Gibt es wirklich Mitochondrien, die man im allgemeinen als nicht primär energieliefernd, sondern als biosynthetisch charakterisieren kann in anderen Fällen als in der Nebennierenrinde? Ist das eine allgemein anerkannte Unterscheidung zweier Mitochondrientypen?

Bücher: Beispielsweise in der Gegenüberstellung von Mitochondrien der Skelettmuskeln und der Leber findet man Unterschiede der Enzymproportionen, welche die Größenordnung betreffen. So ist die Proportion der Glutamatdehydrogenase zu den respiratorischen Pigmenten in Lebermitochondrien hundertmal größer als in Muskelmitochondrien [z. B. Hoppe Seylers Zeitschrift für physiologische Chemie **331**, 180 (1963) und Vortrag von Dr. Pette vor der Hauptversammlung in Weimar]. Natürlich entbehren die Lebermitochondrien nicht der „Kraftwerkfunktion": Die Aktivitäten der DPN spezifischen Dehydrogenasen und anderer Enzyme des Citratzyklus findet man in beiden Typen von Mitochondrien annähernd in denselben gegenseitigen Proportionen und in derselben Proportion zur Konzentration an respiratorischen Pigmenten. Das Plus an Glutamatdehydrogenase in der Leber wird jedoch im Sinne einer hohen Potenz zur reduktiven (TPNH-abhängigen) Aminierung, d. h. im Sinne einer biosynthetischen Funktion gedeutet. Ob diese Deutung allgemein anerkannt wird, ist eine andere Frage. Vielleicht könnte man noch folgendes anfügen: Nach Green und Mitarbeitern kommt Cholesterin nur im mitochondrialen „Envelope" vor und nicht in der Lipidgarnitur der Cristae. Man fragt sich deshalb, ob der relative Reichtum an „Envelope" der sehr kleinen Mitochondrien in der Nebennierenrinde etwas mit der steroidsynthetisierenden Funktion zu tun hat. Aber ich fürchte wirklich, daß wir zu weit vom Thema abkommen.

Komnick: Ich möchte einige Bilder (s. Abb. bei Komnick, Histochemie 3, 1963) zur funktionellen Bedeutung der Bläschen in den Belegzellen zeigen. Man könnte daraus auch Hinweise für die Frage gewinnen, ob die Mitochondrien nur als Energielieferanten dienen oder ob sie noch weiterhin mit der Sekretion dieser Zellen zu tun haben. Wir haben die Magenschleimhaut der Maus vorwiegend aus methodischen Gesichtspunkten untersucht, um die Chloride in den Zellen der Fundusdrüsen zu lokalisieren. Die Mägen wurden in verschiedenen Funktionszuständen — nach Hunger- und Fütterungsversuchen — unter Beachtung des tageszeitlichen Aktivitätszyklus der Tiere in einem Gemisch von Osmiumtetroxyd und Silberlactat fixiert und später mit verdünnter Salpetersäure behandelt, um eventuelle unspezifische Niederschläge wieder herauszulösen. Rein morphologische Untersuchungen nach Osmium-Chrom-Fixation haben die gleichen Ergebnisse erbracht, wie sie Herr Helander geschildert hat: In den Belegzellen der hungernden Maus sind die Sekretkanälchen nicht besonders zahlreich ausgebildet. Sie besitzen ebenso wie die Zellapices nur relativ wenige und kurze Mikrovilli. Das apikale Cytoplasma und die Region um die Sekretkanälchen sind dicht von Vakuolen erfüllt, die wir wohl mit glatten endoplasmatischen Retikulum identifizieren müssen, da sie nach Kaliumpermanganat-Fixation in Form eines tubulären Netzwerkes erscheinen, wie Ito und Sedar gezeigt haben.

Nach der Fütterung bietet sich ein völlig anderes Bild. Die Vakuolen sind bis auf geringe Reste verschwunden; die Sekretkanälchen treten zahlreicher hervor, die Mikrovilli sind deutlich vermehrt und verlängert, d. h. die apikale Zelloberfläche ist vergrößert. Sedar hat an einer anderen Tierart gezeigt, daß sich das endoplasmatische Retikulum der Belegzellen nach der Stimulation zum Drüsenlumen hin öffnet. Es erhebt sich also die Frage, ob die Vergrößerung der apikalen Zellmembranoberfläche primär mit dem Sekretionszyklus bzw. Extrusionszyklus, d. h.

mit dem Transmembrantransport zusammenhängt, oder ist es eine sekundäre Folge
der Extrusion, d. h. ist das Verschwinden des ER damit zu erklären, daß die Membranen bei der Extrusion in die Zellmembran übergehen? Morphologisch denkbar
wäre beides, entscheiden können wir es noch nicht.

In den Belegzellen der hungernden Maus werden während der Fixation durch
Silberlactat Chloride in den Vakuolen niedergeschlagen. Im Grundplasma und in den
Mitochondrien finden sich keine nennenswerten Präzipitate. Offensichtlich sind
demnach die Mitochondrien nicht unmittelbar an dem intrazellulären Chloridtransport und der Chloridspeicherung beteiligt. Nach der Fütterung sind gleichzeitig mit den meisten Vakuolen auch die Chloridniederschläge aus den Zellen verschwunden. Man trifft die Niederschläge jetzt sehr viel häufiger in den Lichtungen
der Fundusdrüsen an und findet sie massiert in den engen Krypten des Magenlumens,
wo sie bei der Entwässerung nicht herausgewaschen werden. Aus diesen Befunden
kann man ableiten, daß — zumindest bei Mäusen — während der Hungerperiode im
endoplasmatischen Retikulum der Belegzellen Chloride gespeichert und nach der
Nahrungsaufnahme in das Lumen der Fundusdrüsen ausgeschüttet werden.

Bücher: Können Sie uns ein Bild geben, welche Vorstellungen Sie für den Mechanismus der Salzsäurebildung erhalten haben?

Komnick: Man kann anhand der histochemischen Befunde nicht sagen, ob diese
Bläschen schon freie Salzsäure enthalten; man kann nur sagen, daß während der
Hungerperiode im endoplasmatischen Retikulum der Belegzelle Chloride gespeichert
werden. Die physiologischen Befunde gehen ja dahin, daß während der Hungerperiode die Chloridspeicherung in der Magenschleimhaut sehr hoch ist, am höchsten
im Vergleich zu allen anderen Geweben des Organismus. Diese Speicherung dürfte
sich nach unseren Befunden hauptsächlich im ER der Belegzellen abspielen und in
etwa auch im intertubulären Bindegewebe, auf das ich hier nicht näher eingegangen
bin. Nach einer ausgiebigen Fütterung verarmt die Magenschleimhaut an Chloriden,
ebenso das Blut. Es erfolgt ein dauernder Nachtransport aus dem Blut, und schließlich werden auch Chloride aus anderen Organen mobilisiert. Nach der Stimulierung,
die durch Nahrungsaufnahme erfolgt, wird die Salzsäure ausgeschüttet. Das stimmt
auch sehr gut mit der Ausschüttung der Chloride überein. Die Salzsäure selbst soll
ja an der Oberfläche der Korbkapillaren frei werden. Die Wasserstoffionen stammen
aus einer Verknüpfung des CO_2- und H_2O-Stoffwechsels. DAVIES konnte an der
isolierten Magenschleimhaut feststellen, daß mit jedem gebildeten Wasserstoffion
ein CO_2-Molekül aufgenommen und ein Bikarbonation freigesetzt wurde. Die
Wasserstoffionen werden durch die Hydrolyse des Wassers freigesetzt, die sich unter
Wirkung der mitochondrialen Stoffwechselenergie vollzieht. Das frei werdende OH
wird unter Mitwirkung der Karboanhydrase auf Kohlendioxyd übertragen, und das
Bikarbonat wird im Austausch gegen Chlorid ins Blut abgeführt. Carboanhydrase-
Aktivität konnte histochemisch von SCHIEBLER und VOLLRATH (1959) in der Region
um die Sekretkanälchen nachgewiesen werden, also dort, wo während der Hungerperiode das ER so stark ausgebildet ist und wo Chloride gespeichert werden. Durch
diese Befunde wird die funktionelle Bedeutung des endoplasmatischen Retikulums
näher erklärt, das ja einen starken funktionellen Wandel in den verschiedenen
Funktionsstadien der Zelle erkennen läßt. Diese Speicherfunktion des ER ist schon
vor dem histochemischen Chloridnachweis allein auf Grund des Strukturwandels
erschlossen worden.

Bücher: So, daß die Chloridionen durch die Parietalzellen durchgeschleust
werden?

Komnick: Letzten Endes ja, aber unter vorübergehender Speicherung im endoplasmatischen Retikulum.

Bücher: Hat diese Speicherung etwas zu tun mit der Salzsäuresekretion, oder
ist das ein Phänomen, das ganz daneben steht?

Komnick: Nach dem Bild in den beiden verschiedenen Funktionsstadien zu
urteilen, muß es etwas mit der Salzsäuresekretion zu tun haben.

Bücher: Ich meine, es ist doch schwer vorzustellen, daß die Parietalzellen nur
soviel Salzsäure ausscheiden können, wie sie Chloridionen in einer Hungerphase
gespeichert haben.

Komnick: Nein. Selbst nach der Nahrungsaufnahme sieht man ja auch Teile des
endoplasmatischen Retikulums mit Chloriden. Das bedeutet also, entweder wird

schon wieder gespeichert, oder, wenn diese Speicherungsphase durch die Stimulierung unterbrochen wird, könnte es sein, daß die Sekretion etwas kontinuierlicher
verläuft. Rein morphologisch ist das auch sehr schön von Sedar nach der Stimulation gezeigt worden. Es ist immer ein Teil von ER da, aber in der Hungerperiode
ist es stärker ausgeprägt als nach der Nahrungsaufnahme. Nun, wir dürfen nicht
vergessen, daß Mäuse einen tageszeitlichen Aktivitätszyklus besitzen. Möglicherweise erfolgt auch die Salzsäureproduktion entsprechend den Freßgewohnheiten
rhythmisch. Diese Fragen haben wir — wie gesagt — nicht im einzelnen weiter
verfolgt, sondern in erster Linie ging es uns hier um methodische Fragen, um den
histochemischen Chloridnachweis auf elektronenoptischer Ebene.

Ullrich: Das Problem des histochemischen Chloridnachweises ist, ob die Chloridionen in vivo so verteilt sind, wie der Niederschlag im histochemischen Bild dies
zeigt. Es ist sehr gut möglich, daß bei der Behandlung mit wäßrigen Medien alle
freien Chloridionen weggeschwemmt und nur die an Anionenaustauschstellen haften
bleibenden Chloridionen gefällt und nachgewiesen werden. In ähnlichen Experimenten mit Nierengewebe hat Nolte eine Silberchloridfällung nur in den interzellulären Spalträumen bekommen und daraus abgeleitet, daß die Chloridionen
nicht trans- sondern interzellulär durch die Nierentubuli transportiert werden. Dies
überrascht, wenn man bedenkt, welch kleine Fläche des Tubulusrohres die Interzellulärräume einnehmen und wie groß die transtubulären Chloridionenflüsse sind.
Diese Einwände schmälern natürlich nicht Ihren Befund, daß in dem einen Funktionszustand der Belegzellen Chloridionen nachweisbar sind, im anderen Zustand
aber nicht. Meine Frage ist, ob das nicht Ausdruck einer physiko-chemischen Änderung von Eiweißkörpern ist, die das eine Mal Chlorid binden und das andere Mal
nicht, so daß alles Chlorid, was in den Zellen ist, ausgewaschen wird.

Komnick: Vor der Fällung kann nichts ausgewaschen werden, weil gleichzeitig
fixiert und gefällt wird in *einem* Gemisch. Daß im Grundplasma keine Chloride
nachweisbar sind, obwohl sie ja von der Basis her wahrscheinlich über die Grundphase des Cytoplasmas in das ER hineintransportiert werden müssen, liegt wahrscheinlich daran, daß die Konzentration dort für eine Fällungsreaktion zu gering ist
oder daß hinterher das Reaktionsprodukt herausgewaschen wird. Wenn Sie einen
Gewebeblock von 1 mm Kantenlänge nehmen, wie das normalerweise in der Elektronenmikroskopie geschieht, dann genügen bereits 5 ml Wasser, um das gefällte
Chlorid restlos wieder auszuwaschen, — wenn Sie die physiologisch ermittelte Cl-
Konzentration der Magenschleimhaut von 600 mmol und die Löslichkeit von
Silberchlorid der Rechnung zugrunde legen. Diesen Verlust können Sie relativ
herabsetzen, wenn Sie größere Gewebescheiben fixieren und hinterher periphere
Bereiche untersuchen. Bemerkenswert ist aber folgendes: Je weiter man in das
Gewebe hineinkommt, desto mehr Diffusionsartefakte findet man. Das spricht
eigentlich dafür, daß im gekühlten Reaktionsmedium die Zelltätigkeit gestört ist,
so daß dieser Anhäufungseffekt im ER nicht mehr erhalten bleibt.

Kuyper: Es kann also eine ganze Menge des Chlorids auch anderswo anwesend
sein, aber wenn die lokale Konzentration zu niedrig ist, können Sie es gar nicht
darstellen.

Komnick: Genau das hatte ich gesagt. Der begrenzende Faktor ist einmal die
Empfindlichkeit der Reaktion und zum anderen beim Entwässern hinterher das
Löslichkeitsprodukt. Quantitativ können diese Befunde nicht ausgewertet werden.

Solomon: Dr. Helander, you said that the ATP-ase was at the basal side of the
cell, and that there were many other enzymes at the apical side of the cell. I was
wondering if you knew what the nature of those other enzymes were?

Helander: I have only made preliminary investigations on the enzymes in
the gastric mucosa, and ATP-ase was the first one I started with. I have not come
to the second one yet, so I do not know which enzymes, but I do believe that there
are enzymes.

Tosteson: The suggestion has been put forward that ATP generated in mitochondria provides energy for the transport of substances across the various membranes between the two sides of the cells, for example the plasma membranes and the
membranes lining the endoplasmic reticulum. I would raise the question as to how
intramitochondrial ATP can be transferred to the membrane sites in order to be

utilized for the transport process. It is generally recognized that intramitochondrial ATP cannot exchange with extramitochondrial ATP.

L. Hokin: It was my understanding that extramitochondrial ATP is synthesized at the surface of the mitochondria.

Tosteson: Well, if it could escape that way —

L. Hokin: I think there is evidence for it?

Tosteson: Yes, if the quantity or amount of ATP so generated is sufficient in relation to the amount of intramitochondrial produced.

L. Hokin: I do not know the quantitative results. Maybe Prof. BÜCHER has them?

Bücher: Allgemeines Einverständnis besteht darüber, daß ATP kompartmentiert ist. Das bedeutet jedoch nicht, daß kein Austausch stattfindet. Wir wissen jetzt etwas mehr darüber, weil im letzten Jahr ein spezieller Inhibitor für den Austausch von Adenosinpolyphosphat gefunden worden ist (Atraktylosid). Von mehreren Arbeitskreisen ist gezeigt worden, daß die Ausschleusung von energiereichem Phosphat aus den Mitochondrien durch einen spezifisch hemmbaren Mechanismus garantiert ist. Ein Glied dieses Mechanismus scheint übrigens auch die Einschleusung von ADP zu sein. Atraktylosid, der spezifische Inhibitor, ist ein wirksames Zellgift. Es hemmt die Zellatmung. Seine Wirkung ist ein Argument für die Existenz „in vivo" des mitochondrial-extramitochondrialen Adenosinpolyphosphatzyklus, wie er „in vitro" an isolierten Mitochondrien ohne weiteres zu beobachten ist. Infolge der engen Kopplung zwischen Atmung und Phosphorylierung ist der zelluläre Sauerstoffverbrauch ein Maß für die Flußrate für diesen Zyklus.

Thoenes: Der Hinweis, daß die apokrine Sekretion vielleicht keine eigene Sekretionsform sei, sondern irgend etwas mit der Fixation zu tun hat, scheint mir recht wichtig. Ich möchte nur die Frage aufwerfen, ob man die blasige Auftreibung des apikalen Cytoplasmas wirklich als reinen Fixationsartefakt auffassen darf, denn die Blasen — wir haben sie am Nierenepithel gefunden — sind sehr oft wohlgeformt und scheinen auch nach Untersuchungen in vivo abgeschnürt zu werden. Ich hätte deshalb die Frage, ob man diese Ausbildung der Blasen vielleicht als ein Supravitalphänomen auffassen muß, als ein Phänomen, das die Zelle hervorruft in der Phase zwischen der Entnahme des Gewebes und dem Eintritt der Fixierung, also in einer Phase, in der sich die Zelle in der Ischämie befindet. Dann wäre dieses Phänomen auf ein aktives Verhalten einer absterbenden Zelle zurückzuführen.

Helander: Yes, I can accept your theory. I think it comes out of anoxy, or something like that, of the dying cell. What I meant to say was that the tissue was poorly preserved, and these structural alterations presumably arise before the fixative has reached that cell.

David: Ich habe eine ganz prinzipielle Frage. Wenn wir annehmen, daß sekretorische Zellen durch folgende Eigenschaften ausgezeichnet sind: Aufnahme von Substanzen, Synthese und Abgabe, möchte ich fragen, welche Zelle eigentlich keine sekretorische Zelle ist, warum z. B. ein Fibroblast keine sekretorische Zelle ist? Wo können wir überhaupt jetzt feststellen, was eine sekretorische Zelle ist und was nicht?

Helander: I refer to the subject of my paper, "Morphology of Animal Secretory Gland Cells". Of course, all the cells will give off certain substances. I have tried to confine my paper to gland cells and what we ordinarily regard as gland cells.

Gersch: Ich weiß nicht, ob es angängig ist, nach dieser Diskussion so spezifischer Fragen noch einmal auf die Grundkonzeption dieser so interessanten Darstellung von Herrn HELANDER einzugehen. Eigentlich berührt das, was ich da sagen möchte, das gleiche, was eben von Herrn DAVID angeschnitten worden ist. Wenn man diese Gegenüberstellung der beiden Typen von Sekretionszellen betrachtet, so erscheint es mir eigentlich wesentlich, die Aufmerksamkeit noch auf einen anderen Zelltyp zu richten, den man ursprünglich nicht in den Typus einer sekretorischen Zelle einbezogen hat. Ich meine die Nervenzellen. Rein morphologisch haben wir bei Nervenzellen genau die gleichen Kennzeichen des Sekretionsablaufes durch eine entsprechende Induktion nach der Arbeitsweise der Stufenuntersuchung von Herrn HIRSCH nachweisen können.

Wenn man die neuen elektronenmikroskopischen Befunde vergleicht, so reiht sich dieser Typus der Zellen ganz in den der proteinhaltigen Sekrete ein. Hier ist nun allerdings das Problem, daß man auch in nicht sekretorischen Nervenzellen ähnliche Vesikelstrukturen gefunden hat. Meine Frage wäre nun, ob Sie bei typischen Drüsenzellen vergleichend in der Ontogenie die Entstehung dieser Sekretionsmuster beobachtet haben und bis zu welchem Zeitpunkt evtl. keinerlei Anzeichen zu erkennen sind. Bei den Nervenzellen jedenfalls hat es nach manchen Befunden den Anschein, daß man keine scharfe Grenze setzen kann. Das ergibt sich auch daraus, daß man vor allem bei wirbellosen Tieren durch besondere äußere Faktoren — ich denke insbesondere an osmotische Belastung — bestimmte Nervenzellen, die normalerweise keine typische Sekretionsleistung zeigen, zu einer Sekretbildung veranlassen kann. Ich möchte allerdings hier nicht ohne weiteres den Begriff Neurosekretion gebrauchen.

Helander: I have just started an investigation into the embryology of the gastric mucosa, and round about the sixteenth day of gestation the animals show secretory granules which are rather similar to the zymogen granules. I have not studied the parietal cells so much, but I measured the stomach pH of new-born mice, and at the fourth day of their lives the pH in their stomachs can get down to below 3, so actually the fourth day I regard them as adult animals in this respect.

Aspects of the Biology of the Animal Cell Secretion[1]

By

LUIZ CARLOS UCHÔA JUNQUEIRA, Sao Paulo

With 1 text-figure

A. Introduction

This review will deal mainly with some aspects regarding the activity of protein secreting cells. Mucous secretions, recently subject of a symposium [51], and lipid secretion will not be considered. There is no intention to cover exhaustively the literature, but a deliberate effort to discuss provocative and developing problems will be made.

Several reviews exist on the subject of cell secretion, covering mainly general aspects [14, 20, 23], its morphology [28, 34], and biochemistry related to protein synthesis [40].

The word secretion will be used throughout as defined by JUNQUEIRA and HIRSCH [23]. Secretion is defined as the cellular process in which macromolecules are, due to cellular activity, synthetized, stored and finally extruded by the cell. This process is divided into the stages of ingestion, synthesis and extrusion. It does not include therefore the transport of ions, water or organic compounds also referred to as secretion in the literature. However, it should be stressed that to consider as secretory cells only cells which have visible accumulations of their products is an arbitrary criterion for it is known that very active protein producing cells do not always have a typical secretory morphology. The liver cell, neuron and plasma cell are typical examples. These cells produce and eliminate proteins continuously and do not have an accumulative phase in their secretory cycle. In this sense they can be differentiated from most secretory cells of the digestive tract.

The study of the process of secretion has received in these last 10 years an enormous impulse with the introduction of modern methods in its analysis. Nevertheless, the problem is a thorny one, due in part to the fact there are few examples of cells having exclusively a protein secretory function. For instance, the pancreas, which is one of the most studied and favorable organs in this respect, is not only composed of two different kinds of secretory cells, but its exocrine portion, in addition to secreting proteins, has distinct water and ion transporting activities.

[1] The author is grateful to the Rockefeller Foundation for support during the initial years of this Laboratory and actually to the National Institutes of Health whose grant (DE 1881) permits the actual work in progress.

Most salivary glands (another frequent subject of secretion studies) have a still more active transport mechanism. This situation is further complicated by the fact that in most salivary glands mucous secretion occurs concomitantly with protein secretion. These activities frequently coexist in the same gland in different cell types or in one cell where both protein and mucous secretion can occur simultaneously. One of several examples of this situation among others are many of the mammalian parotids (including the human) which secrete neutral polysaccharides and amylase.

Apparently protein secretion is a basic cellular activity that appeared early in phylogeny. Although slime secreting elements have been described in the sponges, no recognizable protein secreting cell is known to occur in these animals. In these organisms, despite the existence of special cells that produce a protein (spongin), analogously to what occurs in fibroblasts no visible accumulation of this substance occurs in these cells and they will therefore not be considered as having a typical secretory activity.

However, in the cnidaria distinct cells with the classical secretory morphology appear. Thus the epidermal gland cells of the pedal disk, the so-called enzymatic cells of the gastrodermis of the hydra and the granule type cells of the epidermis of the ctenophora [21] present a typical morphological aspect of protein secreting cells. To my knowledge, no histochemical observations exist on these cells which are of interest due to the fact of being the first to appear in the zoological scale. Protein secretion probably developed initially mainly related to mechanisms of extracellular food digestion that appear first in the coelenterates. Later on in the phylogenetic scale such an activity spreads also to cells of the genital organs, endocrines, *etc.*

Protein secretory cells occur in abundance in the helminths and from then on are present in practically all the animal kingdom. A bird's-eye view of the distribution of these cells in animals discloses a tendency toward a clear predominance of mucous over protein secreting cells. Only in the land-living animals possessing a relatively impermeable tegument such as most arthropodes, reptiles, birds, and mammalia, there is a tendency towards the decrease of the mucous secreting cells as compared to the protein secreting ones.

Evolution and ecology apparently play an important role in the development of secretory cells.

B. Kinetics of secretion

Probably one of the first biochemical approaches to the problem of the duration of the secretory cycle was done by Van Weel and Engel [47], who observed a maximal decrease in the pancreatic carboxypeptidase 3 hours after pilocarpine injection. This period was followed by a wave of synthesis that reached its maximum at 9 hours.

More complete studies on this subject by Daly and Mirsky [9] in the mouse pancreas point toward a 1 hour period of maximal cell degranulation and enzyme expulsion followed by complete restitution of protease, carboxypeptidase, amylase and lipase activities at 6 hours.

Similar timing was observed in the pigeon pancreas amylase by FERNANDES and JUNQUEIRA [13]. A more detailed analysis of the rate of the processes in which ingestion, synthesis and extrusion are occurring was performed by us in 1956 (for literature and discussion see [23]). The conclusion reached at that time was that the overall process of the rat pancreas exocrine protein secretion was rather slow, initiating at 1 hour and attaining its maximum at approximately 3 hours. These initial results obtained with glycine were further confirmed with four other aminoacids [37].

A similar rate of protein production was obtained in several other biological systems including the bovine pancreas [27], blood proteins and antibody appearance in the blood after antigen challenge [44].

These results plus the fact that the times for ingestion and extrusion are very short [23], lead us to postulate that the period called synthesis is the most prolonged and therefore responsible for the greatest part of the time consumed in the secretory process.

Based on recent developments this period might still be subdivided into several steps that include such occurrences as:

a) protein synthesis at the level of the ribosomes,

b) release of the protein from the ribosomes and its passage through the membrane into the cisternae of the endoplasmic reticulum,

c) probable partial dehydration of these proteins in the cisternae, their clumping into granules and migration to the Golgi zone,

d) passage of these granules through the Golgi zone, further dehydration, acquisition of a membrane and formation of a definitive zymogen granule.

The time necessary for all these steps is not known, but data already exist regarding the first steps, i. e., the ribosomic synthesis and passage to the cisternae. Thus SIEKEVITZ and PALADE [41] using C^{14}-leucine administration *in vivo*, observed that in 1 minute the highest specific activity was observed in the chymotrypsinogen of the ribosomes, suggesting that this enzyme synthesis is extremely rapid. In contrast the highest specific activity of the chymotrypsinogen was only observed in the intracisternal granules 15 minutes after the aminoacid administration. Apparently the time necessary to pass through the membranes and form intracisternal granules is larger than that spent in the protein synthesis. To my knowledge no other data exist which permit an accurate evaluation of the length of the other steps.

In collaboration with ROTHSCHILD and FAJER [25] we were able to estimate the average amount of protein present in, and secreted by the rat pancreatic acinar cell. Values of 8.5×10^{-4} µg and 1.3×10^{-5} µg/hr respectively were obtained. This represents in conditions of maximal stimulation in a 1 hour period about 1.5% of the total cell protein.

Finally it should be stressed that in mucous secreting glands the timing of the secretory cycle is very different from the one above mentioned, some glands taking several days for their complete restitution after stimulation [29].

C. Source of energy for cell secretion

It has been known since 1851 [*30*] that cell secretion in the specific case of the submaxillary gland saliva production is an active process not due to filtering of substances from the blood. It must therefore be an energy consuming activity, for the transport of the building blocks of the macromolecules, their synthesis and finally extrusion are active endo-energetic processes. This view was substantiated early this century by the fact that the salivary gland respiration is increased by the organ stimulation [*6*].

The previous literature on gland metabolism has already been reviewed [*14, 23*]. We would like however to comment on the fact that apparently the processes of protein synthesis occurring in cells are not responsible for the major energy consumption. Thus calculations [*5*] suggest that the energy for protein synthesis is less than $^1/_3$ per cent of the total energy liberated in the pancreatic basal metabolism as calculated from its oxygen consumption.

One has the impression that the transport mechanisms occurring in most glands are probably responsible for the consumption of a much larger portion of the cell energy available to them through respiration. As a matter of fact the biochemical and histochemical evidence available suggests that all structures with active ion transport mechanisms have a distinctly higher oxidative metabolism. Typical cases are kidney [*48*] and the striated ducts of the salivary glands [*33, 35*]. This linkage of energetic metabolism and electrolyte transport is also suggested by Thaysen [*46*].

Our previous data [*12*] suggesting that ATP is related to salivary secretion have received support from the data presented by Taro [*45*] who observed a marked increase of secretion when this compound was added to the blood in perfusion experiments.

On the other hand it has been recently described that the nervous supply of a gland has an important role in its metabolism [*32, 42, 43*].

However, a precise analysis of the energetic processes related to true cell protein secretion can only be done on glands in which no ion transport mechanism or other processes coexist.

D. Control mechanisms

Probably the first or earliest mechanism put forth in an organism influencing secretion is tissue interaction. A clear example is described in pancreas tissue cultures by Grobstein [*16*]. This author demonstrated an interaction between the mesenchymal and epithelial components of the pancreatic rudiment of 11 days old mouse embryos. Apparently the mesenchymal cells produce a diffusible substance that influences the epithelial cells orienting their activity toward the formation of secretory acini and synthesis of secretory granules. Unfortunately the author does not present other evidence beside the morphological appearance of zymogenic granules. The profound influence of the connective tissue capsule on the development of the epithelial part of salivary gland of the mouse described by BORGHESE [4] suggests that an analogous phenomenon might be occurring in these glands. Besides this, two other mechanisms are classically considered to control secretion, i. e., nervous and hormonal stimulation. The nervous control of secretion is an universal phenomenon described since LUDWIG [30] who first studied the salivary flow derived from the submaxillary after the stimulation of the chorda-lingual nerve. It is apparently an ubiquitous mechanism occurring in most animal phyla although the literature on the influence of the nervous system on the protein secretory processes of the invertebrates is surprisingly scarce. One of the few examples in the literature is the secretion of digestive enzymes by the crop and intestine of the *Lumbricus* after stimulation of the ventral nerve cord [17].

The action of the nervous system occurs through chemical mediators as can be easily shown by the intravenous injection of these compounds. A still more direct type of approach was performed by EMMELIN, MUREN and STRÖMBLAD [10] by injecting these compounds or their analogues directly into the excretory duct of salivary glands.

The nervous system apparently not only controls the secretory processes of the glands but through this activity also their growth. Thus it is a known fact that denervation of a gland leads to its atrophy and that a constant nervous stimulation promotes phenomena of hypertrophy. This has been clearly shown in the salivary glands where denervation produces atrophy followed by biochemical changes (for literature cf. [7]). In these organs a chronic nervous sympathetic stimulation [50] or a sympathetic drug (isoproterenol) administration [38] can promote considerable hypertrophy. Contrary to what one might expect however, this hypertrophy apparently does not increase the gland efficiency as judged by salivary flow after stimulation [38]. Very little is known on how the chemical mediators act on secretion. Indirect evidences presented some time ago [26] suggest that their action is not on the processes of ingestion or synthesis but on the extrusion phase of the secretory cells and that once this is initiated by some sort of a feed back system the synthetic cycle would be stimulated. The alternation of cycles of synthesis and extrusion has been demonstrated since HIRSCH [18] and followed in detail by several authors since then (for literature see [23]). It is convenient to remember here that by an adequate nervous, pharmacological or hormonal stimulation it is possible to synchronize most of the cellular population of a gland to pass through the various stages of the secretory cycle. In material thus treated quantitative analysis of macromolecular events occurring serially during the processes of ingestion, synthesis and extrusion has been undertaken and yielded very interesting results. The use of adequately labeled precursors and radioautography on the electron microscopic level during the several stages of the secretory cycle is a promising field where very important results have been obtained [8, 49] but we feel that it should be further explored. Thus few results exist comparing quantitatively the intensity of the processes of ingestion and synthesis during the various stages of the secretion cycle.

This technique should also be used in an attempt to try a cytological localization of the sites of action of the chemical mediators or their antagonists. An assay was performed in this direction using C^{14}-methyl-atropine labeled on the methyl group [24]. The sensitivity of the methods then used, permitted no conclusion, for the amount of label present in the pancreas and salivary glands was very close to background. Paradoxically it was the kidney that presented the highest radioactivity. The use of tritiated pancreozymin and secretin coupled with biochemical and radio-autographic studies should also yield interesting results.

Another discussed aspect is if changes in the membrane permeability occur due to cell stimulation. It is pertinent to refer here to the observation of Hirsch [19] that neutral red stains more rapidly and intensely stimulated pancreatic cells than the controls suggesting changes in the ingestion. An analysis of the probable changes in permeability by using substances of different molecular radium would be of interest.

The hormonal mechanism so well exemplified by the action of secretin and pancreozymin on the pancreas has been known for some time and is in this gland very effective. However it does not seem to be so frequent and widespread as the nervous control. The stomach and intestinal secretions are also controlled by hormonal action, but that does not seem to be the case on the salivary glands where so far to our knowledge no specific hormone controlling their secretions has been described [36].

It is interesting to observe that in those glands that have a hormonal control such as the pancreas and thyroid the nervous system apparently does not have a prominent role controlling their secretory activity. One is tempted to speculate on the view set forth in Prosser and Brown [36] who for the digestive system state that a progressive series from nervous to hormonal control is evident. Thus apparently the hormonal mechanism evolved during vertebrate phylogeny as an improvement, substituting to some extent the nervous control.

Referring to hormonal action on glandular activity a clear cut distinction should be made between the effect of specific hormones acting on their target organs and the nonspecific but still important effect of general metabolic hormones on glandular function.

The specific action such as exemplified by the effect of testosterone on the seminal vesicles, oestrogen on the uterus, thyrotropic hormone on the thyroid etc. has received much attention from both the physiological and biochemical point of view. How these substances initiate this stimulation is a mystery. Probably in this field one of the most detailed and promising analyses was the one performed by Mueller (for literature cf. [31]) studying the action on oestrogens on the uterus. In secretory target cells their dependency on the specific hormone is usually almost complete, the glandular secretion depending almost entirely on the presence of the hormone.

On the other hand, emphasis has been given recently to the action of nonspecific hormones on the activity of the pancreas and salivary glands. These organs in mammalia atrophy and present hypofunction after hypophysectomy. This descrease of secretory activity can, and has been treated by replacement therapy with several hormones including thyroxine, cortisone, somatotropine etc. Thus BAKER and co-workers have shown a profound effect of different hormones on the regulation of

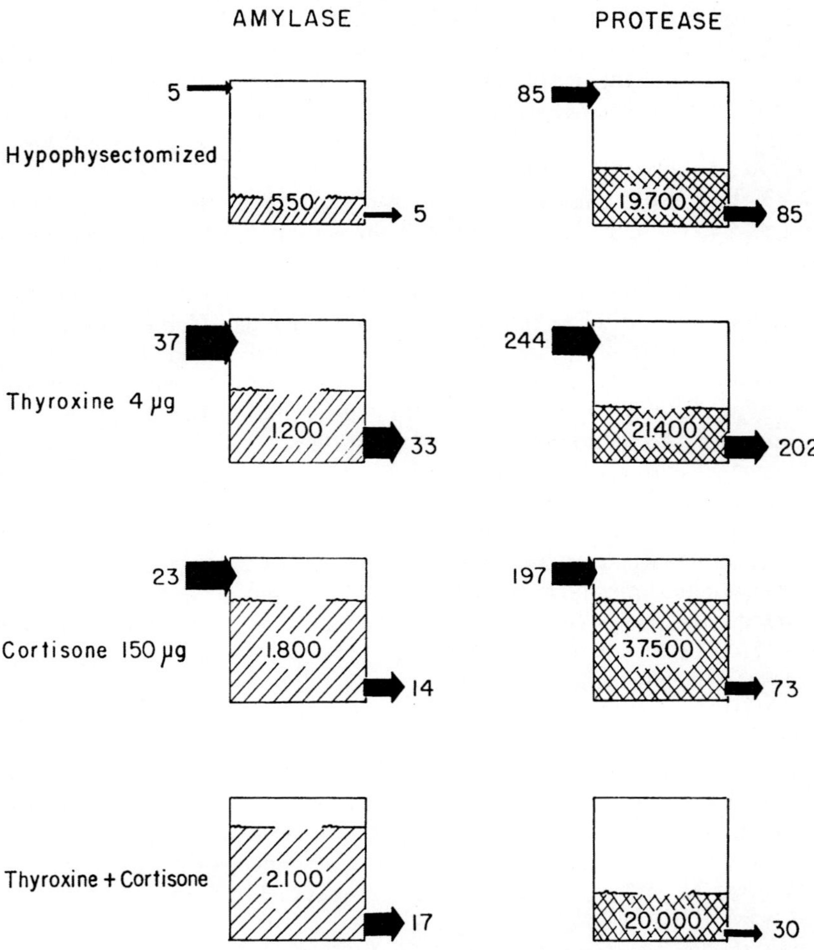

Fig. 1. Action of thyroxine and/or cortisone on the amylase and trypsin activated protease activity of the rat pancreas and pancreatic juice. The arrows to the right represent the enzymes activity per hour in the pancreatic juice and supply an idea of the extrusion process. Observe the considerable increase in the thyroxine treated animals when compared to the other groups. The crossed areas in the quadrangles represent the intracelullar enzymes activities. Observe the accumulating effect of cortisone

zymogenic cells (for literature cf. [1] and [2]). Here contrary to the case of the specific hormone the dependency is not so complete, for secretion still goes on without the action of the hormone.

A more detailed analysis of the action of thyroxine and cortisone on the pancreatic cell has been published recently [39]. By measuring the digestive enzymes activity of the acinar cells and the enzymes secreted into the pancreatic juice of hypophysectomized and thyroxine or/and cortisone injected animals it was possible to conclude that:

1. Thyroxine acts on the acinar cells mainly activating the overall processes of secretion producing therefore an increase in the production of the digestive enzyme. It promotes consequently an increase in the cell efficiency and might be described as having a metabolic effect.

2. The action of cortisone seems to be centered mainly on the processes of ingestion and synthesis with much less effect on the extrusion. The main consequence is therefore an accumulation of protein and enzymes in the cells with no equivalent increase in the overall secretion. The main activity of cortisone therefore is anabolic.

3. Thyroxine plus cortisone have a synergistic effect as one might expect and cellular efficiency is thus increased. It is interesting to observe that this effect depends very much on the dosages of the hormones used.

It must be stressed that the dosages employed by the above mentioned authors were very close to the physiological one as judged from data in the literature. This means that variations in the blood level of these hormones might have an influence on the secretory activity of the pancreas. A summary of these results is presented in Fig. 1.

Recently the action of insulin on the rate of amylase and proteolytic enzyme biosynthesis has been explored by Ben Abdeljlil, Palla and Desnuelle [3].

They demonstrated that in alloxan treated rats, a marked decrease of the pancreatic amylase activity associated with an increase in its proteolytic enzymes level. When insulin is injected in these diabetic rats to restore normal glucose blood levels a reverse of the previous situation was observed with increase of amylase and decrease of protease activities. *In vivo* C^{14}-valine incorporation into pure amylase and chymotrypsinogen showed that the above effect was due to changes in the rate of biosynthesis of these enzymes.

A final comment should be made to the results of Jarett and Lacy [22] who described a catabolic effect of glucagon on the rabbit pancreas. They suggest that this hormone does not act via nervous system, adrenal or pituitary glands and that its most likely mechanism of action is inhibiting the process of synthesis with no parallel decrease of extrusion.

Literature

[1] Baker, B. L.: Ann. N. Y. Acad. Sci. 61, 324 (1955).
[2] —, and E. C. Pliske: Symposium of the Society for Experimental Biology 11, 329 (1957).
[3] Ben Abdeljlil, A., J. C. Palla, and P. Desnuelle: Biophys. Biochem. Res Communic. (in press).
[4] Borghese, E.: J. Anat. 84, 303 (1950).
[5] Borsook, H.: Physiol. Rev. 30, 206 (1950).
[6] Bracroft, J.: J. Physiol. (Lond.) 27, 31 (1901).
[7] Burgen, A. S. V., and N. G. Emmelin: Physiology of the Salivary Glands. London: Edward Arnold Ltd. 1961.
[8] Caro, L. G.: J. biophys. biochem. Cytol. 10, 37 (1961).
[9] Daly, M. M., and A. E. Mirsky: J. gen. Physiol. 36, 173 (1952).

[10] EMMELIN, N., A. MUREN, and R. STRÖMBLAD: Acta physiol. scand. 32, 325 (1954).
[11] FAHRENHOLZ, C.: In: Handbuch der vergleichenden Anatomie der Wirbeltiere, Chapter III, p. 115. Berlin and Vienna: Urban & Schwarzenberg 1937.
[12] FERNANDES, J. F., and L. C. U. JUNQUEIRA: Exp. Cell Res. 5, 329 (1953).
[13] — — Arch. Biochem. 55, 54 (1955).
[14] GABE, M., and L. ARVY: In: The Cell, Chapter V, p. 1. New York-London: Academic Press 1961.
[15] GRASSÉ, P. P.: In: Traité de Zoologie, p. 1789. Paris: Masson et Cie. Ed. 1955.
[16] GROBSTEIN, C.: J. cell comp. Physiol. 60, 35 (1962) (Suppl.).
[17] HERAN, H.: Z. vergl. Physiol. 39, 44 (1956).
[18] HIRSCH, G. C.: Zool. Jahrb. Abt. Physiol. 35, 357 (1915).
[19] — Z. Zellforsch 14, 517 (1931).
[20] — Wiss. Beibl. zur Mat. Med. Nordmark 49, 1 (1964).
[21] HYMAN, L. H.: The Invertebrates. Protozoa through Ctenophora. New York: Hill Book Co. 1940.
[22] JARETT, L., and P. E. LACY: Endocrinology 70, 867 (1962).
[23] JUNQUEIRA, L. C. U., and G. C. HIRSCH: Int. Rev. Cytol. 5, 323 (1956).
[24] —, and H. A. ROTHSCHILD: Unpublished.
[25] — —, and A. FAJER: Exp. Cell Res. 12, 338 (1957).
[26] — —, and I. VUGMAN: Brit. J. Pharmacol. 13, 71 (1958).
[27] KELLER, P. J., E. COHEN, and H. NEURATH: J. biol. Chem. 233, 344 (1958).
[28] KUROSUMI, K.: In: Int. Rev. Cytol. 11, 1 (1961).
[29] LANGSTROTH, G. O., D. R. McRAE, and S. A. KOMAROV: Canad. J. Res. D17, 137 (1939).
[30] LUDWIG, C.: Z. rat. Med. 1, 255 (1851).
[31] MUELLER, G. C.: In: Biological Activities of Steroids in Relation to Cancer. New York: Academic Press 1960.
[32] NORDENFELD, I., P. OHLIN, and B. C. R. STRÖMBLAD: J. Physiol. (Lond.) 152, 99 (1960).
[33] PADYKULA, H. A.: Amer. J. Anat. 91, 107 (1952).
[34] PALAY, S. L.: The Morphology of Secretion. In Frontiers of Cytology, p. 305. Ed. S. L. PALAY. New Haven: Yale Univ. Press 1958.
[35] PERSON, P., R. M. SCHNEIDER, and S. SCAPA: J. Histochem. Cytochem. 9, 197 (1961).
[36] PROSSER, C. L., and F. A. BROWN: Comparative Animal Physiology. Philadelphia: W. B. Saunders, Ed. 1961.
[37] ROTHSCHILD, H. A., G. C. HIRSCH, and L. C. U. JUNQUEIRA: Experientia (Basel) 13, 158 (1957).
[38] SCHNEYER, C. A.: Amer. J. Physiol. 203, 232 (1962).
[39] SESSO, A., V. VALERI, and L. C. U. JUNQUEIRA: Pflügers Arch. ges. Physiol. 277, 473 (1963).
[40] SIEKEVITZ, P.: Exp. Cell Res. Suppl. 7, 90 (1959).
[41] —, and G. E. PALADE: J. biophys. biochem. Cytol. 7, 619 (1960).
[42] STRÖMBLAD, B. C. R.: J. Physiol. (Lond.) 145, 551 (1959).
[43] — Exp. Cell Res. 22, 9 (1961).
[44] TALIAFERRO, W.: J. cell. comp. Physiol. 50 (Suppl. 1), 1 (1957).
[45] TARO, J.: J. Physiol. Soc. Jap. 20, 734 (1958).
[46] THAYSEN, J. H.: Handbuch der exp. Pharmacol. 13, 424 (1960).
[47] VAN WEEL, P. B., and C. ENGEL: Z. vergl. Physiol. 23, 214 (1938).
[48] WACHSTEIN, M.: J. Histochem. Cytochem. 3, 246 (1955).
[49] WARSCHAWSKY, H. C., C. P. LEBLOND, and B. DROZ: J. cell. Biol. 16, 1 (1963).
[50] WELLS, H.: Amer. J. Physiol. 199, 1037 (1960).
[51] WHIPPLE, H. E., S. SILVERZWEIG, and S. JAKOWSKA: Ann. N. Y. Acad. Sci. 106, 1 (1963).

Comparative Aspects
of the Vertebrate Major Salivary Glands Biology*

By

Luiz Carlos Uchôa Junqueira and Flávio Fava de Moraes, São Paulo

With 4 figures

A. Introduction

Interesting developments of the last decade have brought forth results concerning the biology of salivary glands to which hitherto unsuspected functions have been recently ascribed. These glands, much used in the study of secretory processes, present a very interesting distribution and stages of progressive complexity of structure in the vertebrates suggesting that they developed gradually during vertebrate phylogenesis attaining their maximal complexity in the mammals. The vertebrate salivary gland in this sense is quite different from the exocrine pancreas, another structure frequently used for studies on secretion where development occurred very early. Thus while in the lamprey eel the gradual transformation of the intestinal epithelial to exocrine zymogenic cells can be observed [19] in the selachians and teleosts this process is completed with the presence of a ready built pancreatic structure whose basic characteristics will change little during the vertebrate phylogenesis. The evolution of the endocrine portion is apparently a bit slower for its histogenesis from the pancreatic duct cells and can still be observed in the selachians [16].

The salivary glands however have a much more gradual development being absent in the fishes[1], appearing as rudimentary structure in the amphibia, developing further in reptilia and attaining their maximal complexity in the mammalians [5]. There are evidences suggesting that these structures developed from the bucal epithelia of these animals, first by the appearance of secretory granules in the stratified epithelial covering. Secondly unicellular gland cells (usually of the goblet cell type) began to appear among the covering epithelium forming the so-called *diffuse gland* structure.

---

* The work here related was possible due to the grant (DE 1881) from the National Institutes of Health.

[1] To our knowledge the only exception occurs in the lamprey eels (cyclostoma) were a developed bucal gland exists. Very little is known of the biology and structure of this gland. For a description of its anatomical localization and literature on the subject see Fahrenholz [5].

When this process proceeds further these unicellular glands condense in a certain region of the mouth forming a continuous layer. The surface of this layer is then increased by the appearance of folds forming the so-called *crypt fields*. A further development of this structure will result in a circumscribed organ embedded in the mouth wall, called *glandular field*. A progressive deepening of these formations would lead us to glands presenting simultaneously several excretory ducts, the so-called *polystomatic glands*. It is also admitted that a further development of these

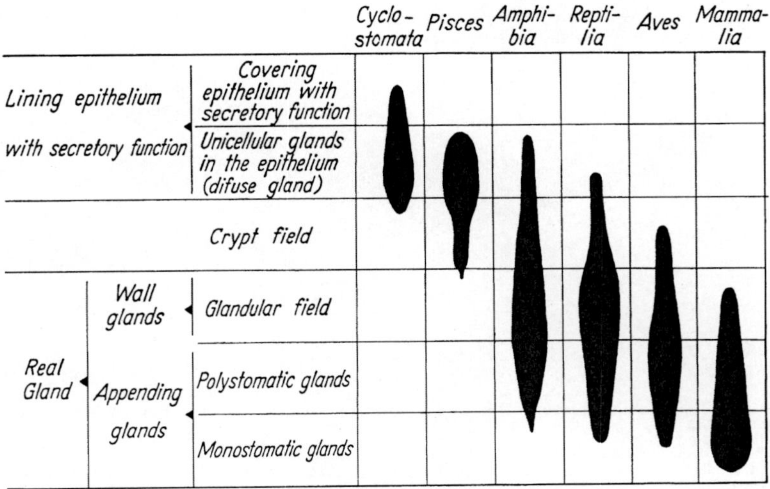

Fig. 1. Graphic representation of the evolution of the salivary glands in the vertebrates. Adapted from FAHRENHOLZ [5]. Observe the gradual appearance of the various stages of development attaining its maximal complexity in the mammals

glands occurs with the fusion of the excretory ducts and final formation of a *monostomatic gland*. It is interesting to observe that a succession of the above described structures can be easily observed during vertebrate phylogeny (Fig. 1).

Salivary glands therefore contrary to the exocrine pancreas developed gradually in the vertebrates. Being essentially from the anatomical, functional and embryological point of view a bucal structure it is to suppose that beside the normal evolutionary modifications above mentioned they have been much influenced by the animals ecology. It seems therefore one of the most favorable materials to study the effect of evolution and ecology on a glandular structure and consequently on cell secretion. To collect further data on this subject a survey of several aspects of mammalian salivary gland biology was initiated one year ago in this laboratory and the initial results are here presented. Special attention is being given to the Chiroptera (bats) for they represent the only mammalian order with six different feeding habits. Thus there are hematofagous, frugivorous, insectivorous, carnivorous, piscivorous and nectarivorous bats [9]. Probably due to the above mentioned facts of gradual development and bucal situation, these structures had the opportunity of being variously influenced during their evolution explaining thus the multiplicity of structure observed and functions attributed to them nowadays.

B. Structure

Anatomically a great variability can be observed on the distribution, variety and size of the salivary glands in mammalia. The extremes are probably on one side the ant eaters (Myrmecophaga) and on the other the Cetacea. In the former the salivary glands not only occupy continuously all the lateral region of these animals' neck but also continue covering the

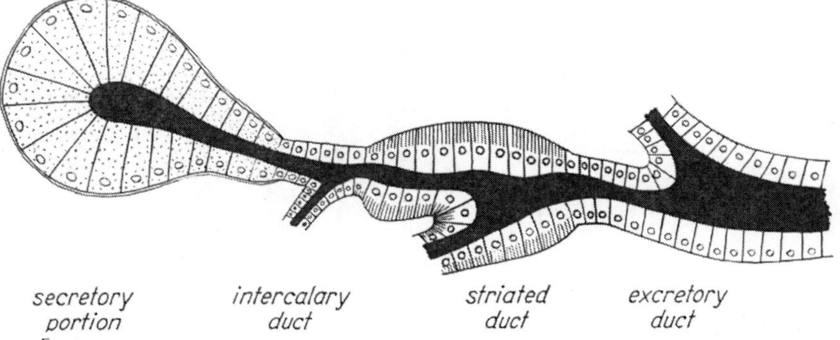

Fig. 2. Graphic representation of the components of a salivary gland adenomere

upper part of the thorax. In the Cetacea however up to now, to our knowledge no definite published description exists of the occurrence of the so-called large salivary glands. Between these both extremes all possible combinations exist. In some animals the submandibular glands are larger than the parotid while in others the opposite is true. As a basis for further discussion a very brief description will be made of the structures that constitute a mammalian salivary gland. These organs are formed by the aggregation of functional units called adenomeres. Each adenomere is composed in the most complete glands of the following portions (Fig. 2):

1) The *secretory portion* usually composed by cells aggregated in acinar or acinotubular form. The cells of this portion can have or not a typical mucous aspect and are usually but not correctly called mucous and "serous" cells. These serous cells have very different morphological and histochemical characteristics and no satisfactory classification for them exists so far and we will therefore provisionally adhere to the above mentioned nomenclature.

2) An *intercalated duct* lined by cuboidal epithelium communicating the preceding structure with the following structure called:

3) *striated duct*, usually composed of mitochondria-rich cylindrical epithelium. The amount of these ducts is extremely abundant in the salivary glands of Monotremes and Marsupials [26].

4) *excretory duct* a structure usually lined by a simple or stratified cuboidal epithelium.

A study of the morphology of these glands in the mammalia shows as compared to the pancreas an enormous variability. Thus the nature and amount of their components changes considerably as one can see from some extreme cases presented in Fig. 3 and 4 and Table 1. To this marked polymorphism expressed by the great variability in the components of a gland other sources of variability may be added such as the changes

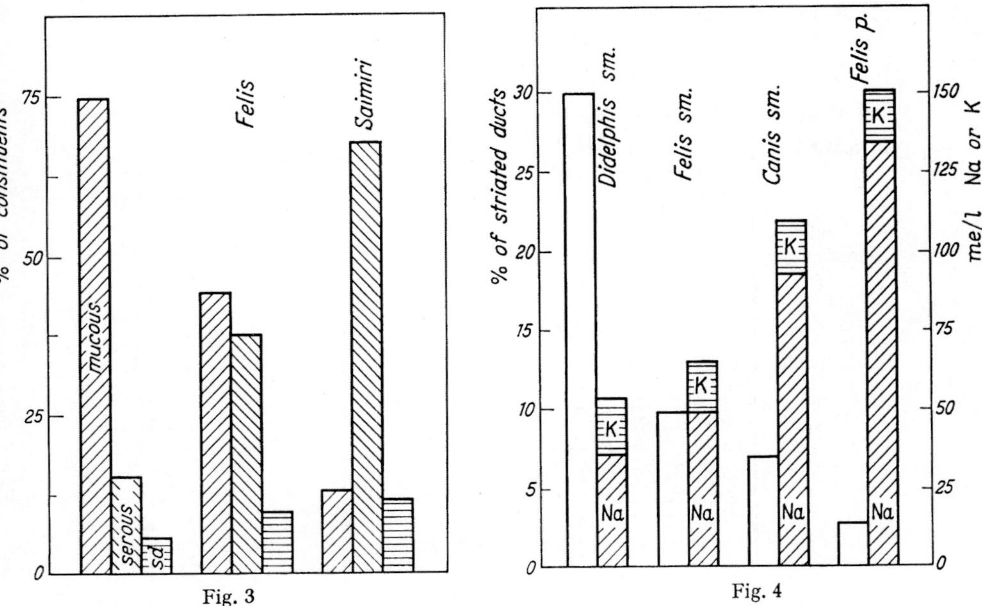

Fig. 3. Relative composition of the constituents in some extreme cases of submandibular glands of different mammals. Observe the clear preponderance of mucous cells in the armadillo *(Cabassous loricatus)* approximately equal amounts of "serous" and mucous components in the cat *(Felis (Felis) domestica)* while the serous portion dominates the picture in a monkey *(Saimiri sciureus)*

Fig. 4. Graphic representation correlating the quantity of striated ducts in salivary glands (open columns) and sodium and potassium content of the saliva obtained from these glands in different mammals. Observe that the amount of sodium present in the saliva is inversely proportional to the quantity of striated ducts. No significant difference could be observed in the potassium content of the different saliva samples. The glands are of the domestic dog and cat and the *Didelphis marsupialis*. *sm* and *p* stand for submandibular and parotid respectively

observed in the striated ducts of some animals who present abundant granules in their cells. This occurs in several animals such as the rat, mouse and hamster. Furthermore this granular change is in some of these animals (rat and mouse) controlled by androgens. The above mentioned are only some of the various sources of variability of these glands here cited as examples to illustrate the possibilities of this polymorphism giving thus morphological basis to a varied function.

C. Function

Several functions have been attributed to mammalian salivary glands and we will analyse those that at the moment seem of greater importance.

Table 1. *Morphological components of the mammalian salivary glands*

Order	Species	Parotid		Submandibular		Sublingual	
		Secretory portion	Modified striated ducts	Secretory portion	Modified striated ducts	Secretory portion	Modified striated ducts
Marsupialia	Didelphis marsupialis	S	—	S—M	—	S—M	—
	Marmosa cinerea	S	—	S—M	—	S—M	—
Insectivora	Blarina brevicaudata			S	+	S—M	—
Chiroptera	Desmodus rotundus	S	+	M	+	M	—
Primates	Callithrix jacchus	S	—	S—M	—	S—M	—
	Cebus sp.	S	—	S—M	—	S—M	—
	Homo sapiens	S	—	S—M	—	S—M	—
	Saimiri sciureus	S	—	S—M	—	S—M	—
Xenarthra	Bradypus tridactylus	S	—	S—M	—	S—M	—
	Cabassous loricatus	S	—	S—M	—	S—M	—
	Dasypus novencinctus	S	—	S—M	—	S—M	—
	Myrmecophaga jubata	S	—	M	—		
Lagomorpha	Lepus cuniculus	S	—	M—S	—		
Rodentia	Cavia porcellus	S	—	S	—	S—M	—
	Citelus beecheye	S	—	S	+	S—M	—
	Citelus tridecilineatus	S	—	S	—	S—M	—
	Cricetus auratus	S	—	S	+	S—M	—
	Cuniculus paca	S	—	S	—	S—M	—
	Dasyprocta azarae	S	—	S	—	S—M	—
	Hydrochoerus hydrochoerus	S	—	S	+		
	Mus musculus	S	—	S	+	S—M	—
	Peromyscus maniculatus austerus	S	—	S	+	S—M	—
	Peromyscus maniculatus maniculatus	S	—	S	—	S—M	—
	Peromyscus maniculatus ontemisae	S	—	S	+	S—M	—
	Rattus norvegicus albinus	S	—	S	+	S—M	—
	Sciurus carolinensis	S	—	S	—	S—M	—
	Tamias striatus	S	—	S	—	S—M	—
Carnivora	Canis brasiliensis	S	—	S—M	—	S—M	—
	Canis familiaris	S	—	S—M	—	S—M	—
	Felis (Felis) domestica	S	—	S—M	—	S—M	—
Artiodactyla	Bos indicus	S	—	S—M	—	S—M	—
	Capra hircus	S	—	S—M	—	S—M	—
	Mazama sp.	S	—	S—M	—	S—M	—
	Ovis aries	S	—	S—M	—	S—M	—
	Sus scrofa domesticus	S	—	S—M	—		

S = Serous cells   M = Mucous cells

I. Production of lubricating substances

utilized as an aid in the process of mastication, capture of insects and swallowing. This seems to be a basic function of this gland as judged by the universal presence of cells producing slimy polysaccharides in all glands studied. Mucous producing cells are particularly developed and dominate the picture in some insect eating animals as the ant eater, armadillo etc. In this last animal even a striated muscle surrounded bladder containing saliva is found in the submandibular gland [5]. In these cases besides aiding in swallowing, these substances have an important role in capturing the insects by adhering them to the tongue surface. The polysaccharides involved in these processes are of three general natures mainly neutral, acid sulfated and sialic acid containing polysaccharides. A survey of these substances in different mammalia so far studied is presented in Table 2. One may observe that all glands

Table 2. *Histochemical distribution of polysaccharides in mammalian salivary glands*

Order	Species	Parotid	Submandibular	Sublingual
Marsupialia	Marmosa cinerea	Np—S	Np—S	Np—S
Chiroptera	Desmodus rotundus	Np—S	Np—S	Np—Sp—S
Primates	Callithrix jacchus	Np—S	Np—S	Np—Sp—S
Xenarthra	Cabassous loricatus	Np—S	Np—S	Np—S
	Dasypus novencinctus	Np—S	Np—S	Np—S
Lagomorpha	Lepus cuniculus	Np—S	Np—S	
Rodentia	Rattus norvegicus albinus	Np—S	Np—S	Np—S
	Mus musculus	Np—S	Np—S	Np—S
	Cricetus auratus	Np—S	Np—S	Np—Sp—S
	Cavia porcellus	Np—S	Np—S	Np—S
	Dasyprocta azarae	Np—S	Np—S	Np—S
Carnivora	Felis (Felis) domestica	Np—S	Np—Sp—S	Np—Sp—S
	Canis familiaris	Np—S	Np—Sp—S	Np—Sp—S
Artiodactyla	Capra hircus	Np—S	Np—S	Np—S

Np = Neutral polysaccharide Sp = Acid sulfated polysaccharide S = Sialic acid

studied present a sialic acid containing and a neutral polysaccharide. Only a reduced number of species present acid sulfated polysaccharide. When this happens it occurs always in the mucous component of the secretory portion. So far, only the carnivora studied present this compound in both submandibular and sublingual glands. In the other species it is only evident in the sublingual. Polysaccharide production seems to be a very basic cellular function as judged by its early appearance in phylogenesis (see preceding paper) and ontogenesis. The observation that the sublingual gland (mainly a mucous producing gland) preceds embryologically the formation of the submandibular and parotid glands in the rat [18] is of interest. Furthermore not only does the sublingual gland initiate its secretion much before the other two [8] but also it has been observed by different authors that several serous glands present a very intense reaction for polysaccharides during the initial stages of development loosing gradually a great part of these characteristics [7]. Further studies on the ontogenesis of these glands is desirable.

The lubricating substances of saliva are produced almost exclusively in the secretory portion as judged from the histochemical evidence available. Only in some species do the intercalary, modified striated ducts and terminal portion of the excretory ducts present mucous producing cells, but the amount however is so reduced that it must contribute very little when compared to the portion originated in the secretory portion. The hamster parotid with abundant mucous secretion in its intercalary duct is the only exception known so far to this rule.

II. Participation in the digestion of food

by secreting enzymes into the saliva. This function so far ascribed only to the amylase is in our opinion underestimated due to insufficient exploration of the field. Thus not only several digestive enzymes exist in the

Table 3. *Enzymatic contents of chiroptera salivary glands*

Species and feeding habit	Sex	No. of animals	Submandibular				Parotid			
			Amylase[1]	Protease[2]	DNAse[3]	Lysozyme[4]	Amylase[1]	Protease[2]	DNAse[3]	Lysozyme[4]
Desmodus rotundus rotundus haematofagous	♂	11	16,62	1,530	0,724	—	24,23	—	—	1,41
	♀	8	—	0,891	0,795	—	9,81	—	—	—
Eumops perotis insectivorous	♂	2	14,43	0,337	—	11,04	19,88	—	—	1,16
	♀	4	20,97	—	—	12,17	30,96	1,045	—	1,16
Mimon bennetti insectivorous	♂	3	11,40	1,023	—	—	10,38	0,620	—	1,68
	♀	3	12,61	0,393	—	—	11,30	0,664	—	1,78
Lonchoglossa caudifera nectarivorous	♂	12	—	0,415	—	—	10,31	—	—	1,37

[1] Amylase = mg glucose/g of fresh weight.
[2] Protease = mg tyrosine/g of fresh weight.
[3] DNAse = Kurnick units/mg of fresh weight.
[4] Lysozyme = mg muramidase/g of fresh weight.
— enzymatic activity if present below methods sensitivity.

saliva, but also their action is possible due to the slow and incomplete mixture of the stomach's global content with the HCl and pepsin secreted. The salivary enzymes therefore have enough time to act efficiently on the stomach content. Calculations of Bergheim suggest that in man up to 60 to 70% of a starch meal can be digested by the salivary amylase [10]. Unpublished results of this laboratory suggest that the same situation exists in the rat regarding the protease activity.

To our knowledge beside amylase, a protease [12], DNAse [24], lysozyme [6] and a fibrinolytic enzyme [4] have been described in the salivary glands and saliva of mammals.

This distribution of enzymes is not restricted to a few exceptional species as one might suppose and as Tables 3, 4 and 5 show. To what extent they participate actively in digestion will have to be worked out in each species. The tables above mentioned present the initial results obtained studying some of these enzymes in different species. It may be seen that so far the species with a richest variety of enzymes present in the salivary glands is the vampire bat *Desmodus rotundus* whose glands contain not only the four enzymes of the table but a strong fibrinolytic

Table 4. *Enzymes in the salivary glands of carnivora*

Species	Submandibular gland				Parotid gland			
	Amylase[1]	Protease[2]	DNAse[3]	Lysozyme[4]	Amylase	Protease	DNAse	Lysozyme
Nasua nasua (8)	58.90	0.90	0.37	—	228.70	—	0.33	—
Felis domesticus (8)	16.70	—	—	—	71.40	—	—	1.18
Canis familiaris (8)	36.60	—	—	—	57.80	—	—	—

[1] mg glucose/g.
[2] mg tyrosine/g.
[3] Kurnick units/mg.
[4] mg muramidase/g.
— enzymatic activity if present below method sensitivity.

activity that according to preliminary tests of this laboratory seems to be different from its caseinolytic (protease) property. Curiously enough in the nectarivorous bat only few enzymes exist and even so with such a

Table 5. *Enzymes distribution in rodentia salivary glands*

Species	Submandibular gland				Parotid gland			
	Amylase[1]	Protease[2]	DNAse[3]	Lysozyme[4]	Amylase	Protease	DNAse	Lysozyme
Rattus norvegicus alb (11)	9.70	8.00	0.696	—	65,782.20	0.918	3.303	
Rattus rattus (6)	6,513.61	29.47	—	—	186,355.00	0.710	2.452	—
Cavia porcellus (8)	55,300.00	1.30	—	—	135,662.00	1.370	—	—
Mus musculus (7)	505.00	20.40	—	—	23,831.00	1.60	—	—
Dasyprocta azarae (4)	159.70	—	0.61	—	19,665.00	—	0.54	—
Cricetus auratus (10)	206.35	1.16	—	—	54,498.17	0.618	1.211	1.702

[1] mg glucose/g.
[2] mg tyrosine/g.
[3] Kurnick units/mg.
[4] mg muramidase/g
— enzymatic activity if present below methods sensitivity.

reduced activity that they have probably no essential physiological significance. The rodentia also have a rich variety of enzymes present in their glands while the carnivora are so far as we know poorly equiped. It is curious to note that the Nasua the most omnivorous of the three animals of this order has also a larger variety of enzymes present. The evidence collected strongly suggests that most of the enzymes studied so far are produced at the secretory portion. An exception is the rat and mouse protease probably secreted in the modified portion of the striated ducts [12].

Enzymes are however not restricted to mammalian salivary glands. It is known that the so-called poison gland of the ophidia contains several enzymes, and enzymes have also been described in amphibian glands [7]. Nothing is known of the importance that these enzymes might have for the digestive process in these animals.

III. Ion transport

The ion transporting activity of some salivary glands seems an established fact. This activity seems to be located in the striated ducts. The main observations that suggest the above mentioned localization of this activity are:

1) Relation between the presence of striated ducts and hypotonic saliva [28]. Recent measurements on saliva ion concentration as related to the amount of striated ducts in different glands of different animals confirm this view (Fig. 4).

2) The fact that experimental cytological changes of the striated ducts influence the saliva ion content [13].

3) Results obtained with kinetic methods also confirm this view [2, 17]. Caution however should be exerted in the interpretation of these results for anatomical data suggest that the salivary ducts are of greatly varying length promoting therefore the mixture of ductal and acinar fluid along the excretory portion [23].

4) Micropunction experiments at the acinar and post striated duct level [29]. These results however should be interpreted with great caution due to the fact that the striated duct very probably secretes ions and water from the capillaries to the lumen as demonstrated by the presence of active salivary secretion in acini deficient glands [23].

IV. Production of pharmacologically active substances

Recently active proteins or polypeptides have been described in the salivary glands of mammals. A bradykinin-like compound was described some time ago [11] in the cat submaxillary secretion while a compound with an angiotensin II activity was also isolated from the mouse submaxillary gland [27]. The same authors isolated kallikrein from submaxillary gland and described a sharp drop of its activity after the ligation of the excretory ducts.

A series of very interesting studies have culminated in the isolation of two active proteins from the mouse submaxillary gland, one affecting the growth of the nervous sympathetic system [21] and the other acting on epidermal keratinization and tooth eruption [3]. An active poison has also been described in the saliva of the shrew *Blarina brevicaudata* [22].

The mammalian salivary gland therefore has revealed so far a surprising amount of pharmacologically active compounds and one feels that further research in this field would be fruitful. Very little is known as to the site of production or accumulation of these compounds. To our knowledge only in the case of the nerve growth factor it has been possible to localize it at the level of the modified striated duct of the mouse submaxillary gland [21]. Doubt however persists if it is produced or accumulated in these structures. If the striated duct is responsible for the production or accumulation of the other compounds as indirect evidences suggest in some cases is still to be determined.

V. Passage of blood proteins

Recently immunological evidence has been presented demonstrating the passage of blood proteins through the salivary glands into the saliva. This has been observed in the man [25] and rat [14]. Furthermore in the last mentioned species the mode of stimulation is important for saliva obtained after adrenergic stimulation contains ten times more blood protein than in saliva obtained after cholinergic stimulation [14]. This confirms the results presented recently showing an increase of glandular permeability after adrenergic stimulation. It is suggested that at least part of this increased permeability occurs at the level of the duct system [20].

VI. Reabsorption of blood proteins

Evidence has been presented recently suggesting that in the rat after cholinergic stimulation blood amylase may accumulate in the salivary gland attaining levels of activity up to twenty times before stimulation. This amylase accumulation probably occurs at the so-called modified portion of the striated duct [15].

VII. Dilution of food

In the ruminants salivary glands are not only developed but their parotids are modified, disappearing the visible secretory granules of the secretory cells and acquiring the capacity of producing large amounts of watery secretion. In the cow this may sum up to about 60 liters per 24 hours. These large quantities of saliva are undoubtedly very important in the complex digestive processes known to occur in these animals stomach. This active fluid transport probably occurs in these animals mainly at the secretory portion of their parotid glands.

C. Correlation between structure and function

From the above mentioned observations one is tempted to locate the site of production of the polysaccharides and of most of the digestive enzymes in the so-called secretory portion. Exceptionally enzymes may be produced in the modified striated duct cells. The function of the intercalary ducts is a mystery. In several species the cells of these structures present abundant secretory granules of unknown significance.

Undoubtedly the striated duct is the most versatile of all structures of the salivary glands for beside producing a digestive enzyme it participates also in the mechanisms of ion and water transport and in the production and or accumulation of at least one pharmacologically active compound. Furthermore the available evidences so far collected suggest that at the rat modified striated duct accumulation of blood amylase occurs after cholinergic stimulation.

D. Final comments

A bird's-eye view of the field suggests to us the following considerations.

In mammalians the salivary glands assume as compared to other organs an exceptional anatomical, histological and functional polymorphism. This probably occurs due to the facts that these glands not only developed slowly during vertebrate phylogenesis but also are closely related to the oral cavity at which external factors such as habitat and feeding habits may act efficiently. The so far collected evidence however does not suggest a strong influence of the terrestrial or aquatic habit on these glands. Thus while certain aquatic mammals such as Cetacea have reduced or no glands at all others such as the Manatee have large glands. Apparently the feeding habit has a stronger influence as suggested by the presence of relatively large glands in the rodents, herbivorous and insectivorous animals while in the carnivora they are less developed. The action of the feeding habit on the digestive function seems however to be much larger as our tables on enzymes suggest. The results with bats are quite clear cut in this connection. It is interesting to observe that in the exclusively carnivorous mammals where mastication is practically inexistant very few enzymes exist and furthermore their activities are so low as to throw doubt on their digestive functions.

The only evolutionary trend observable so far in our preliminary observations is the tendency towards diminution in quantity of the striated ducts. This structure present to our knowledge only in the mammalian salivary glands is very abundant in the glands of the Monotreme (Tachiglossus) and Marsupials (oppossums) studied so far. In the other orders no such richness exists. One might speculate that these structures appeared precociously and florished during mammalian early phylogenesis decreasing posteriorly. It is interesting to observe that the primates' major salivary glands from the morphological and enzymatic point of view do not seem to be the most complex. Apparently these organs attained a more advance stage of development in some rodents, bats and probably insectivora.

It is evident that the speculations above mentioned are provisional and represent only preliminary conclusion based on the animals so far studied. They are therefore subject to later amplifications or modifications and are presented only as a guide to further research in this field.

Literature

[1] BIGNARDI, C.: Riv. Histoch. norm. pat. 7, 231 (1961).
[2] BURGEN, A. S. V.: In: Salivary Glands and its Secretions. vol. III, p. 197. Oxford: Pergamon Press 1964.
[3] COHEN, S., and G. A. ELLIOT: J. invest. Derm. 40, 1 (1963).
[4] DI SANTO, P. E.: J. Morph. 106, 301 (1960).
[5] FAHRENHOLZ, C.: In: Handbuch der vergleichenden Anatomie der Wirbeltiere. Chapter III, p. 115. Berlin-Vienna: Urban & Schwarzenberg 1937.
[6] FLEMING, A., and D. ALLISON: Brit. J. exp. Path. 3, 252 (1922).
[7] FRANCIS, E. T. B.: Proc. Zool. Soc. (Lond.) 13, 453 (1961).
[8] GERSTNER, R., H. FLON, and E. O. BUTCHER: J. Dent. Res. 42, 1250 (1963).
[9] GRASSÉ, P. P.: In: Traité de Zoologie. p. 1789. Paris: Masson et Cie. Ed. 1955.
[10] HAWK, P. B., B. L. OSER, and W. H. SUMMERSON: Practical Physiological Chemistry. p. 354. New York: McGraw-Hill 1954.
[11] HILTON, S. M., and G. P. LEWIS: J. Physiol. (Lond.) 130, 43P (1955).
[12] JUNQUEIRA, L. C. U., A. FAJER, M. RABINOVITCH, and L. FRANKENTHAL: J. Cell. comp. Physiol. 34, 129 (1949).
[13] —, A. M. S. TOLEDO, and F. B. DE JORGE: In: Salivary Glands and its Secretions. vol. III, p. 119. Oxford: Pergamon Press 1964.
[14] — —, and R. G. FERRI: Arch. oral Biol. In press.
[15] — —, and A. SAAD: In: Salivary Glands and its Secretions. vol. III, p. 105. Oxford: Pergamon Press 1964.
[16] KERN, H.: Z. Zellforsch. 63, 134 (1964).
[17] LANGLEY, L. L., and R. S. BROWN: Amer. J. Physiol. 199, 59 (1960).
[18] LEESON, C. R., and W. G. BOOTH: J. Dent. Res. 40, 838 (1961).
[19] LUPPA, H.: Z. mikr. anat. Forsch. 71, 85 (1964).
[20] MARTIN, K., and A. S. BURGEN: J. Gen. Physiol. 46, 225 (1962).
[21] MONTALCINI, R. L., and S. COHEN: Ann. N. Y. Acad. Sci. 85, 324 (1960).
[22] PEARSON, P. O.: Publ. Amer. Ass. Advanc. Sci. 44, 55 (1956).
[23] SCHNEYER, L. H., and C. A. SCHNEYER: Secretion of Saliva. In: Advances in Oral Biology. vol. I, p. 31. New York: Academic Press 1964.
[24] SCHRAMM, M.: In: Salivary Glands and its Secretions. vol. III, p. 315. Oxford: Pergamon Press 1964.
[25] STOFFER, H. R., F. W. KRAUS, and H. C. HOLNOS: Proc. Soc. exp. Biol. (N. Y.) 111, 467 (1962).
[26] TUCKER, R.: Syst. Zool. 7, 74 (1958).
[27] WERLE, E., and I. TRAUTSCHOLD: Ann. N. Y. Acad. Sci. 104, 117 (1963).
[28] WHITE, A. G., P. S. ENTEMACHER, G. RUBIN, and L. LEITER: J. clin. Invest. 34, 246 (1955).
[29] YOSHIMURA, H., T. INONE, T. FUJIMOTO, and M. TOYOKI: Setai no Kagaku 11, 286 (1960).

Discussion

Ullrich: Meine Mitarbeiter, MARTINEZ, HOLZGREVE und FRICK, haben im vergangenen Jahr Mikropunktionsuntersuchungen an der Submaxillardrüse der Ratte durchgeführt und gefunden, daß das Sekret im Schaltstück (intercalated duct) isoton ist und eine Natriumkonzentration von 146 mÄq/l und eine Chloridkonzentration von 121 mÄq/l hat. Im Ausführungsgang am Hilus sind sowohl der osmotische Druck wie auch die Na^+- und Cl^--Konzentration etwas niedriger (225 mosmol/l; Na 117 mÄq/l; Cl 108 mÄq/l). Erst auf der Strecke des Hauptausführungsganges vom Hilus bis zur Ausmündung in die Mundhöhle sinken die Konzentrationen stark ab (Osmolarität auf 89 mosmol/l; Na 21 mÄq/l; Cl 55 mÄq/l). Die Natriumrückresorption, die zum Hypotonwerden des Speichels führt, erfolgt also nicht so sehr in den Streifenstücken, wo man sie bisher vermutet hatte, als vielmehr im Hauptausführungsgang.

Außerdem wurden im Speichel, der durch Mikropunktionen an den genannten Stellen gewonnen worden war, Messungen der Amylaseaktivität durchgeführt (MARTINEZ und EMRICH). Sofern der Einfluß unterschiedlicher Chloridkonzentration bei der Aktivitätsmessung ausgeschaltet war, wurde die Amylaseaktivität überall gleich hoch gefunden. Dies wurde so interpretiert, daß Amylase in den Acini sezerniert wird und das Sekretvolumen bis zur Mundhöhle konstant bleibt, also mit den Elektrolyten nicht auch Wasser rückresorbiert wird.

Junqueira: Die Ratte scheint mir kein sehr günstiges Tier zu sein, um das durchzuführen, weil sich das Streifenstück in verschiedener Hinsicht von dem anderer Tiere unterscheidet, so daß man ein Tier nehmen sollte, das ein echtes Streifenstück hat. Eine Erörterung der Einzelheiten würde hier zu weit führen.

Bücher: Der Mensch hat bekanntlich sehr viel Amylase. Aber die Amylase wird im Magen schnell zerstört, und sie kommt sehr schnell in ein pH-Milieu, in dem sie nicht mehr wirken kann. Ist nicht vielleicht eine wesentliche Funktion dieser Amylase die Reinigung des Gebisses, daß also die Stärke, die sich in den Winkeln der Zähne verkriecht, durch diese Speichelenzyme aufgelöst wird und so neben der mechanischen Reinigung durch den Speichel auch noch eine chemische, enzymatische Reinigungsfunktion bei den höheren Tieren existiert? Wieviel Prozent der Stärke wird schon durch die Speichelamylase verdaut?

Junqueira: Nach Bergheim etwa 50%.

Duspiva: Es ist hier ein Effekt im Spiel, der in erster Linie die Viskositätsherabsetzung eines Stärkebreies betrifft. Wenn man Amylase in einen Stärkebrei gibt, dann sinkt in ganz kurzer Zeit die Viskosität dieses Stärkebreies außerordentlich stark herab, ohne daß man chemisch wesentliche Mengen Maltose dabei findet. Die physiologische Bedeutung dieser Erscheinung scheint damit zu tun zu haben, daß ein in den Magen aufgenommener Stärkebrei, ohne daß er chemisch sehr stark abgebaut wird, den Mageninhalt verflüssigt und damit dem transportierenden Hinterteil des Magens die Arbeit erleichtert. Erst im Darm findet dann die gründliche Hydrolyse der Stärke statt mit den großen Mengen Amylase, die das Pankreas liefert.

Biochemical Aspects of Excitation of Protein Secretion in Pancreas

By

Lowell E. Hokin and Mabel R. Hokin, Madison

With 9 figures

Studies over the past several years have indicated that when secretion is stimulated in a variety of endocrine and exocrine glands there is an increased incorporation of labeled precursors into certain phospholipids (see reviews [5, 6, 8–15]). The phospholipids chiefly affected are phosphatidyl inositol and phosphatidic acid, although certain of the other phosphatides may be affected in certain tissues. Kinetic studies have indicated that the phospholipid effects in different glands are probably not the same phenomenon. For example, in the salt gland the changes in phosphatidyl inositol and phosphatidic acid are quite different than in the pancreas and appear to serve different functions. The changes in the salt gland have been described elsewhere [7, 14–16]. In this presentation we shall confine our discussion primarily to the changes in phospholipid metabolism on stimulating protein secretion in the pancreas.

Our task is to determine the mechanism of this increased incorporation of isotopes into the phospholipids and to determine what role it plays in the overall process of glandular excitation and secretion. We have used several approaches to this problem — kinetic studies, the use of various labeled precursors, differential centrifugation, and tissue radioautography. Unfortunately, we cannot obtain the phospholipid effect in cell free preparations of glands, so we are somewhat limited in our attempts to elucidate the enzymatic mechanisms of the phospholipid stimulations.

Kinetic studies in the pancreas. The rate of uptake of P^{32} into pancreas slices is gradual; several hours of incubation are required before the intracellular inorganic phosphate equilibrates isotopically with the extracellular inorganic phosphate. In the experiment shown in Fig. 1 the slices were incubated for 40 minutes with P^{32} before further additions, so that the changes in specific activity of the intracellular inorganic phosphate over the subsequent short incubation periods were minimized. Under the conditions of these experiments the specific activity of the terminal phosphate of the ATP and the ADP maintains isotopic equilibrium with the intracellular inorganic phosphate. After the initial 40 minutes preincubation the slices were transferred to non-radioactive media without and with pancreozymin. It will be noted that in slices with pancreozymin there was a striking increase in the incorporation of

P^{32} into phosphatidyl inositol. This increase was linear over most of the period after adding pancreozymin. There was a rapid burst in incorporation into phosphatidic acid although of a much smaller magnitude followed by a slow fall. Phosphatidyl ethanolamine showed a linear stimulation like phosphatidyl inositol, but this stimulation was much less. There was no significant stimulation in lecithin.

As we shall show, the rapid increase in incorporation into phosphatidic acid, followed by a fall, was due to a small fraction of the total phosphatidic acid undergoing rapid turnover and rapidly reaching isotopic equilibrium with its precursor. The gradual fall in this particular experiment was due to the fact that the media containing pancreozymin, to which we transferred the slices at 40 minutes, did not contain radioactive phosphate. Thus, the specific activity of the intracellular inorganic phosphate fell slowly due to its replacement by nonradioactive

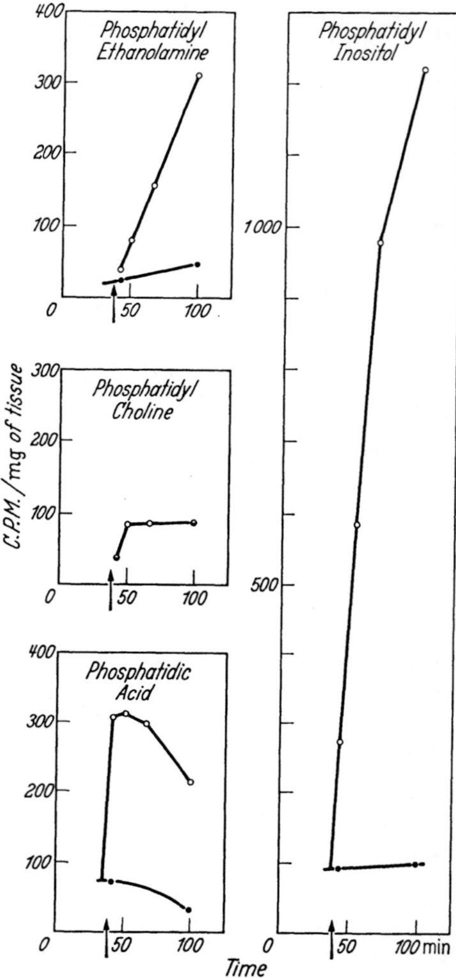

Fig. 1. Time course of the effect of pancreozymin on the amount of P^{32} found in phospholipids in pigeon pancreas slices. Open circles-slices stimulated with pancreozymin. Solid circles-control slices. The arrow indicates the time of transfer to a medium which contained pancreozymin. Approximately 30 mg fresh weight of pigeon pancreas slices were incubated at 38° with shaking, in 1 ml of bicarbonate saline with added glucose (1 mg/ml) and ortho-phosphate-P^{32} (approximately 40 μC/ml) for 40 minutes. They were then transferred to fresh medium of the same composition, except that P^{32} was omitted. The gas phase was 95% O_2 + 5% CO_2 throughout the incubation period. The pancreozymin preparation was Cecekin (Vitrum, Stockholm) and the concentration was 5 Ivy dog units of cholecystokinin per ml. Pancreozymin was present as indicated. Samples were taken at various times after transfer. On removal from the medium the slices were blotted lightly and were then frozen in tubes in an acetone-CO_2 bath; they were kept frozen until workup. Lipids were extracted and separated by paper chromatography [7]. Counts are corrected to a specific activity of 10^6 counts per minute per microgram of phosphorus for the ortho-phosphate of the first medium. The experiment was carried out in triplicate with tissue from one pancreas. Each point is the average of the three individual values obtained

inorganic phosphate from the medium. Due to the rapid turnover of a small fraction of the phosphatidic acid its radioactive phosphate was being measurably replaced by phosphate of progressively lower specific activity. The kinetics with phosphatidyl inositol and phosphatidyl ethanolamine do not indicate a rapid replacement of their phosphate. Rather, the kinetics indicate an accumulative process in which the newly synthesized molecules are not rapidly broken down.

Fig. 2 demonstrates more directly that the leveling off in radioactivity in phosphatidic acid is due to rapid turnover of a small fraction. The slices were incubated without and with pancreozymin for 30 minutes,

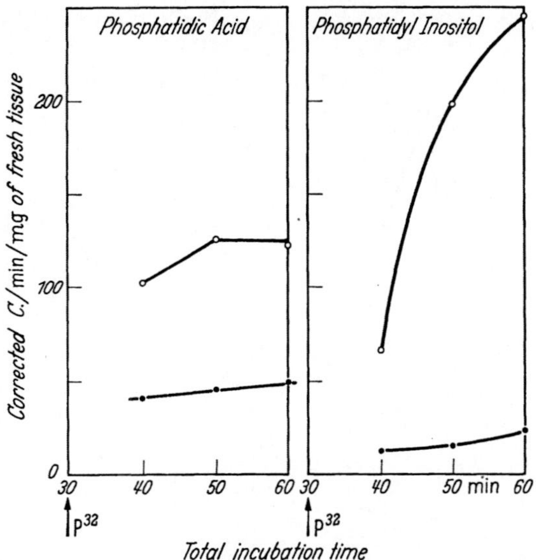

Fig. 2. Time course of the continuing response to pancreozymin. Open circles-slices in contact with pancreozymin from zero time. Solid circles-control slices. Pigeon pancreas slices were incubated in non-radioactive medium in the presence or absence of pancreozymin as indicated for 30 minutes. Ortho-phosphate-P^{32} was then added to the medium and the incubation was continued for a further 10, 20, and 30 minutes. Other conditions were as in Fig. 1. The radioactivity in phosphatidic acid has been corrected to a value of 10,000 counts per minute per mg of fresh tissue for the Norite-adsorbable 7 minute acid-hydrolysable phosphate esters of the tissue; the measured values for this fraction were 1060, 2300, and 3030 counts per minute per mg of tissue at 10, 20, and 30 minutes respectively after addition of P^{32} to the medium. Pancreozymin did not affect these values; it had no effect on the level or specific activity of the Norite-adsorbable, 7 minute acid-hydrolysable phosphate esters under these conditions

P^{32} was then added, and the slices were removed at 40, 50 and 60 minutes. The radioactivities in phosphatidic acid were corrected to a constant radioactivity in the ATP. If the leveling off and gradual fall in radioactivity in phosphatidic acid, seen in Fig. 1, were due to a burst of synthesis of phosphatidic acid, followed by slow breakdown, we should see no more radioactivity in the stimulated slice than in the control slice in the experiment shown in Fig. 2, since the burst of synthesis would have been over long before the 30 minutes exposure to pancreozymin was completed. However, we see that there is more radioactivity in the stimulated slice and that this radioactivity is constant with respect to the specific activity of the ATP. The combined results of Figs. 1 and 2 permit the conclusion that there is a fraction of phosphatidic acid which undergoes rapid and continuous turnover on stimulation with pancreozymin. From the total phosphatidic acid level in pancreas slices [4] and based on the assumption that the specific activity of the precursor of phosphatidic acid is the same as that of ATP, the size of the phosphatidic acid compartment undergoing continuous turnover is calculated to be about 10% of the total phosphatidic acid of the cell. Fig. 2 again shows an accumulative type of labeling in phosphatidyl inositol, further indicating that new molecules of phosphatidyl inositol are being synthesized and that these

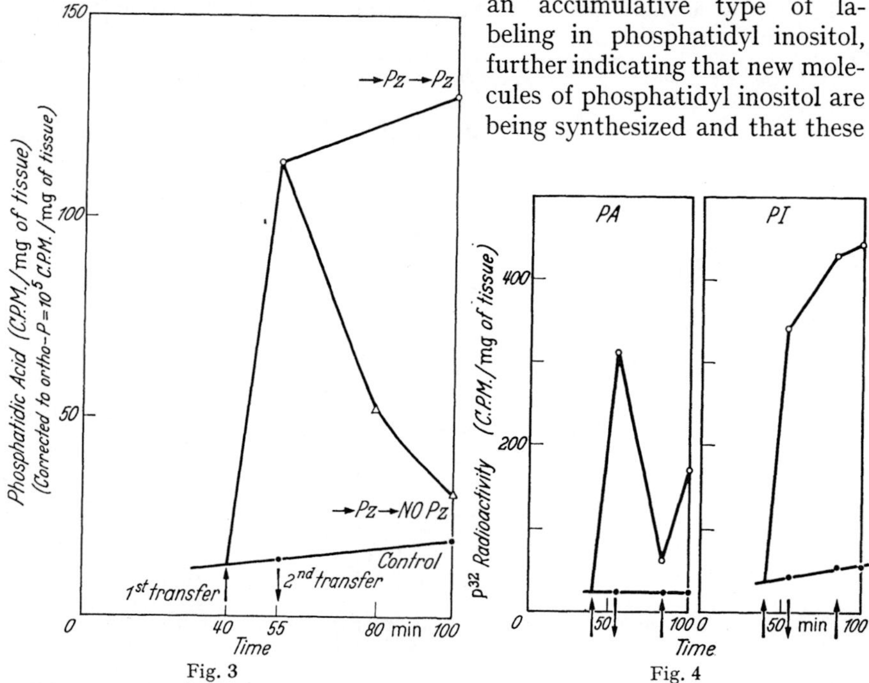

Fig. 3. Loss of P[32] radioactivity from phosphatidic acid on removal of pancreozymin. Open circles-slices were first transferred at 40 minutes to medium with added pancreozymin; they were then either removed for workup at 55 minutes or were transferred to fresh medium with pancreozymin and incubated for a further 45 minutes. Open triangles-slices were transferred at 40 minutes to medium with added pancreozymin and were then transferred at 55 minutes to fresh medium without added pancreozymin and incubated for a further 25 or 45 minutes. Solid circles-slices were transferred each time to medium without pancreozymin. The transfer medium had a specific activity one fifth of that of the first medium. Other conditions were as in Fig. 1

Fig. 4. Radioactivity in phosphatidic acid and phosphatidyl inositol in response to a second exposure to pancreozymin after a recovery period. Open circles-slices were incubated in the presence of orthophosphate-P[32] for 40 minutes; they were then transferred to fresh medium with added pancreozymin (Cecekin 5 Ivy dog units/ml); after a further 15 minutes they were either removed for workup or were again transferred to fresh medium, this time with no pancreozymin; after a further 25 minutes they were either removed for workup or were transferred to fresh medium which contained pancreozymin and incubated for a further 20 minutes. Solid circles-slices were incubated and transferred as above but no pancreozymin was present in any of the vessels. In all cases, transfers were made to medium which contained radioactive phosphate which had a specific activity one fourth the specific activity of the first medium. The specific activity of the tissue orthophosphate under these conditions did not change during the 40 to 95 minute incubation period. The values have been corrected to a specific activity of 10[6] counts/min/µg P for the orthophosphate of the transfer medium, which under these conditions was the same specific activity as that of the orthophosphate of the tissue. The other experimental conditions were as in Fig. 1. The experiment was carried out in triplicate; each point is the average of the three individual values obtained

newly synthesized molecules are not being broken down at a very appreciable rate.

Fig. 3 shows what happens if the pancreozymin is removed from pancreas slices. In this experiment the slices were incubated with P^{32} for 40 minutes. They were then transferred to media containing inorganic phosphate with a specific activity the same as that attained by the intracellular inorganic phosphate at 40 minutes. This technique stabilizes the specific activity of the intracellular inorganic phosphate and the ATP [16]. Some of the media to which the slices were transferred contained pancreozymin and others did not. Those media with pancreozymin showed the usual rapid upshoot in labeling in phosphatidic acid. At 55 minutes some of the slices were transferred to media without pancreozymin and some were transferred to media again containing pancreozymin. It can be seen that in those media without pancreozymin, slices which had previously been exposed to pancreozymin lost radioactivity in phosphatidic acid. This is in spite of the fact that the specific activity of the inorganic phosphate remained constant. This can only mean that on removal of the secretory stimulation the molecules of phosphatidic acid which were undergoing continuous turnover were converted to something else.

Fig. 4 shows that if slices which had lost radioactive phosphatidic acid were again transferred to pancreozymin there was another upshoot in labeling in phosphatidic acid, indicating that the pool of phosphatidic acid which continuously turns over can be recruited again.

These data allow us the following conclusions. When enzyme secretion in the pancreas is stimulated there is an increased synthesis of phosphatidyl inositol and phosphatidyl ethanolamine. Whether these syntheses are matched by an equal and simultaneous breakdown of these lipids is not known. There is a rapid recruitment and continuous turnover of a compartment of phosphatidic acid which is about 10% of the total phosphatidic acid. When stimulation stops this phosphatidic acid disappears. This compartment of phosphatidic acid can be recruited again.

In the salt gland there is a recruitment of a compartment of phosphatidic acid which undergoes rapid and continuous renewal of its phosphate during stimulated sodium secretion [14–16]. In the salt gland, the phosphatidic acid undergoing this rapid turnover is not serving as an intermediate for the synthesis of other lipids [16]. However, we cannot conclude this in the pancreas because of the very rapid synthesis of phosphatidyl inositol taking place under conditions in which phosphatidic acid is undergoing rapid turnover. It is known that phosphatidic acid can serve as a precursor for phosphatidyl inositol [23]. In the pancreas, the shapes of the phosphatidic acid and phosphatidyl inositol curves, as shown for example in Fig. 1, are compatible with a precursor-product relationship between phosphatidic acid and phosphatidyl inositol. We are thus left with two functional possibilities for the compartment of phosphatidic acid which is formed and which undergoes continuous turnover in response to either pancreozymin or acetylcholine in the pancreas: (1) that it functions as a precursor in the synthesis of other lipids, particularly phosphatidyl inositol. (2) that it represents the participation of phosphatidic acid in a special cycle in which phosphatidic acid is continuously regenerated — as in the salt gland [14, 16].

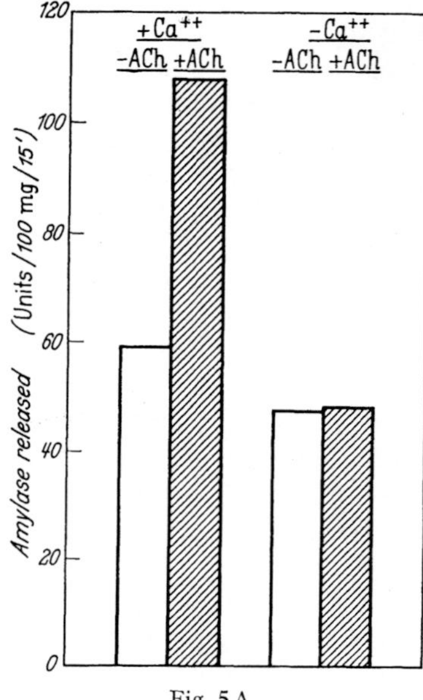

Fig. 5 A

Incorporation of precursors other than P[32]. The increased incorporation of P[32] into phosphatidyl inositol on stimulation of secretion in pancreas is accompanied by an increased incorporation of glycerol-1-C[14] and inositol-2-H[3], and the increased incorporation of P[32] into phosphatidyl ethanolamine is associated with an increased ethanolamine-C[14] incorporation [6]. The increased incorporation of P[32] into phosphatidic acid is also accompanied by an increased incorporation of glycerol-1-C[14]. These observations are all compatible with increased rates of synthesis of these phosphatides.

Is the phospholipid effect in pancreas concerned with reverse pinocytosis? Electron micrographs of Palade [20] as well of those of Fawcett [3] have indicated that the contents of the

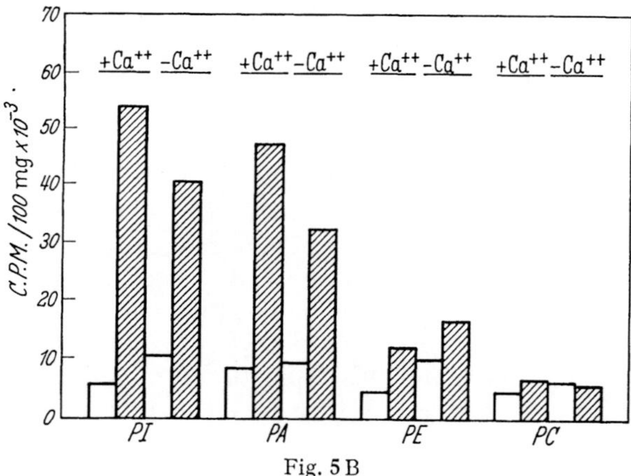

Fig. 5 B

Fig. 5. Effect of acetylcholine on amylase secretion and on the incorporation of P[32] into the phospholipids in the presence and absence of calcium. A. Amylase secretion. B. Radioactivity in the phospholipids. Pigeon pancreas slices were shaken at 0° C for 15 minutes in 0.9% NaCl containing 0.72 mM sodium ethylene diamine tetraacetate. They were then incubated for 40 minutes either in Krebs-Henseleit bicarbonate saline containing 200 mg-% glucose with the usual calcium concentration or with calcium ommitted. P[32]-orthophosphate (specific activity, 1.41 × 10⁶ c.p.m. per μg P) was present in all vessels. The slices were then transferred to identical media but containing P[32]-orthophosphate with one-fifth the original specific activity. Half of each series of vessels with calcium and without calcium contained 10^{-5} M acetylcholine and 10^{-4} M eserine. Incubation was continued for an additional 15 minutes. All other conditions were as described in Fig. 1. *PI* phosphatidyl inositol, *PA* phosphatidic acid, *PE* phosphatidyl ethanolamine, *PC* phosphatidyl choline. Amylase was assayed as described previously [6, 12]

zymogen granules are released by reverse pinocytosis (the term pinocytosis is used since it is assumed that the contents of the zymogen granules are more fluid than solid). In this process the membranes of the zymogen granules fuse with the plasmalemma, the fused membranes pull apart, and the contents of the zymogen granule are discharged into the acinar lumen. We have recently obtained evidence that the phospholipid effect is not correlated with the pinocytosis process *per se*. We have assumed that the release of enzymes on stimulation of pancreas slices with acetylcholine or pancreozymin is a measure of pinocytosis. We have used amylase release as a measure of total enzyme release; it was shown several years ago that the release of amylase by pigeon pancreas slices in response to cholinergic stimulation is paralleled by the release of other digestive enzymes [29]. DOUGLAS and POISNER [7] showed that perfusion of the submaxillary gland with calcium free Ringers solution prevents the secretory response to acetylcholine or adrenaline. As shown in Fig. 5A if pancreas slices were washed with ethylene diamine tetraacetate, rinsed, and then incubated without and with calcium it was found that there was no stimulation of amylase secretion in the absence of calcium. Fig. 5B shows the phospholipid changes under these conditions. Although there was some diminution of the phospholipid effect in the absence of calcium, the stimulation in the phospholipids was still very appreciable. Thus, most of the phospholipid effect can be obtained under conditions in which there is no secretion of enzyme into the incubation medium. It should be pointed out that the effects of calcium ommission cannot be ascribed to the known dependence of amylase activity on calcium [17, 30], since the amylase level in the media of control slices presoaked with ethylene diamine tetraacetate and then incubated without calcium was not very different from the level in the media of control slices presoaked with ethylene diamine tetraacetate and then incubated with calcium. Calcium is removed from amylase only with considerable difficulty [17, 30], and presumably it was not removed from the enzyme under our conditions.

Fig. 6 shows that if amylase secretion and the incorporation of P^{32} into the phospholipids were followed with increasing concentrations of pancreozymin a disproportionality between amylase secretion and the incorporation of P^{32} into phospholipids was observed with higher concentrations of pancreozymin. That is to say, there was no further increase in amylase secretion when the concentration of pancreozymin was increased from 1.5 units per ml to 5 units per ml, but the incorporation of P^{32} into phosphatidyl inositol, phosphatidyl ethanolamine and phosphatidic acid essentially doubled over this concentration range.

We have therefore been able to obtain conditions in which amylase secretion and the phospholipid effect can be dissociated. We feel that this argues strongly against phospholipid metabolism being involved in the final step of the overall secretory cycle; i. e., the coalescence of zymogen granule membrane with plasmalemma and release of zymogen.

Site of the phospholipid effect. The above studies suggest that the phospholipid effect is concerned with a step in the overall secretion process which is earlier than the release of enzymes by reverse pinocytosis. We felt that if the site of the phospholipid effect could be established this might throw light on its physiological significance. Several years ago it was shown that if slices were incubated with P^{32} without and with

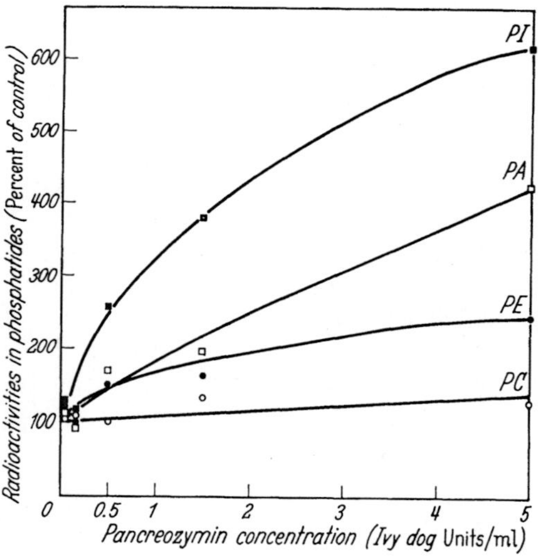

Fig. 6 A

Fig. 6 B

acetylcholine, homogenized in sucrose, and the homogenates resolved by differential centrifugation, most of the phospholipid effect could be recovered in the microsome fraction [25]. Although PALADE and SIEKEVITZ [21] have shown the predominant component of the pancreatic microsome fraction to be vesicles of rough surfaced endoplasmic reticulum it is probable that under our conditions of differential centrifugation smooth membranes were also present. ROTHSHILD [26] found smooth surfaced membranes near the surface region in a microsomal pellet of liver sedimented from 1.31 M sucrose at 105,000 X g. Our microsome fractions were sedimented at 105,000 X g in 0.5 M sucrose; under these conditions sedimentation of smooth membranes would seem to be even more likely. Since the smooth

Fig. 6. The extrusion of amylase and the incorporation of P^{32} into the individual phosphatides in response to increasing concentrations of pancreozymin. A. Amylase secretion. B. Radioactivities in the phosphatides. Pigeon pancreas slices were incubated without pancreozymin for 40 minutes as described in Fig. 1. They were then transferred to a second medium which contained the indicated concentrations of pancreozymin and P^{32}-labeled inorganic phosphate of one-fourth the specific activity of the first medium. The incubation was stopped 30 minutes after transfer. The experiment was carried out with tissue from 3 animals; each value given is calculated from the average of at least 3 observations. All values are expressed as per cent of the control value obtained from slices incubated without pancreozymin throughout

and rough membranes occupy rather characteristic regions in the pancreatic acinar cell, we felt that with the differential centrifugation studies as a background, tissue radioautography might throw light on the site of the phospholipid effect. One problem with radioauto-graphy of lipids is that the usual organic solvents used in embedding the tissue, extract the lipids. This difficulty was overcome by fixing the tissue in 10% neutral formalin and embedding in carbowax, which is a water soluble material. The radioactive tracer which was used was myoinositol-2-H^3 with a specific activity of 150 mc per mmole. A good check that the grains which were counted were due to the tritium which was incorporated into phosphatidyl inositol was shown by the fact that the per cent stimulation by acetylcholine of the incorporation of tritium into the total lipids, as measured in the liquid scintillation spectrometer, was the same as the per cent increase in grain density in the sections (compare, e. g., Figs. 7 and 9). Chromatography of the total lipid extract of pancreas on silicic acid impregnated paper with phenol-ammonia as the developing solvent (this system separates the 3 phosphoinositides [27]), followed by detecting of radioactivity in a strip counter, showed a single radioactive peak coinciding with the phosphatidyl inositol spot, and this peak was higher from the stimulated slice. Diphosphoinositide and triphosphoinositide must therefore incorporate negligible inositol-2-H^3 in pancreas slices under these conditions.

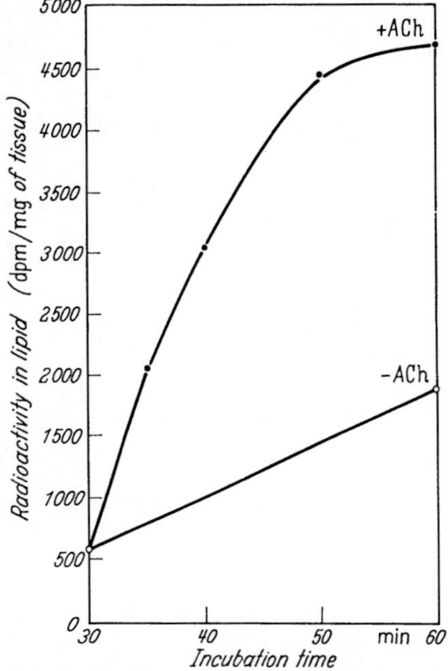

Fig. 7. Incorporation of inositol-2-H^3 into lipid of pigeon pancreas slices incubated without and with acetylcholine. About 50 mg of guinea pig pancreas slices were incubated in 2 ml of KREBS-HENSELEIT bicarbonate saline containing 200 mg per cent glucose and 0.37 mM inositol-2-H^3 (150 mc/mmole) for 30 minutes. 10^{-4} M acetylcholine plus 10^{-4}M eserine were then added to the indicated vessels. Slices were removed at the indicated times and worked up for total radioactivity in the lipid fraction as described previously [7]

Fig. 7 shows the incorporation of tritium into the total lipid extract with time. The slices were incubated for 30 minutes without stimulating agent to allow the inositol-2-H^3 to enter the cell, and acetylcholine was then added. The radioactivity in the total lipid extracts from slices without and with acetylcholine was then followed. The stimulation of inositol-2-H^3 incorporation by acetylcholine can be seen at the shortest sampling period, and it continued more or less linearly until about 20 minutes after adding acetylcholine.

Samples of tissue from this same experiment were also fixed and embedded as described above, 2μ sections were cut and radioautograms were prepared by the stripping film technique. The sections were then stained with toluidine blue. Photographs of sections were taken randomly

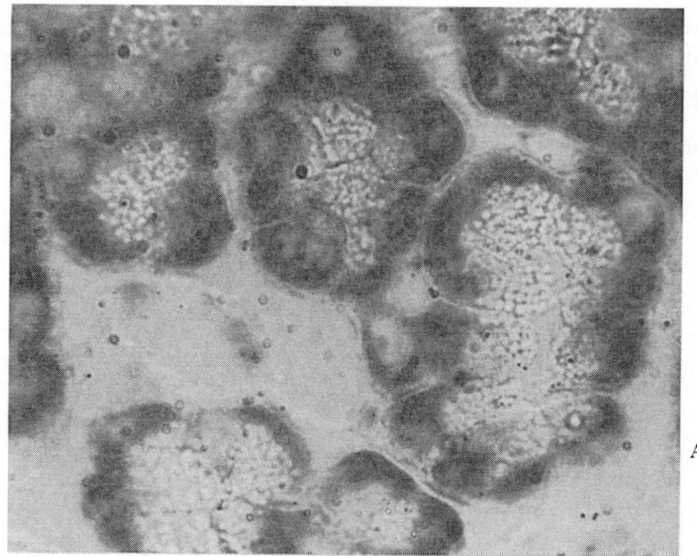

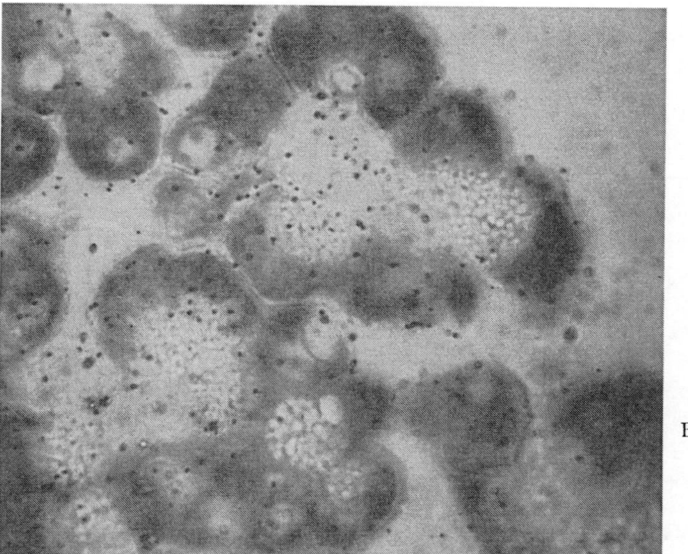

Fig. 8. Radioautograms of pancreas tissue incubated without and with pancreozymin. A. Incubated without pancreozymin. B. Incubated with pancreozymin. Guinea pig pancreas slices were incubated for 30 minutes in 1 ml of KREBS-HENSELEIT bicarbonate saline containing 200 mg-% glucose and 0.37 mM inositol-2-H^3 (specific activity, 150 mc/mmole). Fifty microliters of pancreozymin (100 Ivy dog units per ml) were then added to the stimulated vessels, and 50 microliters of water were added to the control vessels. Incubation was continued for 30 minutes. The tissues were then fixed in 10% neutral formalin and radioautograms were prepared. Incubation with pancreozymin appeared to produce less histological abnormality than did acetylcholine plus eserine

and grains were counted in the nucleus, basophilic cytoplasm, and non-basophilic cytoplasm. At least 200 cells were counted. Only those cells were counted in which there was a fairly clear-cut demarcation between

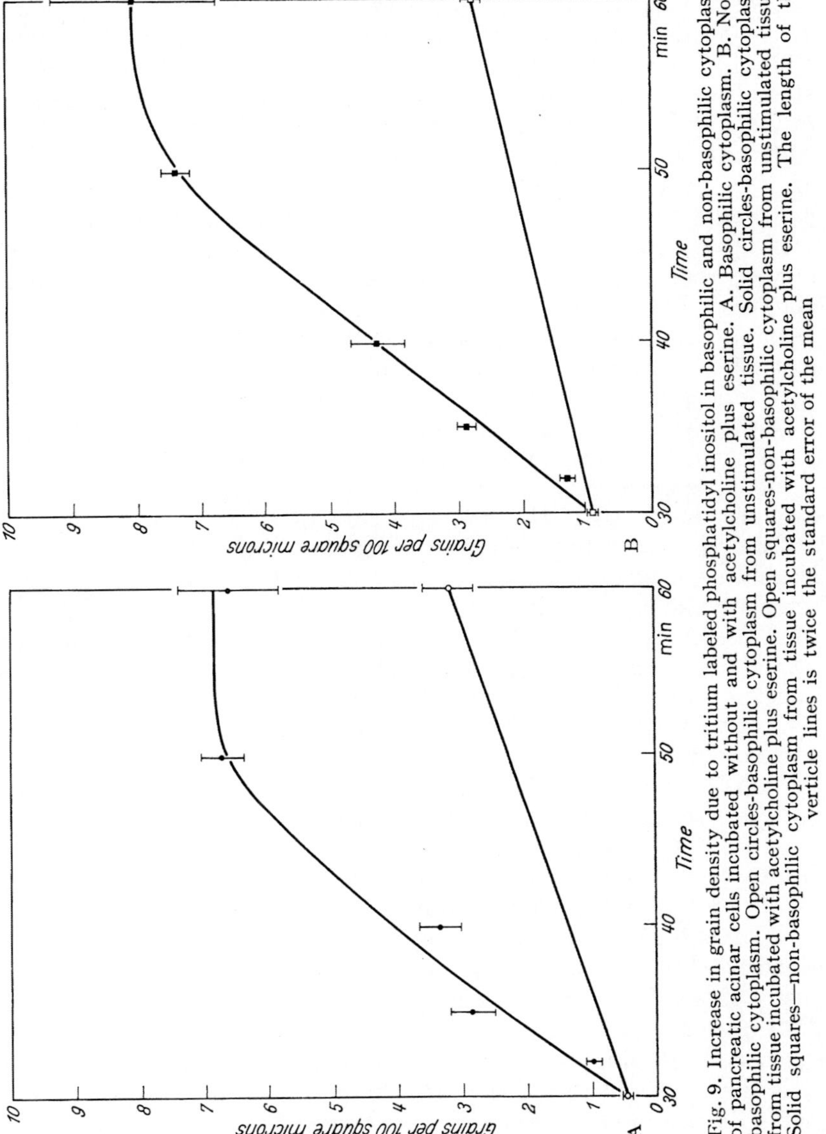

Fig. 9. Increase in grain density due to tritium labeled phosphatidyl inositol in basophilic and non-basophilic cytoplasm of pancreatic acinar cells incubated without and with acetylcholine plus eserine. A. Basophilic cytoplasm. B. Non-basophilic cytoplasm. Open circles-basophilic cytoplasm from unstimulated tissue. Solid circles-basophilic cytoplasm from unstimulated tissue. Open squares-non-basophilic cytoplasm from unstimulated tissue. Solid circles-basophilic cytoplasm from tissue incubated with acetylcholine plus eserine. Open squares-non-basophilic cytoplasm from tissue incubated with acetylcholine plus eserine. The length of the Solid squares—non-basophilic cytoplasm from tissue incubated with acetylcholine plus eserine. The length of the verticle lines is twice the standard error of the mean

basophilic and nonbasophilic cytoplasm. Fig. 8 shows radioautograms from slices incubated without and with pancreozymin. Fig. 9 shows the grain counts in the basophilic and nonbasophilic cytoplasm in slices incubated without and with acetylcholine plus eserine. The grain counts in the nuclei were not significantly above background if one allows for a slight elevation along the edge of those nuclei from stimulated slices; this must have been due to lateral radiation from the basophilic cytoplasm.

The salient feature of Fig. 9 is that there were stimulations in both the basophilic cytoplasm and in the nonbasophilic cytoplasm. This data, in conjunction with the earlier differential centrifugation studies [26], allows us to make certain conclusions. The differential centrifugation studies had already shown the phospholipid effect to be in microsomal membranes, but they did not differentiate between smooth and rough surfaced membranes. If the effect were in the smooth surfaced membranes *only*, we would not expect to see any increase in grains in the basophilic cytoplasm, since in the pancreas this region is essentially devoid of smooth membranes. If the effect were in the rough surfaced membranes *only* we would expect to see some increase in grains in the non-basophilic cytoplasm but not as great as in the basophilic cytoplasm, since the amount of rough surfaced membrane in the non-basophilic region is very much less than that in the basophilic cytoplasm. Since appreciable stimulations were observed in both regions, we are forced to the conclusion that the effect is in both the rough and smooth surfaced membranes.

It could be argued that the phosphatidyl inositol effect occurs at a discrete site in the cell, but that the whole membrane system undergoes dynamic flow so that the increased labeling in phosphatidyl inositol distributes itself throughout the membranes of the cell. If this were so one should have expected to see more localization of graining at early times after adding acetylcholine. However, this was not the case. At the earliest time taken, i. e. 2 minutes, there were statistically significant stimulations in both the basophilic and nonbasophilic regions and of the same magnitude. It seems doubtful that any dynamic flow of membrane — if any such flow exists at all — would have equilibrated all of the cellular membrane by two minutes. These data seem to us to argue fairly strongly for independent stimulations of phosphatidyl inositol synthesis in smooth membranes and in rough surfaced membranes. We might add that the failure to find localization of the increased graining in the region of the apical plasmalemma is further evidence against the effect being concerned with reversed pinocytosis.

What is the physiological significance of the phospholipid effect in the pancreas? In assessing the possible physiological significance of the phospholipid effect it might be useful to refer to a scheme presented by Palade et al. [22] which describes the overall secretion process in the pancreas. These investigators, on the basis of their extensive studies, have divided the overall secretion process into four stages. The *first* stage marks the end of synthesis of enzymes on the ribosomal components of the rough surfaced membranes. The *second* stage consists of the transport of newly synthesized enzymes across the membranes into the lumina of the rough surfaced endoplasmic reticulum. The *third* stage involves segregation of the secretion product into smooth surfaced vacuoles, which appear to arise by budding off from the smooth surfaced end of the endoplasmic reticulum in the centrosphere (Golgi) region of the cell. The *fourth* stage involves storage of the secretion product in mature zymogen granules in the apical region of the cell. The final stage, which PALADE, et al. [22] include as part of the fourth stage, involves discharge of the enzyme contents of the zymogen granules by reverse pinocytosis. That transport occurs across the rough surfaced membranes is shown by the fact that intracisternal granules form when pancreatic secretion is stimulated in the guinea pig by feeding [19]. Since the zymogens are believed to be synthesized on the opposite side of the membranes it would presumably be necessary for them to be transported across the membranes in order to accumulate in the intracisternal spaces. Pertinent to our discussion is the question as to whether transport also occurs across the smooth surfaced membranes in the centrosphere region. The fact that the smooth membrane bounded vesicles contain enzymic material at all stages of filling [20, 22] and that the amount of zymo-genic material in the unfilled vesicles is on average less than in the mature zymogen granules seems to support the view that transport occurs across the smooth membranes from the cell sap. By comparison with the mature zymogen granule the amount of zymogen seen in the smooth surfaced vesicles at early stages of filling with zymogen (see Figs. 5, 6 and 7 in PALADE, et al. [22]) is too small to allow the alternative ex-planation that all of the enzyme of the mature zymogen granule is derived from intracisternal enzyme trapped when the vesicle buds off from the endoplasmic reticulum and that maturation from the partly filled vesicle to mature zymogen granule occurs by resorption of water and shrinkage of membrane. The fact that 85% of the amylase in pigeon pancreas [2] and a fairly high proportion of ribonuclease in mouse [18] and pigeon [2] pancreas are apparently in soluble form in the cytoplasm would be consistent with the view that soluble enzymes are directly transported from the cell sap into the contents of the smooth surfaced vesicles. Further support for this view is the observation of MORRIS and DICKMAN [18] that next to that of the microsome fraction the specific radioactivity of ribonuclease in the soluble fraction shows a more rapid rise than any other fraction after injection of valine-C^{14}. We thus feel that it is highly probable that intracellular transport of enzymes and zymogens occurs across both the rough surfaced membranes and the smooth surfaced membranes.

Since the phosphatidyl inositol effect occurs in rough and in smooth membranes, and since there is good reason to believe that intracellular transport occurs across both these types of membranes, one interpretation of our data would be that the increased synthesis of phosphatidyl inositol on stimulation of secretion is functionally related to the overall process of intracellular transmembrane transport of zymogens and enzymes, which is triggered on stimulation with pancreozymin or acetylcholine. In addition to accounting for the radioautographic data this inter-pretation could account for the lack of parallelism between the phos-pholipid effect and the release of enzyme into the incubation medium, since the release of enzyme into the incubation medium is probably brought about by the reverse pinocytosis process and not by intracellular transmembrane transport (see, however, a discussion of this problem on pages 53—54 of a recent symposium [22]). One difficulty with the postulate that the phospholipid effect is concerned with transmembrane transport is that it is difficult to visualize how the increased synthesis of phosphatidyl inositol, and presumably phosphatidyl ethanolamine, functions in transmembrane transport. One suggestion which has been made is that a "break" in the membrane occurs at a phosphatidyl inositol point, allowing passage of enzyme across the membrane, and resynthesis of phosphatidyl inositol "heals" this break. A difficulty with this interpretation is that one would have anticipated that the "breaks" and the "healing" would occur repeatedly at the same sites, and this would give the rapid equilibration type of kinetics rather than the accumulative type of kinetics observed with phosphatidyl inositol. However, it is possible that a "break" and a "healing" at a site, followed by a "break" at the same site would occur only infrequently in the pancreas so that over the incubation period studied new sites would become progressively involved, which would give the accumulative type of kinetics observed.

An alternative physiological role for the phospholipid effect in pancreas which should be seriously considered is that this effect is concerned with the *transmembrane transport activation* process rather than with the transport process *per se*. In other words the changes in phospholipid metabolism may be part of the mechanism whereby acetylcholine or pancreozymin activate the membrane so as to set the stage for transmembrane transport. The various data presented in this paper are compatible with this hypothesis; i. e., the lack of parallelism between the rate of enzyme secretion and the phospholipid effect, the distribution of the phospholipid effect throughout the smooth and rough membrane system, and the fact that a phospholipid effect could be obtained in the absence of calcium while enzyme secretion could not. The crucial question to be answered in this connection is whether or not the phospholipid effect is paralleled by intracellular transmembrane transport under various conditions. If there is no parallelism we would have evidence that the phospholipid effect is not part of the transmembrane process *per se* and more likely to be concerned with the transmembrane transport activation process. It is possible that combined biochemical and electron microscopic studies could answer this question, since it is possible to follow transmembrane transport by observing the formation of intracisternal granules in the guinea pig pancreas after stimulating secretion by feeding fasted animals [19].

It has been suggested [3] that the increased synthesis of phosphatidyl inositol on stimulation of pancreas slices reflects increased formation of membrane which serves to replace that membrane which is transferred to the surface membrane by reverse pinocytosis. The fact that there is no close correlation between the amount of enzyme extruded, which is presumably a measure of pinocytosis, and the magnitude of the phospholipid effect argues against this explanation, although it does not rigorously disprove it.

One curious fact which must be considered, if one is to espouse the membrane resynthesis hypothesis, and that is the very disproportionate stimulations in the phospholipids. Why, for example, is lecithin not synthesized at an accelerated rate on stimulation, even though it makes up about half of the total phospholipid of most cellular membranes; yet phosphatidyl inositol, which makes up less than 10% of the total phospholipid, shows by far the most striking stimulation in synthesis? It should also be pointed out that SCHUCHER [28] showed several years ago that under conditions in which a maximum phospholipid effect was obtained with acetylcholine there was actually an inhibition of glycine-1-C^{14} incorporation into protein of pigeon pancreas slices. Recently, POORT and SANGSTER [24] found that if fasted rats were pretreated with pilocarpine so as to stimulate release of pancreatic enzymes, and pancreas slices were then prepared and incubated with radioactive leucine, there was much less leucine-C^{14} incorporated into the protein of the microsome fraction as compared to controls. One might have expected that if there were resynthesis of new membrane there would be new synthesis of protein in this membrane, since protein makes up a large proportion of the membrane mass. A counter-argument could be made that in the process of plasmalemma breakdown and synthesis of new intracellular membrane, the major constituents of the membrane such as protein and lecithin, are removed intact from the plasmalemma and are re-introduced into the new intracellular membrane, while phosphatidyl inositol and to a lesser extent phosphatidyl ethanolamine, are degraded to simpler products and must then be resynthesized. The lipids and proteins could be released from the plasmalemma as mixed micelles of characteristic composition and structure by "breakage" of the plasmalemma at phosphatidyl inositol points, and these micelles could then be re-assembled into new membrane as a "package" by resynthesis at phosphatidyl inositol points.

To summarize, we can say that when protein secretion is stimulated in the pancreas, characteristic changes in the metabolism of certain phospholipids occur. There is a formation and a rapid and continuous turnover of a fraction of phosphatidic acid. This fraction of phosphatidic acid amounts to 10% of the total phosphatidic acid of the cell. There is a synthesis of phosphatidyl inositol and phosphatidyl ethanolamine. Lecithin and protein are unaffected. Combined differential centrifugation studies and radioautographic studies indicate that the phosphatidyl inositol effect occurs in smooth and rough surfaced membranes as short as 2 minutes after stimulation. The phospholipid effect is not primarily concerned with the coalescence of zymogen granule membrane with the plasmalemma; i. e., with the extrusion process *per se*. The data point to a participation of the phospholipid effect with some step in the overall process of transmembrane transport of proteins across intracellular membranes during the segregation of these proteins in preparation for zymogen granule formation. This step in the overall process of transmembrane transport may be activation of the membrane by acetylcholine or pancreozymin so as to set the stage for transport, or it may be the transmembrane transport process *per se*. Finally, we should like to emphasize again that the phospholipid effects which we have been talking about in connection with stimulation of secretion in the pancreas appear to be different phenomena from those observed in the salt gland on stimulation of sodium chloride secretion. The effects in the salt gland constitute another story [14–16].

References

[1] DOUGLAS, W. W., and A. M. POISNER: J. Physiol. (Lond.) **165**, 528 (1963).
[2] EGGMAN, L. D., and L. E. HOKIN: Unpublished observations.
[3] FAWCETT, D. W.: In: Symposium on the Plasma Membrane. New York: American Heart Association 1962.
[4] HOKIN, L. E., and M. R. HOKIN: Biochim. biophys. Acta (Amst.) **18**, 102 (1955).
[5] — — Canad. J. Biochem. **34**, 349 (1956).
[6] — — Gastroenterology **36**, 368 (1959).
[7] — — J. gen. Physiol. **44**, 61 (1960).
[8] — — Int. Rev. Neurobiol. **2**, 100 (1960).
[9] — — Symposium on Membrane Transport and Metabolism, by A. KLEIN-ZELLER and A. KOTYK. New York: Academic Press 1960.
[10] — — Lab. Invest. **10**, 1151 (1961).
[11] — — In: Ultrastructure and Metabolism of the Nervous System, ed. by S. R. KOREY, A. POPE, and E. ROBINS. Baltimore: Williams and Wilkins 1962.
[12] — — In: Ciba Symposium on the Exocrine Pancreas, ed. by A. V. S. DE REUCK and MARGARET P. CAMERON, London: Churchill 1962.
[13] — — In: Drugs and Membranes, Vol. 4, ed. by C. A. M. HOGBEN and P. LINDGREN. Oxford: Pergamon 1963.
[14] — — Fed. Proc. **22**, 8 (1963).
[15] — — In: Metabolism and Physiological Significance of Lipids. London: John Wiley and Sons 1965.
[16] — — J. gen. Physiol., Submitted for publication.
[17] HSIU, J., E. H. FISHER, and E. A. STEIN: Biochemistry **3**, 61 (1964).
[18] MORRIS, A. J., and S. R. DICKMANN: J. biol. Chem. **235**, 1404 (1960).
[19] PALADE, G. E.: J. biophys. biochem. Cytol. **2**, 417 (1956).
[20] — In: Subcellular Particles. Washington: American Physiological Society 1959.
[21] —, and P. SIEKEVITZ: J. biophys. biochem. Cytol. **6**, 671 (1956).
[22] — —, and L. G. CARO: In: Ciba Symposium on the Exocrine Pancreas, ed. by A. V. S. DE REUCK and MARGARET P. CAMERON. London: Churchill 1962.
[23] PAULUS, H., and E. P. KENNEDY: J. biol. Chem. **235**, 1303 (1960).
[24] POORT, C., and A. N. SANGSTER: Biochim. biophys. Acta (Amst.) **78**, 741 (1963).
[25] REDMAN, C. M., and L. E. HOKIN: J. biophys. biochem. Cytol. **6**, 207 (1959).
[26] ROTHSCHILD, J.: Biochem. Soc. Symp. **22**, 4 (1963).
[27] SANTIAGO-CALVO, E., S. MULE', C. M. REDMAN, M. R. HOKIN, and L. E. HOKIN: Biochim. biophys. Acta (Amst.) **84**, 550 (1964).
[28] SCHUCHER, R.: Ph. D. Thesis, McGill University 1954.
[29] —, and L. E. HOKIN: J. biol. Chem. **210**, 551 (1954).
[30] STEIN, E. A., J. HSIU, and E. H. FISHER: Biochemistry **3**, 56 (1964).

Discussion

Tosteson: I was interested with regard to the variations in the total amount of phosphatidic acid and phosphatidyl inositol that you might be able to measure under the various conditions. In particular, are there any changes in the total amount of phosphatidic acid, in association with this recruitment phenomenon which you mentioned?

I would also like to hear your comments on the possibility that phosphatidic acid pool might bear a precursor relationship to the phosphatidyl inositol.

L. Hokin: In the pancreas we do not have terribly good quantitative data on the *chemical* levels of phosphatidic acid. Based on the isotopic studies the amount of phosphatidic acid recruited is 10 % of the total phosphatidic acid. The chemical methods are not sufficiently good to demonstrate a 10 % increase in the level of phosphatidic acid. Although we have better data from chemical analyses of the phospholipids in the salt gland, again the increases in the levels of phosphatidic acid and phosphatidyl inositol, based on the isotopic studies, are of such a magnitude that they would not be detected by chemical methods. I think all I can say is that the chemical analysis is fairly compatible with what the isotope data would have predicted.

Diamond: It would appear that, unlike the result with phosphatidic acid which showed an increased turnover, the results with phosphatidyl inositol and phosphatidyl ethanolamine were a continued increased uptake of P^{32} into these compounds. Is it a fact that the pancreas produces a mass synthesis of these two compounds under stimulation, or where is the stuff going?

L. Hokin: With this type of experiment there are two possibilities with phosphatidyl inositol. There is a synthesis, which is certainly going on, as shown by the isotopic studies. You could say that this synthesis is going on unmatched by a breakdown of an equal number of phosphatidyl inositol molecules. The other alternative is that the synthesis is matched by a breakdown of an equal number of phosphatidyl inositol molecules. If you have a large pool of preformed phosphatidyl inositol molecules and you break down molecules randomly you would have essentially the same kinetics as you had if there were only a net synthesis of phosphatidyl inositol unmatched by an equal breakdown. Let me put it another way. Let us say that you have 100 molecules of phosphatidyl inositol, and over the incubation period you have made 5 molecules, and let us say that you have randomly broken down 5 of the 100. The chance of breaking down one of the newly synthesized 5 molecules is very low. So you would not get any different kinetics with that situation than if you just made the 5 molecules and did not break down any. I assume that at some stage or other in the secretory cycle there is a breakdown of molecules matching the rate of synthesis of molecules; otherwise, I presume the cell would eventually burst from filling up with phosphatidyl inositol. However, it is hard to demonstrate breakdown because you have to label a very high percentage of the total phosphatidyl inositol pool, and then stimulate, and maybe you can get a loss of 5 or 10 %. It is not easy.

Diamond: In fact you are saying that the specific activity of phosphatidyl inositol is small throughout this experiment? Small compared with phosphatidic acid?

L. Hokin: Oh yes, probably so.

M. Hokin: The net specific activity of the total phosphatidic acid is misleading because you have actually a small compartment of phosphatidic acid which appears to be in isotopic equilibrium with the ATP^{32} of the tissue, so that the net specific activity of the phosphatidyl inositol would be less than the specific activity of this compartment of phosphatidic acid. I think we did not answer Dr. Tosteson's question about whether the labeling of phosphatidic acid bears a precursor relationship to that of phosphatidyl inositol in the pancreas. It certainly appears to; there is no real evidence that in the pancreas the labeled compartment of phosphatidic acid is not just a precursor for phosphatidyl inositol synthesis but more critical experiments must be done in this area to establish this definitely.

Bücher: May I ask a question concerning your fractionation experiments? With respect to the mitochondria, the bars are very low, but the proportional increase between the dotted bar and the other bar seems to be about the same as in the microsomes. Would this mean an impurity of microsomes going with the mitochondria?

L. Hokin: I think so. Palade and Siekevitz have in fact shown considerable contamination of the mitochondrial fraction by microsomal membranes. If you assume that most of the labelling is in the microsomal membranes, then you would get percentage-wise about the same increase in the mitochondrial fraction as you observe in the microsome fraction — percentage-wise, but not increment-wise.

Niesel: You have correlated your effects with the uptake of protein through the membrane. What is the proof for that? Could it be that you stimulate the membrane flow only, and not the uptake of proteins?

L. Hokin: As I tried to emphasize, all that we can rule out in the overall secretion process as being connected with most of the phospholipid effect is the extrusion process *per se*, since we could dissociate the extrusion process from the phospholipid effect by omitting Ca^{++}. If you want to assume that the phospholipid effect is concerned with the flow of membrane, the calcium experiments would require that the stimulating agent can stimulate this membrane flow directly, since we know that omission of Ca^{++} has stopped the extrusion process. I would have thought that the rate of coalescence of zymogen membrane with plasmalemma — which is part of the extrusion process — would determine the rate at which membrane flowed and the rate at which new membrane was formed. But this may not be the case. It may in fact be that the stimulating agent triggers formation of new membrane and coalescence of zymogen membrane with plasmalemma independently, but this hypothesis requires the assumption that we have thrown the membrane formation step out of gear with the coalescence step when we have omitted Ca^{++}.

Holter: If we accept all your conclusions and all your arguments, and accept your last main conclusion, namely that what you are concerned with is trans-membrane transport of proteins, then the question is, have you found how this membrane transport can go on, and have you formed any hypotheses on just how a high-molecular secretion product is supposed to be transported through the membrane?

Do you assume that there is a formation of a lipoprotein consisting of the enzyme on the one hand, and your lipids on the other hand, which acts as an intermediate carrier?

L Hokin: I must confess that in pancreas we are very much more at a loss to come up with hypotheses that are very meaningful. I think that you have a very special problem to face when you have to move a protein — as a molecular species — across a membrane, and I assume that something like that happens in these intracellular membranes. One hypothesis is that the membrane is some kind of bimolecular leaflet in which phosphatidyl inositol is spaced along the leaflet at intervals. Let us say that when transport is stimulated we break down these phosphatidyl inositol molecules, and this alters the structure of the membrane in some way. That is to say there would be some sort of conformational change or distortion in the membrane which would allow the protein molecule to move through. Perhaps many protein molecules would move through for every phosphatidyl inositol molecule broken. Eventually, the phosphatidyl inositol molecule would be resynthesized. I do not know whether you would call this active transport. We do not know the concentrations of protein on both sides of the membranes. According to this hypothesis one might have expected to find equilibration kinetics with phosphatidyl inositol, because one would have assumed that certain phosphatidyl inositol molecules along the membrane would be broken down and resynthesized rapidly and thus would rapidly reach isotopic equilibrium with their precursor. But this may not be the case, because there may be a very large number of these phosphatidyl inositol molecules, and over the incubation period all of of these molecules may not have turned over more than once.

I might make some more comments about the possibility that the phospholipid effect is concerned with formation of new membrane associated with membrane flow. If this hypothesis is the correct explanation of the data why is lecithin not synthesized at a more rapid rate? After all, lecithin is the main phospholipid of the membranes. Also 50 % of the membrane is protein, but is has been shown that when pancreatic secretion is stimulated there is an inhibition of protein synthesis in the microsome fraction. However, you could argue that when the membrane is broken down it is broken only at phosphatidyl inositol points so that parcels of membrane are released. When new membrane is formed these parcels are reassembled at phosphatidyl inositol points. If the lecithin and protein were all contained within these parcels they would not turn over; only the molecule at the point of breakage and resynthesis would turn over, which would be phosphatidyl inositol.

Tosteson: I should like to comment along very similar lines with regard to the possible mechanism of this effect that Prof. HOKIN has observed. I would like to make a comment about membrane structure, in particular about the so-called lamellar character of the unit membrane.

There is considerable evidence in many sorts of membranes, but in particular in the membranes of red blood cells, that when viewed on the surface plane of the membrane, the membrane is made up of a number of small units; they have been called "plaques". They are approximately 200 Å units across, and about 100 Å units thick, including the two protein coats. Such a structure in section would yield a lamella. Similar type units have been observed in mitochondrial membranes.

Considerable work has been done recently by people interested in the surface chemistry of lipids. I think particularly of the work of BANGHAM in Cambridge, and LUZZATI in Strasbourg and Paris. They have shown that phospholipids can form many different kinds of micelles. BANGHAM has made emulsions of phospholipids in an aqueous solution which look like onions, where each layer is a bi-molecular lamella. He has also shown that the incorporation of ionic detergents into such a lamella tends to convert it to a micellar structure, at least in time average. He has measured the extent of this conversion by measuring the rate at which water soluble substances trapped between the lamellae can leave. In HOKIN's experiment it is interesting that the membrane phospholipids which turn over more rapidly are substances which are anionic detergents; notably ones which are uncharged, like lecithin, are not apparently involved in the effect. One might speculate that the role of these compounds is to tend to convert or to modify the lamella-micelle transition, which probably continually goes on in biological membranes.

Now, it would be very difficult of course to see the micelles in cross-section in the electron-microscope, because of the lack of contrast. It is possible with red blood cell membranes actually to separate those units by breaking up the membranes with ultra-sound, and in fact one can also observe reaggregation of these units under various conditions.

Einleitung

K. Mothes

Über die Beziehungen von Ultrastruktur und Funktion auf dem Gebiet des pflanzlichen Exkretionsstoffwechsels ist bisher recht wenig gearbeitet worden. Ich möchte mir doch erlauben, kurz darauf hinzuweisen, wie groß das Problem „Exkretion im Pflanzenreich" ist.

In der höheren Pflanze liegt zunächst der Tatbestand vor, daß die Differenzierung der Zellen nicht annähernd so weit geht wie im tierischen Organismus. Wir sehen in vielen Fällen mit den bisherigen Methoden der Beobachtung, also der Lichtmikroskopie, den Zellen nicht ohne weiteres an, daß sie spezialisiert sind. Wir können aber in sehr vielen Fällen nachweisen, daß sie es sind.

Zum Beispiel gibt es in einem Blatt, das gerbstoffreich ist, den Gerbstoff oft nur in ganz bestimmten Zellen. Das trifft auch für Alkaloide zu, und es trifft besonders auffällig zu für ätherische Öle, die in Ölzellen vorkommen können. Die Prozesse, die hier bisher in der Hauptsache behandelt worden sind, also z. B. die Ausscheidung in den Verdauungstrakt, fallen für die Pflanze völlig weg. Aber es wird gezeigt werden, daß immerhin ähnliche Phänomene, zellphysiologisch betrachtet, durchaus existieren, z. B. bei den insektivoren Pflanzen, und das sind dann interessante Parallelen. Obwohl die Organismen im großen und ganzen sehr verschieden organisiert sind, gibt es auf zellphysiologischer Basis viele Beziehungen. Die Pflanze hat nicht einen exkretorischen Apparat im Sinne des Nierenapparates, und trotzdem hat sie einen exkretorischen Stoffwechsel in einer viel größeren Mannigfaltigkeit als das Tier, weil alle Stoffe, die als Endprodukte oder Nebenprodukte des Stoffwechsels anfallen, von der Kohlensäure abgesehen, im Organismus verbleiben. Sie hat dafür besondere Räume. Diese Räume sind häufig außerhalb der Zelle, in den Fällen, wo interzellulär ätherische Öle oder Harze gespeichert werden. Die Stoffe können auch in der Zellwand selbst gespeichert werden, wie beim Lignin. Da ist also noch so etwas wie ein Exkretionsvorgang, der die Stoffe aus der Zelle herausbringt, gegeben.

Aber viel komplizierter werden dann die Verhältnisse, wenn diese Exkrete in den Zellsaftraum abgeschieden werden, in die Vakuole, die in der ausgewachsenen und stoffwechselphysiologisch aktiven Pflanzenzelle in geradezu riesigem Ausmaß vorhanden ist. Die Vakuole ist im allgemeinen wenigstens 100 mal so groß im Volumen wie der aktive Plasmabelag einer Pflanzenzelle. Wir können aber nicht generell sagen, Vakuolen sind Exkreträume. In der Vakuole werden ja auch Saccharosen abgestellt, die durchaus wieder in den Stoffwechsel einbezogen werden, und es können genauso in derselben Vakuole Stoffe abgestellt sein, die nicht wieder in den Stoffwechsel einbezogen werden.

Zwischen diesen Exkreten, die stoffwechselinaktiv geworden sind, weil sie durch Ringschlüsse, durch Methylierungen usw. dem Zugriff der Enzyme nicht mehr zugänglich sind, gibt es alle Übergänge. Es ist schwer zu sagen, ob ein Stoff ein endgültiges Exkret im Sinne von FREY-WYSS-LING ist oder ob er noch einmal am Stoffwechsel teilhaben kann.

Ein Beispiel aus meinem eigenen Arbeitsgebiet ist das Nikotin. Es gibt Tabakpflanzen, in denen das Nikotin, einmal gebildet, nicht mehr weiter verändert wird. Jedenfalls geht das eigentliche Ringsystem nicht in irgendeine Art von turn over ein. Das Nikotin wird offenbar in nur wenigen Zellen in der Wurzelspitze gebildet, steigt von dort mit einem allgemeinen Saftstrom in das Blatt und wird im Blatt gespeichert. Dort, wo wir es also finden, ist es nicht gebildet worden. Es ist nun zuerst die Frage: Ist das ein vollkommenes Exkret oder nicht? Es gibt einige Botaniker, die meinen, wenn an einem solchen Exkret noch einmal etwas geschieht, ist es stoffwechselphysiologisch bedeutsam. Und das ist sicher nicht der Fall. Aber man kann an den verschiedensten Tabakpflanzen feststellen, daß dieses Nikotin — das zwei Ringsysteme enthält, also einen Pyrrolidinring, der völlig methyliert ist, und einen Pyridinring —, wenn man das Blatt, in dem es abgelagert ist, einer radioaktiven Kohlendioxydatmosphäre aussetzt, sehr schnell die Methylgruppe austauscht. Also auf dem Weg über Methionin geht dieser radioaktive Kohlenstoff in das Nikotin ein. Das eigentliche Ringsystem wird nicht radioaktiv. Das Blatt kann also kein Nikotin abbauen, kann kein Nikotin aufbauen, aber es kann entmethylieren und wieder methylieren. Das ist kein reversibler Prozeß. Wir wissen genau, daß diese Entmethylierung auf einem anderen Weg geschieht als die Methylierung. Es ist also nicht ein Austausch von Methylgruppen, sondern die Entmethylierung geschieht oxydativ. Solche Prozesse kennen wir heute an einer ganzen Reihe von Alkaloiden und an Nichtalkaloiden, also Steroiden oder Triterpenen. Die Stoffe sind also keineswegs dem Einfluß des Protoplasmas in der Vakuole entzogen, aber im allgemeinen werden sie nicht wieder soweit verändert, daß sie direkt metabolisiert werden können.

Das ist nun keine generelle Aussage in dem Sinne, daß es immer so sein muß. Es gibt durchaus Fälle, wo das gebildete Nikotin nach einiger Zeit wieder abgebaut wird, z. B. wenn das Blatt durch Altern in einen neuen enzymatischen Zustand gerät und im Protoplasma der nikotinspeichernden Zellen Enzyme vorhanden sind, die die Ringsysteme aufzuknacken in der Lage sind. Wir pfropfen z. B. auf eine Wurzel von *Nicotiana tabacum* einen Sproß von *Nicotiana alata*. Der Nikotinstrom geht von der Wurzel in den Sproß von *Alata*. Das Nikotin wird in der ersten Zeit der Vegetation entmethyliert. Es liegt also in *Alata*blättern als Nornikotin vor. Nach einiger Zeit verschwindet das Nikotin vollständig, und es ist nun die Frage, ob man das noch als ein Exkret bezeichnen kann. Irgendeine quantitative oder qualitative Bedeutung hat dieser Abbau des Nikotins wahrscheinlich nicht.

Obwohl es viele solche Beispiele gibt, haben wir keinen Grund zu bezweifeln, daß diese Stoffe Exkrete sind. Nur durch den veränderten Zustand, insbesondere durch einen bemerkenswerten Altersstoffwechsel, werden sie nachträglich wieder einmal angegriffen, ohne daß das eine große physiologische Bedeutung zu haben braucht.

Sehr interessant sind diese Dinge dann bei den Milchsafträhren. In den Milchröhren einiger Pflanzen ist Kautschuk als typisches Exkret in großen Mengen vorhanden, und zwar in dem sog. Milchsaft, der beim Anschneiden aus den Milchröhren durch den Turgor der benachbarten Gewebe austritt. Wir wissen vom Kautschuk absolut sicher, daß er sehr kalorienreich ist. Es ist trotzdem ein echtes Exkret.

Aber wir finden auch Milchsäfte, in denen 50% und mehr des Frischgewichts Eiweiß ist. Diese Eiweiße können ganz spezifische Enzyme darstellen, z. B. das Papain oder Papajodin, das aus dem Milchsaft von *Garica papaja* seit langem gewonnen wird. Es sind also Enzyme konzentriert in diesem Milchsaft, die dort wahrscheinlich gar nichts zu tun haben. Das ist Eiweiß, das irgendwie in diesem Raum sezerniert worden ist und ausfließt. Aber daneben findet man Stoffe, die in jeder sonstigen Zelle einem Stoffwechsel unterliegen könnten, aber in diesem Milchsaftraum unterliegen sie keinem Stoffwechsel.

Zum Beispiel findet man in gewissen *Euphorbia*-Arten Dihydroxyphenylalanine, also Dopa, eine Substanz, die auch den Tierphysiologen bekannt ist, speziell den Pigmentforschern, da sie ein Zwischenprodukt in der Melaninsynthese darstellt. Bei einer Euphorbienart kann Dopa 2% des Frischgewichts des Milchsaftes darstellen, so daß es auskristallisiert. Wenn man die Pflanze mit radioaktivem Tyrosin füttert, das man entweder total oder in der Seitenkette markiert hat, so findet man das radioaktive Tyrosin niemals im Milchsaft, aber man findet dort radioaktives Dopa. Das Tyrosin, dem eine Hydroxylgruppe angebaut wird, geht also in Dihydroxylphenylalanin über. Gibt man radioaktives Tyrosin in den isolierten Milchsaft, kann kein Dopa gebildet werden. Der Milchsaft selbst ist nicht in der Lage, diese einfache Oxydation durchzuführen. Wer eigentlich das Exkret macht, ist bis heute nicht festgestellt. Es ist sehr schwer zu sagen, ob das der wandständige Plasmabelag ist oder ob das die die Milchröhren umgebenden Zellen sind. Jedenfalls finden wir kein Dopa im Blattgewebe. Wir finden es nur im Milchsaft. Und wenn wir radioaktives Tyrosin füttern, finden wir nur Tyrosin im Blattgewebe, aber niemals im Milchsaft. Es müssen also irgendwie spezifische Bildungsorgane vorhanden sein, streng lokalisiert, und es muß irgendwie ein Exkretionsprozeß stattfinden, wobei es zunächst einmal gleichgültig ist, ob dieser vom wandständigen Protoplasma oder von den darum liegenden Zellen ausgeht.

Wie merkwürdig die Dinge sein können, sei noch durch folgendes Beispiel geschildert: Man kann aus diesem Milchsaft durch Fällung und Fraktionierung ein Eiweißpräparat gewinnen, das als Polyphenoloxydase hochaktiv ist und Dopa sofort in Melanin verwandelt. Der intakte Milchsaft ist nicht in der Lage, Dopa in Melanin zu verwandeln. Das ist ein schönes Beispiel dafür, daß die Enzyme, die wir aus den Milchsäften präparieren, unter Umständen im Milchsaft gar nicht als solche vorliegen. Sie sind irgendwie maskiert oder durch irgendeine Strukturveränderung nicht in der Lage, ihre spezifische chemische Funktion auszuüben.

Wir haben zwischen dem Chemismus von Tier und Pflanze viel mehr Analogien als man früher geahnt hat. Aber im Tierreich sind beispielsweise die schon zitierten Steroide fast alle funktionell als Hormone von besonderer Bedeutung. Dafür gibt es in der Pflanze keine Parallele. Wir kennen noch nicht ein einziges Steroid, das irgendeine Stoffwechselfunktion hat. Wir kennen aber Pflanzen, in denen viele Steroide vorliegen und in denen die Steroide 20 bis 30% der Trockensubstanz ausmachen können, beispielsweise die Roßkastanie. Es gibt also ungeheure Mengen von solchen Stoffen, die in der Pflanze als Exkrete vorkommen, im tierischen Organismus bei großer Ähnlichkeit der chemischen Struktur zu wichtigen Funktionen gelangt sind, und obwohl solche Substanzen in riesiger Masse da sind, werden sie eben bei der Keimung der Roßkastanie nicht wieder in den Stoffwechsel einbezogen, sind also echte Exkrete. Es ist eine Überproduktion von Substanzen. Diesen Luxus kann sich die Pflanze leisten.

Elektronenmikroskopisch sind all diese Prozesse noch kaum untersucht, weil wir in den meisten Fällen noch nicht wissen, wo die Substanzen entstehen und wo man nach interessanten Lokalisationen von Prozessen suchen soll. Man ist jetzt erst am Anfang auf diesem Gebiet, und wir werden durch die folgenden Vorträge sehen, was bisher in dieser Richtung getan ist.

Physiologie und Morphologie sekretorischer Pflanzenzellen[1]

Von

EBERHARD SCHNEPF, Göttingen

Mit 8 Abbildungen

Ausscheidungsprozesse spielen bei der Pflanze infolge ihres anders gerichteten Stoffwechsels eine geringere Rolle als beim Tier, wenn man von der „Sekretion" des Zellwandmaterials absieht. Außer diesem und manchen Endprodukten des Atmungsstoffwechsels werden, in vielen Fällen durch spezialisierte Drüsenzellen, Produkte des „sekundären" Stoffwechsels (z. B. ätherische Öle, Harze, Kautschuk, Glykoside), des „primären" Stoffwechsels (z. B. Zucker, Schleime, Enzyme) sowie anorganische Verbindungen, „Rekrete" [13] (Salze, Wasser) ausgeschieden, teilweise in die Vakuolen, z. B. bei Milchröhren, teilweise in die Zellwände, teilweise in die Interzellularen, z. B. bei Harzgängen, teilweise auch aus dem Pflanzenkörper heraus, z. B. durch Drüsenhaare und Drüsenepithelien.

Diese Substanzen lassen sich häufig nicht eindeutig klassifizieren [13, 34]. Die Begriffe „Sekret" und „Exkret" wurden aus der Tierphysiologie übernommen. Durch sie kann man die pflanzlichen Abscheidungen nur dann charakterisieren, wenn ihre Entstehungsgeschichte und ihre Funktion im Stoffwechselgeschehen aufgeklärt ist. Diese Voraussetzungen fehlen jedoch in den meisten Fällen.

Übereinstimmend hält man die zuckerausscheidenden Nektarien sowie Schleim- und Verdauungsdrüsen für die typischsten sekretorischen Drüsen. Nichtspezialisierte sekretorische Zellen müssen im Rahmen dieser Übersicht ebenso unberücksichtigt bleiben wie exkretorische Zellen; einige Befunde über die Sekretion von Zellwandmaterial werden im Referat von SIEVERS angeführt.

A. Physiologie und Morphologie von Schleimdrüsen

I. Die Schleimsekretion von Insektivoren-Drüsen

1. Der Fangschleim

Mehrere Insektivoren fangen ihre Beute mit einem klebrigen Schleim [24, 39]. Er enthält bei *Drosophyllum lusitanicum* etwa 99,75% Wasser, ein saures Polysaccharid (etwa 0,2%), Ascorbinsäure (etwa 0,04%), etwas Calcium, aber keine Verdauungsenzyme. Das Polysaccharid setzt

---

[1] Mit Unterstützung der Deutschen Forschungsgemeinschaft.

sich aus Gluconsäure, Galactose, Arabinose, Xylose und Rhamnose zusammen [44].

2. Die Morphologie der Drüsenzellen

Drosophyllum trägt auf den Blättern 2 Drüsen-Typen, gestielte (Tentakeln), die den Fangschleim ausscheiden, und sitzende, die die Verdauungsenzyme sezernieren [37]. Beide bauen sich aus 2 Lagen von Drüsenzellen auf, die durch Zellen mit cutinisierten Radialwänden vom übrigen Gewebe getrennt werden. Bei *Pinguicula* gibt es ebenfalls beide Drüsentypen. *Drosera* besitzt nur Tentakeln, die ähnlich wie bei *Drosophyllum* strukturiert sind (Übersicht bei Lloyd [24]).

Die plasmatische Oberfläche der Drüsenzellen wird durch Septen und Wandvorsprünge vergrößert (Ausnahme: gestielte Drüsen von *Pinguicula*) [40, 41, 60]. Die Cuticula über den Drüsenzellen ist bei *Drosophyllum* fein perforiert, bei *Pinguicula* dagegen homogen [44,46,60].

Trotz ihrer verschiedenen Funktion ähneln sich die gestielten und sitzenden Drüsen von *Drosophyllum* sowie die äußeren und inneren Drüsenzellen in der Struktur ihres Plasmas [40, 41]. Sie sind nur wenig vakuolisiert und besitzen große Zellkerne, ferner zahlreiche Mitochondrien mit relativ vielen Sacculi, besonders in den inneren Drüsenzellen. Die Plastiden (Leukoplasten) sind wenig differenziert. Das endoplasmatische Reticulum läßt sich nur bei $KMnO_4$-Fixierung in Zisternenform darstellen; bei Osmium-Bichromat zerfällt es in der Regel in einzelne Vesikel [44]. Der Golgi-Apparat besteht aus Dictyosomen und Golgi-Vesikeln. Die Menge der Ribosomen ist nicht auffällig hoch.

3. Die Bildung des Fangschleimes

Der Fangschleim wird bei *Drosophyllum* von den äußeren Drüsenzellen der Tentakeln ausgeschieden. In Dünnschnitten von ihnen findet man etwa doppelt so viele Dictyosomen wie in den äußeren Drüsenzellen der sitzenden Drüsen und etwa 5 mal so viele wie in den inneren Drüsenzellen. Die Golgi-Vesikel sind hier während der Ausscheidungsperiode größer und zahlreicher als in den anderen Zellen und enthalten ein elektronenmikroskopisch erkennbares Material. Dieses fehlt in jungen, noch nicht aktiven Drüsen und ebenso nach Abschluß der Sekretion [41]. Ähnliche Golgi-Vesikel kommen auch in den gestielten Drüsen von *Pinguicula* und bei *Drosera* vor.

Aus einer *Drosophyllum*-Drüse mit relativ schwach elektronenstreuendem Vesikelinhalt stammen die Abb. 1, 2 und 3. Sie zeigen seine Kontrastierbarkeit mit Rutheniumrot[44] undBleihydroxyd.Rutheniumrot ist ein lichtmikroskopisches Reagenz für saure Polysaccharide, der ausgeschiedene Schleim färbt sich damit an. Bleihydroxyd kontrastiert u. a. saure Gruppen und einige Kohlenhydrate [38].

Diese Befunde zeigen, daß sich die ersten erkennbaren Vorstufen des Sekretes im Golgi-Apparat bilden. Die Golgi-Vesikel können daher als Sekretvesikel angesehen werden. Sie entstehen als seitliche Aufblähungen der Golgi-Zisternen (Abb. 5), füllen sich mit einem locker erscheinenden Material und wachsen bis zu einer Größe von etwa $0,3\,\mu$ Durchmesser heran.

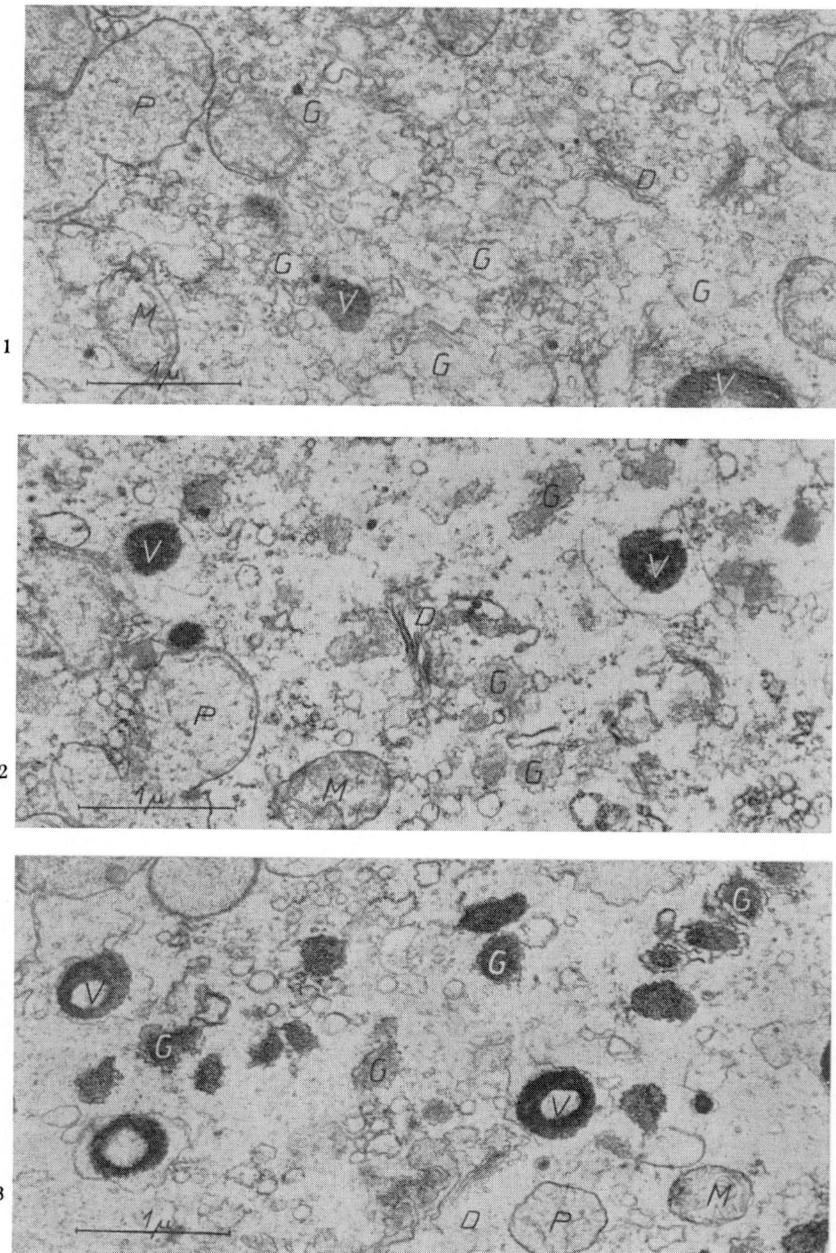

Abb. 1—3. *Drosophyllum*, gestielte Drüse, äußere Drüsenzelle. Schnitte durch dieselbe Drüse ohne Schnittkontrastierung (Abb. 1), nach Kontrastierung mit Rutheniumrot (Abb. 2), nach Kontrastierung mit Bleihydroxyd (Abb. 3). *D* = Dictyosomen, *G* = Golgi-Vesikel, *M* = Mitochondrien, *P* = Plastiden, *V* = Vakuolen mit dunklen Gerbstoffniederschlägen. Abb. 1—9: Fixation: OsO_4—$K_2Cr_2O_7$

Dann lösen sie sich ab. Ihr Inhalt verdichtet sich, verändert seinen Kontrast und läßt sich stärker mit Bleihydroxyd anfärben. Junge Vesikel sind, wohl artifiziell, recht unregelmäßig geformt. Mit zunehmender

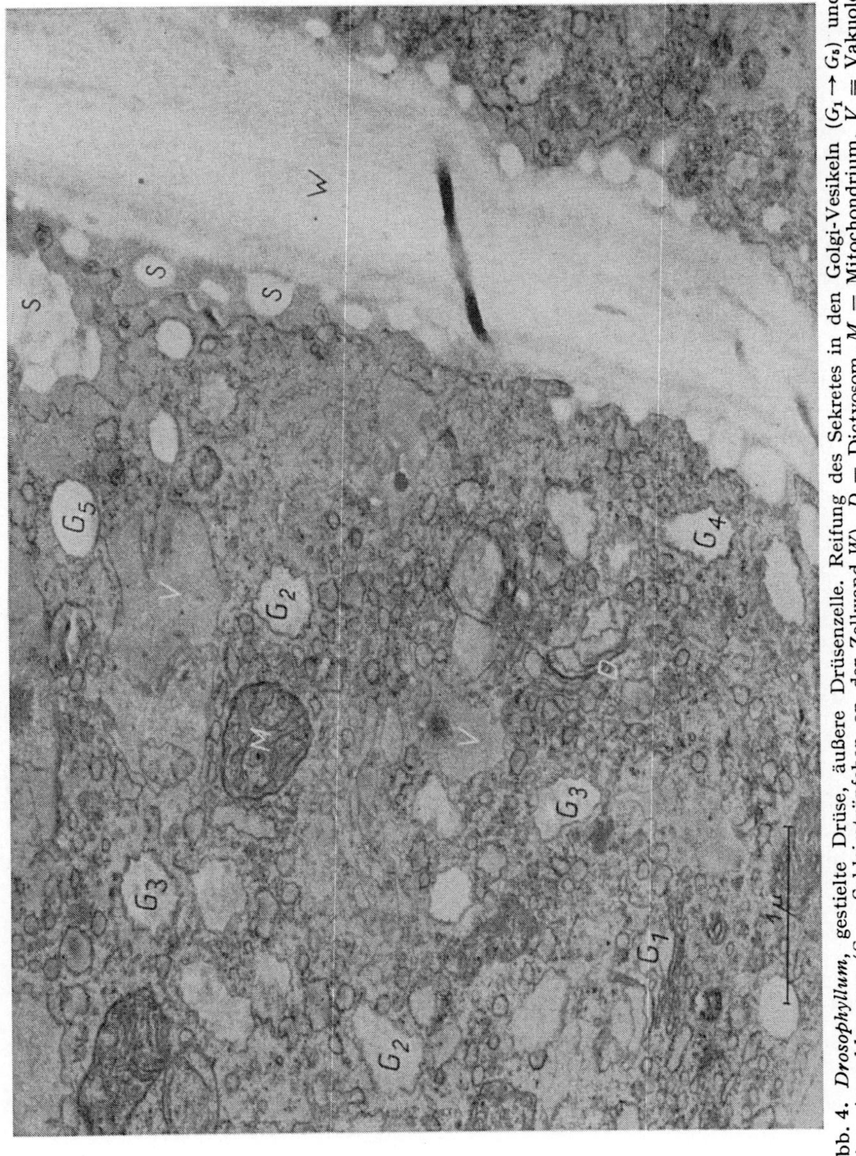

Abb. 4. *Drosophyllum*, gestielte Drüse, äußere Drüsenzelle. Reifung des Sekretes in den Golgi-Vesikeln ($G_1 \rightarrow G_5$) und seine Ausschleusung (S = Schleimtröpfchen an der Zellwand W). D = Dictyosom, M = Mitochondrium, V = Vakuole

Reife werden sie etwas kleiner, ihre Membran glatter (Abb. 4). Das deutet alles darauf hin, daß die Schleimbildung in den Vesikeln auch nach ihrer Abtrennung noch fortgesetzt wird [44]. Dabei scheint es sich um chemische Veränderungen und um eine Konzentrierung des Sekretes zu handeln.

Der Reaktionsverlauf bei der Schleimsynthese ist unbekannt. Die Ausscheidungsintensität steigert sich, wenn die im Sekret nachgewiesenen Zucker zugeführt werden [44]. Sie läßt sich aber auch schon durch die

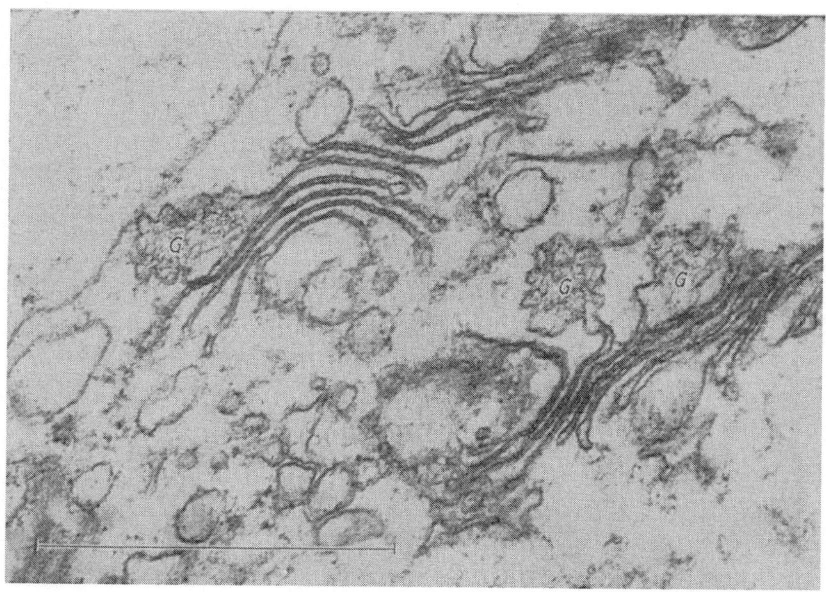

Abb. 5. *Drosophyllum*, gestielte Drüse, äußere Drüsenzelle. Dictyosomen; Golgi-Zisternen zu Vesikeln (*G*) erweitert; diese enthalten ein elektronenmikroskopisch erkennbares Material

Applikation von Glucose oder Glycerinphosphat fördern, Substanzen, die als Ausgangsmaterial für die Synthese oder als Energielieferanten dienen könnten [42].

4. Die Ausscheidung des Fangschleimes

Die freien Golgi-Vesikel wandern durch die Zelle an eine Zellwand. Die Vesikelmembran vereinigt sich, vielleicht nur für kurze Zeit, mit der Plasmamembran und wird dann ab- oder umgebaut. Die Sekrettröpfchen liegen jetzt zwischen Plasmalemma und Zellwand (Abb. 4) [44]. Die Sekretausschleusung ist also ein typisches Beispiel für eine Vesikelextrusion [2].

Die Vesikel werden „ungerichtet" transportiert, die treibende Kraft ist unbekannt. Das Sekret lagert sich bei *Drosophyllum* an keiner Zellwand bevorzugt ab. Bei *Pinguicula* sammeln sich die ausgeschiedenen Tröpfchen dagegen am oberen Teil der Antiklinen. Der Schleim muß dann an den Zellwänden entlang und zuletzt in ihnen wandern, um

nach außen zu kommen. Die cutinisierten Radialwände der Trenn-
schicht erzwingen die Richtung seines Weges [*44*]. Die Cuticula über den
Drüsenzellen ist leicht permeierbar, auch bei *Pinguicula*, wo sie keine
Poren besitzt.

Die Antiklinen sammeln also, von der Drüsenoberfläche aus gesehen,
das meiste Sekret. Dementsprechend tritt es bevorzugt über diesen
Wänden aus. Das kann man bei *Drosera* lichtmikroskopisch im Leben
beobachten [*41*].

Als „Pumpe", die den Schleim aus der Drüse treibt, könnte der Druck
des Plasmas gegen die Zellwände und damit auf die vor diesen liegenden
Schleimtröpfchen wirken; sie würden dadurch herausgepreßt. Sie könnten
aber auch durch eine aktive Wasserausscheidung herausgespült werden
[*44*].

5. Experimentelle Beeinflussung der Sekretion bei *Drosophyllum*

Die Intensität der Schleimsekretion läßt sich leicht durch mikroskopische Messung der ausgeschiedenen, an den Tentakeln haftenden
Sekrettröpfchen ermitteln. Sie ist stark temperaturabhängig. Das Opti-
mum liegt bei etwa 32° C. Der Temperaturkoeffizient Q_{10} beträgt un-
gefähr 2. Die Ausscheidung ist bei 29° C nahezu dreimal so stark wie bei
14° C. In gleichem Maße ist auch die Menge der Sekretvesikel in den
Drüsenzellen bei 29° C gegenüber 14° C erhöht (Abb. 6 und 7) [*42*].
Bei länger dauernden Versuchen verringert sich die Sekretionsintensität.
Sie ist nach 22 Stunden bei 28° C nur noch $^1/_3$ so stark wie zu Beginn. Die
Menge der Golgi-Vesikel sinkt dementsprechend (Abb. 8) [*42*].

Diese engen quantitativen Beziehungen zwischen Struktur und
Leistung ermöglichen eine überschlagsmäßige Berechnung der Umsatz-
geschwindigkeit im Golgi-Apparat. Bei 28° C befinden sich unter 1 μ^2
Drüsenoberfläche insgesamt etwa 3,4 μ^3 Golgi-Vesikel. Diese müssen
sich je Minute 0,4 mal umsetzen, um die Menge von 1,33 μ^3 Schleim
auszuscheiden, die je Minute durch 1 μ^2 Drüsenoberfläche tritt. Das
bedeutet, daß ein Golgi-Vesikel die Lebensdauer von etwa 2,5 min hat.
Die Konzentrationsänderung des Schleimes nach seiner Ausschleusung
aus dem Plasma muß noch berücksichtigt werden. Deshalb bedarf
dieser Wert einer Korrektur, deren Ausmaß jedoch nicht abzuschätzen
ist [*42*].

Die Drüsen sezernieren nur bei ausreichender Wasserversorgung.
Wird diese unterbunden, stellen sie ihre Tätigkeit ein. Dennoch beobachtet man dann in den Drüsenzellen Sekretvesikel. Das bedeutet,
daß das Wasser in erster Linie für die Schleim*extrusion* erforderlich
ist [*43*].

Stoffwechselinhibitoren hemmen die Ausscheidung. Ebenso kommt
sie in einer Stickstoffatmosphäre zum Stillstand. In Drüsen, denen kein
Sauerstoff zugeführt wird, fehlen die Sekretvesikel. Eine Vergiftung mit
10^{-4} m KCN wirkt in gleicher Weise; dabei werden die Zellen nur vorübergehend beeinflußt. Durch die Hemmung des Energiestoffwechsels
wird also in erster Linie die Schleim*synthese* beeinträchtigt [*45*].

Für den Zuckertransport zu den Drüsen hin reichen wahrscheinlich die durch Gärungsvorgänge gewonnenen Energien aus [58, 63]. Die

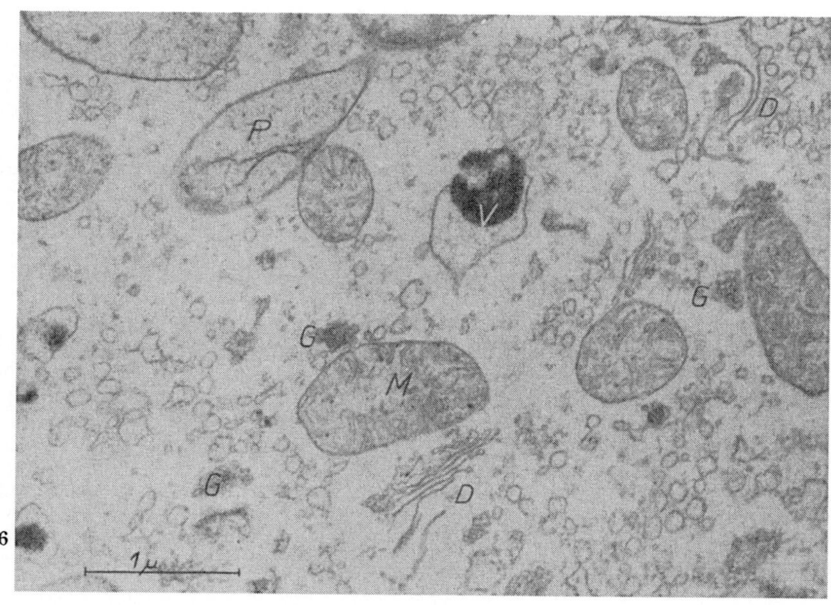

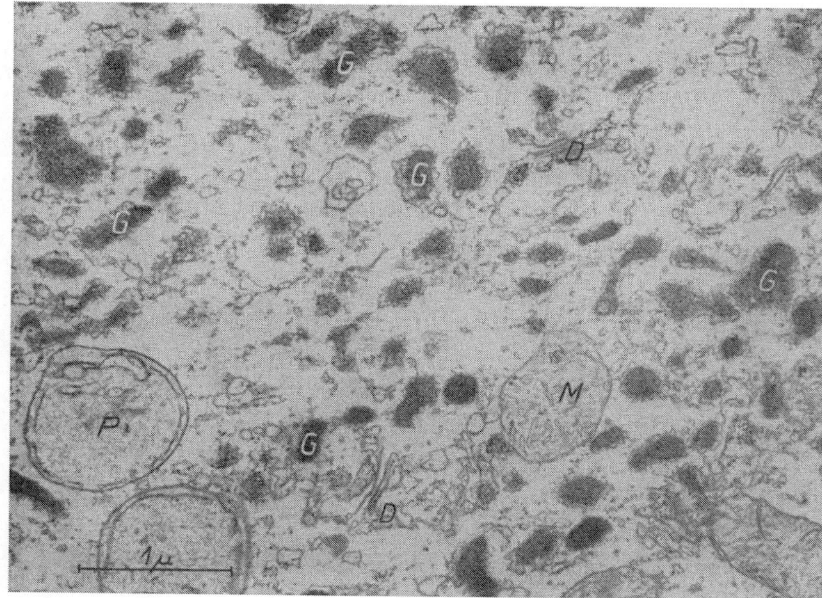

Abb. 6—7. *Drosophyllum*, gestielte Drüsen, äußere Drüsenzellen desselben Blattes. Wirkung der Temperatur auf die Bildung von Golgi-Vesikeln (*G*). Abb. 6 bei 14° C, Abb. 7 2¹/₂ Std bei 29° C. *D* = Dictyosomen, *M* = Mitochondrien, *P* = Plastiden, *V* = Vakuole. Mit Bleihydroxyd kontrastiert

Einschleusung der Zucker in die Kompartimente des Golgi-Apparates und ihre Verknüpfung dürfte ATP verbrauchen. Die große Zahl der Mitochondrien in den Drüsenzellen deutet auf eine hohe Atmungsintensität. Diese wurde bei *Drosera* (allerdings an ganzen Blättern) nachgewiesen [*18*].

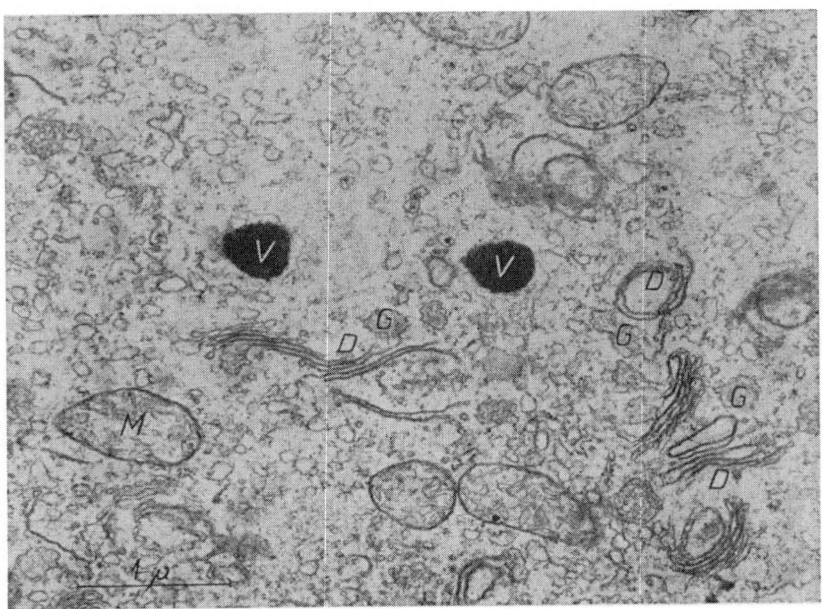

Abb. 8. *Drosophyllum*, gestielte Drüse, äußere Drüsenzelle. In länger dauernden Versuchen bei 29° C nimmt die Bildung von Golgi-Vesikeln (*G*) ab. 22 Std Versuchzeit. *D* = Dictyosomen, *M* = Mitochondrien, *V* = Vakuolen. Mit Bleihydroxyd kontrastiert

II. Andere Schleimdrüsen

Laminaria hyperborea und verwandte Braunalgen besitzen Interzellulargänge, die mit Schleim gefüllt sind. Er wird von kleinen, plasmareichen Zellen ausgeschieden, die in Gruppen an der Innenseite der Kanäle sitzen. In diesen Zellen liegen mehrere große Dictyosomen rings um den Zellkern. Die Golgi-Zisternen bilden relativ große Golgi-Vesikel mit einem elektronenmikroskopisch erkennbaren Inhalt. Ähnlich wird das Material in den Sekretgängen dargestellt. Der Schleim wird offenbar wie bei den Insektivoren erzeugt und ausgeschieden [*47*].

Einige weitere drüsenartige Gebilde, die möglicherweise Schleim sezernieren, enthalten in ihrem Plasma gleichfalls viele Dictyosomen und Golgi-Vesikel, so die Intravaginalschuppen von *Elodea* [*29*], die Drüsenzellen im Thallusinneren der Rotalge *Lomentaria baileyana* [*4*] und das innere Ektalexipulum des Pilzes *Neobulgaria pura* [*31*]. Wahrscheinlich bestehen hier ähnliche Zusammenhänge.

B. Die Sekretion von Verdauungsenzymen

Viele carnivore Pflanzen zersetzen die erbeuteten Insekten mit proteolytischen Enzymen. Diese wirken meist im sauren Bereich und scheinen dem Pepsin nahe zu stehen. Teilweise helfen wahrscheinlich bei der Verdauung Bakterien mit (Übersicht bei [24] und [39]).

Wie bei den anderen Insektivoren ändert sich bei *Drosophyllum* die Struktur der Drüsenzellen kurz nach dem Auftragen von Albumin oder Casein [46]. Die Vakuolen „aggregieren". Das Chromatin kontrahiert sich wie bei einer Prophase zu gröberen Strängen. Granula in der Karyolymphe vergrößern sich. Im Nucleolus entstehen vorübergehend kontrastreiche Partikel (vgl. [23]). Material in den Kernporen deutet auf einen Substanzaustausch zwischen Kern und Cytoplasma. Die Anzahl der Ribosomen scheint zuzunehmen. Das endoplasmatische Reticulum läßt sich jetzt auch nach Osmium-Bichromat-Fixation in Form von Zisternen darstellen. Diese sind teilweise "rough surfaced". Sie zerfallen nicht so stark in einzelne Vesikel wie in ungefütterten Drüsen. Wahrscheinlich vermehrt sich außerdem das Endomembransystem.

In dieser Weise reagieren nicht nur die sitzenden Drüsen, sondern auch die gestielten, deren verdauende Kraft nur gering sein soll [37]. Es ist deshalb nicht sicher, ob die Veränderungen der Plasmastruktur mit der Bildung und Ausscheidung der Verdauungsenzyme oder mit der Resorption zusammenhängen.

Einer Verdauungsdrüse vergleichbar ist das Scutellum des Gras-Embryos. Während der Keimung sezerniert es Enzyme und Enzym-Aktivatoren (Gibberelline) in das Endosperm, mobilisiert die dort abgelagerten Reservesubstanzen, resorbiert sie und führt sie dem wachsenden Keimling zu [8, 32].

Im ruhenden Scutellum, zumindest in den Epithelzellen, ist das endoplasmatische Reticulum nur spärlich entwickelt [32]. Dagegen gleicht die Plasmastruktur während der Keimung eher der einer tierischen, proteinsezernierenden Drüse [3]; ein "rough surfaced" endoplasmatisches Reticulum ist charakteristisch. Zur selben Zeit vermehren sich die Mitochondrien, ihre Aktivität nimmt zu [20].

C. Physiologie und Morphologie der Nektarien

Nektarien scheiden Zucker aus. Nach Frey-Wyssling [12] ist die Sekretion ursprünglich eine Folge mangelnder Korrelation zwischen Zucker mobilisierenden und Zucker verbrauchenden Geweben („Saftventil"). Hierauf deutet auch die Beeinflußbarkeit der Ausscheidung durch Wuchsstoffe [28, 55]. Diese Ansicht ist jedoch nicht unwidersprochen geblieben [25].

I. Das Sekret und seine Herkunft

Die Zuckerkonzentration des Nektars hängt von der Art der Drüse ab [1]; ihre genaue Bestimmung ist wegen der nachträglichen Eintrocknung des ausgeschiedenen Saftes schwierig [11, 67]. Werte um 30%

werden mehrfach angegeben [*1, 22, 25, 71*]. Mikroorganismen, die im Sekret leben, können es verändern und die Angaben über seine Zusammensetzung verfälschen [*25*].

In vielen Fällen scheiden die Nektarien in erster Linie Saccharose aus. Diese wird oft gespalten; dabei entstehen Glucose und Fructose im Verhältnis 1:1 [*25*]. Es können sich aber auch Oligosaccharide bilden [*36, 71, 72*]. Die Invertase-Aktivität zeigt sich auch im isolierten Nektar.

Die Konzentration der ninhydrinpositiven Substanzen ist auffallend niedrig [*67*]. Sie ist in der Regel um so schwächer, je besser die Drüse anatomisch differenziert ist [*25*]. Auch Phosphate treten nur in geringer Menge auf [*25, 67*]. K, Na und Ca kommen etwa im gleichen Verhältnis wie in anderen Pflanzenteilen vor; der Mg-Gehalt ist demgegenüber sehr niedrig [*26*]. Als weitere Komponenten sind verschiedene organische Säuren [*25*], wasserlösliche Vitamine [*61, 69*] sowie in einigen Fällen außer den bereits erwähnten Invertasen andere Enzyme nachgewiesen [*7, 25*].

Als Saccharose werden die Zucker im ausdifferenzierten Phloem transportiert. Nektar wurde deshalb als ausgeschiedener Phloemsaft bezeichnet [*15*]. Befunde über die Phloeminnervierung und Stoffversorgung der Drüsen stützen diese Ansicht [*1, 11, 14, 71*]. Bei den versorgenden Leitelementen handelt es sich um protophloemähnliche Zellen [*11*]. Wenn außerdem noch Tracheiden bis an das Nektarium herantreten, enthält der Nektar viel Wasser [*1*]. Das Sekret ist jedoch, verglichen mit dem Siebröhrensaft, viel ärmer an Nicht-Zucker-Substanzen [*25, 67, 69*]. Es wird daher durch die Drüsen umgewandelt, gefiltert. Dabei können die N-Verbindungen viel stärker zurückgehalten werden als die Phosphate [*67*].

Nektar ist also nicht einfach Phloemsaft, der etwa infolge des Siebröhrenbinnendruckes und polarer Permeabilitätsunterschiede austritt, sondern das Produkt einer spezifischen Drüsentätigkeit. Er hat in vielen Fällen, wenn auch vielleicht erst sekundär durch die Eindunstung, einen höheren osmotischen Wert als der Siebröhreninhalt und der Zellsaft der Drüsenzellen. Die Zucker werden also dann gegen ein Konzentrationsgefälle ausgeschieden [*27*]. Als weitere Zuckerquelle kommt in einigen Fällen die Stärke des Drüsengewebes und der umliegenden Zellen in Betracht.

II. Der Stoffwechsel der Nektarien

Jede Beeinflussung des Kohlenhydratstoffwechsels wirkt sich auf die Ausscheidungsintensität aus [*22, 52, 54, 65*]. Die herangeführten Zucker werden, wie Versuche mit isolierten Nektarien zeigen, in den Drüsenzellen transformiert. Glucose, Fructose und Saccharose wandeln sich ineinander um [*17*]. Lactose und Galactose gehen ebenfalls, wenn auch in geringerem Maße, in diesen Stoffwechsel mit ein. Bei anderen Hexosen sind die Ergebnisse widersprüchlich (objektbedingt?) [*28, 53*]. In der Regel entsteht wohl primär Saccharose. Die an der Synthese mitwirkende Uridindiphosphatglucose ist in Nektarien nachgewiesen [*25*], eine Galactowaldenase wahrscheinlich [*28*]. An weiteren Umsetzungen sind

Transfructosidasen [71] und Transglucosidasen [72] beteiligt. Diese sollen manchmal mit sezerniert werden, sind aber hauptsächlich wohl an der Plasmaoberfläche lokalisiert [28]. Pentosen wie Arabinose und Xylose werden dagegen unverändert ausgeschieden [17, 71].

Die Zucker werden als Phosphate umgesetzt. Als Intermediärformen treten Glucose-Phosphat, Fructose-Phosphat und besonders Saccharose-Phosphat auf [28]. Die hohe Aktivität saurer Phosphatasen in Nektarien hängt offenbar hiermit zusammen [7, 16, 59, 66, 67]. Sie ist so groß, daß meistens der histochemische Phosphatasenachweis zur Identifizierung des Drüsengewebes benutzt werden kann, auch wenn es anatomisch wenig differenziert ist [16]. In einigen Fällen ist nicht so sehr das epidermale Sekretionsgewebe, sondern das weiter innen gelegene Nektarienparenchym phosphataseaktiv [7, 59].

Mg^{++} ist ein Cofaktor bei Phosphorylierungsmechanismen, Ca^{++} hemmt sie (Übersicht bei [21]). Das könnte eine Erklärung dafür sein, daß das Drüsengewebe relativ viel Mg^{++} enthält, während der Nektar daran verarmt ist, und auch für das häufige Auftreten von Ca-Oxalat-Kristallen in Drüsennähe [26, 67] (vgl. auch [56]).

Bei den Phosphorylierungen wird ATP verbraucht. Die Nektarausscheidung ist ein aktiver Prozeß. Die Atmungsintensität ist dementsprechend hoch [30, 66]. Durch die Applikation von Glucose und Fructose steigt der O_2-Verbrauch; ihre Verknüpfung zum Disaccharid erfordert Energie. Galactose und besonders Arabinose, die nur wenig bzw. gar nicht umgeformt werden, wirken nicht in diesem Maße [67]. Substanzen, die die ATP-Produktion der Drüsenzellen beeinträchtigen, hemmen die Ausscheidung des Nektars [5, 28, 55, 67]. Es läßt sich nicht entscheiden, ob nicht auch andere Teile des Sekretionsprozesses gestört werden.

III. Die Morphologie der Nektarien

Nektarien können *sehr* verschieden gebaut sein. Man unterscheidet zwischen „gestalteten" Nektarien mit deutlich ausgebildetem, meist peripherem Ausscheidungsgewebe (Drüsenhaare, -epithel) und „gestaltlosen", denen ein solches fehlt. Bei den letzten tritt das Sekret durch Spaltöffnungen aus [13, 70]. Manche Nektarien sind ähnlich wie die Insektivoren-Drüsen durch eine Scheide aus undurchlässigen Zellwänden isoliert. Dadurch wird verhindert, daß das abgeschiedene Sekret in das Parenchym zurückfließt.

Die Drüsenzellen sind nur relativ schwach vakuolisiert. Sie enthalten nach elektronenmikroskopischen Untersuchungen ein reich strukturiertes Plasma [9, 10, 30, 48, 49, 64]. Das Grundplasma erscheint ungewöhnlich dicht. Die Plastiden sind als Leukoplasten nur wenig differenziert. Entsprechend der hohen Atmungsintensität findet man zahlreiche Mitochondrien. Der Golgi-Apparat ist nicht auffällig gestaltet. Das stark entwickelte endoplasmatische Reticulum kommt in Form von Zisternen und Vesikeln vor. Die Bläschen sind vielleicht zum Teil artifiziell durch den Zerfall von Zisternen entstanden [48], sie lassen sich nicht immer sicher von Golgi-Vesikeln unterscheiden. In *Helleborus*-Nektarien treten kristalline Einschlüsse auf, die mit dem endoplasmatischen Reticulum in Zusammenhang stehen [10].

Mit zahlreichen Plasmodesmen steht das Grundplasma benachbarter Zellen durch die Wände hindurch in offener Verbindung. Durch diese Kanäle zieht außerdem je ein Tubulus des endoplasmatischen Reticulums, so daß dessen Kompartimente interzellulär miteinander kommunizieren [48].

Bei den bisher untersuchten Nektarien ist diese Grundausstattung in verschiedener Weise modifiziert. Sie hängt u. a. stark vom Alter der Zellen ab. Der Beginn der Sekretion fällt mit dem Höhepunkt der cytologischen Entwicklung zusammen [48, 49]. In einigen Fällen kommt es dann in den Drüsenzellen zu einer beträchtlichen Vergrößerung der plasmatischen Oberfläche (Abb. 9) [48, 64] durch Wandprotuberanzen.

IV. Vorstellungen über den Sekretionsmechanismus

Der Saccharosetransport vom Phloem durch das Nektarienparenchym zu den sezernierenden Zellen hin könnte im endoplasmatischen Reticulum erfolgen [30, 48, 49], auch über die Zellgrenzen hinweg. Die Zisternen neigen dazu, während der Sekretionsperiode in Vesikel zu zerfallen. Das deutet auf einen relativ hohen Gehalt an osmotisch wirksamen Substanzen. Die Wanderungsgeschwindigkeit der Saccharosemoleküle könnte — infolge ihrer eingeschränkten Bewegungsfreiheit in den Zisternen — gegenüber der normalen Diffusion erhöht sein. Es ist jedoch zweifelhaft, ob die Kapazität dieses Systems, zumal in den Engpässen (Plasmodesmen) ausreicht, um die Zufuhr zu bewältigen [30]. Da die Nektarien durch protophloemähnliche Elemente versorgt werden, ist es möglich, daß die Zucker in diesen Transportbahnen bereits phosphoryliert sind [28].

Die Feinstruktur der Drüsenzellen von *Euphorbia pulcherrima* verändert sich nicht, wenn die Sekretion durch Abkühlung auf 11° C vorübergehend unterbrochen wird [49]. Es ist also möglich, daß der Nektar nicht wie der Fangschleim der Insektivoren-Drüsen mittels Vesikelextrusion (hier verschwanden die Sekretvesikel bei einer Temperatursenkung), sondern auf andere Weise ausgeschleust wird. Allerdings wird der wirre Verlauf des Plasmalemmas in sezernierenden *Abutilon*-Nektarien von Mercer und Rathgeber [30] als Hinweis auf einen Membranflußmechanismus angesehen. Man kann auch bei den *Euphorbia*-Nektarien nicht ausschließen, daß die Vesikel in den Drüsenzellen das Sekret enthalten, es aber wegen der Unterkühlung nicht ausstoßen [49].

Die eigentümliche Ausbildung der Zellwand in den Drüsenhaaren von *Vicia faba* [64] und besonders bei vielen Septalnektarien (Abb. 9) [48] spricht allerdings für einen anderen Sekretionsmechanismus. Septalnektarien treten in den Fruchtknoten vieler Monokotylen auf. Hier kleidet ein Drüsenepithel die Spalten zwischen den unvollständig verwachsenen Karpellen aus. An der Außenwand der Epithelzellen entsteht bei vielen Pflanzen kurz vor Beginn der Ausscheidung eine „sekundäre" Schicht [51] aus zahllosen sehr feinen Wandprotuberanzen.

Diese bilden ein dichtes Gewirr, dessen Spalten mit Plasma erfüllt sind. Dadurch ist die Plasmaoberfläche und die Substanz des Plasmalemmas

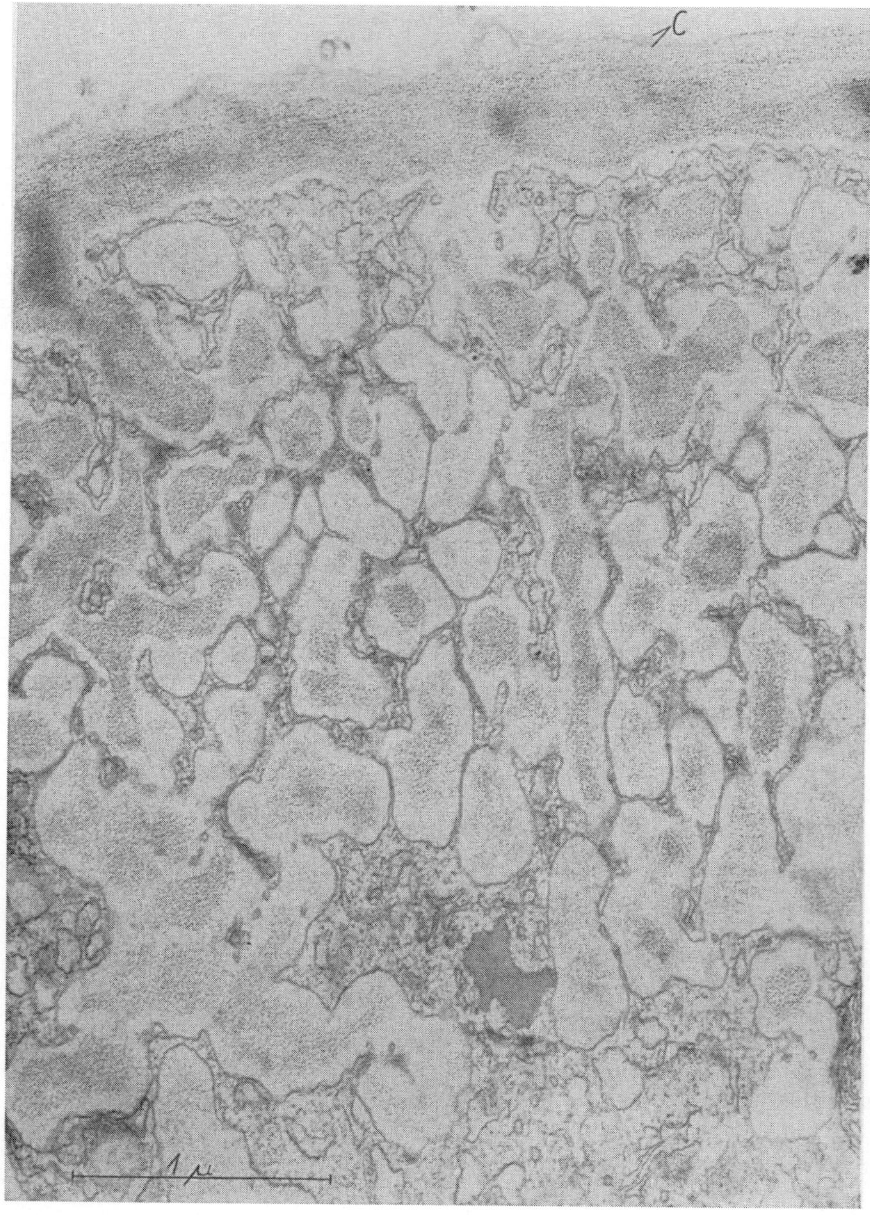

Abb. 9. *Gasteria trigona*, aktives Septalnektarium, „sekundäre Schicht" aus Wandprotuberanzen, durchsetzt mit Plasma. Cuticula (*C*) nur noch stellenweise vorhanden. Mit Bleihydroxyd kontrastiert

während der aktiven Phase um ein vielfaches vergrößert. Nach Abschluß der Sekretion werden diese Strukturen häufig abgebaut [*48*].

In der Plasmamembran können Enzyme lokalisiert sein, u. a. auch Phosphatasen (Übersicht bei [*19*]). Daher liegt die Vermutung nahe, daß die Zucker als einzelne Moleküle mittels aktiver Vorgänge durch das Plasmalemma geschleust werden, wohl durch Carriermechanismen, bei denen Phosphorylierungen und Dephosphorylierungen eine Rolle spielen [*25, 48*]. Die Membran arbeitet als Pumpe, als Energiequelle dient ATP, das an der Membran gespalten wird [*62*]. Vielleicht werden dabei gleichzeitig auch bei Pflanzen Na^+ und K^+ verlagert [*6*].

Solche aktiven Prozesse laufen wohl auch an den inneren cytoplasmatischen Membranen ab. Diese sind dem Plasmalemma in vieler Hinsicht ähnlich und trennen wie dieses stets eine „plasmatische" von einer „wäßrigen Phase" [*50*]. Sie könnten am intra- und interzellulären Zuckertransport zu den sezernierenden Zellen hin beteiligt sein und die Zuckermoleküle in die Kompartimente des endoplasmatischen Reticulums ein- und ausschleusen, gegebenenfalls auch in Sekretvesikel hinein. Eine Verschiebung gegen ein Konzentrationsgefälle ist auf diese Weise möglich.

Es besteht noch keine Klarheit darüber, ob das Wasser unabhängig vom Zucker sezerniert wird. ZIEGLER [*67*] fand, daß bei der Applikation von Stoffwechselgiften die Menge, aber nicht die Konzentration des Nektars beeinflußt wird. Er schloß daraus, daß Zucker und Wasser in ähnlicher Weise und in gleichem Ausmaß durch eine aktive Stoffwechselleistung abgeschieden werden. MATILE [*28*] kam jedoch bei vergleichbaren Versuchen zu anderen, uneinheitlichen Resultaten. MERCER und RATHGEBER [*30*] nehmen an, daß das Wasser teilweise passiv nachströmt.

Bei den großen Unterschieden im Bau [*16*], in der Ausstattung mit Phosphatasen [*7, 59*] und der Leistung [*25*] zwischen verschiedenen Nektarien ist es nicht ausgeschlossen, daß bei der Sekretion verschiedene Wege eingeschlagen werden.

In jedem Fall tritt der Nektar zunächst in die Zellwände. Hier wird er möglicherweise weiter umgestaltet [*28*]. Seine Zusammensetzung verändert sich außerdem durch Rückresorptionen. Nektarien nehmen während der Sekretion Aminosäuren [*68*], Phosphat [*25*], Ca^{++} und SO_4^{--} [*27*] sowie Zucker [*35, 57*] auf. Die Verarmung des Nektars an Nichtzuckern (gegenüber dem Siebröhrensaft) kann jedoch hiermit allein nicht erklärt werden. Die Rückresorption scheint nicht spezifisch genug zu sein [*27*]. Es ist noch ungeklärt, ob sie die in vielen Fällen beobachtete Verschiebung des Verhältnisses von Glucose und Fructose im Nektar verursacht [*28*]. Die Vergrößerung der plasmatischen Oberfläche durch die Protuberanzenbildung erleichtert sicherlich auch solche Prozesse.

Bei den Drüsenhaaren von *Abutilon* sammelt sich das Sekret an der Spitze der Apikalzelle in einem Spalt zwischen der Zellwand und der abgehobenen Cuticula, die hier von feinen Poren durchsetzt ist. Wenn der Druck in diesem Raum durch die Zufuhr immer neuen Nektars einen kritischen Wert übersteigt, öffnen sich die Poren, das Sekret tritt aus [30].

In Septalnektarien ist die Cuticula sehr dünn oder fehlt [48]. Bei anderen Nektarien bildet sie — zusammen mit der Cuticularschicht — häufig eine recht mächtige Lage, in der Poren nicht nachzuweisen sind [9, 49, 64], entgegen älteren Behauptungen [33]. Sie läßt auch in diesen Fällen das Sekret austreten. Gleichzeitig scheint sie aber auch zu verhindern, daß der abgeschiedene, durch die Verdunstung häufig noch eingedickte Nektar die Drüsenzellen plasmolysiert [25].

Literatur

[1] AGTHE, C.: Ber. schweiz. bot. Ges. 61, 240 (1951).
[2] BENNETT, H. S.: J. biophys. biochem. Cytol. 2 (Suppl.), 99 (1956).
[3] BIRBECK, M. S. C., and E. H. MERCER: Nature (Lond.) 189, 558 (1961).
[4] BOUCK, G. B.: J. cell. Biol. 12, 553 (1962).
[5] BROWN, H. D.: Bull. Torrey Bot. Club 86, 290 (1959).
[6] —, R. T. JACKSON, and H. J. DUPUY: Nature (Lond.) 202, 722 (1964).
[7] COTTI, T.: Ber. schweiz. bot. Ges. 72, 306 (1962).
[8] EDELMAN, J., S. I. SHIBKO, and A. J. KEYS: J. exp. Bot. 10, 178 (1959).
[9] EGGMANN, H.: Elektronenoptische Untersuchungen von Nektargeweben. Diplomarbeit Zürich 1962.
[10] EYMÉ, J.: Le Botaniste 46, 137 (1963).
[11] FREI, E.: Ber. schweiz. bot. Ges. 65, 60 (1955).
[12] FREY-WYSSLING, A.: Ber. schweiz. bot. Ges. 42, 109 (1933).
[13] — Die Stoffausscheidungen der höheren Pflanzen. Berlin: Springer 1935.
[14] — Acta bot. neerl. 4, 358 (1955).
[15] —, u. C. AGTHE: Verh. schweiz. Naturf. Ges. 130, 175 (1950).
[16] —, u. E. HÄUSERMANN: Ber. schweiz. bot. Ges. 70, 150 (1960).
[17] —, M. ZIMMERMANN u. A. MAURIZIO: Experientia (Basel) 10, 490 (1954).
[18] GIESSLER, A.: Flora 123, 133 (1928).
[19] GOLDFISCHER, S., E. ESSNER, and A. B. NOVIKOFF: J. Histochem. Cytochem. 12, 72 (1964).
[20] HANSON, J. B., A. E. VATTER, M. E. FISHER, and R. F. BILS: Agron. J. 51, 295 (1959).
[21] HEWITT, E. J.: In: Handbuch der Pflanzenphysiologie, Bd. IV, S. 427. Berlin-Göttingen-Heidelberg: Springer 1958.
[22] HUBER, H.: Planta 48, 47 (1956).
[23] LAFONTAINE, J. G.: J. biophys. biochem. Cytol. 4, 229 (1958).
[24] LLOYD, F. E.: The carnivorous plants. New York: Ronald Press Comp. 1942.
[25] LÜTTGE, U.: Planta 56, 189 (1961).
[26] — Planta 59, 108 (1962).
[27] — Planta 59, 175 (1962).
[28] MATILE, P.: Ber. schweiz. bot. Ges. 66, 237 (1956).
[29] MENKE, W.: Z. Naturforsch. 15b, 800 (1960).
[30] MERCER, F. V., and N. RATHGEBER: In: Electron Microscopy, Vol. 2, WW 11. New York and London: Academic Press 1962.
[31] MOORE, R. T., and J. H. McALEAR: J. cell. Biol. 16, 131 (1963).
[32] NIEUWDORP, P. J.: Acta bot. neerl. 12, 295 (1963).
[33] NIEUWENHUIS-VON UEXKÜLL-GÜLDENBRAND, M.: Rec. Trav. bot. neerl. 11, 291 (1914).
[34] PAECH, K.: Biochemie und Physiologie der sekundären Pflanzenstoffe. Berlin-Göttingen-Heidelberg: Springer 1950.
[35] PEDERSEN, M. W., C. LeFEVRE, and H. H. WILBE: Science 127, 758 (1958).
[36] PERCIVAL, M. S.: New Phytolog. 60, 235 (1961).
[37] QUINTANILHA, A.: Bol. Soc. Brot. 4, 44 (1926).

[38] REYNOLDS, E. S.: J. cell. Biol. **17**, 208 (1963).
[39] SCHMUCKER, T., u. G. LINNEMANN: In: Handbuch der Pflanzenphysiologie, Bd. XI, S. 198. Berlin-Göttingen-Heidelberg: Springer 1959.
[40] SCHNEPF, E.: Planta **54**, 641 (1960).
[41] — Flora **151**, 73 (1961).
[42] — Z. Naturforschg. **16**b, 605 (1961).
[43] — Planta **57**, 156 (1961).
[44] — Flora **153**, 1 (1963).
[45] — Flora **153**, 23 (1963).
[46] — Planta **59**, 351 (1963).
[47] — Naturwissenschaften **50**, 674 (1963).
[48] — Protoplasma (Wien) **58**, 137 (1964).
[49] — Protoplasma (Wien) **58**, 193 (1964).
[50] — Arch. Mikrobiol. **49**, 112 (1964).
[51] SCHNIEWIND-THIES, J.: Beiträge zur Kenntnis der Septalnektarien. Jena: Fischer 1897.
[52] SHUEL, R. W.: Canad. J. agricult. Sci. **35**, 124 (1955).
[53] — Canad. J. Bot. **34**, 142 (1956).
[54] — Canad. J. Plant. Sci. **37**, 220 (1957).
[55] — Canad. J. Bot. **37**, 1167 (1959).
[56] — Canad. J. Plant Sci. **41**, 50 (1961).
[57] — Plant Physiol. **36**, 265 (1961).
[58] ULLRICH, W.: Planta **57**, 402 (1961).
[59] VIS, J. H.: Acta bot. neerl. **7**, 124 (1958).
[60] VOGEL, A.: Beih. Z. schweiz. Forstverw. **30**, 113 (1960).
[61] WEBER, F.: Protoplasma **36**, 613 (1942).
[62] WILBRANDT, W.: In: Funktionelle und morphologische Organisation der Zelle, S. 219. Berlin-Göttingen-Heidelberg: Springer 1963.
[63] WILLENBRINK, J.: Planta **48**, 269 (1957).
[64] WRISCHER, M.: Acta botan. croat. **20/21**, 75 (1962).
[65] WYKES, G. R.: New Phytolog. **51**, 294 (1952).
[66] ZIEGLER, H.: Naturwissenschaften **42**, 259 (1955).
[67] — Planta **47**, 447 (1956).
[68] —, u. U. LÜTTGE: Naturwissenschaften **46**, 176 (1959).
[69] —, —, u. U. LÜTTGE: Flora **154**, 215 (1964).
[70] ZIMMERMANN, J. G.: Beih. bot. Zbl. **49**, I. Abt., 99 (1932).
[71] ZIMMERMANN, M.: Ber. schweiz. bot. Ges. **63**, 402 (1953).
[72] — Experientia (Basel) **10**, 145 (1954).

Summary

The carnivorous plant Drosophyllum captures insects by means of a mucilage, which consists mainly of an acid polysaccharide and of ascorbic acid. It is produced by the Golgi apparatus in the outer gland cells of the stalked glands. The Golgi cisternae dilate marginally and form Golgi vesicles which contain the mucilage; it stains in the electron microscope with ruthenium red and lead hydroxide. The Golgi vesicles become free and extrude their content when their membranes fuse with the plasmalemma.

The rate of secretion depends on the temperature. It decreases in experiments of long duration with isolated leaves. These effects can just as well be observed directly in the Golgi apparatus. Interruption of the water supply inhibits the extrusion of secretory vesicles. A disturbance of the energy supply stops the formation of secretory vesicles. In Drosophyllum gland cells the structure of the nucleus and of the endoplasmic reticulum changes markedly during digestion. There is still uncertainty about the secretion of the digestive enzymes. In mucilage glands of other plants the Golgi apparatus also participates in secretion.

Nectar is basically a solution of sucrose. Glucose and fructose are often formed by invertases, and there are also many other components. The amount of ninhydrin-positive substance depends on the anatomical differentiation of the glands. In some cases sugar is secreted against a concentration gradient. Phosphorylating processes are involved in secretion. There are very different types of nectaries. The morphology of secretion has not yet been completely explained. There are indications of a membrane flow mechanism in *Abutilon* nectaries, but experiments with *Euphorbia* glands and the peculiar structure of the epithelial cells in septal nectaries indicate secretion by molecular processes in the plasma membrane. In these epithelial cells the protoplasmic surface is considerably enlarged by pectic cell wall protuberances, which are only found in "ripe" glands. This enlargement may likewise facilitate reabsorption. However, these processes are not specific enough to explain the peculiar composition of the nectar.

Funktion des Golgi-Apparates in pflanzlichen und tierischen Zellen[1]

Von

ANDREAS SIEVERS, Bonn

Mit 11 Abbildungen

A. Einleitung und Definition

In den letzten zehn Jahren hat die elektronenmikroskopische Analyse des pflanzlichen und tierischen Protoplasmas erwiesen, daß der Golgi-Apparat in der Regel eine obligatorische Komponente der Zelle darstellt. Mit Ausnahme der Bakterien und blaugrünen Algen ist im Pflanzenreich z. B. sein Vorkommen inzwischen so vielfältig überprüft worden, daß sein Fehlen als Ausnahme gelten kann. In Anbetracht der großen Anzahl und Verschiedenartigkeit der Untersuchungen über den Golgi-Apparat soll in diesem Referat versucht werden, anhand einiger ausgewählter Ergebnisse seine Funktion in der Zelle zu erörtern, soweit wir sie bereits heute als gesichert ansehen können.

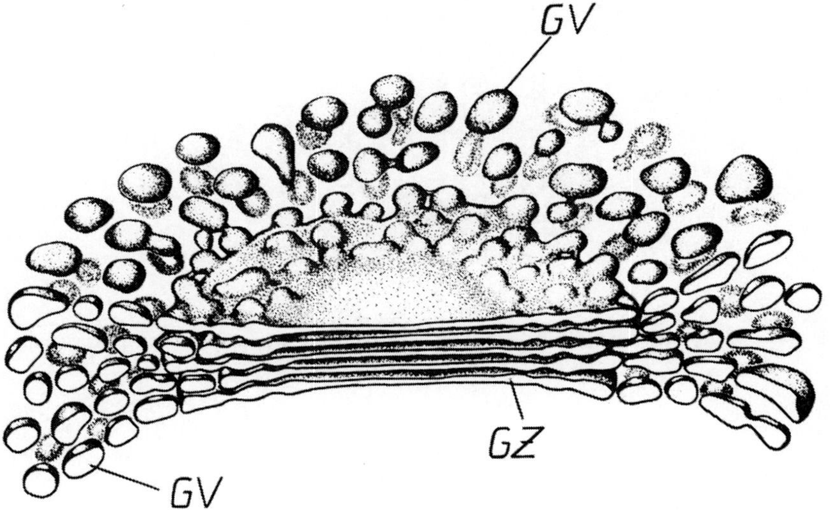

Abb. 1. Schema eines Dictyosoms mit fünf Golgi-Zisternen (*GZ*) und zahlreichen Golgi-Vesikeln (*GV*). (Aus DRAWERT und MIX [*19*])

Ohne in die Diskussion über terminologische Fragen eingreifen zu wollen, soll jedoch vorweg hinreichend geklärt werden, was hier unter dem Begriff *Golgi-Apparat* verstanden wird. Als strukturelle Einheit gelte in der Regel das *Dictyosom*, das als

---

[1] Mit Unterstützung durch die Deutsche Forschungsgemeinschaft.

Stapel mehrerer flacher Membranen auf elektronenmikroskopischem Wege sichtbar gemacht werden kann. Ein in sich geschlossenes Membranpaar aus einem Dictyosom heiße *Golgi-Zisterne*, die an seinem Rande durch Abschnürung gebildeten Bläschen *Golgi-Vesikel* (Abb. 1, 2 und 4a). Gelegentlich wird es zweckmäßig sein, ein
Dictyosom mit den umliegenden Golgi-Vesikeln als *Golgi-Feld* zu bezeichnen.
(Der Terminus Golgi-Feld erweist sich besonders dort als brauchbar und notwendig,

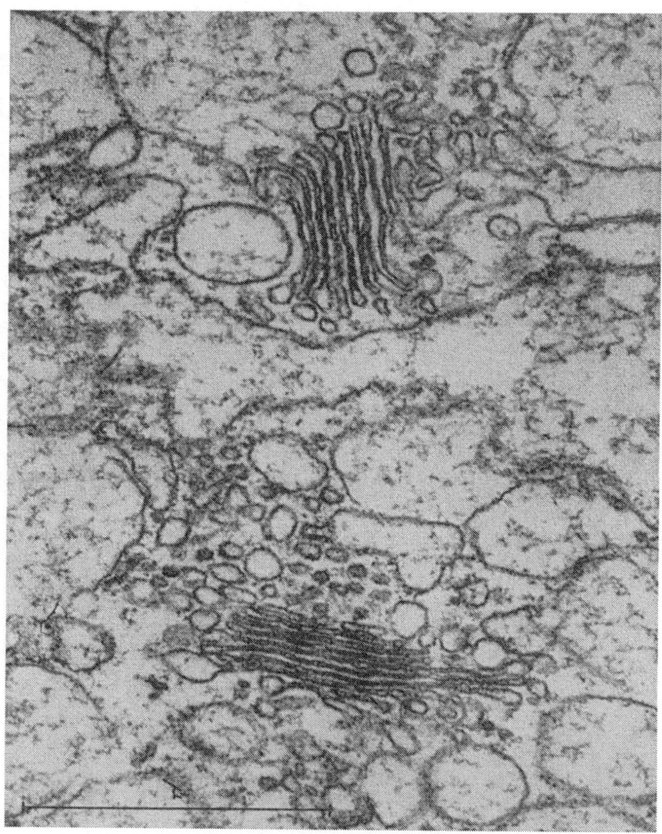

Abb. 2. Zwei Dictyosomen aus der Speicheldrüse von *Mycus persicae* mit Golgi-Vesikel-
Bildung. OsO$_4$-Fixation. (Mit freundlicher Genehmigung aus WOHLFARTH-BOTTERMANN
[*104*])

wo ein Dictyosom nicht zu erkennen ist, wohl aber auf Grund topographisch eindeutiger Beziehungen eine besondere Ansammlung von Vesikeln mit ihm homologisiert werden kann.) Bei einer derartigen Begriffsbestimmung bleibt dann der
Terminus *Golgi-Apparat* der Gesamtheit der genannten Strukturen innerhalb einer
Zelle vorbehalten.

Mit dieser Definition ist nun allerdings eine Erörterung über die
Funktion des Golgi-Apparates in die Dimension der Ultrastruktur der
Zelle verwiesen, wie sie nur mit dem Elektronenmikroskop dargestellt
werden kann. Schließlich ist jedweder Strukturwandel dieser Zell-
komponente — und nur ein erkennbarer Strukturwandel erlaubt den
Schluß auf die Funktion — zurückführbar auf Formveränderungen von

Membranen mit einer Dicke von etwa 75 Å. In dieser Größenordnung versagt das Lichtmikroskop. Ich werde daher über den Golgi-Apparat in diesem engeren Sinne und seine Funktion berichten, ohne dabei die Vielfalt von Strukturen berücksichtigen zu können, die früher — also allein mit Hilfe des Lichtmikroskopes — als Golgi-Apparate bezeichnet worden sind. Nur an den Stellen, wo eine exakte Homologisierung möglich ist, wird sich im Rahmen dieses Themas ein Rückblick als lohnend erweisen.

B. Funktion des Golgi-Apparates in pflanzlichen Zellen

Durch HODGE, MCLEAN und MERCER [42] wurde 1956 zum ersten Male auf Strukturen in der Pflanzenzelle aufmerksam gemacht, und zwar in der Armleuchteralge *Nitella*, die wir heute als Dictyosomen bezeichnen. Ein Jahr später beschrieben mehrere Autoren gleichzeitig dieselbe Struktur in verschiedenen Pflanzen und identifizierten sie in Anlehnung an die tierische Zelle als Golgi-Strukturen (nähere Einzelheiten darüber z. B. bei [98] oder [7]). SCHNEPF [81] erbrachte 1960 die ersten Hinweise auf eine Funktion des Golgi-Apparates in pflanzlichen Drüsenzellen, wo der Transport von Sekret via Golgi-Vesikel nach außen vermutet wird. Von MOLLENHAUER, WHALEY und LEECH [60] wurde 1961 zum ersten Male gezeigt, daß in Wurzelhaubenzellen Zellwandmaterial vom Golgi-Apparat abgegeben wird, und vermutet, daß außerdem die Membran von Golgi-Vesikeln dem Plasmalemma inkorporiert wird. Damit sind nun aber bereits die Grundzüge der Funktion des Golgi-Apparates in pflanzlichen Zellen angedeutet, auf die in der Folge näher eingegangen werden soll.

I. Der Golgi-Apparat in pflanzlichen Drüsenzellen

Über die Rolle des Golgi-Apparates in pflanzlichen Drüsen wird im Referat von SCHNEPF ausführlich berichtet. Es kann demnach als erwiesen gelten, daß der Golgi-Apparat bei einer Reihe von Drüsen zum mindesten die letzten Teilschritte der Synthese oder die Kondensation des Sekretes übernimmt und daß er das Sekret — im wesentlichen Polysaccharide — in Form von Golgi-Vesikeln zum Plasmalemma befördert. Man kann also in diesen Fällen Golgi-Vesikel und Sekret-Vesikel miteinander identifizieren.

II. Golgi-Apparat und Zellwandbildung

1. Zellwand-Wachstum

a) Kalyptra

Wie bereits erwähnt, wiesen MOLLENHAUER, WHALEY und LEECH [60] bei *Zea mays* nach, daß im Laufe der Differenzierung der Kalyptrazellen der Golgi-Apparat eine vorübergehende hohe Aktivität zeigt. Es werden zahlreiche Golgi-Vesikel von 0,3–0,5 μ Durchmesser gebildet (und zwar vom Rande der Zisternen und durch Hypertrophieren ganzer Zisternen), die offenbar zur Plasmamembran wandern und dort ihren Inhalt abgeben. Das Innere der Vesikel und die Zellwand sind nach $KMnO_4$-Fixation und nachfolgender Schnittkontrastierung mit $Ba(MnO_4)_2$ dunkel kontrastiert. Es wird vermutet, daß auf diese Weise Zellwandmaterial nach außen befördert wird. Spätere Untersuchungen zeigten nach OsO_4-Fixation [58], daß im Zuge der Abschnürung in den Vesikeln eine in zunehmendem Maße mit OsO_4 kontrastierbare Substanz von flockig-fibrillärer Konsistenz gebildet wird, vergleichbar den Verhältnissen in Insektivoren-Drüsen [82]. Über die Natur der in den Kalyptrazellen ausgeschiedenen Substanz liegen keine Untersuchungen vor. Da die Zellwand nicht nur während der Differenzierung zunimmt, sondern auch während der Isolierung verschleimt (vgl. [88]), ist es nicht möglich, in diesem Falle beide Vorgänge voneinander zu trennen. Es ist jedoch bekannt, daß diese Zellwände weitgehend aus Pektinstoffen aufgebaut sind [14].

b) Rhizodermis

α) Dermatogen

Eine recht interessante Aktivität zeigt der Golgi-Apparat in den Epidermiszellen der Wurzel. Der Golgi-Vesikel-Inhalt hat zunächst nahe

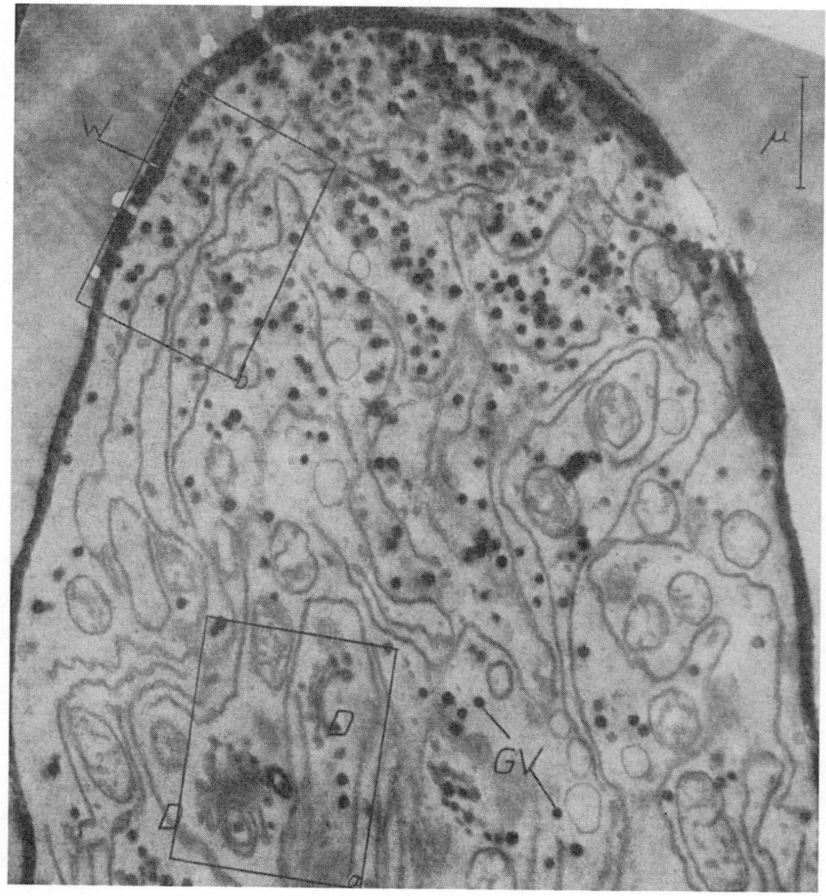

Abb. 3. Längsschnitt durch die Spitze eines wachsenden Wurzelhaares von *Zea mays* mit deutlicher Anreicherung der Golgi-Vesikel (*GV*) am Zellapex. *D* = Dictyosom, *W* = Zellwand. $KMnO_4$-Fixation. (Aus SIEVERS [86])

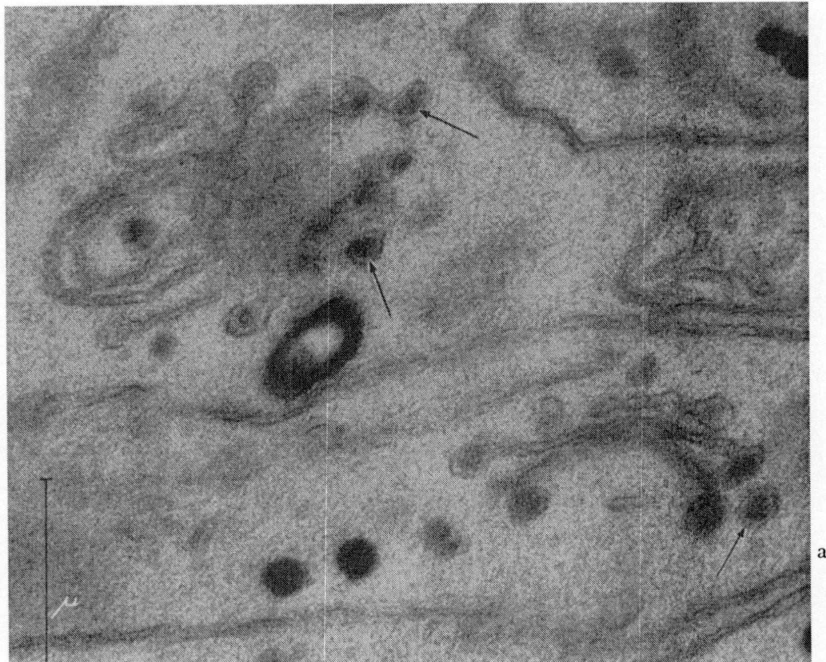

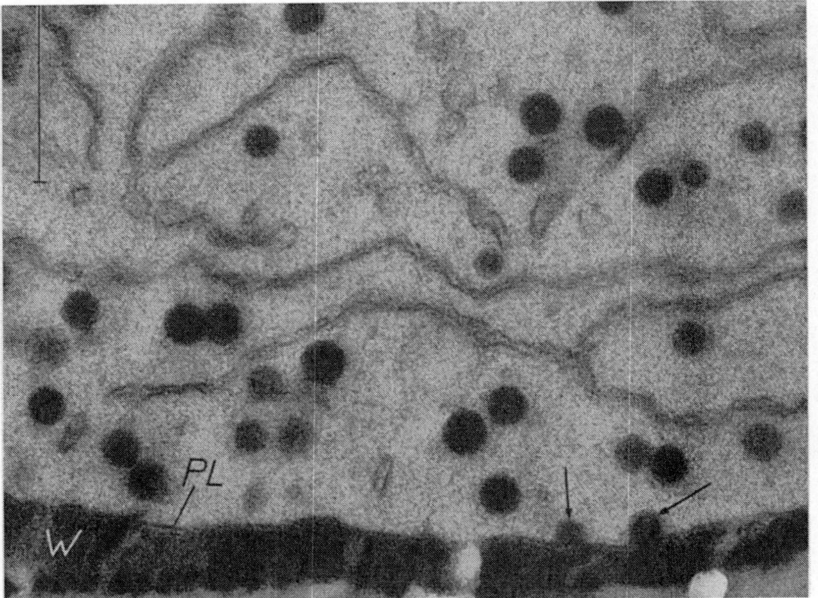

Abb. 4a u. b. Ausschnittvergrößerungen aus Abb. 3. — Abb. 4a zeigt die Genese von Golgi-Vesikeln. Der Kontrast im Innern der Vesikel nimmt mit zunehmender Entfernung von der Golgi-Zisterne zu (Pfeile). — In Abb. 4b sind durch Pfeile zwei Golgi-Vesikel markiert, die im Zeitpunkt der Extrusion ihres Inhaltes fixiert worden sind. PL = Plasmalemma, W = Zellwand. (Aus Sievers [86])

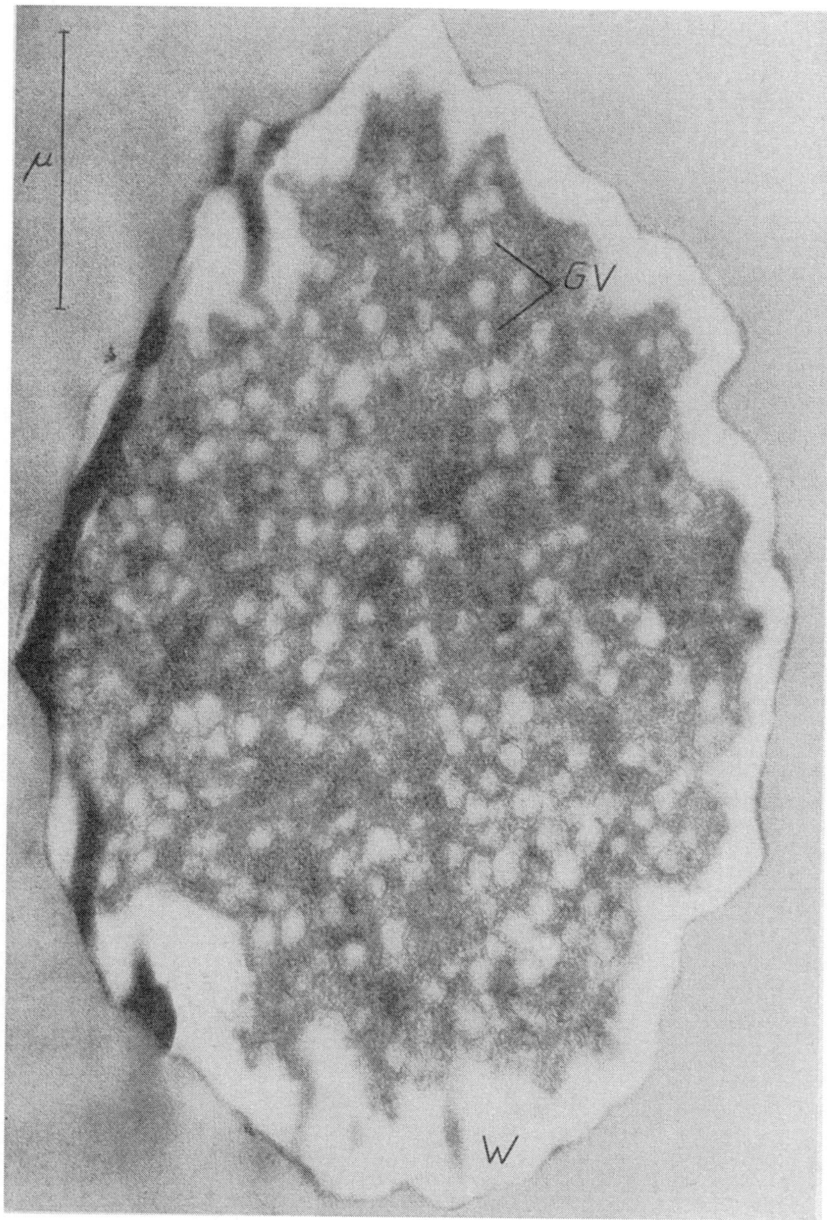

Abb. 5 u. 6. Zwei basipetale Querschnitte durch die Spitze wachsender Wurzelhaare von *Zea mays* nach OsO₄-Fixation und bei schwach hydratisiertem Cytoplasma. Man beachte den Mangel an Kontrast in Golgi-Vesikeln (*GV*) und der Zellwand (*W*) sowie die Anreicherung der Golgi-Vesikel an der Spitze (Abb. 5). Abb. 6 zeigt zahlreiche Dictyosomen (*D*) und Golgi-Vesikel (*GV*). *ER* = Endoplasmatisches Retikulum. (Abb. 5 und 6 SIEVERS, unveröffentlicht)

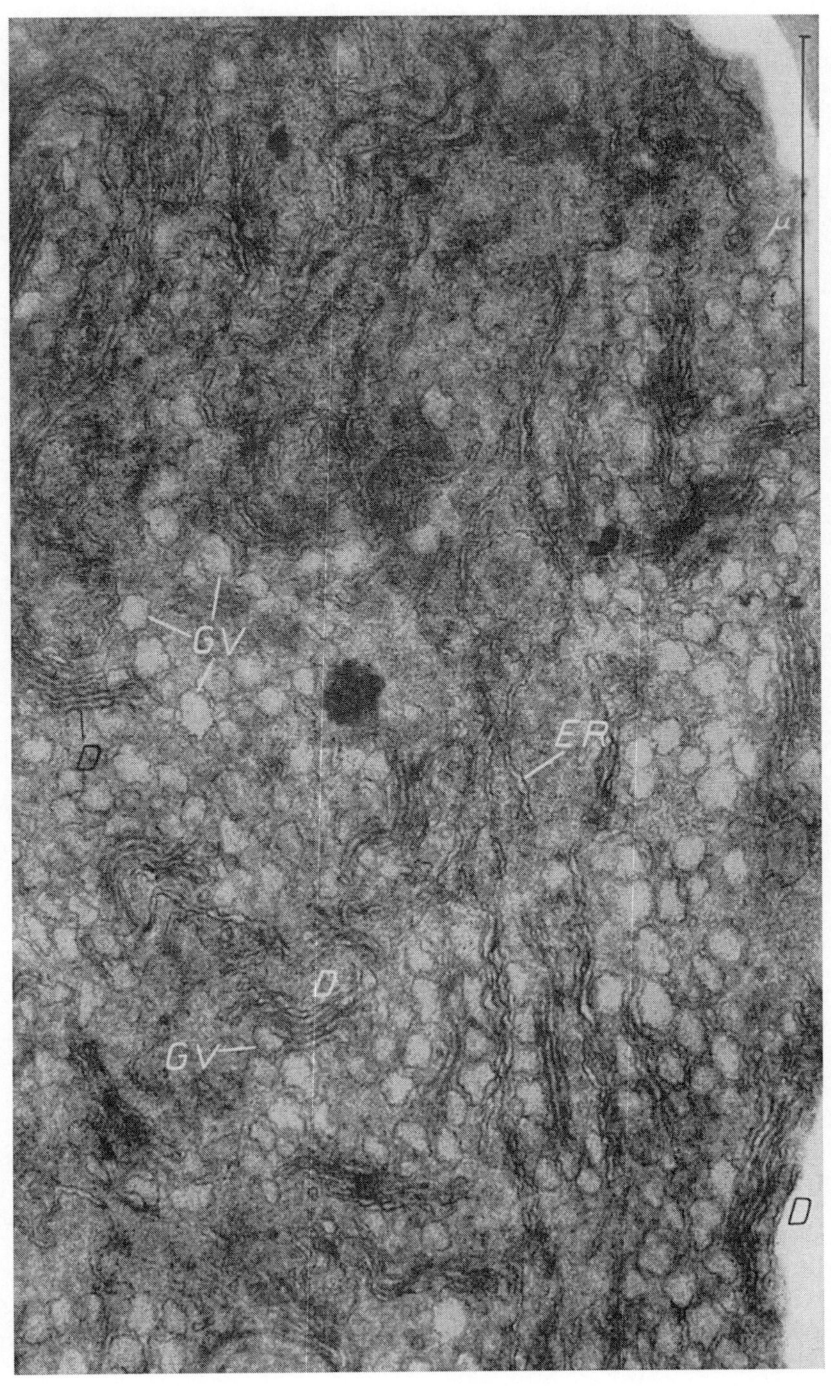

Abb. 6

dem Vegetationspunkt keinen Kontrast, und die Vesikel sind verhältnismäßig klein. Etwa von der zehnten Zelle an erreichen sie einen konstanten Durchmesser von etwa 0,17 μ und sind stark kontrastiert nach KMnO$_4$-Fixation [101]. Golgi-Vesikel-Substanz wird nun in die Wände der Rhizodermiszellen eingebaut, und zwar in die Außenwand sowie in die periklinen und antiklinen Innenwände [99].

β) Wurzelhaare

Eine besondere Differenzierung der Rhizodermiszelle ist das Wurzelhaar, das sich papillenartig aus der Außenwand herauswölbt und als schlauchförmige Ausstülpung von etwa 10 μ Durchmesser mehrere Millimeter lang werden kann. Die Oberfläche der Rhizodermis, eine für die Stoffaufnahme der Pflanze wesentliche Zellschicht, wird auf diese Weise um ein Vielfaches vergrößert (Zusammenfassung u. a. bei [15]). Wurzelhaare weisen bekanntlich Spitzenwachstum auf, d. h. der Einbau von Zellwandstoffen ist auf die äußerste apikale Rundung begrenzt. Diese schon lange bekannte Tatsache wurde in jüngster Zeit durch elektronenmikroskopische Untersuchungen an der Textur der Zellulosemikrofibrillen in der Wand erneut bestätigt [5, 17, 25, 46].

Erweitert man nun aber die Feinstrukturanalyse auch auf das Protoplasma, so ergibt sich sowohl bei *Zea mays* als auch bei *Tradescantia albiflora*, daß die bereits für die Dermatogenzelle beschriebene Aktivität des Golgi-Apparates im Wurzelhaar noch eindrucksvoller in Erscheinung tritt [86, 87]. Im gesamten Wurzelhaar befinden sich zahlreiche aktive Dictyosomen (Abb. 3, 4a und 6), deren Golgi-Vesikel von etwa 0,1 μ Durchmesser offenbar in steigender Konzentration an der Wurzelhaarspitze angereichert werden (Abb. 3) und die dort Substanzen in die wachsende Zellwandpartie ergießen (Abb. 4a und b, 7). Der starke Kontrast, den die Golgi-Vesikel-Substanz mit KMnO$_4$ als Fixans abgibt, wird auch hier erst im Zuge der Isolierung der Vesikel gebildet (Abb. 4a) und läßt auf die im werdenden Vesikel stattfindenden Prozesse (Synthese, Kondensation?) schließen (vgl. – nach OsO$_4$-Fixation – [82], ferner [58]). Nach OsO$_4$-Fixation bleibt dagegen der Inhalt der Vesikel bei Wurzelhaaren i. a. unkontrastiert [86]. Basipetale Querschnitte (nach OsO$_4$-Fixation und im schwach hydratisierten Plasma) sind in den Abb. 5 und 6 wiedergegeben. Im außerordentlich dichten Plasma fallen die zahlreichen unkontrastierten Golgi-Vesikel besonders deutlich auf. Die Zellwand hat stets dieselben Kontrast-Eigenschaften wie der Inhalt der Golgi-Vesikel.

Obwohl auch die Rhizodermis sowie die Wurzelhaare selbst mit einer Schleimschicht überzogen sind, wird man dennoch annehmen dürfen, daß an jeder durch Zellwand-Wachstum ausgezeichneten Wandpartie, in die extrem viel Golgi-Vesikel-Substanz abgegeben wird, auch normale Zellwandbaustoffe „sezerniert" werden [87]. Es ist ferner plausibel, daß dort, wo die Membran eines Golgi-Vesikels und das Plasmalemma sich berühren und beide an tangierenden Punkten sich öffnen müssen, um dem Vesikel-Inhalt den Eintritt in den extraplasmatischen Raum zu ermöglichen, auch die beiden genuin verschiedenen Membranen miteinander verschmelzen. Abb. 7 gibt nach OsO$_4$-Fixation einen Ausschnitt aus dem (stark hydratisierten, vgl. [89]) apikalen Plasma eines

Wurzelhaares wieder. Die Golgi-Vesikel sind in diesem Falle ausnahmsweise – wie die Zellwand – dunkel kontrastiert, das Plasmalemma hat sich etwas wellig durch einen hellen Saum von der Wand abgehoben. An

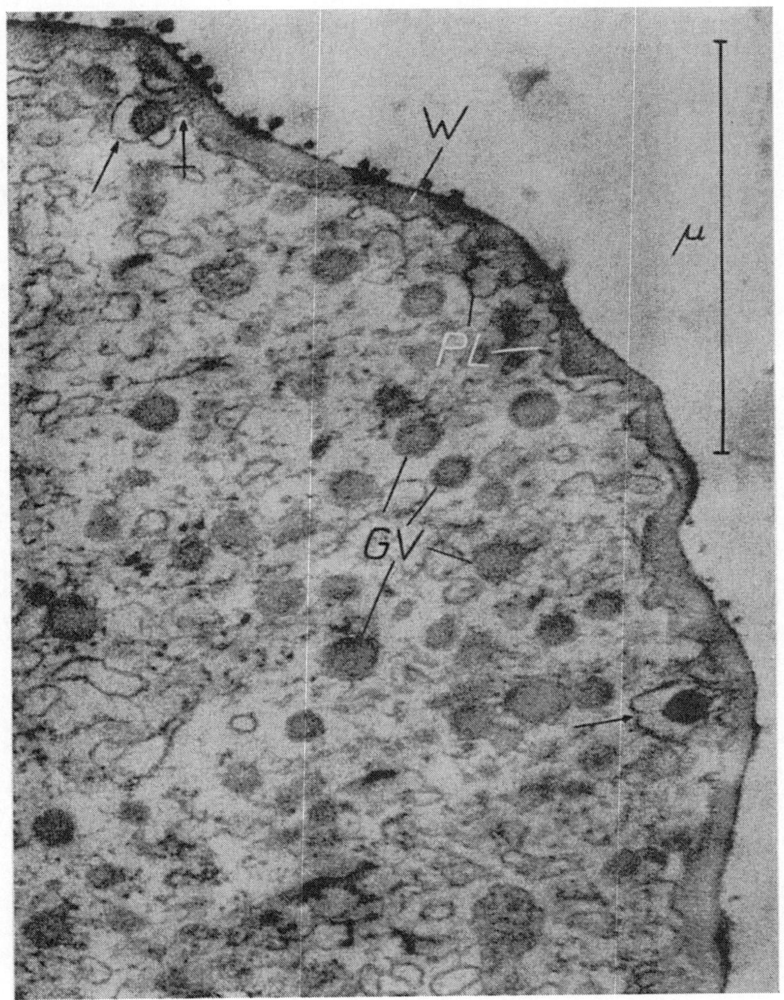

Abb. 7. Detail aus der Spitze eines wachsenden Wurzelhaares von *Tradescantia albiflora* nach OsO$_4$-Fixation. Stark hydratisiertes Cytoplasma. Golgi-Vesikel (*GV*) mit kontrastiertem Inhalt, ebenso Zellwand (*W*) kontrastiert. Plasmalemma (*PL*) wellig von der Wand abgehoben. Die beiden Pfeile weisen auf zwei Golgi-Vesikel, deren Membran sich vom kontrastierten Inhalt abgesetzt hat und die wahrscheinlich im Zeitpunkt der Berührung von Vesikel- und Plasma-Membran fixiert wurden. Am oberen Vesikel scheinen Plasmalemma und Vesikel-Membran fusioniert zu sein (gekreuzter Pfeil). (SIEVERS, unveröffentlicht)

zwei Stellen (Pfeile) sieht man je ein Golgi-Vesikel in unmittelbarer Wandnähe. Die Vesikel-Substanz ist deutlich von der Vesikel-Membran abgesetzt, die sich offenbar gedehnt hat. Wahrscheinlich handelt es sich hier um zwei Vesikel, die genau im Zeitpunkt der Berührung mit der Plasmamembran fixiert worden sind. Es hat den Anschein, als ob bei dem oberen Vesikel bereits ein Zusammenhang zwischen den beiden Membranen hergestellt worden sei. Jedoch fehlen hierzu noch detailliertere Untersuchungen, die mit hoher Auflösung und in Folgeschnitten exakt die fraglichen Membranen vermessen lassen (vgl. dazu [21], Fig. 8).

c) Pollenschlauch

In kürzlich erschienenen Arbeiten [73, 74] wird auch für einen weiteren Fall von pflanzlichem Spitzenwachstum, nämlich dem von Pollenschläuchen *(Lilium)*, eine starke Ansammlung von Vesikeln an der Spitze der wachsenden Zelle angegeben. Auch hier findet man — durchaus vergleichbar mit den Verhältnissen im Wurzelhaar — neben größeren Vesikeln von etwa 0,1 μ Durchmesser, die für Golgi-Vesikel gehalten werden, noch kleinere, die ich als Elemente des endoplasmatischen Retikulums (ER) ansprechen möchte, und außerdem Ribosomen. Zellwand und Vesikel-Inhalt sind nach OsO_4-Fixation unkontrastiert und lassen sich beide mit Wolframatophosphorsäure am Schnitt nachkontrastieren. Auf Grund ihrer Befunde vermuten die Autoren sowohl Zellwand- als auch Plasmalemma-Bildung des Pollenschlauches auf dem Wege der Golgi-Vesikel. Zu den gleichen Ergebnissen kamen bereits LARSON und LEWIS [54] an Pollenschläuchen von *Ranunculus macranthus*, ohne jedoch die Befunde im einzelnen mit photographischen Aufnahmen zu belegen; SASSEN [79, 79a] bestätigt das Ergebnis an Pollenschläuchen von *Petunia*.

d) Rhizoide

An jungen Rhizoiden der Armleuchteralge *Chara foetida* konnte inzwischen ebenfalls eine besondere Konzentrierung von Golgi-Vesikeln im Zellapex und die Einschleusung von Vesikel-Substanz in die apikale Zellwand beobachtet werden [89a].

2. Zellwand-Neubildung

a) Zentrifugale Wandbildung

Während bisher Fälle von Wandzuwachs betrachtet wurden, soll nunmehr die Neubildung von Zellwänden im Zuge der Zellteilung näher analysiert werden. Die Abb. 8 zeigt in der frühen Telophase der Karyokinese den Beginn der Bildung der Zellplatte, also der frühesten Zellwandpartien, in der Wurzelrinde von *Zea mays*. Man erkennt deutlich eine Vielzahl von dunklen Vesikeln, die sich in einer Linie zwischen den beiden Tochterkernen angesammelt haben und deren Herkunft von Dictyosomen unzweifelhaft ist. In Abb. 9 ist eine etwas spätere Telophase aus der Rhizodermis dargestellt, wo die Vesikel bereits in Richtung der Zellplatte zusammengeflossen sind und wo ebenfalls entsprechende Vesikel von Golgi-Zisternen abgeschnürt werden. Nach diesen neuesten Ergebnissen halten es WHALEY und MOLLENHAUER [100] für höchst wahrscheinlich, daß die Wandsubstanz der Zellplatte jedenfalls zunächst aus Golgi-Material aufgebaut wird, daß ferner das neue Plasmalemma von der Golgi-Vesikel-Membran gebildet wird und daß das ER die Zellplatte in ausgesparten kleinen Räumen von der Anlage her bereits „überbrückt" (vgl. zum letzten Punkt auch [8] sowie [68]).

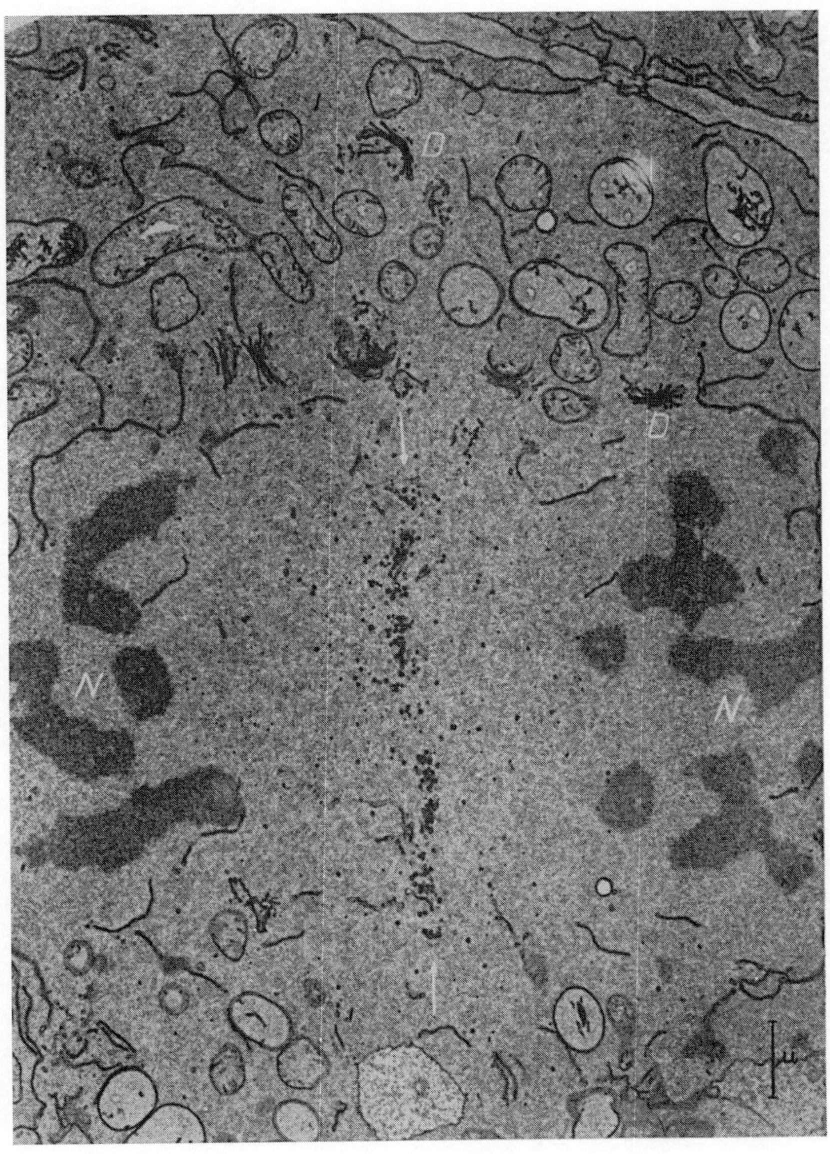

Abb. 8. Frühe Telophase der Karyokinese und Beginn der Zellplattenbildung in der Wurzelrinde von *Zea mays*. Die beiden Tochterkerne (*N*) sind durch eine lineare Anordnung von Golgi-Vesikeln getrennt (Pfeilpaar). *D* = Dictyosom. KMnO₄-Fixation. (Mit dankenswerter Genehmigung aus WHALEY und MOLLENHAUER [100])

7*

Eine Zunahme der Dictyosomenzahl in Zellplattennähe wurde schon von BUVAT und PUISSANT [8] und von WHALEY, MOLLENHAUER und LEECH [101] beobachtet. BUVAT und PUISSANT [8] vermuten jedoch, daß die „vacuoles pectiques", die die Zellplatte bilden, aus dem ER entstehen, was jedoch aus den Abbildungen nicht zu entnehmen ist. PORTER und MACHADO [68] weisen ebenfalls auf kleine, distinkte Vesikel von unbekannter Herkunft hin — abgesehen von den sog. „Phragmo-

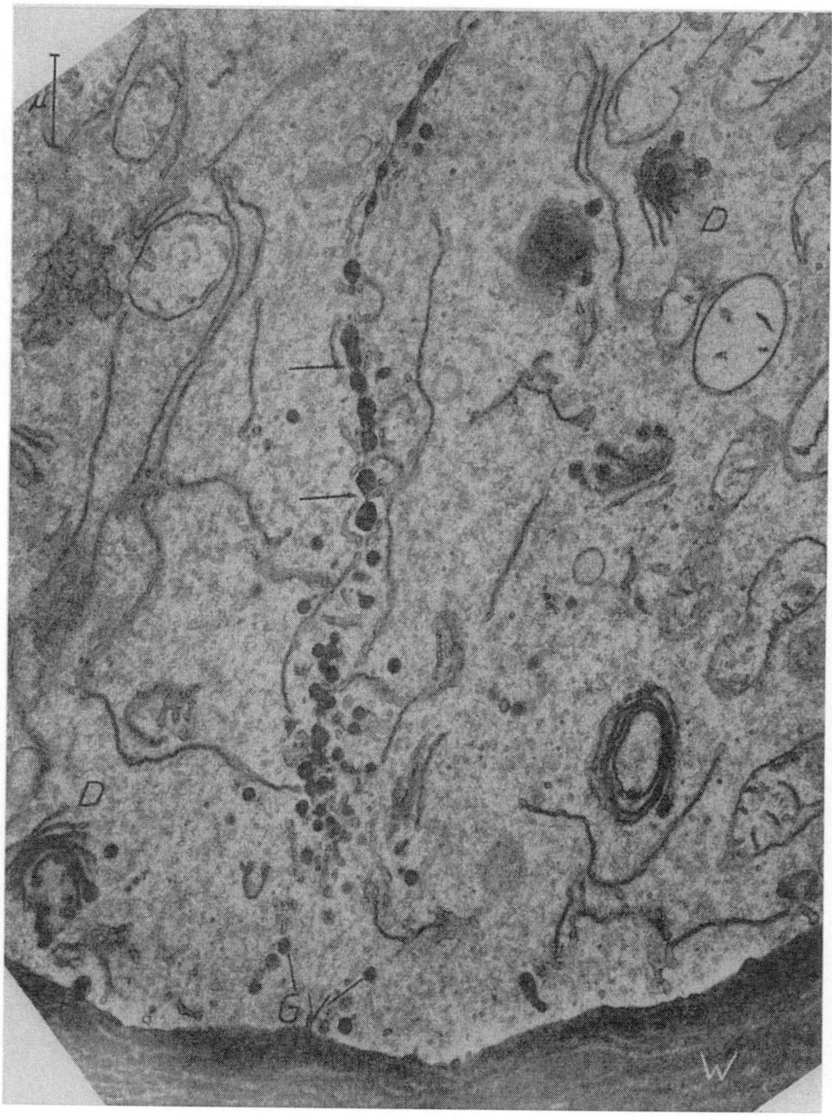

Abb. 9. Spätere Telophase aus der Rhizodermis von *Zea mays*, die teilweise das „Zusammenfließen" der Golgi-Vesikel (*GV*) im Bereich der Zellplatte zeigt (Pfeile). *D* = Dictyosom, *W* = Außenwand. KMnO₄-Fixation. (Aus WHALEY und MOLLENHAUER [100])

somen" —, die durch Verschmelzen die Zellplatte bilden. Ohne daß die Autoren darauf näher eingehen, soll an dieser Stelle doch festgestellt werden, daß die fraglichen Vesikel — nach einem Vergleich der verschiedenen Abbildungen miteinander — zum mindesten in der Größenordnung der dort abgebildeten Golgi-Vesikel liegen. Es ergibt sich also daraus ebenfalls kein Widerspruch zu den neuesten Befunden von WHALEY und MOLLENHAUER [100].

Nach diesen ersten Ergebnissen, die noch weiterer Bestätigung bedürfen, kann man also damit rechnen, daß der Golgi-Apparat auch die frühesten Elemente der die Tochterzellen trennenden Zellwand bildet[1].

b) Zentripetale Wandbildung

Die soeben geschilderte Neubildung der Trennwand zwischen zwei Tochterzellen geht bei den höheren Pflanzen i. a. zentrifugal vor sich. Bei den meisten niederen Pflanzen wird die entsprechende neue Zellwand jedoch irisblendenartig von außen nach innen angelegt, also zentripetal. Auch zu dieser andersartigen Wandbildung liegen Befunde vor, die im Falle der Zieralge *Micrasterias rotata* zeigen, daß das Wandmaterial von Golgi-Vesikeln geliefert wird [18, 19].

3. Weitere Bemerkungen zur Zellwandbildung

Gerade die letzten Beobachtungen zur Neubildung der Zellwand nach Kernteilung zeigen nun aber sehr deutlich, daß der Golgi-Apparat tatsächlich Stoffe zum normalen Wandwachstum liefert. Bereits früher [87] wurde schon anhand der Beobachtungen an Wurzelhaaren gefolgert, daß er die Grundsubstanz (= Matrix) der Primärwand ausscheidet (vgl. auch [63]), die im wesentlichen — wie die der Zellplatte — aus Pektinstoffen und Hemizellulosen besteht und die in ihrer Konsistenz *in statu nascendi* im Gegensatz zur starren Zellulose viskös ist. Schematische Übersichten über die bisherigen Ergebnisse hinsichtlich der Funktion des Golgi-Apparates bei der Zellwandbildung sind in den Abb. 10 und 11 wiedergegeben.

Obwohl noch keine chemischen Analysen der ausgeschiedenen Stoffe vorliegen, lassen sich die vorliegenden Beobachtungen doch unter dem Gesichtspunkt betrachten, daß der Golgi-Apparat der Bildner der Pektinstoffe ist. Auch die Tatsache, daß in jungen Kollenchymzellen besonders viele Dictyosomen auftreten [43], spricht dafür, ebenfalls die vermehrte Zahl der Dictyosomen nach Verletzung [59]. Eine bemerkenswerte Aktivität des Golgi-Apparates finden BUVAT [7] in sich entwickelnden Siebröhren- und Gefäßgliedern, WHALEY, KEPHART und MOLLENHAUER [99] sowie KOLLMANN und SCHUMACHER [52] ebenfalls in Siebzellen, was vielleicht mit den in diesen Zellen besonderen Wandstrukturen in Zusammenhang zu bringen ist.

Aber auch die Schleimstoffe, die von der Kalyptra, der Rhizodermis und den Wurzelhaaren ausgeschieden werden, sind von ähnlicher chemischer Beschaffenheit wie die Matrix der Primärwand und verdienen noch besondere Aufmerksamkeit. Wahrscheinlich wird sich gar keine exakte Trennung zwischen Schleim-Sekretion und Matrix-Bildung ergeben — zum mindesten in den Außenwänden der jungen Wurzel. Es wäre nämlich denkbar, daß die von der Wurzel auf dem dargelegten

---

[1] Eine Bestätigung erfolgte kürzlich [25a].

Wege an ihrer Oberfläche ausgeschiedenen Substanzen sowohl als plastische Matrix für das noch zu bildende Zellulose-Gerüst als auch als extrazellulärer Ionenaustauscher für die Stoffaufnahme fungieren, für die Pektinstoffe infolge ihres Gehaltes an Carboxylgruppen in besonderem Maße in Frage kommen [71]. Unter diesem Aspekt läßt sich auch eine plausible Erklärung für die Verschleimung und die damit verbundene Isolierung sowie der Golgi-Aktivität der Kalyptrazellen geben: Möglicherweise „machen" sie nicht nur den Boden „geschmeidig" für das Vordringen der Wurzel und „schützen" das Meristem, sondern produzieren ebenfalls bereits aktiv Material (u. a. Pektinstoffe), das dann als Ionenaustauscher von großer physiologischer Wichtigkeit im absorbierenden Teil der Wurzel sein kann.

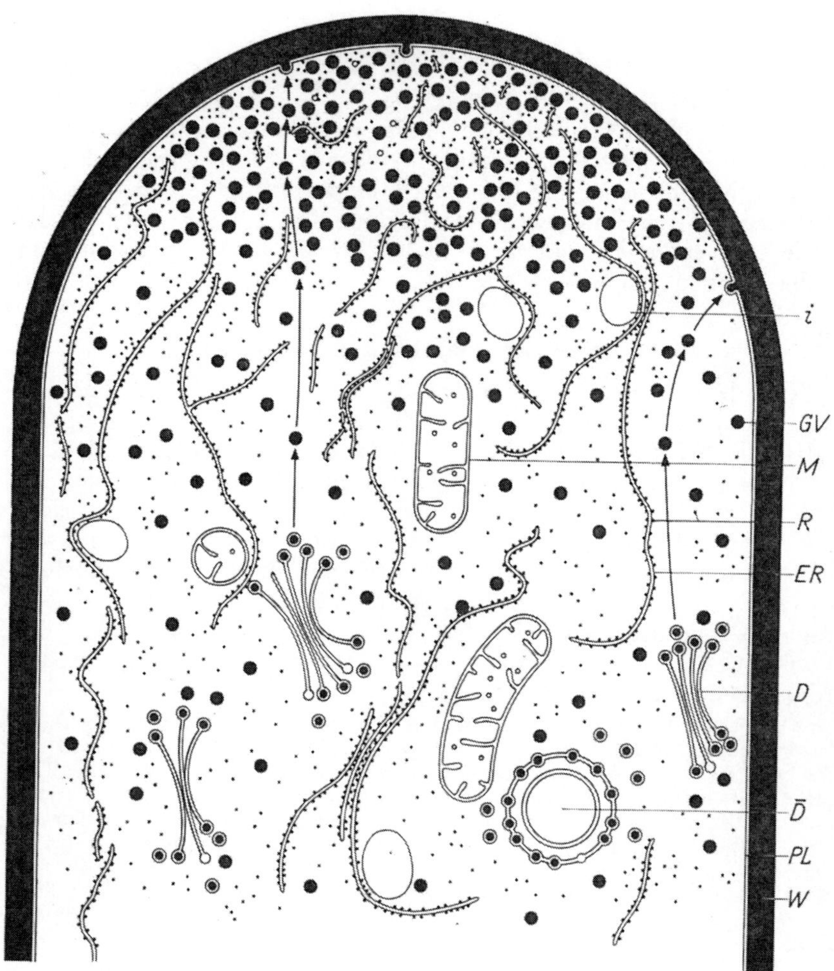

Abb. 10. Schematischer Längsschnitt durch den Apex einer pflanzlichen Zelle mit polarem Spitzenwachstum. Bildung von Golgi-Vesikeln (*GV*) am Rande der Golgi-Zisternen mit kontrastierbarer Substanz. Der Weg der Golgi-Vesikel von den Dictyosomen zur wachsenden Zellwandpartie ist durch Pfeile gekennzeichnet. *D* = Dictyosom mit teilweise gebogenen Zisternen, *D̄* = Flachschnitt durch zwei gebogene Golgi-Zisternen eines Dictyosoms, deren äußere durch den vesikelbildenden Rand geschnitten ist, *PL* = Plasmalemma, *W* = Zellwand, *ER* = endoplasmatisches Retikulum, *R* = Ribosomen, *M* = Mitochondrien, *i* = nicht identifiziertes Körperchen. (Aus SIEVERS [89a])

Abb. 11. Schema der Zellplattenbildung. (In Abb. 11a ist der Ausschnitt aus den beiden Tochterzellen, den Abb. 11 darstellt, markiert.) ZP = Zellplatte, N = Tochterkerne, sonst wie Abb. 10. Die Hüllmembran der freien Golgi-Vesikel wurde in Abb. 10 und 11 nicht eingezeichnet, da sie sich bei kontrastiertem Vesikelinhalt nicht deutlich abhebt. Sie ist jedoch stets vorhanden, wie Bilder mit nicht kontrastiertem Vesikelinhalt (vgl. Abb. 6) zeigen. (Sievers, unveröffentlicht, nach elektronenmikroskopischen Aufnahmen aus [68] und [100])

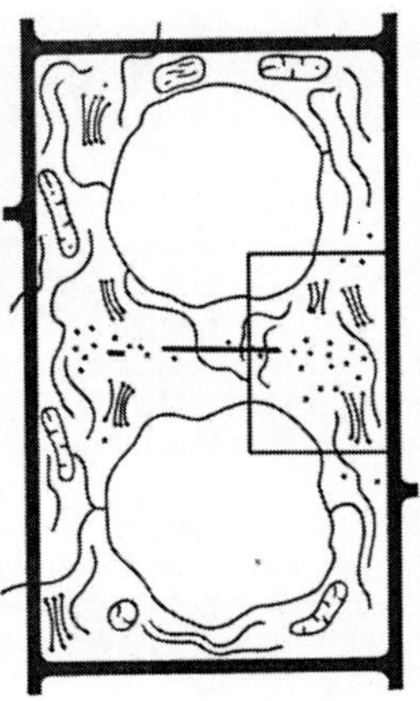

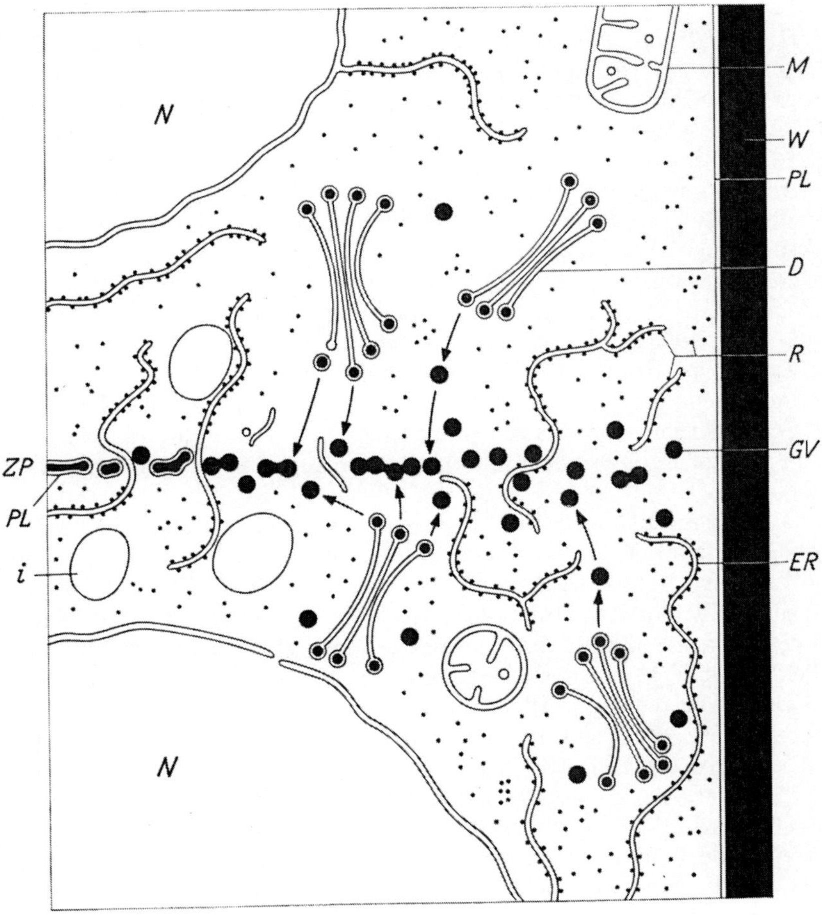

C. Funktion des Golgi-Apparates in tierischen Zellen

Während der Golgi-Apparat der pflanzlichen Zelle ein Novum darstellt, ist er in der tierischen Zelle schon lange bekannt und allerdings auch Gegenstand mancher Kontroversen gewesen. Die heute als Golgi-Apparat bezeichneten Strukturen wurden 1954 von DALTON und FELIX [16] in Epithelzellen des Nebenhodens sowie von SJÖSTRAND und HANZON [91] in exokrinen Pankreaszellen zuerst beschrieben und mit dem „klassischen" Golgi-Apparat homologisiert. Zwei Dictyosomen aus der Speicheldrüse einer Blattlaus sind in Abb. 2 dargestellt. Sie mögen die Identität der Strukturen zwischen pflanzlicher und tierischer Zelle demonstrieren. In Anbetracht der Mannigfaltigkeit der Gebilde, die sich unter dem klassischen Golgi-Apparat verbargen, hat es nicht an Zweifeln gefehlt, ob die Homologisierung legitim ist. Auf eine derartige Frage einzugehen, bedeutet, den Rahmen dieses Themas zu sprengen. Es sei jedoch auf eine Arbeit [37] verwiesen, die die Möglichkeit einer· solchen Homologisierung mit der von GOLGI gemeinten Struktur in der Purkinje-Zelle durchaus wieder gestattet.

I. Der Golgi-Apparat in der exokrinen Pankreas-Zelle

Die zahlreichsten Untersuchungen am Golgi-Apparat sind in der tierischen Zelle an der Bauchspeicheldrüse vorgenommen worden. Es fehlt daher nicht an zusammenfassenden Darstellungen (z. B. [*39, 40, 41, 65, 66, 90*].) Ferner wird auch in dem Referat von HELANDER die Morphologie der exokrinen Pankreaszelle behandelt.

An der Sekretion der Verdauungsenzyme ist der Golgi-Apparat unmittelbar beteiligt, und zwar ist er der Lieferant der seit langem bekannten Zymogen-Granula, die am Apex der Drüsenzelle angesammelt werden und nach Fütterung des Tieres verschwinden, um dann neu gebildet zu werden. Die elektronenmikroskopischen Untersuchungen haben ergeben, daß die Zymogen-Granula als Vesikel von Golgi-Zisternen abgeschnürt werden und daß ihr Inhalt — nach OsO_4-Fixation — in zunehmendem Maße im Zuge der Isolierung und des Transportes zum Zellapex „reift", d. h. elektronendichter wird (loc. cit.). Von EKHOLM, ZELANDER und EDLUND [*21*] liegen Aufnahmen vor, wie bereits erwähnt, die im Augenblick der Extrusion des Granulum-Inhaltes in den Drüsengang die Fusion von Plasmalemma und Vesikel-Membran beweisen. Golgi-Vesikel und Zymogen-Granula sind also auch im Falle der Pankreaszelle identisch.

Die rein morphologischen Untersuchungen sind mit physiologischen kombiniert worden. Ernährt man Ratten eiweißfrei, so verschwinden die Zymogen-Granula. Allerdings normalisieren sich nach zehn bis zwölf Tagen die Strukturen wieder [*97*]. Nach Injektion von Äthionin, einer körperfremden Aminosäure, degenerieren die Strukturen der Drüsenzellen. Zwei Tage nach der letzten Injektion werden die Zellen neu gebildet, wobei der Golgi-Apparat früher als die Zymogen-Granula wieder erscheint [*20, 35*]. Eine totale Hemmung der Zymogen-Granula-Bildung und eine Hyperplasie des Golgi-Apparates kann erzielt werden, wenn β-3-Thiophen-Alanin, ein Analogon eines natürlichen Bausteines, verabreicht wird [*47*]. Schließlich läßt sich das Pankreas durch die Hormone Sekretin und Pankreozymin stimulieren. In der Tat verschwinden nach einer derartigen Stimulierung die Zymogen-Granula [*36*]. Die genannten Beispiele zeigen die enge Verknüpfung zwischen der Funktion der Drüsenzellen und ihrer Feinstruktur.

Besonders eindrucksvoll sind die Ergebnisse, die durch Kombination von autoradiographischen und elektronenmikroskopischen Methoden gewonnen wurden. Wird eine markierte Aminosäure, z. B. DL-Leucin-4,5-H³, injiziert, so kann ihr Weg in der exokrinen Pankreaszelle verfolgt werden [9, 10, 11]. Er beginnt im Ergastoplasma und geht über das Golgi-Feld, über Golgi-Vesikel mit schwachem Kontrast zu den Zymogen-Granula am Zellapex und endet im Lumen des Drüsenganges (Tab. 1).

Tabelle 1. *Verteilung von Korn-Zahlen in* % *über verschiedene Zellstrukturen zu unterschiedlichen Zeiten nach Injektion von* DL-*Leucin-4,5-H³* (nach CARO und PALADE [*11*])

Zeit (min)	ER, Ribosomen und umgebendes Grundplasma	Golgi-Feld	Zymogen-Granula	Zellkern	Mitochondrien
4	67	27	1	2	3
6	53	39	2	5	1
20	11	73	10	3	3
240	11	10	73	4	2

Entsprechende Befunde liegen von WARSHAWSKY, LEBLOND und DROZ [*94*] vor, die die kinetische Seite dieses Prozesses mit Hilfe lichtmikroskopisch-autoradiographischen Untersuchungen bestätigen. Diese Art von Untersuchungen ist noch beweiskräftiger als bereits früher vorgenommene Zellfraktionierungen [*11*], da der Golgi-Apparat z. B. in keiner Fraktion eindeutig zu isolieren ist.

Die Untersuchungen an der exokrinen Pankreaszelle haben mit neueren Methoden die Rolle des Golgi-Apparates bestätigt, die bereits früher von HIRSCH [*38*] auf Grund lichtmikroskopischer und cytochemischer Methoden in einer Theorie zusammengefaßt worden ist. Die Pankreaszelle stellt also insofern den Prototyp für die Funktion des Golgi-Apparates dar, zumal auch infolge der mikroskopisch beobachtbaren Zymogen-Granula Abläufe in den Zellen als Funktion der Zeit unmittelbar zur Erkenntnis gelangen können [*38*]. Zugleich wird hier aber auch die Frage nach der Herkunft des Sekretmaterials gestellt. Was bei der pflanzlichen Zelle zunächst noch völlig offen ist — nämlich die Herkunft der Polysaccharid-Bausteine — weist hier im Falle der Proteine zurück auf die Rolle der Ribosomen und des ER. Daß die für den Pankreassaft notwendigen Proteine an den Ribosomen synthetisiert werden, scheint sicher zu sein. Einigkeit ist jedoch noch nicht über die strukturellen Zusammenhänge zwischen dem ER und dem Golgi-Apparat erzielt (vgl. [*90*] mit [*66*]).

II. Der Golgi-Apparat in weiteren sekretorischen Zellen

1. Milchdrüse

Anhand einer Reihe von Ergebnissen läßt sich die Tatsache, daß der Golgi-Apparat in der tierischen Zelle nach unseren bisherigen Kenntnissen vorwiegend eiweißhaltige Sekrete ausscheidet, noch erhärten. Bei verschiedenen laktierenden Tieren wurden die Milchdrüsen elektronen-optisch untersucht. Die Milch besteht aus Fettkügelchen von etwa 4 μ Durchmesser, Eiweißpartikeln von 0,15–3 μ Durchmesser und weiteren, morphologisch nicht faßbaren Komponenten des Milchplasmas [4]. Das Milcheiweiß wird vom Golgi-Apparat in Vesikel geformt, deren Inhalt am Zellapex ausgeschüttet wird [3, 4, 45, 95]. WELLINGS und PHILP [96] haben DL-Leucin-4,5-H³, also die Aminosäure, die bereits im Pankreas zur Markierung des Protein-Weges benutzt wurde, laktierenden Mäusen injiziert. Dreißig Minuten nach der Injektion erschien die höchste Radio-aktivität im Bereich des Golgi-Apparates, nachdem sie zuvor in der Zone des Ergastoplasmas am stärksten war. Das Ergebnis zeigt also, daß kein wesentlicher Unterschied zwischen dem Weg der Aminosäure in der Pankreaszelle und dem in der Milchdrüse besteht.

2. Weitere sezernierende Organe mit auffallender Aktivität des Golgi-Apparates

Außer Pankreas und Milchdrüsen ist in jüngster Zeit eine große Anzahl von Organen elektronenmikroskopisch untersucht worden, die alle sekretorisch tätig sind und die die Formung von deutlich kontrastier-

Tabelle 2. *Übersicht über einige weitere Organe mit Zellen von besonderer Aktivität des Golgi-Apparates*

Organ	Autoren
Nervenzellen (Neurosekretion)	[1, 6, 33, 62, 67, 72, 75, 78, 80]
Hypophyse	[23, 55, 77]
Schilddrüse	[102, 103]
Nebenschilddrüse	[61]
Nebennierenmark	[12, 13]
Leber	[50]
Speicheldrüse	[57]
Magenschleimhaut	[34, 49]
Ovarium und Ovidukt	[2, 32, 48, 105]
Plazenta	[70, 92]
Prostata	[31, 84]
Chondrocyten[1]	[27, 69, 85]
Seidendrüse[1]	[93]

[1] In Chondrocyten und in der Seidendrüse ist nicht nur die Bildung von Golgi-Vesikeln mit kontrastreichem Inhalt, sondern auch seine Extrusion aus der Zelle nachgewiesen.

baren Granula in Golgi-Vesikel erkennen lassen. In einer Übersicht soll daher die Vielfalt dieser Organe hier einmal zusammengestellt werden (Tab. 2).

Die Tabelle zeigt in aller Deutlichkeit, daß tierische Zellen mit einer wie auch immer gearteten Sekretions-Tätigkeit durch Kontrast-bildung mit dem Fixans Substanzen in Golgi-Vesikeln vermuten lassen. Es wird sich in allen hier erwähnten Fällen in erster Linie um vorwiegend eiweißhaltige Substanzen handeln, die Enzym- (Pankreas), Hormon- (z. B. Plazenta), Nahrungs- (Milchdrüse) und Gerüst-Funktion (Chondrocyten) haben können.

III. Der Golgi-Apparat in Spermatozoen und Protozoen

1. Spermatozoen

Während die bisher geschilderten tierischen Zellen durchweg in erster Linie jeweils überwiegend bestimmte Eiweißstoffe in Form von Golgi-Vesikeln durch eine Membran abgrenzen, sollen schließlich noch einige interessante Fälle erwähnt werden, deren Golgi-Apparat Polysaccharide produziert. Das Akrosom, ein dem Zellkern von Spermatozoen anliegender membranbegrenzter Körper, wird während der Spermiogenese vom Golgi-Apparat gebildet [24, 26, 29, 44, 51]. Das elektronendichte Material in seinem Innern wird vom Golgi-Apparat abgesondert; es enthält wahrscheinlich Polysaccharide [44].

2. Protozoen

Der Parabasalkörper einiger Zooflagellaten ist strukturell identisch mit dem Golgi-Apparat der Metazoen [28]. Für *Trichonympha* wies GRIMSTONE [30] nach, daß er Polysaccharide produziert. Zum Schluß sollen jedoch noch zwei weitere Sonderfälle von Golgi-Apparat-Aktivität geschildert werden. Deutlich geformte Schuppen, die die Oberfläche des Flagellaten *Halosphaera* bedecken, werden vom Golgi-Apparat gebildet [55a]. Ferner finden sich in dem Radiolar *Aulacantha scolymantha* außerordentlich große Dictyosomen [76], die strukturierte Stäbchen von Polysaccharidnatur intrazisternal bilden. Sie werden in den extraplasmatischen Raum ausgeschieden und sind offenbar maßgeblich am Aufbau der Hüllmembran der Zentralkapsel beteiligt: Ein Fall von „Zellwandbildung" auf Polysaccharid-Basis bei Protozoen.

D. Schlußbemerkungen

Versucht man, das Fazit aus den bisherigen Untersuchungen über die Funktion des Golgi-Apparates zu ziehen, so leuchtet eine seiner möglichen Rollen in der Zelle ohne weiteres ein, nämlich die, *Stoffe durch eine Membran vom Grundplasma zu trennen*, um sie dann i. a. vom intra- zum extraplasmatischen Raum zu befördern.

Diese Tatsache löst aber eine Reihe von Fragen aus, auf die wir zunächst nur unbefriedigend antworten können und die hier zum Teil genannt werden sollen:

a) Welche Art von Stoffen ist hier gemeint? Woher kommen die Stoffe und in welchem Zustand übernimmt sie der Golgi-Apparat? Werden sie im Golgi-Apparat noch chemisch verändert, und — wenn ja — wie und in welcher seiner strukturellen Komponenten?

b) Woher stammt das Material zum ständigen Membran-Neubau? Wird möglicherweise die Fähigkeit des Golgi-Apparates zur Membranbildung von der Zelle nicht nur zum genannten Zweck, sondern auch für die Bildung anderer membranöser Komponenten des Cytoplasmas eingesetzt? Wie wird die Anzahl der Dictyosomen, wie die der Golgi-Zisternen vermehrt? Welche Art von Vorgängen spielt sich in der Zisterne ab beim Abschnürungsprozeß der Vesikel? Gibt es eine Konvertierbarkeit von Membranen im weiteren Sinne, als es bisher für Golgi-Vesikel-Membran und Plasmalemma gezeigt wurde?

c) Welche Kräfte spielen beim — zumeist gerichteten — Transport der Golgi-Vesikel eine Rolle? Ein Dictyosom ist nicht von einer geschlossenen Membran umgeben. Welche Kräfte halten die Zisternen zu einem derartigen Stapel zusammen?

d) Schließlich ergibt sich noch die Frage, ob nicht auch die Intrusion von Stoffen in die Zelle hinein — also der Prozeß in umgekehrter Richtung — über den Golgi-Apparat ablaufen kann.

Zur letzten Frage d) wird sicher eine Antwort im Referat über Pinocytose gegeben werden, zu den Fragen c) und b) lassen sich noch keine sicheren Aussagen machen. Auf die Fragen a) wissen wir bisher zu antworten, daß die Stoffe sowohl Polysaccharid- als auch Protein-Natur besitzen; der Syntheseort der Proteine liegt im Ergastoplasma, speziell in den Ribosomen. Der Transport der neugebildeten Proteine zum Golgi-Apparat in Form von „Prosekreten" wird dem ER zugeschrieben (vgl. u. a. [41]). Gegen eine ausschließlich mechanisch zu deutende „Verpackung" der Substanzen im Golgi-Apparat und für eine weitere chemische Veränderung sprechen die zahlreichen Beobachtungen, aus denen ersichtlich ist, daß der Vesikel-Inhalt i. a. an Kontrast zunimmt im Zuge der Vesikel-„Reifung", woraus man auf seine zunehmende Reaktionsbereitschaft mit den kontrastgebenden Fixantien bzw. „Färbemitteln" schließen darf. Ferner sprechen auch die — allerdings noch spärlichen, heterogenen und nicht streng lokalisierbaren — Befunde über die Fermentausstattung des Golgi-Apparates [53, 56, 64] und über die Sauerstoffabhängigkeit der Vesikel-Bildung dafür [22, 83].

Im Golgi-Apparat hat die Zelle ein System zur Verfügung, das sie einsetzen kann, um gewisse — chemisch keinesfalls identische — Substanzen bzw. Substanz-Gemische bereits intrazellulär durch eine Membran vom übrigen Plasma abzusondern. Die Absonderung gipfelt i. a. in eine Extrusion der Substanzen aus der Zelle heraus. Es gibt aber auch Fälle, in denen sie bei der Ausschleusung von Stoffen sicherlich auf eine derartige intrazelluläre Membranbindung verzichtet, wie z. B. bei der Zellulose-Synthese oder in Nektarien, wo die Sekretion wahrscheinlich molekular über das Plasmalemma erfolgt und nicht via vorfabrizierten, separaten Raum, d. i. Golgi-Vesikel (zur Nektar-Sekretion vgl. Referat von Schnepf).

Worin liegt diese „Doppelwegigkeit" begründet? Vielleicht läßt sich der Grund in der Natur der Stoffe finden, die ausgeschleust werden sollen: Einfache, neutrale Zucker — wie im Falle der Zellulose-Synthese oder aber auch bei der Nektar-Sekretion — stören auch in größeren Mengen nicht den Stoffwechselhaushalt der Zelle; saure Polysaccharide — also z. B. Pektinstoffe — bzw. auch deren Monomere dürften infolge ihrer Carboxylgruppe pro Monomer in starker Anreicherung eine Gefahr für die Zelle darstellen. Sogar die Gefahr der Autolyse der Zelle dürfte durch die Verdauungsenzyme in den Zymogen-Granula gegeben sein. In den hier herausgegriffenen Beispielen der sauren Polysaccharide und des Pankreassaftes isoliert die Zelle die Substanzen durch eine Golgi-Vesikel-Membran. Sie schützt sich auf diese Weise vor Substanzen, die sie zwar einerseits im Sinne des Organismus sezernieren muß, die ihr selbst jedoch andererseits in derartigen Konzentrationen eine Gefahr bedeuten müssen. Möglicherweise lassen sich darin Sinn und Funktion des Golgi-Apparates in der Zelle am ehesten erklären.

Literatur

[1] AFZELIUS, B. A., u. G. FRIDBERG: Z. Zellforsch. **59**, 289 (1963).

[2] ANDERSEN, E., and H. W. BEAMS: J. Ultrastruct. Res. **3**, 432 (1960).

[3] BARGMANN, W., H. FLEISCHHAUER u. A. KNOOP: Z. Zellforsch. **53**, 545 (1961).

[4] —, u. A. KNOOP: Z. Zellforsch. **49**, 344 (1959).

[5] BELFORD, D. S., and R. D. PRESTON: J. exp. Bot. (Oxford) **12**, 157 (1961).

[6] BERN, H. A., R. S. NISHIOKA, and I. R. HAGADORN: J. Ultrastruct. Res. **5**, 311 (1961).

[7] BUVAT, R.: In: International Review of Cytology, Vol. 14, S. 41. New York and London: Academic Press 1963.

[8] —, et A. PUISSANT: C. R. Acad. Sci. (Paris) **247**, 233 (1958).

[9] CARO, L. G.: J. biophys. biochem. Cytol. **10**, 37 (1961).

[10] —, et G. E. PALADE: C. R. Soc. Biol. (Paris) **155**, 1750 (1961).

[11] — — J. Cell Biol. **20**, 473 (1964).

[12] CLEMENTI, F., and G. P. ZOCCHE: In: Electron Microscopy, Vol. 2, S. YY—9. New York and London: Academic Press 1962.

[13] — — J. Cell Biol. **17**, 587 (1963).

[14] CORMACK, R. G. H.: Science **122**, 1019 (1955).

[15] — Bot. Rev. **28**, 446 (1962).

[16] DALTON, A. J., and M. D. FELIX: Amer. J. Anat. **94**, 171 (1954).

[17] DAWES, C. J., and E. BOWLER: Amer. J. Bot. **46**, 561 (1959).

[18] DRAWERT, H., u. M. MIX: Planta (Berl.) **58**, 448 (1962).

[19] — — Sitzungsber. Ges. Beförder. ges. Naturwiss. Marburg **83/84**, 361 (1962).

[20] EKHOLM, R., Y. EDLUND, and T. ZELANDER: J. Ultrastruct. Res. **7**, 102 (1962).

[21] —, T. ZELANDER, and Y. EDLUND: J. Ultrastruct. Res. **7**, 61 (1962).

[22] FALK, H.: Z. Naturforsch. **17 b**, 862 (1962).

[23] FARQUHAR, M. G.: Trans. N. Y. Acad. Sci., Ser. 2, **23**, 346 (1960).

[24] FAWCETT, D. W., u. R. D. HOLLENBERG: Z. Zellforsch. **60**, 276 (1963).

[25] FREY-WYSSLING, A., u. K. MÜHLETHALER: Mikroskopie (Wien) **4**, 257 (1949).

[25a] —, J. F. LÓPEZ-SÁEZ, and K. MÜHLETHALER: J. Ultrastruct. Res. **10**, 422 (1964).

[26] GATENBY, J. B., T. N. TAHMISIAN, R. DEVINE, and H. W. BEAMS: Cellule **59**, 27 (1959).

[27] GODMAN, G. C., and K. R. PORTER: J. biophys. biochem. Cytol. **8**, 719 (1960).

[28] GRASSÉ, P. P.: C. R. Acad. Sci. (Paris) **242**, 858 (1956).

[29] —, and N. CARASSO: Nature (Lond.) **179**, 31 (1957).

[30] GRIMSTONE, A. V.: J. biophys. biochem. Cytol. **6**, 369 (1959).

[31] GROTH, D. P., and D. BRANDES: J. Ultrastruct. Res. **4**, 166 (1960).

[32] HADEK, R.: J. Ultrastruct. Res. **9**, 445 (1963).

[33] HAGADORN, I. R., H. A. BERN u. R. S. NISHIOKA: Z. Zellforsch. **58**, 714 (1963).

[34] HELANDER, H. F.: J. Ultrastruct. Res., Suppl. 4 (1962).

[35] HERMAN, L., and P. J. FITZGERALD: J. Cell Biol. **12**, 277 (1962).

[36] HERMODSSON, L. H.: In: Electron Microscopy, Vol. 2, S. YY—11. New York and London: Academic Press 1962.

[37] HERNDORN, R. M.: J. Cell Biol. **18**, 167 (1963).

[38] HIRSCH, G. C.: Protoplasma-Monographien, Bd. 18. Berlin: Gebr. Bornträger 1939.

[39] — Naturwissenschaften 47, 25 (1960).

[40] — In: Biological Structure and Function, Vol. 1, S. 195. New York and London: Academic Press 1961.

[41] — Wissenschaftl. Beibl. Materia Medica Nordmark, Nr. 49. Uetersen 1964.

[42] HODGE, A. J., J. D. MCLEAN, and F. V. MERCER: J. biophys. biochem. Cytol. 2, 597 (1956).

[43] HOHL, H. R.: Ber. schweiz. bot. Ges. 70, 395 (1960).

[44] HORSTMANN, E.: Z. Zellforsch. 54, 68 (1961).

[45] HOLLMANN, K. H.: J. Ultrastruct. Res. 2, 423 (1959).

[46] HOUWINK, A. L., en P. A. ROELOFSEN: Acta bot. neerl. 3, 385 (1954).

[47] HRUBAN, Z., H. SWIFT, and R. W. WISSLER: J. Ultrastruct. Res. 7, 359 (1962).

[48] HSU, W. S.: Cellule 62, 147 (1961).

[49] ITO, S., and R. J. WINCHESTER: J. Cell Biol. 16, 541 (1963).

[50] KARRER, H. E.: J. Ultrastruct. Res. 4, 191 (1960).

[51] KAYE, J. S.: J. Cell Biol. 12, 411 (1962).

[52] KOLLMANN, R., u. W. SCHUMACHER: Planta (Berl.), 63, 155 (1964).

[53] KUFF, E. L., and A. J. DALTON: In: Subcellular Particles, S. 114. New York: Ronald Press 1959.

[54] LARSON, D. A., and C. W. LEWIS JR.: In: Electron Microscopy, Vol. 2, S. W—11. New York and London: Academic Press 1962.

[55] MAILLARD, M.: J. Microscopie (Paris) 2, 81 (1963).

[55a] MANTON, I., K. OATES, and M. PARKE: J. mar. biol. Ass. U. K. 43, 225 (1963).

[56] MEEK, G. A., and S. BRADBURY: J. Cell Biol. 18, 73 (1963).

[57] MOERICKE, V., u. K. E. WOHLFARTH-BOTTERMANN: Z. Zellforsch. 59, 165 (1963).

[58] MOLLENHAUER, H. H., and W. G. WHALEY: J. Cell Biol. 17, 222 (1963).

[59] — —, and J. H. LEECH: J. Ultrastruct. Res. 4, 473 (1960).

[60] — — — J. Ultrastruct. Res. 5, 193 (1961).

[61] MUNGER, B. L., and S. I. ROTH: J. Cell. Biol. 16, 379 (1963).

[62] MURAKAMI, M.: Z. Zellforsch. 59, 684 (1963).

[63] NORTHCOTE, D. H.: In:Cell Differentiation, Symposia of the Society for Experimental Biology, Nr. 17, S. 157. Cambridge: University Press 1963.

[64] NOVIKOFF, A. B., and S. GOLDFISCHER: Proc. nat. Acad. Sci. (Wash.) 47, 802 (1961).

[65] PALADE, G. E.: In: Subcellular Particles, S. 64. New York: Ronald Press 1959.

[66] —, P. SIEKEWITZ, and L. G. CARO: In: Exocrine Pancreas, Ciba Foundation Symposium, S. 23. London: J. A. Churchill Ltd. 1962.

[67] PETZOLD, H.: Protoplasma (Wien) 56, 553 (1963).

[68] PORTER, K. R., and R. D. MACHADO: J. biophys. biochem. Cytol. 7, 167 (1960).

[69] REVEL, J. P., u. E. D. HAY: Z. Zellforsch. 61, 110 (1963).

[70] RHODIN, J. A. G., and J. TERZAKIS: J. Ultrastruct. Res. 6, 88 (1962).

[71] ROBERTSON, R. N.: In: Handbuch der Pflanzenphysiologie, Bd. 4, S. 243. Berlin-Göttingen-Heidelberg: Springer 1958.

[72] RÖHLICH, P., B. AROS u. B. VIGH: Z. Zellforsch. 58, 524 (1962).

[73] ROSEN, W. G.: In: Pollen Physiology and Fertilization, S. 159. Amsterdam: North-Holland Publishing Company 1964.

[74] —, S. R. GAWLIK, W. V. DASHEK, and K. A. SIEGESMUND: Amer. J. Bot. 51, 61 (1964).

[75] ROSENBLUTH, J.: Z. Zellforsch. 60, 213 (1963).

[76] RUTHMANN, A., u. K. G. GRELL: Z. Zellforsch. 63, 97 (1964).

[77] SANO, M.: J. Cell Biol. 15, 85 (1962).

[78] —, u. A. KNOOP: Z. Zellforsch. 49, 464 (1959).

[79] SASSEN, M. M. A.: In: Pollen Physiology and Fertilization, S. 167. Amsterdam: North-Holland Publishing Company 1964.

[79a] — Acta bot. neerl. 13, 175 (1964).

[80] SCHARRER, E., u. S. BROWN: Z. Zellforsch. **54**, 530 (1961).
[81] SCHNEPF, E.: Planta (Berl.) **54**, 641 (1960).
[82] — Flora (Jena) **151**, 73 (1961).
[83] — Flora (Jena) **153**, 23 (1963).
[84] SCHRODT, G. R.: J. Ultrastruct. Res. **5**, 485 (1961).
[85] SHELDON, H., and F. B. KIMBALL: J. Cell Biol. **12**, 599 (1962).
[86] SIEVERS, A.: Protoplasma (Wien) **56**, 188 (1963).
[87] — Z. Naturforsch. **18 b**, 830 (1963).
[88] — Planta (Berl.) **61**, 97 (1964).
[89] — Z. Zellforsch. **64**, 280 (1964).
[89a] — Vortr. Tagung d. Deutsch. Bot. Ges. München 1964. Ber. Deutsch. Bot. Ges. **77**, 388 (1964).
[90] SJÖSTRAND, F. S.: In: Exocrine Pancreas, Ciba Foundation Symposium, S. 1. London: J. A. Churchill Ltd. 1962.
[91] —, and V. HANZON: Experientia (Basel) **10**, 367 (1954).
[92] TERZAKIS, J. A.: J. Ultrastruct. Res. **9**, 268 (1963).
[93] VOIGT, W. H.: Zur funktionellen Morphologie der Fibroin- und Sericin-Sekretion der Seidendrüse von *Bombyx mori* L. Dissertation Bonn 1964.
[94] WARSHAWSKY, H., C. P. LEBLOND, and B. DROZ: J. Cell Biol. **16**, 1 (1963).
[95] WELLINGS, S. R., and K. B. DEOME: J. biophys. biochem. Cytol. **9**, 479 (1961).
[96] —, u. J. R. PHILP: Z. Zellforsch. **61**, 871 (1964).
[97] WEISBLUM, B., L. HERMAN, and P. J. FITZGERALD: J. Cell Biol. **12**, 313 (1962).
[98] WHALEY, W. G., J. E. KEPHART, and H. H. MOLLENHAUER: Amer. J. Bot. **46**, 743 (1959).
[99] — — — In: Cellular Membranes in Development, S. 135. New York and London: Academic Press 1964.
[100] —, and H. H. MOLLENHAUER: J. Cell Biol. **17**, 216 (1963).
[101] — —, and J. H. LEECH: Amer. J. Bot. **47**, 401 (1960).
[102] WISSIG, S. L.: J. biophys. biochem. Cytol. **7**, 419 (1960).
[103] — J. Cell Biol. **16**, 93 (1963).
[104] WOHLFARTH-BOTTERMANN, K. E.: Naturwissenschaften **50**, 237 (1963).
[105] ZEIGEL, R. F., and A. J. DALTON: J. Cell Biol. **15**, 45 (1962).

Summary

The function of the Golgi apparatus in the cell, insofar as it seems established, is discussed on the basis of some examples. Its main role is evidently to separate substances within the cell from the ground cytoplasm by a membrane and to localize them in vesicles. Often the vesicles are transported to the plasma membrane, where they discharge their content into the extracellular space. In this process the vesicle membrane is incorporated into the plasma membrane, or else a new plasma membrane may arise from Golgi membranes by fusion of individual vesicles, as in the case of formation of new cell walls after karyokinesis in plant cells. The substances transported out of the cell by the Golgi apparatus are heterogeneous. Polysaccharides probably predominate in protozoa and plants, while proteins predominate in animals. It is discussed whether the Golgi apparatus particularly secretes substances whose concentration and chemical properties would render them harmful to the cell if they were not separated from the ground cytoplasm by a membrane. Examples of this might be pectins in plants and pancreatic juice in animals.

Diskussion

Hirsch: Ich darf Herrn SIEVERS beglückwünschen zu seinem klaren Referat. Im Anschluß an die noch offenen Fragen am Ende des Vortrages möchte ich zur Terminologie und zu den Verhältnissen in der exokrinen Pankreaszelle einige Bemerkungen machen.

Zunächst muß ich auf die leidige Terminologie des „Golgi-Feldes" eingehen. Schon 1939 habe ich in meiner Monographie über „Golgi-Körper" 125 Bezeichnungen aufgezählt. In meiner zweiten Monographie 1955 — mittlerweile war die Anzahl der Namen auf 155 angestiegen — habe ich das fragliche System nach J. R. BAKER

„osmiophile Körper" genannt, nachdem in Frage gestellt war, ob die mit Metallen imprägnierbaren Körper von GOLGI 1898 entdeckt wurden. In einer dritten Monographie 1962 habe ich versucht, einen objektiven Ausdruck zu wählen, und habe als Oberbegriff „*Lamellen-Vakuolen-Feld*" vorgeschlagen. Damit sollte vor allem gesagt werden, daß es sich bei den fraglichen „Körpern" nicht handelt um zirkumskripte, von einer Hüllmembran umgebene Körper, sondern um ein *freies Feld* im Cytoplasma, das vor allem Lamellen-Systeme und Vakuolen beherbergt (Abb. 1, 16—22). Diese neutrale Bezeichnung hat jedoch die meisten Cytologen nicht überzeugen können. Dafür wird jetzt der alte Terminus „Dictyosomen" (PERONCITO 1909) — zu deutsch „Netzkörper" — ausgegraben, der in der unzutreffenden Annahme erfunden wurde, daß die mit OsO$_4$ imprägnierten Strukturen ein Netzwerk bilden.

Es liegt mir fern, meinen Ausdruck „Lamellen-Vakuolen-Feld" durchsetzen zu wollen. Aber auf *einen* objektiven Punkt muß ich Wert legen, auf den Ausdruck „Feld". Ob wir dieses offene Feld nun nach seinen Hauptteilen, den Lamellen und Vakuolen, ein „Lamellen-Vakuolen-Feld" oder ein „Golgi-Feld" nennen, ist von untergeordneter Bedeutung. In Wahrheit enthält dieses Feld zahlreiche Strukturen, die physiologisch in einer Kette — wie bei einem Fließband — zusammenhängen. Unsere Aufgabe ist es, diese Kette in jeder der sehr zahlreichen Zellformen mit einem solchen Felde zu erforschen und dabei logisch begründete Bezeichnungen zu gebrauchen, die dem Leser verständlich sind.

Die exokrine Pankreaszelle

Zunächst eine kurze Vorbemerkung zu den „Intrazisternen Granula", die von PALADE 1956 in den ER-Kanälchen des Meerschweinchen-Pankreas nach langem Hunger und plötzlich gegebenem Futter beschrieben wurden. Sie stimmen in Größe und Wanderungszeit überein mit den lichtmikroskopisch im lebenden Pankreas der Maus nach Pilokarpin-Reizung von mir 1931 gefundenen Granula A und mit den von SUZUKI 1959 gefundenen Granula nach Unterbindung des Ductus pancreaticus. Also werden solche Granula im ER nur gebildet durch künstliche Verhältnisse. Diese Granula sind rundliche Konglomerate von Proteinen, die von Ribosomen gebildet und in das Kanallumen des ER durch aktiven Transport eingeschleust werden. Sie sind nicht identisch mit den Zymogen-Granula, sondern nur künstlich hervorgerufene Stauungen, die sich vor Erreichen des Golgi-Feldes wieder auflösen. Ich habe daher vorgeschlagen, sie „Kondensgranula" zu nennen.

Jetzt möchte ich Ihnen die kettenförmigen Prozesse im Golgi-Feld der exokrinen Pankreaszelle schildern, wie sie von SLUITER 1944 lichtmikroskopisch-statistisch gefunden, von WARSHAWSKY, LEBLOND und DROZ 1963 durch Autoradiographie statistisch bestätigt und von CARO und PALADE 1964 ergänzt wurden. Ich selbst trug meine Ergebnisse 1963 in Japan und Sofia vor, publizierte sie 1964 in der Materia Medica Nordmark; 1965 werden sie in extenso erscheinen (Z. Zellforsch.).

Das gezeigte Schema von PALADE 1962 ist richtig in folgenden Punkten: Bei der Ingestion werden nur Aminosäuren aufgenommen, keine Blutplasma-Proteine; Proteinsynthese an den Ribosomen; Einschleusen der Proteine in die Lumina des ER; Transport dieser Proteine apikalwärts. Alles weitere an PALADEs Schema ist unrichtig. Richtig ist vielmehr: Die im ER-Kanal apikalwärts transportierten Proteine gelangen nach wenigen Minuten zum Golgi-Feld; als gelöste Proteine bewegen sie sich in 11, 7 min von der Basis zum Golgi-Feld, als Kondensgranula in 7—13 min. pro μ und werden vor Erreichen des Golgi-Feldes wieder aufgelöst.

Statische Beschreibung des Feldes. Bei der exokrinen Pankreaszelle, auf die wir uns hier beschränken wollen, besteht — statisch gesehen — das Golgi-Feld aus folgenden Teilen (Abb. 1):

1. In der Mitte des Feldes liegt ein Stapel von 4—6 platten, geschlossenen *Säcken* (Dictyosom im Sinne von SIEVERS). Die Wände der Säcke, deren Inhalt ich den *intralamellären Raum* nenne, sind etwa 60 Å dick. In der Mitte, wo die Säcke sich streng parallel aufeinanderlagern, ist dieser Raum von einer dichten, osmiophilen Substanz erfüllt. An den Rändern aber ist er zu je einer Vakuole (19) stark aufgeweitet. Die dunkle Substanz in der Mitte und die großen, hellen Aufblähungen am Rande liegen also im gleichen intralamellären Raum. Die Vakuolen, die weithin „Golgi-Vakuolen" genannt werden, bezeichnet Herr SIEVERS als „Golgi-Vesikel", wogegen ich Bedenken anmelde.

Ferner gibt es zwischen den parallel gelagerten Säcken einen *interlamellären Raum*, sehr hell, fast ohne Substanz, mit einer auffallend gleichbleibenden Lichtung von etwa 150 Å.

2. *Die Vakuolen* (19) entstehen aus dem intralamellären Raum durch seitliche Ausweitung der Säcke und entfernen während ihres Wachstums deren Wände

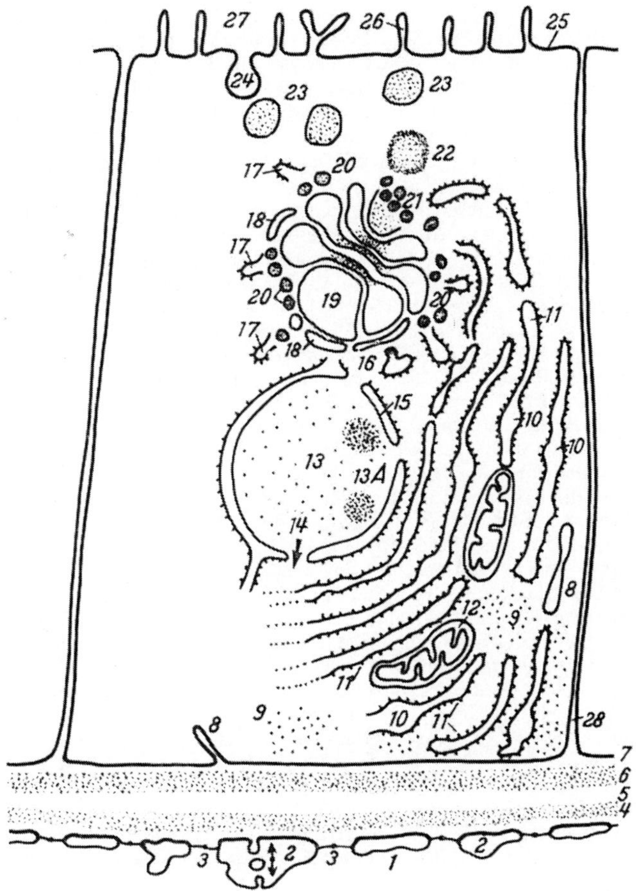

Abb. 1. Schema einer exokrinen Pankreaszelle mit darunter liegender Blutkapillare. (Es werden hier nur die Strukturen des Golgi-Feldes erläutert.) 10 Lumen des Endoplasmatischen Retikulums, 11 membranständige Ribosomen; 16 Entstehung eines X-Körpers aus einer ER-Knospe; 17 Freie X-Körper vor dem „offenen Mund" des ER; 18 Entstehung eines kleinen Sackes aus den X-Körpern und deren Anlagerung an einen alten Sack; 19 große Vakuole, entstanden aus einer peripheren Ausweitung eines Sackes; 20 X-Körper; 21 Intermediär-Körper mit einwandernden X-Körpern; 22 Intermediärkörper; 23 Zymogengranula; 24 deren Extrusion

immer mehr voneinander. Entgegen meiner ursprünglichen Annahme gibt es nicht zwei Arten von Vakuolen, sondern nur eine Form, nämlich die peripheren Aufblähungen der Säcke. Das System der Säcke mit ihren Vakuolen entsteht zweifellos durch Anlagerung kleinerer platter Säcke an die größeren, alten. Dabei bleibt es ein Rätsel, warum diese Anschmiegung immer in einem so regelmäßigen Abstand von 150 Å erfolgt.

3. *X-Körper* (im amerik. Sprachgebrauch ,,Golgi-vesicles") sind rundliche bis längliche Körper von 160—1300 Å Längsdurchmesser, die sich in allen Golgi-Feldern finden. Unschwer lassen sich mehrere Typen von X-Körpern unterscheiden, die anscheinend in einem genetischen Zusammenhang stehen. Auf die Frage nach ihrer Entstehung komme ich später zurück.

4. *Intermediärkörper* sind rundliche Gebilde (21, 22), die in ihrer Größe zwischen den Vakuolen und den Zymogengranula stehen. Sie sind stets größer und dichter als die Vakuolen. Dort, wo die Hüllmembran stellenweise fehlt, legen sich X-Körper den Intermediär-Körpern an (21). Die Intermediärkörper stimmen wahrscheinlich mit den ,,Prozymogengranula" nach Thienylamin überein.

5. *Zymogengranula* haben eine neue, relativ starke Hüllmembran (23). Diese entsteht vermutlich aus dichter werdenden Randsubstanzen und bildet einen Schutz gegen die Aktivität einiger innerer Enzyme.

6. Das *Endoplasmatische Retikulum* (ER) umgibt ringförmig das gesamte Golgi-Feld, indem es sich nach außen an die Schicht der X-Körper anschließt (10, 17). Besonders auffallend ist nun, daß die ER-Kanäle sich in der Außenschicht des Golgi-Feldes ausbuchten oder gar öffnen (17). Diese ,,Münder" sind alle nach dem Zentrum des Golgi-Feldes zu gerichtet. Neben der örtlichen Gemeinschaft dieser ,,Münder" und der X-Körper fällt ferner die starke Ähnlichkeit der kleineren X-Körper mit den Ausbuchtungen oder *Knospen* des ER auf (16), so daß man auf den Gedanken kommt, daß entweder die X-Körper aus den ER-Kanälen entstehen oder umgekehrt. Aber das ist bereits eine Frage der dynamischen Interpretation.

Versuch einer Dynamik. Da bei den elektronenoptischen Bildern der Strom der Geschehnisse sowohl in der einen Richtung als auch in der entgegengesetzten verlaufen kann, ist das Ergebnis einer dynamischen Interpretation zunächst eine *Hypothese.* Die Dynamik des Golgi-Feldes besteht zweifellos aus einer Zusammenarbeit aller 6 hier geschilderten Teile. Dabei hat der Strom der Vorgänge offenbar zwei Endziele: die Kondensation der Proteine in den Zymogengranula (wahrscheinlich in Verbindung mit der Entstehung zumindest einiger Enzyme) und den Wiederaufbau verbrauchter Teile der Säcke.

Kondensation und Transport der Proteine. Die an den Ribosomen gebildeten Proteine werden durch die ER-Kanäle in Richtung Golgi-Feld transportiert und können dort theoretisch auf verschiedenen Wegen die ER-Kanäle verlassen. Doch von allen Möglichkeiten, die hier nicht näher erörtert werden sollen, ist mir die folgende am wahrscheinlichsten: Die X-Körper schnüren sich wie Knospen aus den ER-Kanälen ab und wandern zur Mitte des Golgi-Feldes. Wenn diese Deutung richtig ist, dann würden die X-Körper ein *Vehikel der Proteine* sein, die vom ER direkt zu den Golgi-Säcken transportiert werden. In diesem Falle verliefe also die Integration: Protein der Ribosomen → ER → X-Körper → Golgi-Feld. Ob diese Annahme stimmt, hängt vom Weg der X-Körper ab, d. h. ob sie aus dem ER entstehen und Säcke werden oder umgekehrt. Ich glaube: ER → X-Körper → Säcke. Eine sorgfältige Analyse elektronenoptischer Bilder über Lage, Größe und Übergangsformen von X-Körpern führt zu dem Ergebnis, daß neue kleine Säcke, die sich den bereits vorhandenen Säcken anlegen, aus X-Körpern durch Anschwellung hervorgehen (20). Auf diese Weise könnten die Proteine der ER-Kanäle direkt den neuen Säcken zugeführt werden. Auf diesem Wege erfolgt gleichzeitig aber auch — wie auch aus ontogenetischen Studien hervorgeht (FERREIRA 1959, ZAGURY 1962) — die Restitution der mit der Sekretextrusion verbrauchten Säcke und zwar auf dem Wege ER → X-Körper → kleine Säcke → große Säcke. Aber eine Kommunikation nach dem Prinzip der kommunizierenden Röhren besteht bei der Pankreaszelle zwischen ER und den Säcken sicher nicht.

Eine weitere Möglichkeit, Proteine durch die X-Körper vom ER in die aus den Säcken entstehenden Vakuolen zu transportieren, könnte folgender Weg sein: Es ist schon 1958 von mir beobachtet worden, daß X-Körper sich den Intermediärkörpern anlegen und mit ihnen verschmelzen (21). Die Intermediärkörper erhalten demnach von 2 Seiten Proteine aus dem ER: 1. durch X-Körper, welche primär Proteine aus dem ER zu den Säcken bringen — wobei physikalisch auch noch die Lamellen durch aktiven Transport gelöster Proteine beteiligt sein könnten — und 2. durch die direkte Verschmelzung von X-Körpern mit den Intermediärkörpern (21). Die Intermediärkörper entstehen durch Ablösung der Vakuolen vom Rande der Säcke, was schon lichtmikroskopisch beobachtet werden konnte, und entwickeln sich weiter zu Zymogengranula (23), in deren Innerem mitunter Strukturen auftreten, die größenmäßig an X-Körper erinnern.

Der Weg der Proteine: Ergastoplasma → Golgi-Feld → Zymogengranula wurde bereits durch Lebendbeobachtung (HIRSCH 1931) nachgewiesen und neuerdings durch autoradiographische Studien bestätigt (CARO 1961, WARSHAWSKY, LEBLOND und DROZ 1963, vgl. a. SIEKEVITZ und PALADE 1960). Nach dem heutigen Stand unseres Wissens kann dieser Weg spezifisch wie folgt angegeben werden: Ribosomen → ER → X-Körper → Golgi-Sack →

→ Golgi-Vakuole → Intermediärkörper → Zymogengranulum.

Zusammenfassend darf festgehalten werden, daß zahlreiche Untersuchungen der Ontogenie und Physiologie des Pankreas für die Existenz von zwei „blasenförmigen Gebilden" im Arbeitszyklus des Golgi-Feldes sprechen, deren Herkunft und Bedeutung ich Ihnen geschildert habe. Diese beiden sehr verschiedenartigen Gebilde müssen jedenfalls terminologisch streng unterschieden werden, wenn nicht eine heillose Sprachverwirrung entstehen soll. Der allgemeine Sprachgebrauch unterscheidet: 1. Die Abschnürungen des ER als *X-Körper* oder Golgi-Vesikel, die in verschiedenen Formen zwischen dem ER und dem System der Säcke liegen und die dem Transport der Proteine aus dem ER und der Restitution der verbrauchten Säcke dienen — und 2. Die *Golgi-Vakuolen*, die aus den Säcken abgeschnürt werden, bestimmte Produkte des Golgi-Feldes enthalten und über Intermediärkörper sich zu Zymogengranula entwickeln.

Ich bitte also, diese traditionsgemäßen Unterscheidungen zwischen zwei ganz verschiedenen „blasenförmigen Gebilden" im Raume des Golgi-Feldes beizubehalten, auch wenn die Botaniker sich an dem Ausdruck „Vakuole" stoßen.

Ich glaube, daß die zahlreichen Strukturen des Golgi-Feldes im exokrinen Pankreas in einem integrierten physiologischen Zusammenhang stehen und daß mein obiger Versuch, diese Integration zu finden, den momentan höchsten Grad von Wahrscheinlichkeit besitzt, aber die hier dargestellte Kette noch der statistischen Untermauerung bedarf.

Sievers: In der Pflanzenzelle haben wir keine morphologischen Hinweise dafür, daß uns bekannte Strukturen etwas zum Golgi-Apparat hin liefern. Nach den bisherigen Befunden sieht es nicht so aus, als ob in der Pflanzenzelle etwa das ER den Golgi-Apparat beschickt. Wir müssen vielmehr annehmen, daß das Material aus dem Grundplasma kommt. Daraus kann der Schluß gezogen werden, daß es den Golgi-Strukturen möglich ist, durch die Zisternenmembranen hindurch Stoffe aus der Umgebung aufzunehmen. Das dürfen wir von der pflanzlichen Zelle sicherlich ableiten. In gleichem Sinne hat ja heute auch Prof. HOKIN seine Befunde am Pankreas interpretiert.

Was ist bislang in der tierischen Zelle bewiesen? Ich glaube, am ergebnisreichsten ist die Kombination von Autoradiographie und Elektronenmikroskopie, wie ich sie in der Tabelle geschildert habe. Mir scheint *bewiesen* der Gang der Radioaktivität vom Ergastoplasma über das Golgi-Feld zu den Zymogengranula am Apex der Zelle und schließlich in den Drüsengang hinein. Die Frage, was die Membranen des Ergastoplasmas und der Golgi-Zisternen im einzelnen leisten und leisten können, kann man aus den bisherigen Befunden noch nicht exakt beantworten. Weitere Folgerungen sind also eigentlich zunächst nur *Hypothesen*, die sicherlich weitertragen werden.

Solomon: Dr. SIEVERS, you presented a table, which summarizes statistically the results of grain counting from the autoradiographs of CARO and PALADE. With regard to the Golgi apparatus, are you able to tell us where the grains lie exactly? Are they inside or outside the Golgi cisternae?

Sievers: Because of the large size of the silver grains, this question is hard to decide. They lie neither inside nor outside the cisternae. You can only say, they lie in the ergastoplasm resp. in the Golgi field. Only with respect to the zymogen granules you can say, they are intragranular, for the zymogen granules are of lightmicroscopic dimensions. Therefore in my reply on Prof. HIRSCHs remark I said, that the only conclusion, we can draw from the current results in autoradiography, is the main route of the secretory products through the cell. From these results, we can not say with sureness, what really happens in detail, when the proteins are going from the ergastoplasma to the Golgi field.

P. Sitte: Ich wollte darauf hinweisen, daß die Situation im Pankreas und in den Pflanzenzellen ganz verschieden ist. Im Pankreas werden immerhin Enzyme gebildet, die also spezifische Makromoleküle sind, zu deren Charakterisierung es nicht genügt, das Molekulargewicht zu kennen, sondern zu deren Charakterisierung auch die Primärstruktur bekannt sein muß. In der Pflanzenzelle handelt es sich um die Bildung der Grundsubstanz der Zellwand. Während also Proteine oder Enzymproteine nur an Matrizen gebildet werden können — und das mag sicher den Zusammenhang mit dem ER bedingen oder zumindest voraussetzen —, so ist die Situation in der Pflanzenzelle doch wohl anders: Hier werden Polysaccharide gebildet. Daß Polysaccharide in den Dictyosomen selbst gebildet werden können, das ist auch für tierische Zellen gezeigt worden.

Bücher: Herr SITTE, das Pektin wird doch wahrscheinlich aus Uridindiphosphatgalakturonsäure gebildet. Das muß die Vorstufe sein. Wenn wir annehmen, daß die Polymerase aus dem ER in die Dictyosomen geht, hätten wir in den Golgi-Zisternen das Enzym, und von außen — aus dem Grundplasma — kommen die monomolekularen Uridindiphosphatgalakturonsäuremoleküle. Die Synthese des Pektins ist ein relativ einfacher chemischer Prozeß, im einfachsten Falle wäre vielleicht nur ein Enzym notwendig. Ein Teil der Galakturonsäuremoleküle ist vielleicht schon methyliert oder wird es vielleicht erst später.

P. Sitte: Sie haben es so gut gesagt, daß ich nichts dagegen hätte. Aber dennoch bleibt ein wesentlicher Unterschied: Im Falle des Pankreas wird das Protein selbst, das dann sezerniert wird, wenn vielleicht auch in einer einfacheren Form, vom endoplasmatischen Retikulum den Golgi-Zisternen zugeliefert, während im Falle der Pflanzenzelle doch nur die Enzyme zur Bildung der Pektine zugeliefert werden.

Bücher: Wer weiß, ob nicht im Pankreas bestimmte enzymatisch gerichtete Reaktionen in den Golgi-Zisternen ablaufen?

Thoenes: Es wäre doch vielleicht auch daran zu denken, daß die Polymerase mit Hilfe von freien Ribosomen gebildet werden könnte, die in der Pflanzenzelle vorkommen, so daß man in der Pflanzenzelle das ER nicht unbedingt — weder direkt noch indirekt — mit der Synthese von Polysacchariden in Verbindung bringen muß.

Ich darf noch eine Bemerkung anschließen, die die grundsätzliche Rolle des Golgi-Apparates im Sekretionszyklus betrifft. Ich möchte gern auf die Leberzelle hinweisen, bei der ja erwiesenermaßen das Albumin im Ergastoplasma gebildet wird. Das Golgi-Feld liegt hier ausschließlich im Bereich des Gallenpols der Zelle, so daß man sich schwer vorstellen kann, daß Albumin, das zweifellos an die Blutbahn abgegeben wird, hier auch über den Golgi-Apparat ausgeschieden wird.

Mothes: Vielleicht darf ich zur Frage der Polymerase einen Analogiefall schildern. Es handelt sich dabei um die Kautschuksynthese im Milchsaft von *Hevea* nach den Untersuchungen von LYNEN. Wir wissen heute genau, daß die Kautschuksynthese im isolierten Milchsaft stattfindet aus Stoffen, die von außen in den Milchsaft abgeschieden werden — die gilt zumindest für den Milchsaft von *Hevea*. Dazu gehören zwei Enzymsysteme. Das eine ist ein Multienzymsystem, das aus Azethylkoenzym A das Isopentenylphosphat macht. Das polymerisierende Enzymsystem, das dann die lange Kautschukkette knüpft, wurde lange gesucht. Schließlich konnte LYNEN zeigen, daß das Multienzymsystem ganz woanders sitzt als das polymerisierende System. Dieses polymerisierende Enzymsystem ist bei der Zentrifugation des Milchsaftes in der leichtesten, oben schwimmenden Fraktion. Wir wissen, daß die Kautschuktröpfchen schon bei Beginn des Entstehens mit einem Eiweißfilm umgeben sind. LYNEN ist der Meinung (ohne daß er elektronenmikroskopisch etwas untersucht hat), daß diese Eiweißhülle das polymerisierende Enzymsystem enthält. Jedenfalls gelingt es ihm ohne weiteres, mit Isopentenylphosphat und der schwimmenden Fraktion Kautschuk in beliebigen Mengen herzustellen. Da haben wir so eine Trennung von zwei Enzymsystemen. Isopentenylphosphat ist gebunden an ein Multienzymsystem, das selbst in hohem Maße strukturgebunden ist und beim Zentrifugieren in der schwersten Fraktion sitzen bleibt, und das polymerisierende System ist in einer leichten Fraktion, weil es einen Eiweißfilm darstellt, der die leichten Kautschuktröpfchen umgibt. Die Vorstufe, also das Isopentenylphosphat, muß in ein Bläschen abgeschieden werden, wo dann das endgültige Exkret, der Kautschuk, gebildet wird. Das könnte also eine gewisse Versöhnung der hier geäußerten Auffassungen darstellen.

Holter: Ich wollte einmal die Frage aufwerfen, ob etwas darüber bekannt ist, welche Kräfte oder Strukturen dafür sorgen, daß alle diese Stoffe immer an die richtige Stelle gelangen. Was verbirgt sich eigentlich hinter den Pfeilen in den Schemata von Herrn SIEVERS, oder auch in dem Wort „gerichtet"? Gibt es da irgendwelche Leitstrukturen, Mikrofibrillen oder dergleichen? Ich denke ganz besonders an die Fälle, wo das Sekret nicht nur eine Zellgrenze zu erreichen hat. Wenn zwischen zwei Tochterzellen eine neue Wand aufzubauen ist und die Sekrettröpfchen in der Mitte der Mutterzelle zum Stillstand zu bringen sind, finde ich, daß die Frage nach Leitstrukturen sehr dringlich wird.

Sievers: Da stimme ich Ihnen zu. Gerade das Beispiel, daß inmitten einer Zelle eine neue Zellplatte aufgebaut wird, beweist recht anschaulich, daß es einen gerichteten Transport gibt. Allerdings wurden bisher noch keine Strukturen gefunden, denen mit Sicherheit eine Bedeutung als Leitstrukturen zugeschrieben werden kann. Aber vielleicht sollte ich in diesem Zusammenhang die sog. microtubules erwähnen, die 1963 von LEDBETTER und PORTER nach Glutaraldehydfixation in Pflanzenzellen gefunden wurden und die nach Ansicht der Autoren vielleicht etwas mit der Plasmaströmung zu tun haben. Möglicherweise stellen sie auch eine Leitstruktur oder ein Richtungsmoment für den Vesikeltransport dar.

Komnick: Ich habe eine etwas allgemeinere Frage. Wir haben hauptsächlich über die Funktion des Golgi-Apparates in Drüsenzellen und in anderen Zellen gesprochen, die geformte Produkte ausscheiden. Wir wissen aber, daß der Golgi-Apparat eine ubiquitäre Strukturkomponente der Zellen ist, vielleicht mit Ausnahme der Säugererythrocyten. Auch wenn wir die Sekretion im weitesten Sinne auffassen — d. h. im Sinne einer Abgabe einzelner Moleküle — so erhebt sich doch vor allen Dingen im Hinblick auf die funktionelle Deutung, die Herr SIEVERS gegeben hat, daß also der Golgi-Apparat dazu da ist, Stoffe räumlich abzutrennen, die Frage: Welche Funktion kann ihm in Zellen zugeschrieben werden, die keine geformten Sekrete produzieren? In solchen Zellen — ich denke auch an bestimmte Epithelzellen — ist er ja häufig in sehr starkem Maße ausgeprägt. Gibt es da irgendwelche gesicherten Anhaltspunkte? Hypothesen gibt es eine ganze Reihe.

Schmidt: Bei resorbierenden Zellen, wie der Darmepithelzelle, ist die Aufgabe des Golgi-Apparates offensichtlich auch die Verarbeitung der Resorbate. Außerdem hat jede Zelle einen zelleigenen Stoffwechsel. Und dieser zelleigene Stoffwechsel einer Epidermiszelle, z. B. von Amphibien, spielt sich zum großen Teil in den Vakuolen des Golgi-Apparates ab. Wenn wir z. B. eine Zelle mit irgendeinem Vitalfarbstoff behandeln und sie schwer schädigen, dann werden die Abbauprodukte in Vakuolen abgelagert, die sie zum großen Teil in den Golgi-Apparat ableiten.

David: Ich glaube, man sollte die Rolle des Golgi-Feldes nicht auf die Frage der Sekretion beschränken. Zum Beispiel wissen wir von verschiedenen tierischen Zellarten sicher, daß der Golgi-Apparat bestimmte Beziehungen zum Zentrosom hat, daß er wahrscheinlich auch eine wichtige Rolle bei bestimmten Zellteilungsvorgängen spielt.

Gersch: Was weiß man über die Entstehung dieser Dictyosomen beispielsweise in den Wurzelhaarzellen? Aus welchen Elementen entstehen sie im Verlauf der Ausbildung des Wurzelhaares?

Sievers: Darüber ist nichts Sicheres bekannt.

P. Sitte: Bei Diatomeen ist durch Vitalbeobachtung der Dictyosomen festgestellt, daß sie sich synchron mit dem Kern teilen. Wie das im Elektronenmikroskop aussieht, weiß man noch nicht. Aber sie teilen sich in diesen Organismen, das steht fest.

Sievers: Ist es bekannt, woher die Membranen stammen, die nach der Mitose zwei tierische Tochterzellen trennen?

Schmidt: In einigen Fällen ist das bekannt, z. B. bei der ersten Furchungsteilung. Es ist hier ähnlich, wie in der pflanzlichen Zelle; die neuen Zellmembranen entstehen durch Fusion von Vesikeln. Wir wissen nicht, ob diese Vesikel aus dem ER oder aus dem Golgi-Apparat stammen. Bei der tierischen Zelle gibt es keine Wandsubstanzen, die von den Vesikeln mitgeliefert werden müßten und den Herkunftsort so deutlich verraten könnten wie bei der Pflanzenzelle.

Hier erhebt sich nun eine andere Frage: Von SJÖSTRAND u. a. sind kürzlich neue Messungen an Zytoplasmamembranen vorgenommen worden. Sie ließen zumindest vermuten, daß verschiedene Dicken bei den verschiedenen Membranen vorliegen. Spricht nun die Möglichkeit des Einbaus einer Golgi-Membran in das Plasmalemma dafür, daß verschieden dicke Membranen vorliegen?

Helander: I would like to comment on your question. Indeed, the different cytomembranes vary in thickness, and the plasma membrane, which is usually thicker than the Golgi-membranes, varies in thickness at different sites of the cell circumference. But I think it is quite possible that these membranes of the Golgi vacuoles may ultimately become incorporated with the plasma membrane, and at the same time these membranes are rebuilt, or slightly rebuilt, so that they increase in size. As you know from the investigations of SJÖSTRAND, the plasma membrane is not symmetrical, but asymmetrical, so that one side of it is a little thicker than the other. He believes that this is because of the protein layers on each side of the lipid layer being thicker on one side than on the other, and this might be the solution to this question. The protein layer might be remodelled as the membranes move from one part of the cell circumference to another part of it.

Bücher: One of the general phenomena we are confronted with in nearly all the contributions of today is the dynamic state of subcellular membranes, the rapid formation, and I personally believe in the 'specificity' of all these membranes. The question is how this rapid formation of membrane material goes on, and what is the mechanism of that.

There is a question which I would like to put, as there are so many morphologists here. Has any of you ever seen an end of such a membrane? That is to say, is the mechanism for formation of the membrane a building at the end of the membrane, or is the mechanism of the sort that it is introduction of material into an existing membrane?

Sievers: Ein Membranende ist bisher nie gesehen worden, es sei denn bei artefiziell gerissenen Membranen. Die Membranen sind immer in sich geschlossen und wachsen wohl durch Einfügung neuer Bausteine. SCHNEIDER konnte allerdings 1963 bei der Konjugation von Paramecium zeigen, daß Teile der Zellmembran in Strukturen des Grundplasmas zerfallen und sich daraus wieder neu aufbauen. Aber auch da waren nie Membranenden zu sehen.

Wohlfarth-Bottermann: Ich darf zu der Frage der Schnelligkeit der Membranbildung ein altes Experiment erwähnen, das wir nur nebenbei publiziert haben, das aber ganz eindrucksvoll ist. Sie kennen sicher die Plasmaschläuche von Schleimpilzen, die im Plasmodium in einem Netzwerk wachsen. Wenn man einen solchen Protoplasmafaden mit einer Nadel ansticht, dann tritt blitzartig ein Protoplasmatropfen aus, in einem Augenblick, wo das Protoplasma in dem Schlauch wirklich strömte. Wir dachten, das wäre ein gutes Objekt, um die Frage der Membranbildung, der Bildung des Plasmalemmas, zu untersuchen, und haben dann im Augenblick unmittelbar nach dem Anstich sofort mit Osmiumtetroxyd fixiert. Es ist uns nicht gelungen, zu irgendeinem Zeitpunkt elektronenmikroskopisch etwa einen nackten Protoplasmatropfen nachzuweisen. Es ist ganz sicher, daß die Neubildung der Zellmembran keine für uns meßbare Zeit beansprucht. Man findet kein nacktes Protoplasma.

Physiologie der Pinocytose bei Amöben

Von

Heinz Holter, Kopenhagen

Mit 13 Abbildungen

A. Einleitung

Die ursprüngliche Definition der Pinocytose als das „Trinken" der Zellen, also die tropfenweise Aufnahme von Flüssigkeit in diskreten, vakuolisierten Quanten oder Schlucken, hat durch die Arbeiten der letzten Jahre einen gewissen Bedeutungswandel erfahren. Es soll daher schon einleitungsweise hervorgehoben werden, daß nach dem gegenwärtigen Stand unseres Wissens das Wesentliche der Pinocytose nicht die Aufnahme von *Flüssigkeit* ist, sondern die Aufnahme der in der Flüssigkeit enthaltenen, gelösten oder suspendierten *Substanzen*, deren Anreicherung durch Adsorption eine wichtige Komponente im Pinocytosemechanismus darstellt.

Diese Bedeutungsverschiebung ist natürlich von großer Wichtigkeit für die physiologischen Aspekte des Pinocytoseprozesses, berührt aber die morphologische Definition der Pinocytose viel weniger. In morphologischer Beziehung können wir daran festhalten, daß das entscheidende Kennzeichen der Pinocytose die *Invagination* und Abschnürung der Zellmembran ist, wodurch unter Umschließung kleiner Volumina des Außenmediums gewisse Teile der Zelloberfläche als Vakuolenwand ins Zellinnere befördert werden.

Diese Betrachtungsweise führt es mit sich, daß eine scharfe begriffsmäßige Sonderung der Pinocytose von anderen, verwandten Formen der zellulären Substanzaufnahme immer schwieriger wird. Sowohl METCHNIKOFFs [30] klassische Phagocytose wie eine lange Reihe verwandter Prozesse dienen dem gleichen Zweck, der Substanzaufnahme, mit den gleichen Mitteln, der Vakuolenbildung nach Invagination. Eine Untereinteilung in Athrocytose [21], Kolloidopexie [9], Ultraphagocytose [22], Mikropinocytose [36], Rhopheocytose [40], u. a., je nach der Größe der aufgenommenen Teilchen oder der Größe der aufnehmenden Vesikel ist zweifellos praktisch, morphologisch und historisch gerechtfertigt. Doch erscheint es wichtig, daran festzuhalten, daß das zellphysiologisch entscheidende Charakteristikum allen diesen Prozessen gemeinsam ist: es ist die Verlegung der Membranbarriere von der Oberfläche in das Zellinnere.

NOVIKOFF [*35*] hat 1961 vorgeschlagen, diese Sachlage durch einen gemeinsamen Namen anzuerkennen, und DE DUVE hat auf einem 1963 in London abgehaltenen Symposium [*19*] das Wort „Endocytose" als Sammelbegriff in Vorschlag gebracht, zugleich mit „Exocytose" als Sammelname für den umgekehrt gerichteten exkretorischen Prozeß. Trotz einiger etymologischer Bedenken [*38*] hat der Vorschlag seiner praktischen Anwendbarkeit wegen Anschluß gefunden.

Das Thema meines Referates ist somit die Besprechung eines Spezialfalles der Endocytose, nämlich der Pinocytose, unter besonderer Berücksichtigung der zellphysiologischen Aspekte dieses Prozesses. Da aus rein praktischen Gründen die experimentelle physiologische Bearbeitung der Pinocytose hauptsächlich an Amöben geschehen ist, wird vor allem von diesen Organismen die Rede sein. Dies scheint berechtigt auch in Anbetracht des Umstandes, daß gerade bei der Arbeit mit Amöben der Kontakt zwischen den mehr morphologisch und mehr physiologisch orientierten Arbeitsrichtungen besonders gut ist (siehe z. B. WOHLFARTH-BOTTERMANN [*49*]). Es ist daher zu hoffen, daß die so gewonnenen Resultate auch bei der Erforschung anderer Zelltypen von Nutzen sein werden.

B. Initialphase der Pinocytose:
Adsorption und Invagination

Das charakteristische Merkmal der Pinocytose ist die Kombination von Oberflächenadsorption und Invagination, die von BENNETT [*4*] in seiner Hypothese über Membranfluß und Vesikulation dargestellt wurde. Diese Hypothese hat WITTEKIND [*47*] in seinem Vortrag auf der 102. Versammlung Deutscher Naturforscher und Ärzte besprochen; dort ist auch BENNETTs figürliche Darstellung als Abb. 5 wiedergegeben.

BENNETTs Hypothese bezieht sich in erster Linie auf die von PALADE [*37*] elektronenoptisch nachgewiesene und von vielen Seiten bestätigte „mikropinocytische" Invagination und Vakuolenbildung. Bei Amöben spielt sich die Pinocytose im lichtmikroskopischen Bereich ab, zumindest in der Hauptsache und unter den Bedingungen unserer Versuche. Abgesehen von den Dimensionen entspricht jedoch auch bei *Amoeba proteus* der Invaginationsprozeß ganz dem einen der beiden von BENNETT skizzierten Fälle: Die mit der adsorbierten Substanz beladene Zellmembran kleidet einen durch Invagination gebildeten Kanal aus (nach BENNETT „fließt" sie in den Kanal hinein), von dessen Grund dann durch Wandverschmelzung kleine (wenige μ im Durchmesser) Vakuolen abgeschnürt werden, die durch die Cytoplasmaströmung über das Endoplasma verteilt werden.

Die initiale Phase der Pinocytose hat somit 3 Aspekte: den adsorptiven Stimulus durch spezifische Induktoren, die Invaginationsreaktion und die Vakuolenbildung. Aus praktischen Gründen wollen wir uns mit diesen drei Aspekten gesondert beschäftigen. Später wird von den nachfolgenden Phasen der Pinocytose die Rede sein, nämlich den Veränderungen der Vakuolen im Zellinneren, ihrer Eingliederung in das Stoffwechselsystem der Zelle, und der Exkretion des Residualmateriales.

I. Adsorption

Zu Beginn der experimentellen Untersuchungen der Amöbenpinocytose ist man davon ausgegangen [24, 25, 28], daß Pinocytose nur als Reaktion auf eine spezifische Stimulation zustande käme, herbeigeführt durch die Anwesenheit bestimmter induzierender Substanzen im Außenmedium. In letzter Zeit mehren sich die Anzeichen dafür [48, 50], daß eine Pinocytosewirksamkeit geringerer Intensität zu den permanenten Lebensäußerungen der Amöben gehört, vielleicht als ein Prozeß der Membraneinschmelzung im Zusammenhang mit der amöboiden Bewegung.

Sehen wir von dieser „Residualpinocytose" vorläufig ab, bleibt die Tatsache bestehen, daß intensive Pinocytoseprozesse, die im primitiven physiologischen Experiment gemessen werden können, einer Induktion bedürfen. Die Induktion besteht, nach BENNETTs Hypothese, in der Adsorption der Induktorsubstanz an geeignete „Haftpunkte" der Oberflächenmembran. In Übereinstimmung mit dieser Annahme fanden BRANDT [6] und SCHUMAKER [45] beim Studium der Lokalisation fluorescierender und radioaktiver Induktoren (Proteine), daß die markierte Substanz als der Oberfläche aufgelagerte Schicht nachgewiesen werden kann.

Der Adsorptionsprozeß wurde dann von BRANDT und PAPPAS [7, 8], NACHMIAS und MARSHALL [32], und CHAPMAN-ANDRESEN [12] mehr eingehend studiert. Es hat sich gezeigt, daß die „Haftsubstanz" die mucoide Außenschicht der Zellmembran ist, die als Deckschicht auf dem eigentlichen Plasmalemma von LEHMANN [27] und SCHNEIDER und WOHLFARTH-BOTTERMANN [44] nachgewiesen wurde. Die chemische Zusammensetzung dieser Deckschicht ist noch unbekannt, doch dürfte ein saures Polysaccharid [2, 6, 46] aller Wahrscheinlichkeit nach eine wesentliche Komponente sein.

Eine solche Mucoidschicht ist elektronen- und lichtoptisch nicht leicht nachzuweisen und es ist daher nicht bekannt, ob es sich um den Amöben eigentümliche Zellorganellen handelt; doch mehren sich in letzter Zeit Anzeichen für eine sehr allgemeine Verbreitung dieser Struktur (siehe z. B. BELL [3]), und man ist versucht anzunehmen, daß alle der Pinocytose fähigen Zelltypen sie besitzen.

Bei Amöben ist sie jedenfalls in imponierender Entwicklung anzutreffen. MERCER [29], PAPPAS [39] und WOHLFARTH-BOTTERMANN [48] haben nachgewiesen, daß die Mucoidschicht eine komplizierte Struktur besitzt, indem eine dem Plasmalemma anliegende amorphe etwa 200 Å dicke Grundschicht an ihrer Außenseite etwa 2000 Å lange Papillen oder Haare trägt (Abb. 1). Es ist nicht bekannt, ob diese Struktur ein Spezialorgan der pinocytischen Adsorption ist, oder ob sie auch andere Funktionen hat; nachgewiesen ist lediglich [8, 15], daß elektronenoptisch sichtbare Induktoren wie Protein oder Thoriumdioxyd an die Papillen und die Außenfläche der amorphen Zone adsorbiert werden (Abb. 2).

Die adsorptive Beladung des Mucoids muß irgendwie die invaginatorische Reaktion des Plasmalemmas oder der Ektoplasmaschicht verursachen, doch die Überführung des Stimulus, sowie der Mechanismus der Invagination selbst, sind unbekannt.

Etwas besser steht es mit dem Studium der Induktionsspezifität, mit dem sich CHAPMAN-ANDRESEN [*10, 11, 12*] recht eingehend beschäftigt hat.

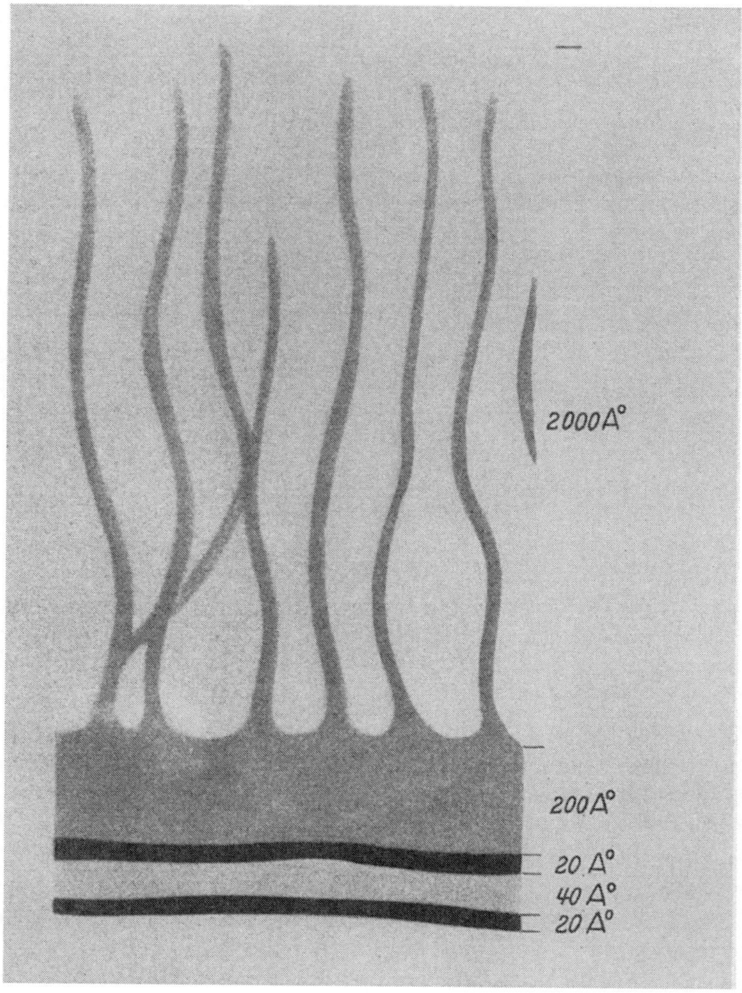

Abb. 1. Diagramm der Zellmembranstruktur von *Chaos chaos*. Die 2000 Å dicke „Papillenschicht" ist eine Fortsetzung der unterliegenden 200 Å dicken amorphen Zone. Darunter liegt das eigentliche Plasmalemma, eine "unit membrane"-Struktur bestehend aus zwei dichten Linien von je 20 Å, geschieden von einer 40 Å weiten, weniger dichten Zone [*8*]

Das Hauptresultat dieser Untersuchungen war, daß eine Reihe von Salzen, von Aminosäuren, Proteinen und Farbstoffen Induktionswirkung zeigen, alle abhängig vom pH der induzierenden Lösung. Viele elektrisch neutrale Substanzen, wie Kohlenhydrate, sind wirkungslos. Hieraus

scheint hervorzugehen, daß die elektrische Ladung des Induktors eine
wesentliche Rolle für seine Adsorption an die Mucoidschicht spielt.
Es gibt aber auch Anzeichen für eine mehr spezifische Wechselwirkung
zwischen der Mucoidschicht und der adsorbierten Substanz. So scheinen
z. B. Globuline im selben pH-Bereich wirksamer zu sein als Albumine.
Solange aber Zusammensetzung und chemische Eigenschaften des Mucoi-
des unbekannt bleiben, scheinen detaillierte Überlegungen solcher Probleme verfrüht.

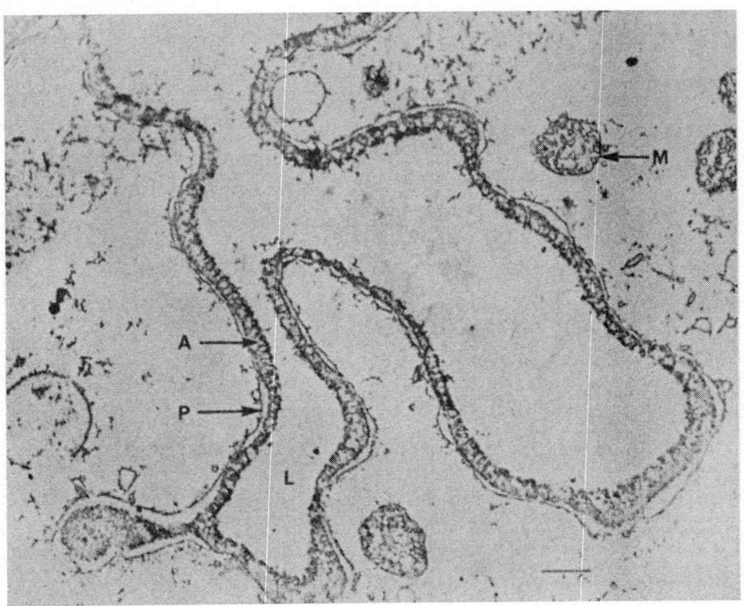

Abb. 2. Längsschnitt durch eine Pinocytoseinvagination von *Amoeba proteus*. Die Schicht
der Albumin-beladenen Papillen (*A*) setzt sich von der Oberfläche in den Kanal fort. Das
Lumen (*L*) ist optisch leer. Plasmalemma (*P*) und Albuminschicht an einigen Stellen getrennt. Mitochondrien (*M*) in der Nähe des Kanals. Vergrößerung: 8000mal [*15*]

CHAPMAN-ANDRESEN [*12*] fand auch beträchtliche Unterschiede
in der Reversibilität der adsorptiven Bindung. Salze werden leicht aus
der Mucoidschicht ausgewaschen. Auch Proteine und basische Farbstoffe lassen sich leicht wieder entfernen, wenn das pH der Waschflüssig-
keit entsprechend gewählt wird; im allgemeinen genügt eine Erhöhung
um etwa 2 pH-Einheiten gegenüber dem Induktionsoptimum (bei den
erwähnten Proteinen etwa 4,5; Wasch-pH 6,5). Es gibt aber auch Induktoren, z. B. die „schleimfärbenden" Farbstoffe der Cu-phthalocyanin-
Gruppe (Alcian Blue), welche ganz irreversibel gebunden zu werden
scheinen, so daß bis jetzt kein Weg zu ihrer Entfernung gefunden werden
konnte. Im Elektronenmikroskop geben diese Farbstoffe Bilder (Abb. 3),
die recht verschieden von den nach Proteinadsorption erhaltenen sind
(Abb. 2). Es ist nicht bekannt, ob sich dieser Befund generalisieren läßt,
und ob er mit dem Umstand der Irreversibilität der Adsorption im
Zusammenhang steht.

Bei den Untersuchungen von CHAPMAN-ANDRESEN handelte es sich darum, die Pinocytose-Reaktion auf den Adsorptionsstimulus soweit möglich quantitativ zu messen. Die von CHAPMAN-ANDRESEN zu diesem Zweck ausgearbeitete Methode [12] beruht auf der Zählung der per Oberflächeneinheit der Amöbe gebildeten Invaginations-Kanäle. Diese Messung der Pinocytose-Intensität gibt natürlich angenäherte, aber

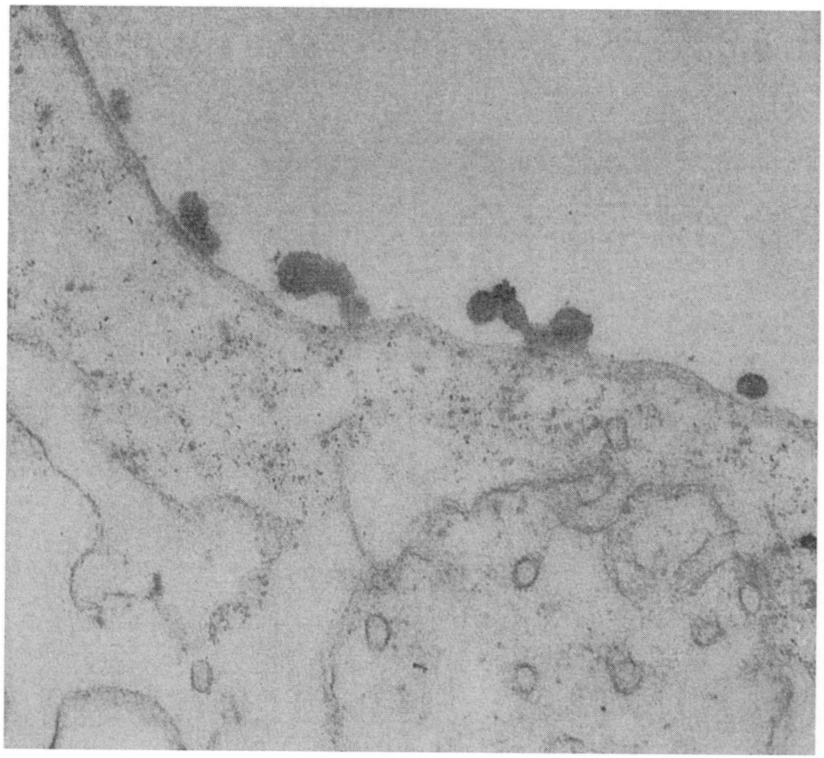

Abb. 3. Zellmembran von *Amoeba proteus* 10 min nach Immersion in Alcian Blue. Der Alcian Blue-Mucoidkomplex der Papillen ist zu Körnchen kontrahiert. Vergrößerung: 43000mal [23]

doch überraschend reproduzierbare (Variationsbreite unter 20%) Resultate. Abb. 4 zeigt ein Histogramm der Durchschnittszahl von beobachteten Kanälchen nach Pinocytoseinduktion mit 0,125 n NaCl-Lösung unter standardisierten Versuchsbedingungen. Abb. 5 gibt die mit Hilfe dieser Methode bestimmte Temperaturvariation der Pinocytose-intensität, als Beispiel für die Art der Messungen, für welche sich die Methode eignet.

Die von der Mucoidschicht aufgenommenen Substanzmengen können erstaunlich groß sein. Einen qualitativen Eindruck hiervon vermittelt schon die Betrachtung der Bilder, die nach der Adsorption von fluorescierenden Proteinen [6] oder Elektronenkontraststoffen (z. B. [7]) erhalten werden. Quantitativ wurde die Frage meines Wissens nur von CHAPMAN-

ANDRESEN und HOLTER [15] bearbeitet. Wir bestimmten die Menge des von *Amoeba proteus* bei pH 4,5 adsorbierten und bei pH 6,5 abwaschbaren Serumalbumins (mit ^{131}I markiert) und fanden unter optimalen Bedingungen $9 \times 10^{-2}\,\mu g$ pro Amöbe. Dies entspricht etwa 60% des Trockengewichtes der Amöbe. Die Dicke der adsorbierten Schicht kann

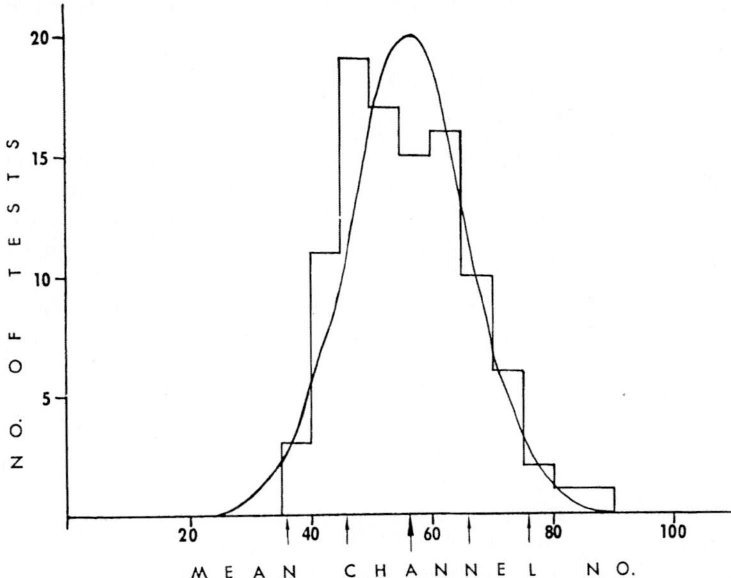

Abb. 4. Histiogramm der Verteilung der durchschnittlichen Kanalzahl per *Amoeba proteus* nach Immersion in 0,125 n NaCl korrigiert auf Standard-pH 6,5 und Standardtemperatur 21, 5° C. Die Kurve bezeichnet die berechnete Normalverteilung der Werte [12]

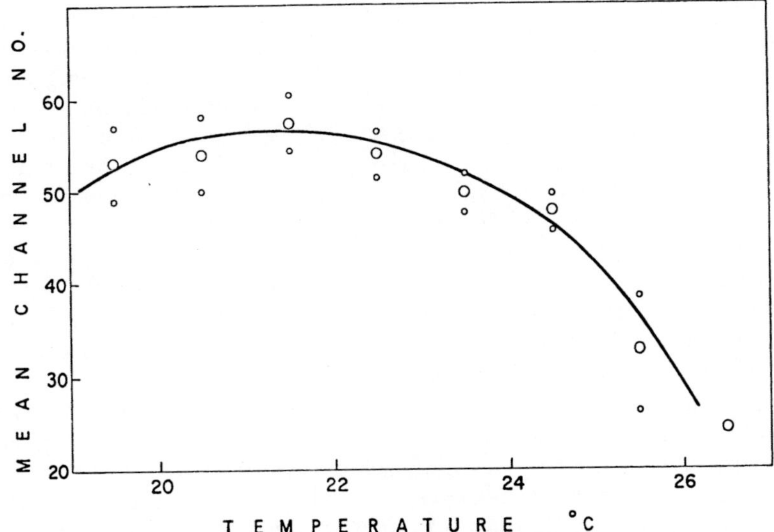

Abb. 5. Temperatur-Abhängigkeit der Zahl von in 0,125 n NaCl gebildeten Pinocytose-kanälen per *Amoeba proteus*, korrigiert auf pH 6,5. Große Kreise: Durchschnittswert; kleine Kreise: Standardstreuung [12]

auf etwa 0,3 μ geschätzt werden. Eine solche Schätzung beruht natürlich auf sehr vereinfachenden, wahrscheinlich abwegigen, Annahmen über Proteinhydratation und Oberflächenareal der Amöben und muß daher mit Skepsis betrachtet werden. Immerhin aber stimmt sie größenordnungsmäßig überein mit der anscheinenden Dicke der proteinbeladenen Mucoidschicht, die man von Bildern wie Abb. 2 ablesen kann.

II. Invagination und Vakuolenbildung

Es wird allgemein angenommen, daß die Adsorption eine notwendige Vorbedingung der Invagination sei. Ob sie aber auch eine ausreichende Bedingung hierfür ist, wissen wir nicht. Es ist mit anderen Worten unbekannt, ob jede adsorptive Beladung der Mucoidschicht Invaginationen auslöst, oder ob dies nur geschieht, wenn bestimmte, „invaginationsspezifische" Substanzen adsorbiert werden.

Auch die Annahme, daß Invaginationen *nur* nach vorhergehender Adsorption einer spezifischen Induktorsubstanz auftreten, läßt sich möglicherweise nicht aufrechterhalten. WOHLFARTH-BOTTERMANN [48] hat gefunden, daß die Pinocytose bei der Limax-Amöbe *Hyalodiscus* kontinuierlich, ohne experimentelle Induktion vor sich geht, wenn auch wahrscheinlich mit geringerer Intensität als im offensichtlich induzierten Prozeß. Wir müssen also entweder annehmen, daß im normalen Medium von *Hyalodiscus* Induktoren in geringen Mengen vorhanden sind (vielleicht von der Amöbe selbst gebildet werden) oder wir müssen die Hypothese fallen lassen, daß die adsorptive Veränderung der Mucoidschicht (welche *Hyalodiscus* in schöner Ausbildung besitzt) eine notwendige Bedingung des Invaginationsmechanismus sei.

Die letztere Annahme ist mehr oder weniger stillschweigend von mehreren Autoren gemacht worden. Es ist ja verhältnismäßig leicht, sich eine Veränderung der Oberflächenspannung als Folge der Adsorption vorzustellen; wie diese veränderte Oberflächenspannung aber zur Invagination führen sollte, leuchtet nicht unmittelbar ein. — Ausführlicher ist ein Vorschlag von BRANDT [6], in dem angenommen wird, daß das Plasmalemma nur in gewissen Punkten fest mit dem unterliegenden Ektoplasma (Plasmagel) verbunden bleibe. Die Invagination käme dann zustande entweder durch Vorfließen von Plasmasol am fixierten Areal vorbei, oder durch Retraktion des Plasmagels, oder eine Kombination dieser Effekte. HAYWARD [23] hat am Grunde von Invaginationen Zonen von besonders dichtem, granulärem Cytoplasma beobachtet, die vielleicht mit BRANDTs Vorschlag in Beziehung stehen könnten.

Wollen wir aber ehrlich sein, so müssen wir zugeben, daß die Anhaltspunkte für auch nur einigermaßen wohlfundierte Annahmen über den Invaginationsmechanismus gegenwärtig einfach noch nicht existieren.

Auch über den Energieverbrauch der Invaginationsreaktion wissen wir sehr wenig. SCHUMAKER [45] und DE TERRA und RUSTAD [20] haben nachgewiesen, daß die Invagination (nicht aber die Adsorption) vom oxydativen Amöbenstoffwechsel abhängig ist und durch Stoffwechselgifte gebremst werden kann. Damit würde übereinstimmen,

daß gelegentlich [16] eine räumliche Assoziation von Mitochondrien
und Pinocytosekanälen beobachtet wurde. Weitgehende Rückschlüsse
auf den Invaginationsmechanismus lassen aber auch diese Befunde
nicht zu.

Die Invagination in Amöben ist morphologisch recht variabel [17],
abhängig von der Art des verwendeten Induktors (Abb. 6). Am häufigsten sind 1–2 μ weite Kanäle mit einigermaßen parallelen Wänden

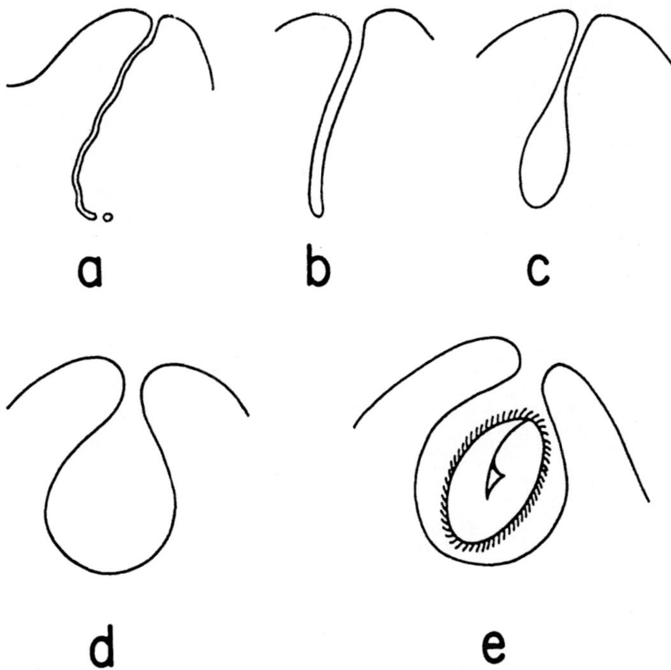

Abb. 6. Diagramm verschiedener Invaginationstypen, erhalten in verschiedenen Induktoren mit *Amoeba proteus*. a Pinocytosekanal in Salzlösung; b Pinocytosekanal in Tabakmosaikvirus; c Flaschenförmiger Hohlraum in Tabakmosaikvirus; d Flaschenförmiger
Hohlraum in Methionin; e Phagocytische Invagination [17]

(Type a und b; in Salz- und Proteinlösungen) während die in Virus-
und gewissen Farbstofflösungen erscheinenden „Flaschen"-Kanäle
(Type c und d) sich in extremen Fällen in Form und Dimensionen den
phagocytischen Invaginationen nähern können (Type e).

Die in Abb. 6 schematisch dargestellten Formen dürfen jedoch nicht
als feste Strukturen angesehen werden. Bei Lebendbeobachtung und
besonders im Zeitrafferfilm bemerkt man peristaltische Bewegungen [12],
welche bläschenartige Erweiterungen des Kanals zelleinwärts zu befördern scheinen. Diese Peristaltik ist schon von MAST und DOYLE [28],
den Entdeckern der Amöben-Pinocytose, beobachtet worden, wie aus
ihrer von WITTEKIND [47] als Abb. 6 wiedergegebenen Zeichnung
klar hervorgeht.

Der Mechanismus dieser Peristaltik ist ebenso unbekannt wie der
Mechanismus der Invagination selbst. Doch sei darauf hingewiesen, daß

das von HAYWARD [23] beobachtete dichte, granuläre Cytoplasma (Abb.7)
bisweilen auch in der Halszone der Kanäle auftritt und vielleicht mit den
peristaltischen Kontraktionen in Verbindung gebracht werden könnte.
REICHSTEIN-NILSSON [41] hat ähnliche Cytoplasmastrukturen in der
Mundregion von *Tetrahymena* nachgewiesen, dort wo die Nahrungsvakuolen abgeschnürt werden.

Die Lebenszeit der einzelnen Invagination ist bei *Amoeba proteus*
im Durchschnitt etwa 2 min [12]. Auch die Gesamtdauer der Pinocytose-

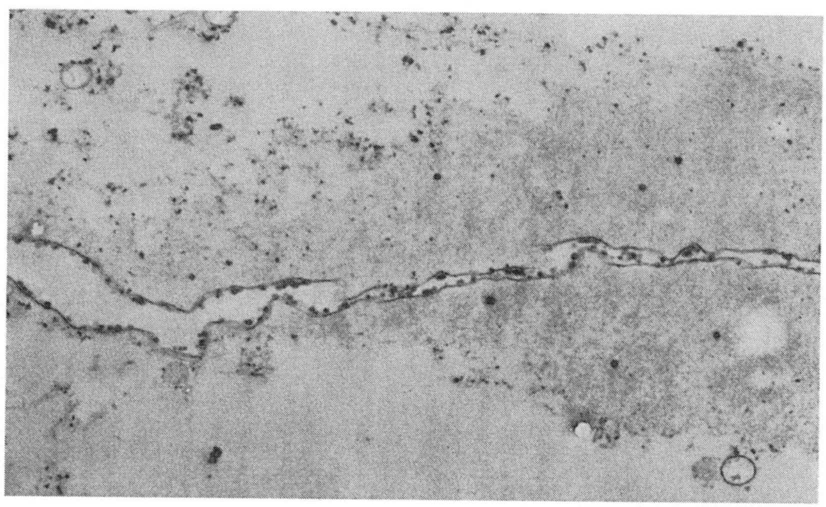

Abb. 7. Kanalhals in *Amoeba proteus* nach Alcian Blue. Der Farbstoff-Mucoidkomplex
an der Innenwand erkennbar. Die betreffende Kanalstrecke ist umgeben von dichtem,
granulärem Cytoplasma, wie es auch am in Abb. 10 dargestellten Kanalboden auftritt.
Vergrößerung: 13000mal [23]

reaktion ist begrenzt. Sie dauert etwa 20 min [12], gerechnet vom Zeitpunkt der Einbringung der Amöbe in die Induktorlösung. Beim Ablauf
eines solchen „Pinocytosezyklus" scheint die Amöbe ein Sättigungs-
stadium zu erreichen, und auch die Überführung in neue Induktorlösung
verursacht nicht die Wiederaufnahme der Pinocytoseaktivität, wenn
man der Amöbe nicht dazwischen Zeit gibt, sich zu erholen.

CHAPMAN-ANDRESEN [13] hat dieses Sättigungsphänomen genauer
untersucht und ist zu dem Schluß gekommen, daß die Erschöpfung des
für die Pinocytose verfügbaren Membranareales der begrenzende
Faktor ist. Ihre Ergebnisse deuten darauf hin, daß nur etwa die Hälfte
der gesamten Amöbenoberfläche (gemessen im monopodialen, kriechen-
den Stadium) durch Invagination und Vakuolenabschnürung ins Zellinnere überführt werden kann. Wenn dieser Grenzwert erreicht ist,
kann kein weiterer Induktionsreiz die Wiederaufnahme der Pinocytose
erzwingen. Unterbricht man vor Erreichen dieser Grenze den aktiven
Pinocytosezyklus durch kurzes Waschen und Überführung in neue
Induktionslösung, so bleibt die Summe der in beiden Perioden gebildeten

Kanäle stets gleich, unabhängig davon, in welchem Verhältnis die Gesamt-
immersionszeit aufgeteilt wurde (Tab. 1).

Der so erreichte Grenzwert der Gesamtkanalzahl ist nach CHAPMAN-
ANDRESEN auch weitgehend unabhängig von der Art des verwendeten
Induktors. In mehr qualitativer Weise geht dies sogar soweit, daß
Phagocytose und Pinocytose einander ersetzen können. Eine zu voller
Sättigung mit *Tetrahymena* gefütterte *Amöbe* reagiert praktisch nicht
auf die Einbringung in eine Pinocytose-Induktorlösung und umgekehrt.

Tabelle 1. *Amoeba proteus, Pinocytose in 0,125 n NaCl. Wirkung der Dauer der
ersten Induktionsperiode auf die Pinocytoseintensität während der zweiten Induktionsperiode*

Gruppe Nr.	Immersionszeit		Kanäle pro Amöbe		
	1. Periode	2. Periode	1. Periode	2. Periode	Gesamtzahl
1	21	20	63,7	3,5	67,2
2	12	20	39,0	9,8	48,8
3	7	20	8,6	48,6	57,2
4	4	20	2,5	63,2	65,7

4 Gruppen von Amöben wurden die in Kolonne 2 gezeigte Zeit in der Induktionslösung belassen, 20 min in Normalmedium übergeführt und dann wieder 20 min
in die Induktionslösung eingebracht. Kanalzahlen sind Durchschnittswerte der
Zählung an 10 Amöben [13].

Diese Versuche besagen nichts darüber, ob der Grund für das Sättigungsphänomen mechanischer oder physiologischer Art ist; doch kann
die Annahme der für die Pinocytose nur in begrenzten Mengen „zur
Verfügung stehenden" Membran natürlich nur einen Sinn haben, wenn
von einer eventuellen Neubildung von Membran während des Pinocytosezyklus quantitativ abgesehen werden kann. Dies ist einer der
Gründe für die Vermutung, daß die permanent verlaufende „Residualpinocytose" von wesentlich geringerer Intensität sein muß als die in-
duzierte Pinocytose. Außerdem ist zu beachten, daß *Amoeba proteus*
und *Chaos chaos* in intensiver Pinocytose die Kriechbewegung einstellen
und eine typische stationäre Rosettenform annehmen.

Unter der Annahme also, die Membranbildung vernachlässigen
zu können, hat CHAPMAN-ANDRESEN auf Grund der bekannten Kanalzahl und der geschätzten Zahl der pro Kanal abgeschnürten Vakuolen
das Gesamtareal der während des Pinocytosezyklus ins Zellinnere über-
führten Oberflächenmembran berechnet, wobei sie zu dem oben genannten
Wert von etwa 50% gelangte.

Die Erholungszeit, d. h. die Zeit, nach welcher die Amöbe auf einen
neuen pinocytischen Reiz mit annähernd der ursprünglichen Intensität
reagiert, wurde ebenfalls von CHAPMAN-ANDRESEN [13] untersucht. Bei
Zimmertemperatur und in normalem Kulturmedium ist diese Zeit
etwa 4 Std, während welcher die Amöbe ihren Habitus von der Pino-
cytoserosette durch eine Reihe von Übergangsstadien zurück zu der
monopodialen Form umwandelt (Abb. 8). Morphologisch und pinocytisch scheint der Erholungsvorgang also parallel abzulaufen.

Den gemachten Annahmen zufolge würde dieser Befund also besagen, daß die von der Zelloberfläche verschwundene Membran während 4 Std resynthetisiert oder in anderer Weise an die Oberfläche zurückgeliefert werden kann. Es ist natürlich nicht leicht, schlüssige Beweise für die Richtigkeit dieser Auslegung zu finden, doch sei darauf hingewiesen, daß WOLPERT und O'NEILL [50], die ebenso wie WOHLFARTH-BOTTERMANN [48] eine ständige pinocytische Membraneinschmelzung

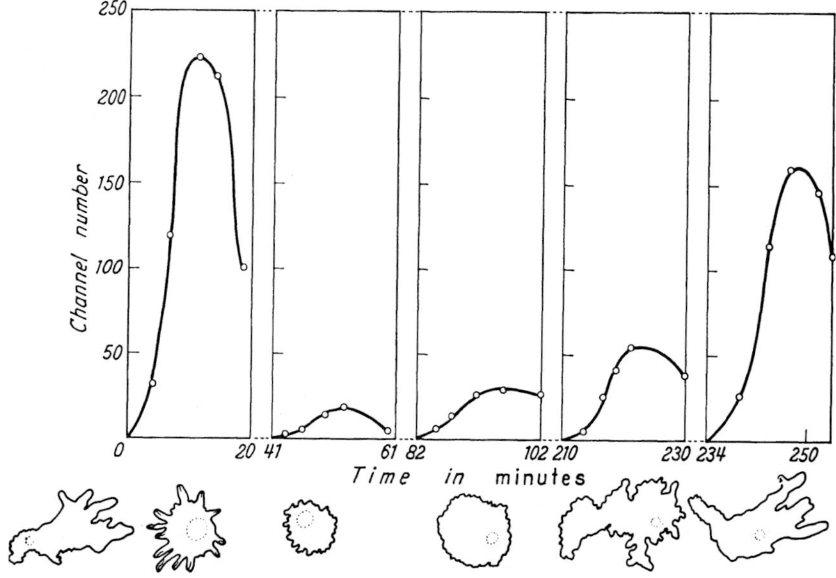

Abb. 8. Erholung von *Amoeba proteus* nach intensiver Pinocytose in 0,125 n NaCl. Ordinate: Kanalzahl per Versuchsgruppe von 10 Amöben. Abscisse: Zeit nach Immersion in die erste Induktorlösung. Die Umrißzeichnungen von Amöben sind nach Photographien schematisiert. Die erste Kurve gibt die Pinocytoseaktivität während der ersten Induktionsperiode. Die folgenden Kurven zeigen die Reaktionen während der jeweiligen zweiten Immersion in den Induktor. Morphologisch illustrieren die Zeichnungen den Übergang vom monopodialen Stadium (links) durch Pinocytose- und Verdauungsstadien zurück zum monopodialen (rechts) [13]

während der amöboiden Bewegung annehmen, und zwar bei *Amoeba proteus* in der Größenordnung von 0,2% pro Minute, auf Grund dieser Annahme eine Restitutionszeit von 4,2 Std für 50% der Amöbenmembran berechnen, also denselben Wert, den CHAPMAN-ANDRESEN von ganz anderen Voraussetzungen her fand.

Die für die Abschätzung der eventuellen physiologischen Bedeutung des Pinocytoseprozesses interessanteste Frage ist die nach den Stoffmengen, die in die Pinocytosevakuolen aufgenommen werden. CHAPMAN-ANDRESEN und HOLTER [15] haben dieses Problem an *Amoeba proteus* untersucht und haben dabei, um zwischen adsorbiertem Material und miteingeschlossener Lösung unterscheiden zu können, zwei Isotopmarkierte Substanzen angewendet, nämlich ^{131}I-Serumalbumin und ^{14}C-Glukose. Das Albumin ist der Induktor und wird als solcher an das Mucoid der Zellmembran gebunden. Glukose ist, wie wir früher [14] gefunden haben, kein Induktor und wird nicht adsorbiert. Wir nahmen daher an, daß die zusammen mit Albumin aufgenommene Glukosemenge als Maß für das gleichzeitig mit dem Induktor aufgenommene Lösungsvolumen angesehen werden könne. Durch Bestimmung des Verhältnisses ^{131}I/^{14}C im induzierenden Medium und in der Amöbe (nach Abwaschen der nur an die Oberfläche adsorbierten Substanzmengen) erhält man somit ein Maß für die Effektivität der Pinocytose in der Abtrennung und Anreicherung der induzierenden Substanz.

Wie Tab. 2 zeigt, ist die Pinocytose in dieser Beziehung bemerkenswert effektiv. Der relative Albumingehalt des von der Amöbe aufgenommenen Materials ist mindestens 10mal höher als der der zur Induktion verwendeten Lösung.

Auch in absolutem Maß ist die aufgenommene Substanzmenge erstaunlich groß, indem das im 20 min langen Pinocytosezyklus in die Amöbe beförderte Protein 40% des Trockengewichtes der Amöbe entspricht. Im Gegensatz dazu steht das miteingeschlossene Flüssigkeitsvolumen von nur etwa 2% des Amöbenvolumens.

So, wie die Pinocytose unter den hier gewählten, für massive Adsorption und Induktion optimalen

Tabelle 2. *Amoeba proteus*, Pinocytose in 2% ^{131}I-Serumalbumin, 0,04% ^{14}C-Glukose bei pH 4,4. Pinocytoseintensität (für 1 Amöbe) bestimmt durch Kanalzählung an 10 Amöben durch 20 min (Kolonne 8). Werte für Protein- und Glukoseaufnahme gelten für 1 Amöbe (gemessen an 30 Amöben). Die Induktorlösung enthielt pro μl: $5,8 \times 10^2 \mu$c ^{131}I, $7,7 \times 10^2 \mu$c ^{14}C. Zählerwerte gegeben in c.p.m. (counts per min)

Temperatur des Induktors	Kanalzahl	pH der Waschlösung	Immersionszeit (min)	Proteinaufnahme			Glukoseaufnahme		
				c.p.m.	μc $\times 10^6$	μg $\times 10^4$	c.p.m.	μc $\times 10^6$	μg $\times 10^4$
24.1	47	4.4	1	188 ± 28	325 ± 42	1118 ± 166	3.6 ± 0.6	14 ± 4.4	0.70 ± 0.12
			15	353 ± 65	620 ± 56	2100 ± 386	11 ± 5.6	41 ± 8	2.15 ± 1.09
		6.2	1	54 ± 11	93 ± 18	321 ± 65	0.3 ± 0.2	1.2 ± 0.9	0.06 ± 0.04
			15	156 ± 32	270 ± 38	928 ± 190	7.2 ± 0.2	27 ± 6	1.40 ± 0.04
3.8	0	4.4	1	166 ± 12	287 ± 39	987 ± 71	3.3 ± 1.2	12 ± 4	0.64 ± 0.23
			15	200 ± 26	346 ± 43	1190 ± 155	6.4 ± 1.1	24 ± 6	1.25 ± 0.22
		6.2	1	17 ± 3.5	29 ± 6	101 ± 21	1.7 ± 0.7	6.5 ± 4.2	0.33 ± 0.14
			15	60 ± 18	104 ± 19	367 ± 110	4.3 ± 0.5	16 ± 5	0.88 ± 0.10

^{131}I/^{14}C im Induktor: 0,755.

An Oberfläche adsorbiert (μc $\times 10^6$): ^{131}I 270, ^{14}C 10; ^{131}I/^{14}C: 27; Protein-Anreicherungsfaktor: 36x. Durch Pinocytose aufgenommen (μc $\times 10^6$): ^{131}I 169, ^{14}C 16; ^{131}I/^{14}C: 11; Protein-Anreicherungsfaktor: 14 $\times$ [15]

Versuchsbedingungen verläuft, entspricht sie jedenfalls nicht ihrem Namen; sie ist kein „Trinken der Zelle".

In der Bewertung dieses Befundes muß man sich jedoch vor Augen halten, daß die Pinocytose unter den Bedingungen dieser Versuche kaum mehr als normale Zellenaktivität betrachtet werden kann. Die Ingestion so großer Proteinmengen in 20 min hat leicht beobachtbare pathologische Wirkungen, welche nicht alle Zellen überleben. Solche „Spitzenpinocytose" ist durch hochkonzentrierte Induktorlösungen mit pH-Werten außerhalb des physiologischen Bereiches bedingt. Andererseits kennen wir, wie schon erwähnt, eine „Residualpinocytose", die ohne spezifische Induktion zu verlaufen scheint, vielleicht nur der Membraneinschmelzung dient und jedenfalls als permanent verlaufender, höchst normaler Lebensprozeß gedeutet werden muß. Wo zwischen diesen beiden Extremen der eventuelle Bereich einer physiologisch signifikanten, von der Amöbe gut und permanent vertragenen pinocytischen Substanzzufuhr zu suchen ist, wäre noch durch geeignete Versuche zu prüfen.

Vorläufige und unsystematische Fütterungsversuche mit ausschließlich pinocytischer Nahrungszufuhr haben jedenfalls ergeben, daß die typische Hungerdegeneration der Amöben beträchtlich hinausgeschoben werden kann. Es ist aber bis jetzt noch nie gelungen, durch Pinocytose ernährte Amöben zur Teilung zu bringen.

C. Die endoplasmatische Vakuole

Sobald die pinocytischen Vakuolen sich von der Oberflächenmembran gelöst haben und ihr „Eigenleben" im Cytoplasma beginnen, erleiden sie eine Reihe von Veränderungen, von denen nicht alle morphologisch nachweisbar sind. Vom physiologischen Standpunkt aus gesehen, muß diese vakuoläre Phase wohl als der entscheidende Abschnitt der Endocytose angesehen werden. Während dieser Phase muß die invaginierte Zellmembran, die jetzt die Vakuolen auskleidet, irgendwie so umgewandelt werden, daß sie ihre ursprüngliche weitgehende Impermeabilität verliert und den Kontakt zwischen den vakuolisierten Substanzen und dem Stoffwechselsystem der Zelle ermöglicht.

Der anscheinend am nächsten liegende Weg zur Realisierung dieses Kontaktes wäre das Verschwinden der Membran durch enzymatische oder andere Einflüsse, wodurch der Vakuoleninhalt ohne weiteres bloßgelegt würde. Ein solcher Vorgang ist denn auch von BENNETT [4] in seiner oben erwähnten Hypothese angenommen worden. Die einzige mir bekannte wohlbelegte experimentelle Bestätigung dieser Annahme rührt von WOHLFARTH-BOTTERMANN [48] her, der in *Hyalodiscus simplex* zwar nicht das Verschwinden, wohl aber eine Fragmentierung der Membran pinocytischer Vakuolen nachgewiesen hat. Ähnliche Befunde an anderen Objekten wurden auf Kongressen häufig diskutiert mit dem Ergebnis, daß es sehr schwierig zu sein scheint, die Möglichkeit von Artefakten bei solchen Befunden gänzlich auszuschließen. Eine endgültige Entscheidung dieser ungemein wichtigen Frage wäre äußerst wünschenswert.

Bei *Amoeba proteus* und der ihr nahestehenden *Chaos chaos* ist es uns nie gelungen, wirklich überzeugende Belege für Verschwinden oder Fragmentierung der Vakuolenmembran zu finden, wohl aber haben wir eine Reihe von Indikationen dafür, daß die digestiven Veränderungen pinocytisch zugeführter Substanzen sich im allgemeinen innerhalb der Vakuolen abspielen. Wir neigen daher bis auf weiteres zu der Anschauung, daß die zellphysiologisch sehr weitreichende Annahme des gänzlichen Verschwindens der Membranbarriere noch verfrüht wäre, und beschränken uns auf die Betrachtung der anderen Umwandlungen, welche an Pinocytosevakuolen beobachtet worden sind. Es handelt sich um folgende: 1. Sichtliche Veränderungen der Vakuolenwand und ihrer Auskleidung. — 2. Oberflächenvergrößerung durch Unterteilung und Verzweigung. — 3. Vakuolenkoaleszenz. — Alle drei Erscheinungen sind wichtig für das Problem des Substanztransportes durch die Vakuolenmembran.

I. Vakuolenwand

Wie schon früher erwähnt, ist das Plasmalemma der Amöben eine dreischichtige Membran, welche — wenngleich möglicherweise mit einiger Vergröberung [49] — im wesentlichen der Definition einer „unit-membrane" entspricht. Alle Beobachter sind sich darüber einig, daß das Plasmalemma zumindest während der initialen Invaginationsphasen unverändert bleibt, so daß zwischen dem „Außenplasmalemma", und der Bekleidung der Invaginationskanäle sowie der frisch abgeschnürten Vesikel elektronenoptisch keine Unterschiede nachgewiesen werden können [8, 48].

Dasselbe gilt bis zu einem gewissen Grad für das elektronenoptische Bild der dem Plasmalemma überlagerten Mucoidschicht. Auch hier findet man oft, und anscheinend besonders dann, wenn keine massive Adsorption stattgefunden hat, Aussehen und Dichte des Mucoid-Haarkleides in Kanälen und Vesikeln unverändert bewahrt [8, 48]. Ist die Mucoidschicht aber durch Adsorption mit Fremdsubstanzen beladen, so scheint sie instabil zu werden, und man beobachtet oft ein Abstoßen der Mucoidschicht vom Plasmalemma. Diese „Häutung" kann schon kurz nach der Adsorption eintreten, noch bevor die Invagination stattgefunden hat [7], öfter geschieht sie im Invaginationskanal [23], und zur Regel wird sie nach der Abschnürung der Pinocytosevakuolen [8, 32]. Der Häutungsprozeß hinterläßt die Vakuolenmembran nackt, aber sonst anscheinend unverändert. Da es unbekannt ist, ob die Gegenwart der Mucoidschicht die Permeabilitätseigenschaften der Zellmembran beeinflußt, kann auch nichts darüber ausgesagt werden, ob die Permeabilität durch die Häutung verändert wird. Die abgestoßene Schicht von Mucoid und Induktor liegt frei im Lumen der Vakuole, oft unter Beibehaltung der ursprünglichen Flächenstruktur (Abb. 9).

Die Abstoßung der Mucoidschicht scheint zeitlich mit dem von Müller und Rappay [31] berichteten Verlust der PAS-Reaktion zusammenzufallen.

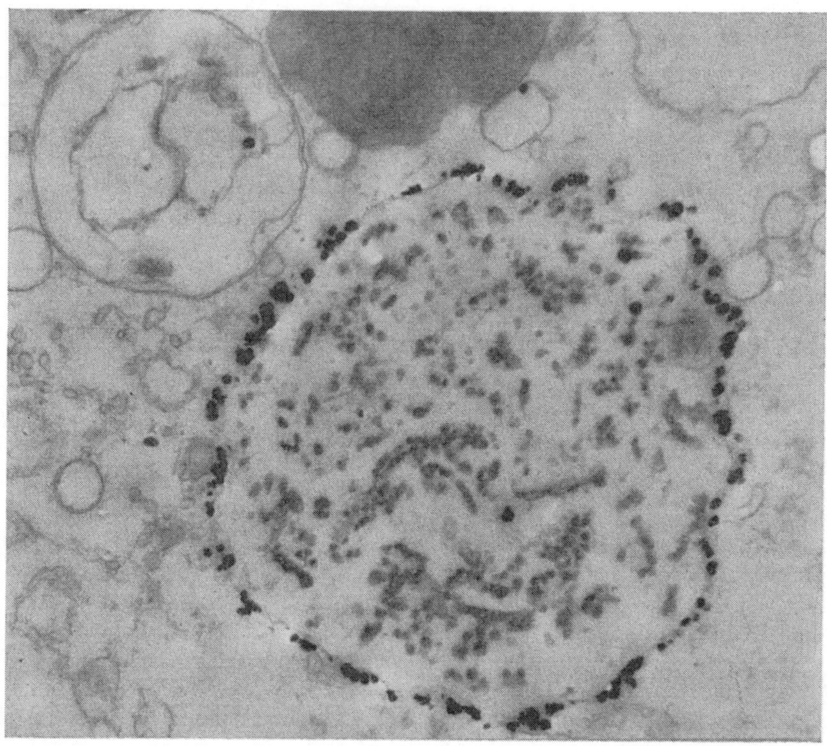

Abb. 9. Vakuole 1 Std nach Alcian Blue-Immersion. Das Lumen enthält große Mengen des Induktor-Mucoidkomplexes, zum Teil noch linear geordnet, wie er von der Membran abgestoßen wurde. Die Außenseite der Vakuolenwand ist besetzt mit den von vielen Autoren beschriebenen dichten Körnchen unbekannter Zusammensetzung. Vergrößerung:12000 mal [23]

II. Vergrößerung der Vakuolenoberfläche

Brandt und Pappas [8] und Hayward [23] beschreiben die Bildung von Verzweigungen und fingerförmigen Erweiterungen der Pinocytosevakuolen, die zu einer beträchtlichen Vergrößerung der Vakuolenoberfläche führen. Diese Digitation kann, wie Abb. 10 zeigt, ebenso wie die Mucoidhäutung schon vor der Abschnürung der Vakuolen am Grunde des Invaginationskanals auftreten.

In späteren Stadien kann dieses Phänomen zu labyrinthartigen Strukturen führen [8, 23], in welchen die interdigitierten Lumina der Pinocytosevakuolen nur erkannt werden können, wenn sie durch geeignete Induktorsubstanzen gekennzeichnet sind.

Ein anderer zur Oberflächenvergrößerung führender Vorgang ist die Unterteilung der Vakuolen durch Bildung kleiner (0,2—0,5 μ Durchmesser) Sekundärvakuolen. Sie entstehen an der Peripherie der ursprünglichen Vakuolen durch einen Prozeß, der von MERCER [29] „Knospung", von ROTH [43] „Mikropinocytose" genannt wurde. ROTH hat diese Erscheinung an den Nahrungsvakuolen von Amöben studiert, NACHMIAS und MARSCHALL [32] haben ihn an Pinocytosevakuolen

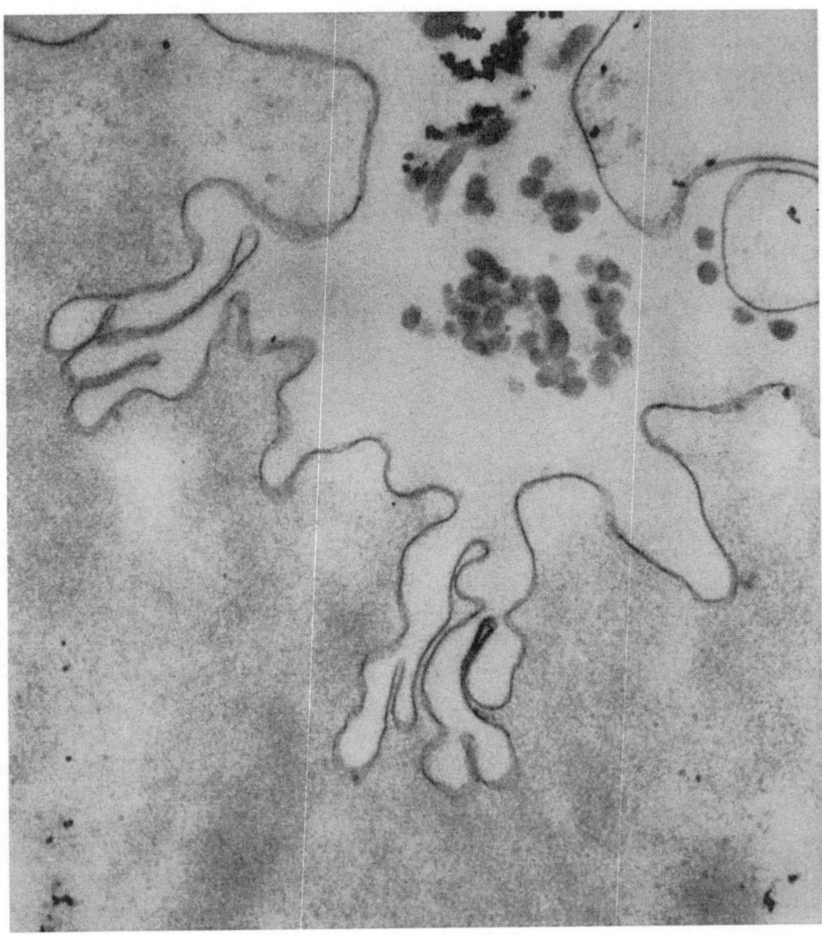

Abb. 10. Boden eines Pinocytosekanales, 5 min nach Immersion in Alcian Blue. Die finger-förmigen Fortsätze sind vom früher erwähnten dichten Cytoplasma umgeben. Im Lumen Induktor-Mucoidkomplex. Vergrößerung: 33000mal [23]

wiedergefunden. CHAPMAN-ANDRESEN und NILSSON [16] meinen, daß solche Mikropinocytosevakuolen schon an den Invaginationskanälen auftreten können.

ROTH [43], der sich am eingehendsten mit der Mikropinocytose beschäftigt hat, kommt auf Grund seiner Berechnungen zu der Ansicht, die dadurch herbeigeführte Oberflächenvergrößerung der Vakuolen sei so groß (ein Faktor von etwa 300), daß sie allein genüge zur

Erklärung einer wesentlichen Erhöhung der Permeation des Vakuoleninhaltes. Seiner Meinung nach ist es also unnötig, spezifische, zu erhöhter Permeabilität führende Veränderungen der Membranstruktur anzunehmen.

Derartige Veränderungen sind jedoch morphologisch gefunden oder zumindest angedeutet worden. WOHLFARTH-BOTTERMANN [48] schreibt (von *Hyalodiscus*): „Nicht selten findet man aber auch bei deutlich quer geschnittenen Membranen von Pinocytose-Bläschen einen kontinuierlich abnehmenden Elektronenkontrast, wobei gleichzeitig die Doppelschichtigkeit der Membranen immer undeutlicher wird. Es muß daher mit der Möglichkeit gerechnet werden, daß die durch Pinocytose ins Endoplasma gelangten Zellmembranen hier in einfache cytoplasmatische Membranen mit gleichem oder geringerem Elektronenkontrast übergehen."

Derartige Beobachtungen nähern unser Problem dem umfassenden Fragenkomplex der intracellulären Membranenbildung, der ja für das Verständnis von so beträchtlichen Oberflächenvergrößerungen wie den hier angenommenen, von entscheidender Bedeutung wäre. Leider wissen wir sehr wenig darüber [49], und es kann daher nicht näher darauf eingegangen werden.

Eine alternative Deutung der Mikropinocytose [23] wird im nächsten Abschnitt erwähnt.

III. Vakuolenverschmelzung

Ebenso physiologisch wichtig und morphologisch auffallend wie die eben besprochenen Vorgänge des Oberflächen*wachstums* ist die Erscheinung der Vakuolenverschmelzung, die zu einer Oberflächen*reduktion* führt. Die Verschmelzung von membranbegrenzten Cytoplasma-Inklusionen ist nicht auf Endocytose-Vakuolen beschränkt und ist meiner Meinung nach ein ganz wesentlicher Vorgang der intracellulären Dynamik, da sie möglicherweise einen der wichtigsten Verbindungspfade im Transport der Metaboliten darstellen könnte. Die Verschmelzung von Vakuolen und Organellen unterschiedlicher Herkunft wird später besprochen werden, im Zusammenhang mit der Beziehung zwischen Endocytosevakuolen und Lysosomen. Vorläufig wollen wir uns auf die Verschmelzung der Pinocytosevakuolen untereinander beschränken.

Gleich nach der Aufnahme sind die Pinocytosevakuolen bei *Amoeba proteus* und *Chaos chaos* alle ziemlich gleich groß, etwa 2 μ im Durchmesser. Diese Gleichartigkeit dauert aber nicht lange; wenn die Vakuolen durch den Cytoplasmastrom in engen Kontakt miteinander gebracht werden, verschmelzen sie oft, und das Ergebnis ist das Bild von Vakuolen sehr verschiedener Größe, welches man bei der Betrachtung von mit fluorescierenden Proteinen gefütterten Amöben normalerweise sieht [25]. Im Zeitrafferfilm kann die Verschmelzung sowohl bei Amöben wie bei anderen Zellen direkt beobachtet werden.

Wir sehen also, mehr oder weniger gleichzeitig, zwei entgegengesetzt gerichtete Prozesse im Endoplasma verlaufen: der eine führt zum Wachs-

tum, der andere zur Verminderung der Oberfläche der Pinocytose-
vakuolen.

Mit dem letzteren Vorgang, dem der Oberflächeneinschmelzung,
hat sich HAYWARD [23] gelegentlich der Diskussion seiner elektronenoptischen Befunde recht eingehend beschäftigt. Er betrachtet drei

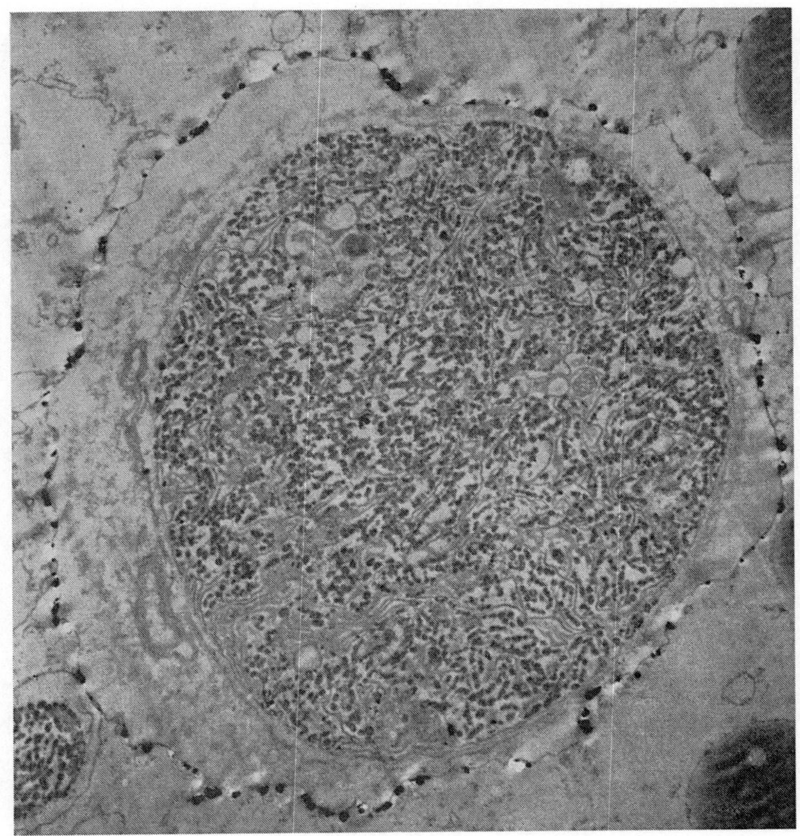

Abb. 11. Große Verschmelzungsvakuole 6 Std nach Alcian Blue-Immersion. Die zentrale
sphärische Masse besteht aus Induktor-Mucoidkomplex und Membran. Die Außenwand
besetzt mit den dichten Körnchen (vgl. Abb. 9). Vergrößerung: 9000mal [23]

verschiedene Erscheinungen als wahrscheinliche Folgen der Vakuolenverschmelzung:

Eine ist die wiederholte Einfaltung und Überführung der Membran
in das Lumen der Vakuole, wodurch Bilder von der Art des in Abb. 11
gezeigten entstehen.

Die zweite Möglichkeit ist die Abstoßung überschüssiger Membran
durch Bildung mikropinocytischer Vesikel (Abb. 12). Diese Interpretation
des Mikropinocytoseprozesses ist also sehr verschieden von ROTHs [43]
Anschauungen.

Die dritte Möglichkeit der Umwandlung überschüssiger Membran ist nach Haywards Vorschlag die des enzymatischen Abbaues. Hayward begründet diese Vermutung mit dem Auftreten gewisser laminärer Strukturen in den Vakuolen, welche er als freie Phospholipide ansieht. Hayward nimmt an, daß diese Strukturen von aus Protein und Phospholipid bestehenden Membranen herstammen, und durch selektive intravakuoläre Verdauung des Proteinanteils gebildet sein mögen.

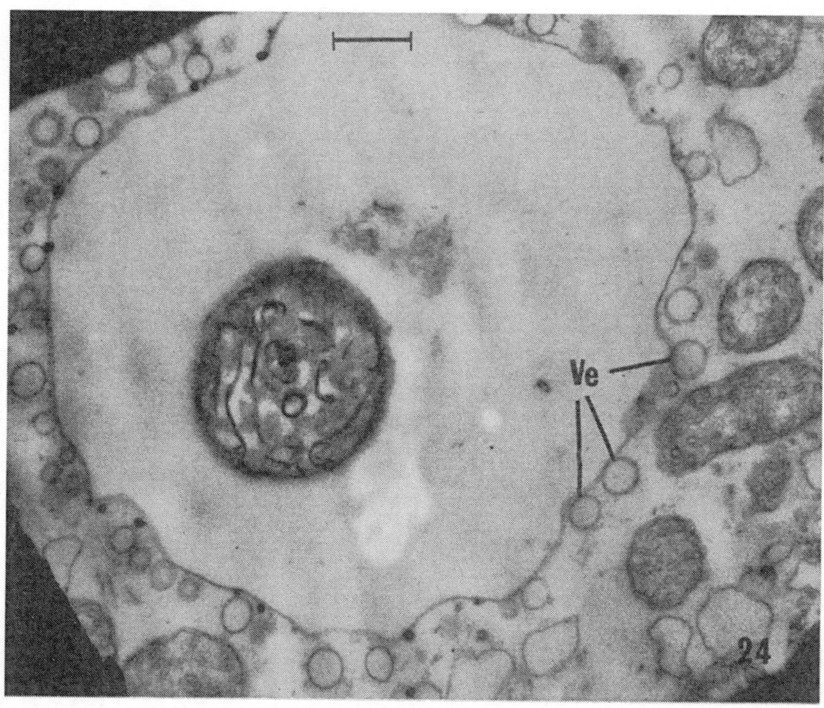

Abb. 12. Verschmelzungsvakuole 1 Std nach Induktion mit NaCl. Kleiner Zentralkörper mit membranösen Residuen. Die Vakuole ist umgeben von kleinen Vesikeln, die sich auch um Nahrungsvakuolen zu scharen pflegen. Vergrößerung: 14000mal [*23*]

Wie man sieht, sind alle diese Vermutungen basiert auf der Grundannahme, daß die Membranbarriere zwischen Vakuoleninhalt und Cytoplasma bestehen bleibt. Auch in dem Fall eines vermuteten enzymatischen Membranabbaues wird der Prozeß als intravakuolär angesehen.

Wenn man an dieser Grundannahme festhält, ist man also nicht in der Lage, die Pinocytose als Alternative zu den verschiedenen Prozessen des Transmembran-Transports durch die Zelloberfläche aufzufassen. Eher wäre die Pinocytose als Hilfsmechanismus des Membrantransports zu betrachten, welcher durch die Translokation großer Anteile der Zelloberfläche ins Zellinnere und darauf folgende Membranumwandlungen die Transportprozesse in ein anderes zelluläres Niveau verlegt, aber die entscheidenden physiologischen Merkmale des Membrantransportes unverändert läßt, nämlich vor allem die Selektivität und die energetische Kopplung an den Zellenstoffwechsel. Hierin unterscheidet sich diese Betrachtungsweise wesentlich von der früher erwähnten andern Alternative, welche die einfache Freisetzung des Vakuoleninhaltes durch Membranverdauung oder Membranfragmentierung annimmt.

Welche dieser Annahmen sich auch schließlich als richtig erweisen möge, so existieren jedenfalls Anzeichen dafür, daß die pinocytische Einfuhr von Material ins Zellinnere metabolisch wirkungsvoll sein kann.

CHAPMAN-ANDRESEN und HOLTER [14] haben die metabolische Erfassung pinocytisch zugeführter Glukose an *Chaos chaos* untersucht. Wie schon früher angeführt, ist diese Amöbe für Glukose sehr impermeabel, und Glukose wird, da sie kein Pinocytoseinduktor ist, normal auch nicht pinocytisch aufgenommen. Bietet man aber Glukose zusammen mit einem induzierenden Protein, so geht, wie früher beschrieben, eine gewisse Menge davon zusammen mit dem Protein in die Pinocytosevakuolen ein. Mit Hilfe einer autoradiographischen Methode untersuchten CHAPMAN-ANDRESEN und HOLTER die intrazelluläre Lokalisation so dargebotener, mit ^{14}C markierter Glukose. Sie fanden, daß die Radioaktivität schon sehr bald nach der Pinocytose auch außerhalb der Vakuolen im Cytoplasma nachgewiesen werden kann. In Respirationsversuchen im Cartesianischen Taucher wurde die Menge von ausgeschiedenem $^{14}CO_2$ gemessen, und es wurde festgestellt, daß die Amöbe imstande war, die Glukose ungefähr gleichzeitig mit dem Auftreten im Cytoplasma metabolisch zu erfassen. Hieraus wurde geschlossen, daß die Vakuolenmembran im Endoplasma in der Richtung erhöhter Glukosepermeabilität verändert worden sein mußte.

In Anbetracht unseres heutigen Wissens über Mikropinocytose und Digitation ist diese Annahme nicht mehr einwandfrei. Erstens würde eine Glukosedistribution in vielen mikropinocytischen Vesikeln autoradiographisch von einer echten diffusen Verteilung im Cytoplasma nicht unterscheidbar sein, und zweitens macht, wie von ROTH [43] angeführt, die Oberflächenvergrößerung die Annahme spezifischer Permeabilitätsveränderungen überflüssig. Außerdem ist zu bedenken, daß die Glukose, selbst wenn sie zweifellos in den glykolytischen Stoffwechsel eingegangen ist, nicht notwendigerweise die Vakuolenmembran als Glukose passiert hat; es kann dies auch in Form eines Abbauproduktes geschehen sein.

Versuche dieser Art bieten also nur wenig Information in bezug auf den Mechanismus der Membranpassage .Sie zeigen aber eindeutig, daß die Passageverlegung ins Zellinnere die erstaunlich rasche Ausnützung einer Substanz vermittelt, die für die Zelle ursprünglich fast unzugänglich war.

D. Enzymatische Aspekte der Vakuolenverschmelzung

Der im vorigen Abschnitt genannte Fall der Glukoseausnützung wirft die folgende Frage auf: Wenn wir den Verdacht hegen müssen, daß nicht Glukose selbst, sondern z. B. ein phosphoryliertes Zwischenprodukt tatsächlich die Membran passiert; oder, wenn im Falle von hochmolekulären Stoffen wie Proteinen das Molekül erst teilweise abgebaut werden muß, um passieren zu können — wie kommen diese Abbaureaktionen zustande? Es gibt keine experimentellen Anhaltspunkte dafür, daß die Zellmembran der Amöben mit einer Vielfalt katabolischer Enzyme ausgerüstet sei (nur in wenigen Fällen wird von einer solchen Lokalisation berichtet; vgl. [5]), und dennoch scheint, sobald dieselbe Membran sich im Endoplasma befindet, eine lebhafte enzymatische Wirksamkeit eingeleitet zu werden. Dieses Problem ist natürlich nicht neu; es handelt sich essentiell um die alte Frage, wie die Verdauung des Inhaltes der Nahrungsvakuolen zustandekommt.

Die Antwort auf diese Frage scheint in den Beziehungen zwischen endocytischen Vakuolen und Lysosomen zu liegen. Neuere Untersuchungen der bemerkenswerten morphologischen Variabilität der Lysosomen (vgl. Übersicht von DE DUVE [19]) haben etwa folgenden Sachverhalt ergeben: Die „ursprünglichen" Lysosomen, dichte strukturlose Körper, welche DE DUVEs [18] ursprünglicher Definition entsprechen (vgl. auch NOVIKOFF [33, 34]), haben die Tendenz mit einer Reihe von anderen cytoplasmatischen Organellen zu verschmelzen. Hierdurch entstehen morphologisch sehr verschiedenartige, „zusammengesetzte" Inklusionskörper, die jedoch alle die enzymatischen Charakteristika der Lysosomen zeigen und daher unter den erweiterten Lysosomenbegriff einzuordnen sind. Zu den häufigsten Verschmelzungsrecipienten der Lysosomen gehören die Endocytosevakuolen, und man kann daher annehmen, daß sie auf diesem Wege ihre Ausrüstung mit hydrolytischen Enzymen empfangen. Soweit ich weiß, war ROSE [42] der erste, der diese wichtige Rolle der Lysosomen und der Organellenverschmelzung erkannte.

Diese ganze Hypothese führt zu einer Reihe von sehr interessanten Fragestellungen in bezug auf die intrazelluläre Physiologie der Enzyme, auf die hier nicht näher eingegangen werden kann. Nur auf zwei Aspekte sei besonders hingewiesen.

Der eine betrifft den Umstand, daß die Organellenverschmelzung einen Weg anzeigt, auf dem hochmolekulare Substanzen (in diesem Fall Enzyme) unter Umgehung der Membranbarriere von einem Zellort zum andern transportiert werden können. Der Vorgang scheint von großer prinzipieller Bedeutung auch für wichtige Phasen des Exkretionsmechanismus und verdient eingehendes Studium. Soweit ich weiß, nehmen die gangbaren Membranmodelle keine spezielle Rücksicht auf eventuelle Membraneigenschaften, die eine strukturelle Grundlage für die Tatsache der Membranverschmelzung abgeben könnten.

Der andere Hinweis betrifft die Frage, ob die Pinocytose von bestimmten Substanzen, z. B. Proteinen, als spezifischer Stimulus für die Lysosombildung der Zelle wirken kann. In Anbetracht der großen Ansprüche an das lysosomale Enzymsystem, welche die pinocytische Substanzzufuhr darstellen muß, war eine solche Annahme naheliegend, und sie ist von LAGUNOFF [26] experimentell geprüft worden. Die Untersuchung wurde an *Chaos chaos* mit Serumalbumin, Natriumchlorid und Alcian Blue als Pinocytoseinduktoren angestellt, und das Ergebnis war, daß in keinem Fall eine markante spezifische Neubildung hydrolytischer Enzyme nachweisbar war. Es scheint also, daß die Amöben mit einem mehr oder weniger konstanten „Vorrat" lysosomaler Enzyme ihr Auslangen finden, und daß trotz weitgehender lokaler Umdisponierung durch Organellenverschmelzung die totale vorhandene Enzymmenge nur langsam durch langfristige Synthese verändert wird.

E. Die Exkretion bei Amöben

Eine ähnliche Sequenz von Vakuolenverschmelzungen kennzeichnet die Bildung jener großen zusammengesetzten Vakuolen, in welchen verschiedene Abfallsprodukte als Vorstufe zur Exkretion aufgesammelt werden. Dieser Vorgang wurde vor vielen Jahren von ANDRESEN und HOLTER [1] beobachtet, und schon damals wurde darauf hingewiesen, daß es sich um eine Verschmelzung nicht nur von verschiedenen Vakuolentypen, sondern auch von Vakuolen und nichtvakuolisierten Organellen handle. Die so gebildeten Mischvakuolen wurden „Defäkationsvakuolen" genannt.

Heute wissen wir, daß die Defäkationsvakuolen aller Wahrscheinlichkeit nach identisch sind mit den "compound lysosomes" oder "residual bodies" [19], die nicht nur in Amöben, sondern in einer ganzen Anzahl anderer Zelltypen beschrieben worden sind. Wir wissen auch, daß solche Sammelvakuolen nicht nur durch

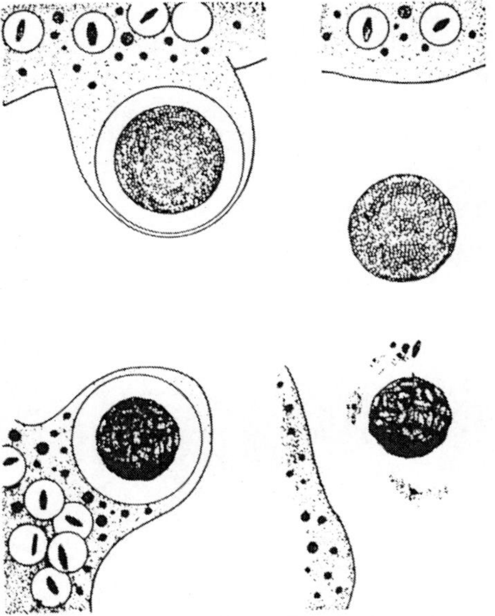

Abb. 13. Zwei Fälle von Defäkation in *Chaos chaos* (schematisch). Links: Die Defäkationsvakuolen liegen in kleinen Pseudopodien. Die obere enthält Nahrungsreste, die untere ein Krystall-Konglomerat, hervorgegangen aus der Verschmelzung von Krystallvakuolen (wie solche an der Basis der Pseudopodien dargestellt sind). Rechts: Der Vakuoleninhalt ist ausgestoßen und liegt frei im Medium [1]

Verschmelzung entstehen können, sondern auch durch die Bildung von Einschlußmembranen in einem Prozeß, der von HAYWARD [23] „Sequestration" genannt wurde.

Das endliche Schicksal solcher Vakuolen ist in Amöben ihre Ausstoßung durch den von ANDRESEN und HOLTER [1] beschriebenen Defäkationsvorgang (Abb. 13). Die Vakuole wird durch einen kräftigen Cytoplasmastrom gegen das Plasmalemma geführt, oft an der Spitze eines kleinen Pseudopodiums. Wenn Vakuolenmembran und Plasma-

lemma in engen Kontakt gebracht werden, geschieht eine im Lichtmikroskop nicht genau zu beobachtende Verschmelzung, wodurch der
Inhalt der Vakuole freigelegt wird, ohne daß gleichzeitig das Cytoplasma
gegen das Außenmedium entblößt wird. Vermutlich durch Oberflächenspannung wird der Vakuolengrund sehr rasch auf das Niveau der Zelloberfläche gebracht und der Membraninhalt dadurch ausgestoßen.

Oberflächlich gesehen erinnert der Defäkationsvorgang an eine in-
vertierte Invagination, und die Defäkation der Amöben kann sohin wohl
als ein infolge seiner verhältnismäßigen Heftigkeit vielleicht etwas
spezieller Fall von Exocytose betrachtet werden.

In vielen anderen Zelltypen sind Fälle von „umgekehrter Pinocytose"
beschrieben worden, entweder als postulierter Sekretionsmechanismus
hochmolekulärer Zellprodukte, oder als Transportprozeß, bei welchem
Substanzen an einer Zelloberfläche durch Pinocytose aufgenommen und
auf der Gegenseite durch Exocytose wieder ausgeschieden werden. Die
Beschreibung gründet sich in den meisten dieser Fälle auf die statischen
Bilder des Elektronenmikroskopes, und man kann daher kaum beurteilen,
ob der eigentliche Ausscheidungsvorgang an die rasch und heftig verlaufende Amöbendefäkation erinnert, oder ob es sich um eine echte In-
vertierung der viel langsamer ablaufenden Invagination und Vesikulation handelt.

Literatur

[1] ANDRESEN, N., and H. HOLTER: C. R. Carlsberg, Ser. chim. **25**, 107 (1945).
[2] BAIRATI, A., and F. E. LEHMANN: Exp. Cell Res. **5**, 220 (1953).
[3] BELL, L. G. E.: J. theoret. Biol. **3**, 132 (1962).
[4] BENNETT, H. S.: J. biophys. biochem. Cytol. **2**, Suppl. 99 (1956).
[5] BIRNS, M.: Exp. Cell Res. **20**, 202 (1960).
[6] BRANDT, P. W.: Exp. Cell Res. **15**, 300 (1958).
[7] —, and G. D. PAPPAS: J. biophys. biochem. Cytol. **8**, 675 (1960).
[8] — — J. cell. Biol. **15**, 55 (1962).
[9] BRATIANU, S., et A. LLOMBART: C. R. Soc. Biol. (Paris) **101**, 299 (1929).
[10] CHAPMAN-ANDRESEN, C.: C. R. Carlsberg, Ser. chim. **31**, 77 (1958).
[11] — Soc. Exptl. Biol. Joint British-Dutch-Scandinavian Meeting, Copenhagen (1959).
[12] — C. R. Carlsberg **33**, 73 (1962).
[13] — Progress in Protozoology. Publ. House Czechoslovak Acad. Sci. Eds.
Ludvik, Lom und Vavra 1963.
[14] —, and H. HOLTER: Exp. Cell Res. Suppl. **3**, 52 (1955).
[15] — — C. R. Carlsberg **34**, 211 (1964).
[16] —, and J. R. NILSSON: Exp. Cell Res. **19**, 631 (1960).
[17] —, and D. M. PRESCOTT: C. R. Carlsberg, Ser. chim **30**, 57 (1956).
[18] DE DUVE, C.: Subcellular Particles. New York: Ronald Press (1959).
[19] — Ciba Foundation Symposium "Lysosomes", S. 126 u. 411, London:
J. A. Churchill Ltd. 1963.
[20] DE TERRA, N., and R. C. RUSTAD: Exp. Cell Res. **17**, 191 (1959).
[21] GERARD, P., and R. CORDIER: Biol. Rev. **9**, 110 (1934).
[22] GOSSELIN, R. E.: J. gen. Physiol. **39**, 625 (1956).
[23] HAYWARD, A. F.: C. R. Carlsberg **33**, 535 (1963).
[24] HOLTER, H.: Int. Rev. Cytol. **8**, 481 (1959).
[25] —, and J. M. MARSHALL JR.: C. R. Carlsberg, Ser. chim. **29**, 7 (1954).
[26] LAGUNOFF, D.: C. R. Carlsberg **34**, in press.
[27] LEHMANN, F. E.: Ergebn. Biol. **21**, 88 (1958).
[28] MAST, S. O., u. W. L. DOYLE: Protoplasma (Wien) **20**, 555 (1934).
[29] MERCER, E. H.: Proc. roy. Soc. B. **150**, 216 (1959).

[30] METCHNIKOFF, E.: Biol. Zbl. 3, 560 (1883).

[31] MÜLLER, M., és G. RAPPAY: Magy. Tud. Akad., Biol. orv. Tud. Osztal. Közl. 3, 81 (1959).

[32] NACHMIAS, V. T., and J. M. MARSHALL JR.: IUB/IUBS Symposium "Biological structure and function" II, 605 (1962).

[33] NOVIKOFF, A. B.: Biol. Bull. 119, 287 (1960a).

[34] — Developing Cell Systems and their Control. New York: Ronald Press (1960b).

[35] — The Cell II. New York. Acad. Press 1961.

[36] ODOR, L. D.: J. biophys. biochem. Cytol. 2, Suppl. 105 (1956).

[37] PALADE, G. E.: J. biophys. biochem. Cytol 2, Suppl. 85 (1956).

[38] — Ciba Foundation Symposium "Lysosomes", S. 412. London: J. A. Churchill Ltd. 1963.

[39] PAPPAS, G. D.: Ann. N. Y. Acad. Sci. 78, 448 (1959).

[40] POLICARD, A., and M. BESSIS: C. R. Acad. Sci. (Paris) 246, 3194 (1958).

[41] REICHSTEIN NILSSON, J.: Unveröffentl. Beobachtung.

[42] ROSE, G. G.: J. biophys. biochem. Cytol. 3, 697 (1957).

[43] ROTH, L. E.: J. Protozool. 7, 176 (1960).

[44] SCHNEIDER, L., u. K. E. WOHLFARTH-BOTTERMANN: Protoplasma (Wien) 51, 377 (1959).

[45] SCHUMAKER, V. N.: Exp. Cell. Res. 15, 314 (1958).

[46] SPEK, J., u. G. GILLISSEN: Protoplasma (Wien) 37, 259 (1943).

[47] WITTEKIND, D.: Naturwissenschaften 50, 270 (1963).

[48] WOHLFARTH-BOTTERMANN, K. E.: Protoplasma (Wien) 52, 58 (1960).

[49] — Forschungsber. des. Landes Nordrhein-Westfalen, Nr. 1247, Köln: Westdeutscher Verlag 1963.

[50] WOLPERT, L., and C. H. O'NEILL: Nature (Lond.) 196, 1261 (1962).

Summary

The initial phase of pinocytosis has two steps: 1. adsorption of the substance to be ingested (the pinocytosis-inducer) to the mucoid layer of the plasmalemma; 2. invagination of the loaded membrane, leading to the formation of pinocytosis channels and the detachment of pinocytic vacuoles which contain the adsorbate and fluid medium. Adsorption leads to a selective accumulation of the inducer. The ingested vacuoles, therefore, contain considerable amounts of solid, but relatively little fluid.

Pinocytosis does not continue indefinitely under a prolonged pinocytic stimulus. According to CHAPMAN-ANDRESEN this is due to an exhaustion of surface membrane available for invagination. The limiting value seems to be reached when about 50% of the surface has been internalized.

The pinocytic vacuoles undergo morphological changes which lead to an increase of the vacuolar surface, partly by digitation, partly by the detachment of secondary "micropinocytic" vesicles. These processes seem to be accompanied by a permeation of vacuolar contents into the groundplasm.

Pinocytic vacuoles coalesce with each other and with other components of the cytoplasm. Probably the most significant of these processes is the fusion with lysosomes which provides the vacuoles with an equipment of digestive hydrolases. It is not known whether high molecular substances always have to be broken down by such enzymes before they can permeate, or whether there also exists a mechanism for the direct transfer of unchanged large molecules. On the basis of present evidence it seems most probable that a membrane barrier is retained throughout the intracellular cycle. If this is assumed one is led to regard pinocytosis as being a form of transport not essentially different from the transport mechanisms of the cell surface, but rather as an auxiliary transferring surface processes to internalized membrane areas.

The final fate of pinocytic vacuoles seems to be excretion. Indigestible remnants of pinocytically ingested substances are observed as components of compound debris vacuoles. They must be assumed to have entered such vacuoles by means of vacuolar fusion, and are expelled from the cell by defaecation.

Diskussion

Wohlfarth-Bottermann: Ich danke Herrn Holter für seinen Vortrag, der uns den letzten Stand der Kenntnis über die Pinocytose vermittelt hat.

Ich darf vielleicht gleich auf Ihre Frage antworten, ob sich die Residualpinocytose auch morphologisch von der induzierten Pinocytose unterscheidet. Mit der Einschränkung, daß wir nicht die großen Amöben untersucht haben, sondern Amöben vom Limax-Typ, kann man das eigentlich bejahen; denn wir haben immer wirklich runde, abgeschnürte Blasen gesehen und nie diese typischen Pinocytoseblasen mit den zusammengedrückten Membranen, wie Sie sie gezeigt haben. Wie es bei den großen Amöben ist, können wir leider nicht sagen, weil die noch immer die nichtbewältigte Schwierigkeit der Fixation bieten.

Staubesand: Erlauben Sie mir, daß ich meinem eigenen Referat schon den ersten Teil in der Diskussion vorwegnehme; denn Ihre Frage nach Unterschieden zwischen der induzierten Pinocytose und der Residualpinocytose berührt sehr stark die Dinge, über die ich hier zu sprechen die Absicht hatte. de Duve hat ja den Vorschlag gemacht, alle Begriffe wie Phagocytose, Pinocytose, Mikropinocytose zusammenzufassen. Es kann keinem Zweifel unterliegen, daß ein wesentliches Merkmal, nämlich die Vereinnahmung von Oberflächenmembran in das Innere der Zelle, bei allen diesen Prozessen vorhanden ist. Wir glauben aber doch, daß zwischen der Pinocytose und der Mikropinocytose, d. h. der Pinocytose elektronenmikroskopischer Dimensionen, Differenzen bestehen, und zwar nicht nur morphologische Differenzen, sondern auch physiologische. Dazu darf ich drei Hauptpunkte nennen:

1. Zur Pinocytose lichtmikroskopischer Dimension sind erfahrungsgemäß in erster Linie freie Zellen, Zellen mit der Fähigkeit zur Pseudopodienbildung, befähigt.

2. Wenn man die Pinocytose dadurch aktiviert, daß man Zellen — wir haben mit Leukozyten gearbeitet[1] — in hochprozentige Eiweißlösungen bringt, dann wird die lichtmikroskopische Pinocytose aktiviert, nicht jedoch die Mikropinocytose.

3. Eines der wesentlichen morphologischen Merkmale der lichtmikroskopischen Pinocytose ist, daß sich unter Vermittlung von Pseudopodien am Rande der Zellen Vakuolen ausbilden, die nach und nach in das Zellzentrum wandern und sich dabei konzentrieren — oder sagen wir: die sich verändern — während die mikropinocytotischen Vesikel praktisch keinerlei Veränderung ihres Inhalts aufweisen.

Zwischen der lichtmikroskopischen Pinocytose und der Mikropinocytose bestehen sicherlich funktionelle Wechselbeziehungen. Sie haben Untersuchungen erwähnt, die zeigen, daß sich von den lichtmikroskopisch entstandenen Vakuolen Sekundärvesikel abschnüren. Das ist aber unseres Wissens bislang nur an Amöben beobachtet worden. Wir haben dasselbe jetzt auch bei menschlichen Leukozyten beobachtet.

Holter: Im wesentlichen sind wir ganz einig. Es ist sehr erfreulich, daß diese Phänomene, die wir bislang nur bei Amöben gefunden haben, jetzt von Ihnen auch bei Leukozyten gefunden wurden. Was die Eindickung betrifft, so können wir Ihnen umgekehrt mitteilen, daß wir das auch wiederum bei Amöben gefunden haben. Marshall und ich haben schon vor zehn Jahren durch Zentrifugierungsversuche in verschiedenen Stadien nach der Pinocytose nachgewiesen, daß die Pinocytosevakuolen immer schwerer werden. Daraus haben wir geschlossen, daß in den Pinocytosevakuolen der Amöben eine Eindickung vor sich geht.

Die Abschnürung der mikropinocytotischen Vesikel von der Hauptvakuole macht mir großes Kopfzerbrechen. Sie ist ohne jeden Zweifel da. Es erhebt sich wieder die Frage nach der Spezifität. Wenn man nämlich annimmt, wie Marshall es tut, daß die mikropinocytotischen Vesikel die seien, die den erwähnten Ionenaustausch, also den aktiven Transport, in der Amöbe besorgen, dann haben wir wiederum die Frage: Wie geschieht eigentlich die Abscheidung zwischen dem Gesamtinhalt der lichtmikroskopischen Großvakuole von dem Inhalt der mikropinocytotischen Vesikel? Wir haben, wenn sich das mikropinocytotische Vesikel aus der großen Vakuole durch eine neue Invagination bildet, keinen Mechanismus,

---

[1] J. Staubesand und D. Wittekind: Dtsch. med. Forsch. 2, 203 (1964).

der uns einen Teil des Vakuoleninhalts hier zurückhält und nur einen Teil — nämlich den Teil, der nachher für den aktiven Transport in Betracht kommt — in das mikropinocytotische Vesikel aufnimmt.

Darüber wissen wir noch gar nichts. Daß das Phänomen auftritt, ist ohne jeden Zweifel. Nach allem, was wir heute wissen, dürfen wir annehmen, daß die Pinocytose, die zur Bildung von lichtmikroskopisch nachweisbaren Vesikeln führt, im wesentlichen eine Stoffzufuhr darstellt. Die Probleme, die wir dann zu behandeln haben, sind in dieser großen Vakuole im wesentlichen dieselben wie bei der Phagocytose: Enzymzufuhr, Verdauung und die Fragen, ob die Verdauungsprodukte durch die Vakuolenmembran herausdiffundieren oder ob die Diffusion erleichtert wird durch die Oberflächenvergrößerung oder durch die Bildung mikropinocytotischer Vesikel.

Gersch: Kann man nicht an einem anderen Beispiel als Modell diese Unterscheidung zwischen der induzierten Pinocytose und der Residualpinocytose studieren? Ich denke an die Bildung der Nahrungsvakuole von Paramecium, bei der man auf der einen Seite durch stoffliche Gaben eine Vakuole induzieren kann, aber in der gleichen Weise auch durch reine Flüssigkeit, beispielsweise durch Indikatoren. Meine Frage ist, ob man von dieser Seite im Mechanismus der Bildung der Vakuolen elektronenmikroskopisch evtl. Unterschiede oder Gleichheiten feststellen könnte, um den Mechanismus bei einem bestimmten induktionsfähigen Vorgang genauer zu studieren.

Schneider: Mir ist nicht bekannt, daß Paramecium Nahrungsvakuolen nur in reinen Flüssigkeiten, d. h. ohne Induktionsreiz bildet. Es muß anscheinend ein — mechanischer oder chemischer — Reiz vorliegen. Dabei ist es gleichgültig, ob die induzierenden Partikel hinterher verdaubar sind; es können z. B. auch Latex-Kügelchen sein. Bei den Farbstoffversuchen, die Sie erwähnten, besteht ja die Möglichkeit, daß sie entweder einen chemischen Reiz darstellen oder auch einen mechanischen, wenn sie kolloidal gelöst sind.

Duspiva: Die im Süßwasser lebenden Protisten lassen sich im allgemeinen schwer mit irgendwelchen Tracern markieren, wenn man Tritium-Thymidin oder etwas ähnliches in die Tiere hineinbringen will. Mit der Zeit gehen diese Tracerstoffe doch in die Zelle ein. Es ist die Frage, ob sie nicht auch über einen Pinocytosemechanismus in die Zelle eingeschleust werden, da die Pellicula so impermeabel ist. Dann könnte man nämlich mit einem zusätzlichen Induktor mehr von dem Tracerstoff in die Zelle hineinbekommen.

Holter: Aus eigenen Erfahrungen kann ich Ihnen nichts darüber sagen, aber mit meinem Kollegen ZEUTHEN, der mit synchronisierten Tetrahymenakulturen arbeitet und sich sehr dafür interessiert, alle möglichen Antimetaboliten in die Tetrahymena hineinzubringen, habe ich diese Frage sehr oft diskutiert. Es ist ohne jeden Zweifel bei Tetrahymena möglich, induktionsartige Reaktionen herbeizuführen. Die Anzahl der Vakuolen in den Tetrahymena erhöht sich sehr bedeutend, wenn man sie in Proteinlösungen setzt. Das ist von CHAPMAN-ANDRESEN gefunden und von SEAMAN bestätigt worden. ZEUTHEN denkt daher daran, diesen Mechanismus als Einschleusungsmechanismus zu benutzen, also durch Zugabe von Protein usw. eine künstliche — nennen wie sie einmal — Endocytose herbeizuführen, um die Antimetaboliten hineinzubringen, für die seine Zellen sonst ganz impermeabel sind. Wie das gehen wird, weiß ich noch nicht. Aber die Frage ist ohne jeden Zweifel aktuell und evtl. auch von pharmakologischem Interesse.

Duspiva: Es gibt einen wunderschönen Film aus dem Pasteur-Institut, der die Acanthamöbe beinhaltet. Das ist eine winzige Amöbe, die Bakterien frißt. Die Bakterien werden am physiologischen Vorderende der Amöbe agglutiniert und an der Amöbenoberfläche bis an das physiologische Hinterende befördert, wo dann erst eine Nahrungsvakuole gebildet wird.

Holter: Wir haben diese Frage diskutiert und sind uns darüber einig geworden, daß in der Acanthamöbe die pinocytotische Tätigkeit anscheinend an das Uroid lokalisiert ist, zum Unterschied von den anderen Frischwasseramöben, wo das über die ganze Oberfläche vor sich zu gehen scheint. Daran anschließend, möchte ich fragen, ob es Anzeichen dafür gibt, daß das Uroid in der Residualpinocytose besonders aktiv wäre?

Wohlfarth-Bottermann: Bei den Limax-Amöben sicher.

Holter: I should like to ask Dr. Tosteson whether there is anything in the existing membrane models that does account for the possibility of membrane fusion. Membrane fusion obviously plays a very important role indeed in many of these processes, not only in the fusion of vacuoles but also in the pinching off of vacuoles and in the extrusion. I have always regretted very much that in the various membrane models that have been put forward so far, this very important property of membranes has not been taken account of.

Tosteson: Well, I think that most of the existing membrane models are science fiction, but I think there are some observations on the behaviour of so-called liquid crystals of phospholipids in aqueous media that do give reason to believe that such a process could occur with materials of the sort that make up biological membranes.

Morphologische Aspekte der Stoffaufnahme und intrazellulären Stoffverarbeitung

Von

Walter Schmidt, München

Mit 5 Abbildungen

Vorbemerkungen und Begriffsbestimmungen
(zur Diskussion gestellt)

Der Terminus *„intraplasmatisch"* wird im folgenden auf das zum *„inneren Milieu"* der Zelle gehörige Gebiet angewandt, *„extraplasmatisch"* auf den vom inneren Milieu durch das Plasmalemm oder durch eine homologe Membran abgegrenzten Raum. Er gehört dem *„äußeren Milieu"* an, kann aber sehr wohl intrazellulär liegen, wie dies nach Abnabelung der Pinocytosebläschen der Fall ist. Eine strikte Grenzziehung zwischen beiden Begriffen ist oft nicht möglich. Als *„Permeation"* wird die aktive oder passive Passage des Plasmalemms oder einer vergleichbaren Membran bezeichnet, also der Übertritt von einem Milieu in das andere und von einer Phase (im Sinne Ruskas) zur anderen. Die Bezeichnung *„Stoffaufnahme"* = Ingestion wird ganz allgemein für die Hereinnahme oder das Eindringen von Stoffen in die Zelle verwandt.

A. Die Stoffaufnahme

Nach obiger Definition können Stoffe nur nach *Permeation* in das Grundplasma gelangen. In welchem Umfang Konzentrationsgefälle und aktive Zelltätigkeit hierbei wirksam sind, vermag der Morphologe nicht zu entscheiden. Vitalmikroskopische Untersuchungen mit Fluorochromen zeigten, daß lipoidlösliche Farbstoffe schneller permeieren als wasserlösliche. Die hieraus gefolgerte Lipoidlöslichkeitstheorie der Permeation gewinnt aber darüber hinaus auch für die Aufnahme von Fettstoffen im physiologischen Zellstoffwechsel an Bedeutung. Man hat die Vorstellung, die Fettstoffe würden sich auf Grund ihrer physikalisch-chemischen Eigenschaften durch die Lipoidkomponente des Plasmalemms in bisher noch unsichtbarer Form hindurchlösen [42]. Die Pinocytose spielt bei der Fettresorption in den bisher untersuchten Zellen keine oder eine nur zufällige Rolle [4, 36]. Außerdem ist nachgewiesen, daß auch hochmolekulare wasserlösliche Stoffe [30, 33] zu permeieren vermögen. Zur Ultrafilter- und Porentheorie gestatten unsere derzeitigen Kenntnisse vom Feinbau des Plasmalemms noch keine Stellungnahme. Ob der sich ständig und anscheinend unvorstellbar schnell vollziehende Umbau des Molekulargefüges des Plasmalemms [40] auch für die Hereinnahme von Stoffen verantwortlich gemacht werden kann, bleibt vorerst eine

10*

Theorie (Schmidt). Auch müßte die Frage entschieden werden, ob die unter dem Stoffangebot beobachteten Inkontinuitäten des Plasmalemms, die das „Einsinken" von Stoffen [16] direkt in das Grundplasma ermöglichen sollen, dem vitalen Zustand entsprechen. Die Beobachtungen Wohlfarth-Bottermanns [40] beweisen eindringlich, daß das Plasmalemm einem steten Strukturwandel unterliegt und die Vorstellung, man könne es mit einer semipermeablen Kollodiumhaut analogisieren, revisionsbedürftig ist.

Der 2. Mechanismus, der im Dienst der Stoffaufnahme steht, ist die *Pinocytose*. Unter dem Bild der Plasmalemmvesikulation werden an der Zelloberfläche Stoffe vereinnahmt und in das Innere des Zelleibes transportiert. Der Vakuoleninhalt liegt damit intrazellulär — jedoch nicht intraplasmatisch — umschlossen von einer aus dem Plasmalemm gebildeten Membran. Um in das Grundplasma zu gelangen, muß er durch diese Membran permeieren. Hieran ändert grundsätzlich auch die Vermutung nichts, ihre Permeabilität würde sich wie bei der Amöbe [8] erhöhen. Zu einem anderen Gesichtspunkt führt die Hypothese, die Membran der Pinocytosebläschen löse sich in der Zelle auf [3] und der Inhalt würde dem Grundplasma direkt einverleibt. Für die Aufnahme von Stoffen durch das in *Plasmalemmporen* mündende endoplasmatische Retikulum fehlt bisher jeglicher Anhalt.

Prädilektionsstellen für die Permeation sind Mikrovillirasen und dem Stoffwechselstrom zugekehrte Plasmalemmflächen. Vermutlich beteiligt sich die gesamte Zelloberfläche an der Stoffaufnahme- und abgabe, wie auch die Pinocytose durch besondere Auslösungseffekte anscheinend an jeder Stelle des Plasmalemms angeregt werden kann. Die an der Ingestion beteiligten Abschnitte zeichnen sich durch einen mehr oder weniger dichten Besatz mit neutralen Mukopolysacchariden, Phosphatasen und anderen Fermenten aus.

Die Geschwindigkeit, mit der die Stoffe in die Zelle gelangen, ist überraschend hoch. Basische lipophile Fluorochrome sind bereits nach $^1/_2$—1 sec im Cytoplasma nachweisbar. Die ersten Spuren des wasserlöslichen Ferrlecits finden sich 2—10 sec nach der Applikation im Grundplasma, während die Aufnahme von Goldsolteilchen durch Pinocytose ungefähr 1 min dauert [33].

B. Die intrazelluläre Stoffverarbeitung

Mit den in den Zelleib gelangten Stoffen setzt sich die Zelle auseinander; sie werden verarbeitet. Das morphologische Bild und das Ergebnis dieses Vorganges ist bei den verschiedenen Zelltypen sehr unterschiedlich. Sie werden bestimmt von den Aufgaben und Möglichkeiten, die der Zelle im Rahmen ihrer Funktion innerhalb des Gesamtorganismus zur Verfügung stehen und ihrer funktionsbedingten inneren Organisation. Eine resorbierende, an das äußere Milieu des Körpers grenzende Dünndarmepithelzelle verhält sich denselben Stoffen gegenüber anders als eine resorbierende Nierentubuluszelle, die sich nur mit den bei der Rückresorption anfallenden Stoffen auseinanderzusetzen hat. Außerdem verfährt ein und dieselbe Zelle mit Fremdstoffen oft anders als mit physiologischen Resorbaten, weshalb die meisten Vitalfarbstoffe zur Markierung des normalen Transportweges ungeeignet sind [30]. So wird Trypanblau von der Nierentubuluszelle aufgenommen, im supra-nucleären Bereich zurückgehalten und wieder ausgeschieden, während die Dünndarmepithelzelle den Farbstoff überhaupt nicht resorbiert, die Amnionepithelzelle ihn ohne erkennbare Reaktion passieren läßt. Verarbeitung und intrazellulärer Transport sind eng miteinander verbunden; sie werden in meinen Ausführungen getrennt abgehandelt.

I. Morphologisches Bild des trans- und intrazellulären Transportes

Kleinste Partikelchen permeationsfähiger Stoffe wurden bereits von mehreren Untersuchern in diffuser Verteilung im Grundplasma aufgefunden und ihr *Transport im Grundplasma* nachgewiesen. Außerdem muß gefolgert werden, daß auch der Austausch von Ionen und niedermolekularen Stoffen zwischen den Zellorganellen auf diesem Weg erfolgt.

Als besondere Einrichtungen für den intrazellulären Transport fungieren *Vakuolen*. Man beobachtet allgemein die Tendenz der Zelle, die im Grundplasma diffus verteilten Stoffe zur weiteren Verarbeitung in Vakuolen zu segregieren und mit deren Hilfe an den Bestimmungsort zu transportieren. Durch Pinocytose aufgenommene Stoffe liegen bereits primär in transportablen Vakuolen. In sie können noch sekundär Stoffe aus dem Grundplasma segregiert und eingelagert werden. Außer Vakuolen dienen *Leitstrukturen*, die selbst keine Vehikelfunktion haben, dem intrazellulären Transport. Meist sind in ihnen die vermuteten Stoffe schwer nachweisbar: Entweder wurden sie bei der Präparation herausgelöst oder sie liegen in einer nicht mehr erfaßbaren Konzentration vor. Die Funktion von Leitstrukturen übernehmen vermutlich die *mikrotubulären Bildungen* der Mikrovilli. Das *endoplasmatische Retikulum* scheint für die intrazelluläre Leitung gelöster Stoffe geradezu prädestiniert. Es wurde denn auch von zahlreichen Untersuchern in den Lumina des Retikulums ein verdichteter Inhalt abgebildet, der den gesuchten Stoffen entsprechen dürfte. Sehr überzeugend sind die Beobachtungen über seine Zubringerfunktion im Zusammenhang mit der kontraktilen Vakuole bei Paramecium [31]. In einigen Zelltypen findet man das endoplasmatische Retikulum stellenweise zu großen Blasen aufgetrieben. In ihnen ist die Verdichtung und Anreicherung des in den Schläuchen schwach konzentrierten Inhaltes zu beobachten [18, 38]. Diese Erweiterungen lösen sich anscheinend nicht als selbständige Vakuolen ab, sondern bleiben im Zusammenhang mit dem Retikulum. Die *basalen Plasmalemmeinfaltungen* und das *basale Labyrinth* erfüllen die Aufgabe extraplasmatischer Leitstrukturen. Sie dienen in der Nierentubuluszelle der Ableitung von Flüssigkeit bei der Konzentrierung des Primärharnes [27] andererseits auch der Zuleitung von Flüssigkeit bei der tubulären Ausscheidung [25]. In der Salzdrüse bestehen vergleichbare

Plasmalemmeinstülpungen [20], die jedoch die Zelle bis zur Oberfläche hin durchziehen und in deren Lumina extraplasmatisch der Transport von Chlorionen erfolgt. Anscheinend sind alle diese Plasmalemminvaginationen viel dynamischer als man zu vermuten gewohnt ist, und es ist sehr wahrscheinlich, daß sie unter der Notwendigkeit der momentanen Stoffwechselsituation entstehen und wieder verschwinden können. Schließlich seien als Leitstrukturen noch die *intrazellulären Sekretkapillaren* genannt. Auch sie formieren sich in Abhängigkeit von der Funktion durch Fusion zahlreicher Bläschen, um ihren Inhalt auszuschütten, verschwinden wieder und bilden sich auf gleiche Weise neu.

Der Transportweg ist in den verschiedenen Zelltypen entsprechend der inneren Organisation unterschiedlich. In einer polar differenzierten, hochprismatischen Zelle verläuft er zwangsläufig anders als in einer kugeligen oder flach ausgebreiteten. Es ist eine zu schematische Darstellung, wenn nur ein Weg berücksichtigt wird: bei der resorbierenden Zelle von der Oberfläche zur Basis, bei der sezernierenden in umgekehrter Richtung. An Mesothelzellen konnte bereits experimentell ein gegenläufiger Transport nachgewiesen werden [33], und sicherlich trifft dies auch für andere Zelltypen zu. Außerdem weiß man schon seit geraumer Zeit, daß die Tubulus- wie die Dünndarmepithelzelle auch sekretorisch tätig sein können. Vollständig entziehen sich unseren Nachweismethoden Stoffe, die für den zelleigenen Energie- und Betriebsstoffwechsel verwendet werden. Ihr Weg endet in der Zelle.

Die *Geschwindigkeit* des transzellulären Transportes beträgt in Mesothelzellen von Amphibien mit einer Dicke von ungefähr 1 μ 2–3 min, in der Pankreasdrüsenzelle während der Sekretbereitung 4 Std [7, 33].

II. Über die Herkunft der Vakuolen,

in denen Transport, Stoffanreicherung- und Verarbeitung erfolgen, sind wir durch zahlreiche Beobachtungen und experimentelle Untersuchungen an verschiedenen Zelltypen hinreichend orientiert. Die Abschnürung durch *Plasmalemmvesikulation* soll in diesem Zusammenhang nur der Vollständigkeit wegen erwähnt werden (Abb. 1a–f). Die Oberfläche der Bläschen ist zunächst glatt, ihr Durchmesser nach dem Loslösen vom Plasmalemm gering. Ein Großteil der Vakuolen entsteht aus dem *Golgi-Apparat* (Abb. 2a–c) durch Abschnüren aus den Membranpaketen, seltener durch Entfaltung und Aufweitung ganzer Säcke [1, 30]. Die abgeschnürten, meist relativ kleinen Vesikel liegen zunächst gehäuft im Bereich der Membranpakete. Ob jedoch alle diese Vesikel aus den Golgilamellen stammen, wird z. Z. noch diskutiert [14]. Die Entfaltung der Membransäcke ist offenbar die Folge einer hochaktiven oder überstürzten Zelltätigkeit. Da die meisten Zelltypen über mehrere Lamellenpakete verfügen, erfolgt die Vesikelbildung gleichzeitig an mehreren Stellen. Der Stoffnachweis gelingt meist erst in den abgeschnürten Bläschen und nur vereinzelt in den Lumina der Membransäcke. Die Abschnürung von Vakuolen aus dem *endoplasmatischen*

Retikulum (Abb. 3a—d) erfolgt an den ribosomenfreien phiolenartig aufgetriebenen Enden der Schläuche oder durch Querteilung der zisternalen Form in zahlreiche (ergastoplasmatische) Bläschen. In einigen

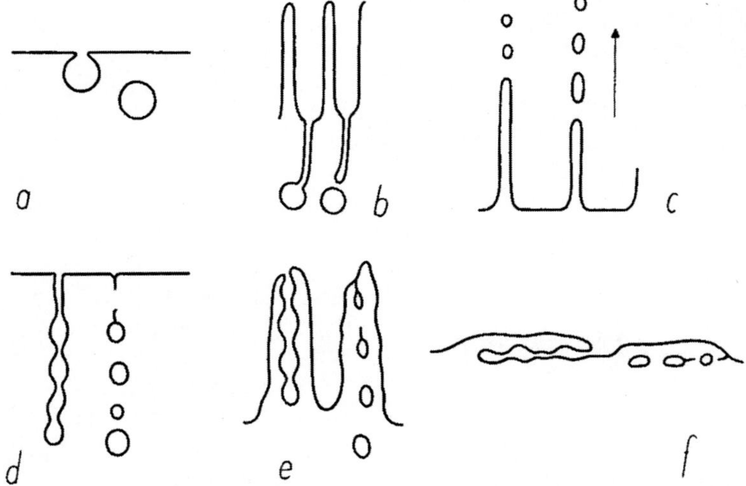

Abb. 1. Beispiele für verschiedene Möglichkeiten der Plasmalemmvesikulation

Zelltypen (Malpighische Gefäße [*38*], Fibroblasten [*18*], Oocyte [*2*]) spielt sich die Anreicherung in blasenförmig erweiterten Abschnitten des endoplasmatischen Retikulums ab (Abb. 3d), die jedoch offensichtlich nicht zu selbständigen Vakuolen werden. Auch aus den *annulated*

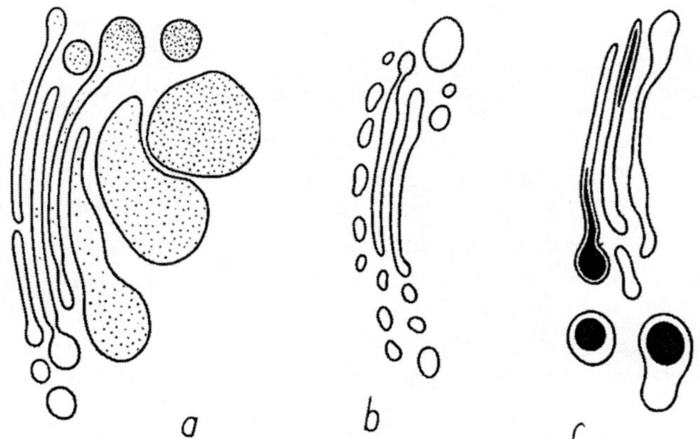

Abb. 2. Bildung von Vakuolen aus dem Golgi-Apparat und Stoffanreicherung in den Golgi-Vakuolen

membranes (Abb. 3e) können sich Vesikel bilden. Ebenso sind blasenförmige Auftreibungen der *perinukleären Zisterne* beobachtet worden [*18, 30*], die sich zu selbständigen Vakuolen entwickeln. In welchem Umfang *Multivesikulärkörper* an der Vesikelbildung beteiligt sind, ist nicht ausreichend geklärt.

Außerdem sind die *Mitochondrien* einiger Zelltypen, vielleicht auch nur unter besonderen Bedingungen, zur Stoffanreicherung und Vakuolenbildung befähigt (Abb. 4a—d), indem sich zwischen äußerer und innerer Lamelle oder im Innern der Cristae Blasen ausbilden. Nach anderen Untersuchungen [32] ist in der Nebennierenrinde die vakuoläre Abhebung der Mitochondrienhülle (Abb. 4c) ein physiologischer Vorgang. Nach Vitalmarkierungsversuchen gelang es, in blasig aufgetriebenen Mitochondrien mit deutlichem Verlust an Binnenstruktur die angebotenen Stoffe in granulärer Form wiederzufinden (Abb. 4d) [16]. Anscheinend übernehmen jedoch im normalen Stoffwechselgeschehen die Mitochondrien der meisten Zelltypen keine Speicherfunktion.

Waren bei den bisher aufgeführten Beispielen die Vakuolen aus einer morphologisch definierten Membranstruktur abzuleiten, so ist auch die Neubildung von Vakuolen aus dem *Grundplasma* hinreichend gesichert [40].

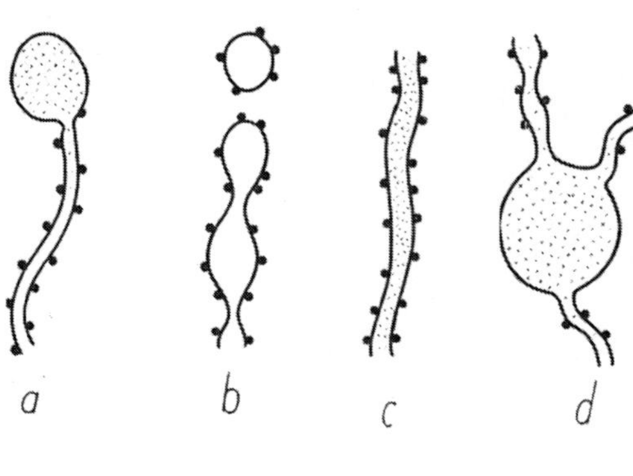

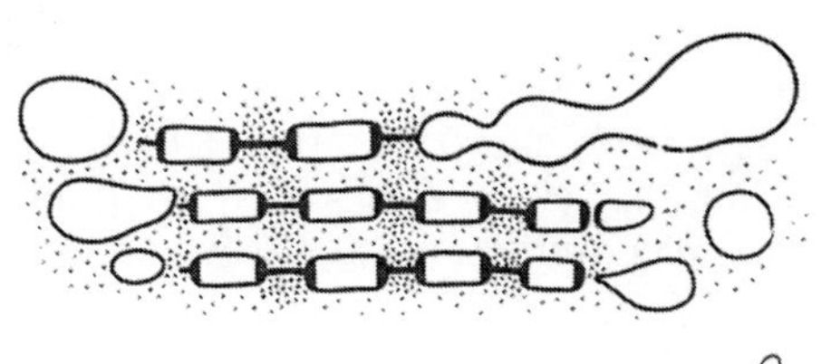

Abb. 3. Vesikelbildung und Anreicherung von Stoffen im endoplasmatischen Retikulum. 3e Vakuolenbildung aus annulated membranes

Keinesfalls sind die verschiedenen Bildungsmodi von Vakuolen für alle Zelltypen von gleicher Bedeutung. In dem einen steht die Plasmalemmvesikulation im Vordergrund, in einem anderen die Abschnürung von Vakuolen aus dem Golgi-Apparat oder aus dem endoplasmatischen Retikulum. Ohne experimentelle Markierung ist es so gut wie unmöglich, festzustellen, von welcher Matrix eine Vakuole abstammt. Vielleicht gestatten neuere Meßergebnisse [28, 29], die eine unterschiedliche Dicke der verschiedenen Cytoplasmamembranen wahrscheinlich gemacht haben (Plasmalemm 95 Å, Ergastoplasmamembran 55 Å, Golgimembran 70 Å), auch die von ihnen gebildeten Vakuolen zu unterscheiden. Der Oberflächenbesatz mit Ribosomen ist kein sicherer Anhalt für ihre Herkunft; denn während der Verarbeitung des Inhaltes werden auch an ursprünglich glattwandige Vakuolenmembranen Ribosomen angelagert [32]. In zahlreichen Arbeiten findet man Vakuolen abgebildet, deren Membran nicht vollständig geschlossen ist. Sind es Vakuolen, deren Membran sich soeben aus dem Grundplasma formiert oder in Auflösung begriffen ist [3], oder ist es ein Fixierungsartefakt?

Während des Transportes und während der Verarbeitung des Vakuoleninhaltes erfährt der Umfang der Vakuole sehr oft eine Veränderung.
Seine *Vergrößerung* erfolgt entweder durch schrittweisen Konflux (Fusion)

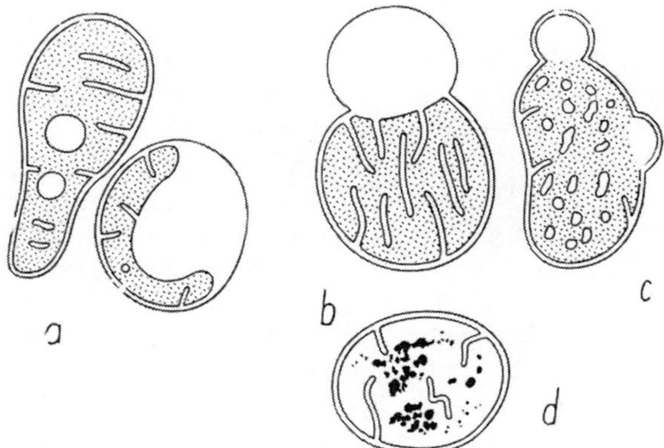

Abb. 4. Entstehung von Vakuolen in Mitochondrien. 4d Speicherung in Mitochondrien

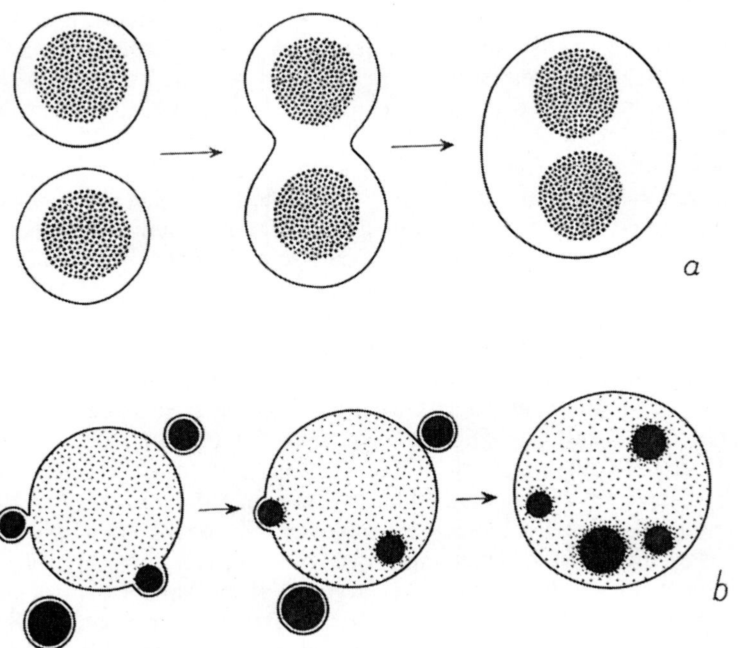

Abb. 5. Konflux von Vakuolen und Inkorporation ihres Inhaltes

annähernd gleichgroßer Vakuolen oder durch Einbeziehen mehrerer
kleiner Vakuolen (Abb. 5). Die Membranen werden dabei lückenlos in
die der nächst größeren Vakuole eingebaut, der Inhalt inkorporiert
(Abb. 5b). Eine andere Möglichkeit, den Vakuolenumfang zu vergrößern,

ist der successive Einbau von Lipoiden und Proteinen aus dem Grundplasma. Die *Verkleinerung* von Vakuolen geht meist mit einer Verdichtung ihres Inhaltes einher. Zur Verringerung des Inhaltes und der Membranoberfläche trägt die Mikropinocytose unter Abschnürung zahlreicher kleiner Bläschen aus der Vakuolenwand bei. Schließlich kann gefolgert werden, daß auch durch die allmähliche Herausnahme von Material aus der Membran während der Stoffausschleusung in das Grundplasma der Umfang der Vakuole herabgesetzt wird.

Völlig unzureichend sind unsere derzeitigen Kenntnisse vom Mechanismus, der für den Transport der Bläschen verantwortlich ist. Die Beobachtung eines gegenläufigen Transportes erschwert die Vorstellung, als treibende Kraft könne ein Flüssigkeitsstrom wirksam sein.

III. Morphologischer Aspekt der intrazellulären Verarbeitung

Die Verarbeitung der aufgenommenen Stoffe, ihr Ab- und Umbau spielt sich, soweit wir heute wissen, weitgehend in Vakuolen ab. Hierbei fällt besonders das Phänomen der *Anreicherung* im Vakuolenlumen auf. Sie besteht in einer Konzentrierung des Inhaltes durch Wasserentzug, so daß bisweilen das Bild eines Granulums zustande kommt. Durch die Zugabe von Stoffen, die für die Verarbeitung und gegebenenfalls für die Resorption aus der Vakuole notwendig sind, wie Fermente und Mukopolysaccharide, wird die Dichte im elektronenmikroskopischen Bild noch erhöht. Bei sezernierenden Zellen folgt auf die anfängliche Konzentrierung meist wieder eine Quellung, bis die notwendige Konsistenz des Sekretes erreicht ist. Diese Vorgänge sind mit dem Konflux von Vakuolen und einer damit verbundenen Zusammenfassung und lokalen Konzentrierung vergesellschaftet. Infolgedessen zeigen oft größere Granula einen inhomogenen Bau, ganz besonders dann, wenn in eine Vakuole Stoffe verschiedener Zusammensetzung oder Dichte inkorporiert werden (Abb. 5b). Unter „*Speicherung*" soll eine Sonderform der intrazellulären Anreicherung verstanden werden, bei der die Zelle Stoffe in größeren Mengen und über längere Zeit zurückhält, um sie bei Bedarf oder Gelegenheit einer weiteren Verwendung zuzuführen oder als unverwertbar auszustoßen. Bekanntlich verfügt nicht jede Zelle über die Fähigkeit zu speichern. Dies schließt jedoch nicht aus, daß sie unter bestimmten Bedingungen diese Fähigkeit erlangt (z. B. Skelettmuskelfaser). Einblicke in den molekularen Bereich des Umbaues sind dem Morphologen bisher so gut wie verschlossen. Zu beobachten sind lediglich Verdichtung und Quellung sowie die Umwandlung ursprünglich homogener Stoffe in feinkörnige (Myofer nach Abbau der Dextrankomponente, Ferritinbildung). Der Austritt der umgebauten Stoffe aus dem Vakuolenlumen in das Grundplasma entzieht sich gleichfalls noch der sicheren Identifizierung im elektronenmikroskopischen Bild.

IV. Histochemischer Aspekt der intrazellulären Verarbeitung und Bedeutung der Cytosomen

Das Plasmalemm ist an den Prädilektionsstellen für die Permeation und Pinocytosetätigkeit, an Mikrovilli und basalen Plasmalemmeinfaltungen mit neutralen Mukopolysacchariden und Fermenten (Phosphatasen, Esterase, Aminopeptidase, ATPase) besetzt. Die PJS-positiven Mukopolysaccharide bilden meist einen Rasen zarter fadenförmiger Auflagerungen. Über die Bedeutung dieser Besätze für den Mechanismus und Chemismus der Permeation ist, abgesehen von der spaltenden Tätigkeit der Fermente, vorerst wenig bekannt. Die Mukopolysaccharide übernehmen anscheinend das Haften der Partikelchen an der Plasmalemmoberfläche [6, 30, 39]. Bei der Plasmalemmvesikulation werden diese Besätze mithereingenommen und lassen sich im Bläscheninhalt nachweisen [21]. Ihre Konzentration ist, gemessen an der Intensität der chemischen Reaktion, in den tiefer im Cytoplasma gelegenen Bläschen meist höher, so daß angenommen werden darf, sie habe durch Konflux und Verdichtung und außerdem durch weitere Fermentsegregation zugenommen. Im licht- und elektronenmikroskopischen Bild erscheinen diese Stoffe als Matrix, in die Fremdstoffpartikelchen und Vitalfarbstoffe eingelagert sind. Durch weitere Fusion entstehen allmählich umfangreiche dichte Gebilde, die als Cytosomen, Phagosomen oder Pinosomen bezeichnet werden. Steht die Einlagerung von Siderin im Vordergrund, trifft die Bezeichnung Siderosomen [22] zu. Für die Speicherung von Trypanblau gilt grundsätzlich der gleiche Mechanismus [30]. Es entstehen Trypanblaugranula oder anders formuliert: Trypanblau-Cytosomen. Hier muß die Frage gestellt werden, inwieweit sind diese Gebilde mit den bio-cytochemisch genau definierten Lysosomen identisch? Lysosomen enthalten [12] neben PJS-positiven Mukopolysacchariden Lipide und vor allem hydrolytische Enzyme wie Phosphatasen, DNase, RNase, Proteasen, Kathepsin, β-Glukuronidase, α-Glukosidase. Die stets in hoher Konzentration nachweisbare Phosphatase gilt geradezu als Kriterium für die Identifizierung von Granula als Lysosomen. Die gesamte Fermentgarnitur ist ein sicherer Beweis, daß die Lysosomen mit dem Abbau von Stoffen in Zusammenhang stehen. Im elektronenmikroskopischen Bild umschließt den dichten fermentreichen Inhalt eine einschichtige Membran, die sich von der "unit membrane" einer Vakuole nicht unterscheidet. Pinosomen und Lysosomen zeigen damit ein morphologisch gleiches Bild. Auch gelang bereits in größeren Pinosomen der Nachweis der oben erwähnten Fermente. Die Frage, ob Pinosomen und Lysosomen wesensgleich seien, findet durch neue histochemische Untersuchungen [34] und durch den Vergleich mit der Nahrungsvakuole der Amöbe eine einleuchtende Erklärung: Zum Abbau der aufgenommenen Stoffe werden in die Pinocytosevakuole die notwendigen Fermente abgegeben und lassen sich dann in den Pinosomen nachweisen. Auf welchem Weg die Fermente in das Pinosom gelangen, ist nur teilweise gesichert. Vielleicht kommen sie aus dem Grundplasma, wie vermutlich die Enzyme an der Plasmalemmoberfläche. Eine zweite Möglichkeit ist heute so gut wie nachgewiesen [34]. Die Fermente werden in kleinen Vakuolen als „Verdauungssaft" an die Pinosomen herangebracht und durch Fusion in deren Innenraum entleert. Es bleibt jetzt die Frage, ist nur das fermentführende Bläschen als Lysosom zu bezeichnen oder auch das Pinosom? Allem

Anschein nach entstehen in Vakuolen verschiedener Genese Stoffeinlagerungen, die abgebaut werden sollen, und damit auch Lysosomen. In der Darmepithelzelle beteiligen sich Pinocytosebläschen zum Beispiel nicht an der Bildung solcher Körper. Welche Fermentgarnitur Vakuoleninhalte besitzen, in denen anabole Vorgänge ablaufen, scheint mir noch nicht ausreichend geklärt, ebensowenig wie die Stellung der Microbodies. Nach allen bisherigen Befunden und Überlegungen hat man in den Lysosomen die Zellorte gefunden, in denen der Abbau der aufgenommenen Stoffe erfolgt. Grundsätzlich wesensgleich sind Vakuolen, in denen Abraumstoffe aus dem Cytoplasma im Anschluß an Zelldifferenzierungsvorgänge oder infolge toxischer Schädigungen bisweilen unter dem Bild des Krinoms [30] gespeichert werden. Der elektronenmikroskopische Aspekt der Cytosomen, ob fein- oder grobgranulär, ob homogen oder myelinfigurenähnlich geschichtet oder komplex zusammengesetzt, wird weitgehend von der chemischen Natur und der quantitativen Zusammensetzung der eingelagerten Stoffe, dem Verarbeitungsmodus des jeweiligen Zelltyps und nicht zuletzt von dem Grad des erfolgten chemischen Abbaues zum Zeitpunkt der Fixation bestimmt werden. Die chemische Natur der in die Vakuole segregierten Stoffe dürfte die Zusammensetzung der Fermentgarnitur entscheidend bestimmen.

Die Zellorte, in denen die anabolen und katabolen Vorgänge ablaufen, sind membranbegrenzte Räume, letztlich Vakuolen. Es scheint so, als ob für einen bestimmten Verarbeitungsvorgang Vakuolen einer bestimmten Matrix herangezogen würden. In einem anderen Zelltyp, vielleicht auch schon unter veränderten Bedingungen, übernehmen Vakuolen anderer Herkunft die gleiche Aufgabe. Untersuchungen mit Vitalfarbstoffen veranlaßten mich [30], deshalb die Gesamtheit der zur Stoffanreicherung befähigten Vakuolen ungeachtet ihrer verschiedenen Herkunft und ihres Inhalts unter der Bezeichnung „Vakuolärer Apparat" zusammenzufassen. Hiermit wurden bewußt diese Vakuolen als der eigentlich funktionell wichtige Apparat einem Zellorganell gleichgesetzt.

V. Cytopempsis, Speicherung, Resorption, Sekretion

Die morphologisch faßbaren Vorgänge bei der Verarbeitung und Bereitung von Stoffen, wie Speicherung, Resorption und Sekretion sollen anschließend an einigen charakteristischen Beispielen umrissen und unter Berücksichtigung der Vielfalt der in den einzelnen Zelltypen eingeschlagenen Wege einander gegenübergestellt werden.

1. Über die *Cytopempsis*, die Durchschleusung von Stoffen wird Herr Staubesand ausführlicher berichten (S. 162).

2. Zellen mit *Speicherfunktion* lagern die aufgenommenen Stoffe vorwiegend in Vakuolen ab. In der Nierentubuluszelle sind es bei der Trypanblauspeicherung ausschließlich Pinocytosebläschen, die durch Fusion zu lichtmikroskopisch sichtbaren, den Farbstoff enthaltenden Vakuolen heranwachsen [30]. Histiocyten und Makrophagen bedienen sich gleichfalls der Plasmalemmvesikulation zur Aufnahme kolloider Stoffe und speichern sie auch in diesen Vesikeln. Die Bildung von Dottermaterial in der Oocyte erfolgt bei Crustaceen in blasigen Auftreibungen

des endoplasmatischen Retikulums [2], bei der Moskito-Mücke in Pinocytosevakuolen [26] und beim Frosch [35] in Mitochondrien. Vakuolen
des Golgi-Apparates sind als Speicherorte für Hyaluronidase im Acrosom
nachgewiesen. Die Stapelung von Neutralfetten findet in der Amnionepithelzelle in Vakuolen nicht bekannter Genese statt, im Fettkörper
der *Drosophila* in cystisch erweiterten Abschnitten des endoplasmatischen
Retikulums und in den Fettzellen der Säuger im Grundplasma ohne
membranöse Abgrenzung [24, 36]. Glykogen wird anscheinend gleichfalls nur im Grundplasma in Form kleinster Tröpfchen oder Körnchen
ohne Membran abgelagert.

3. Die *resorptive Leistung* ist an der Dünndarmepithelzelle für
Fettstoffe [28] und hochmolekulare Kohlenhydrate [30] genauer untersucht. Die Aufnahme erfolgt durch Permeation im Bereich des Bürstensaumes; Pinocytosevorgänge sind ohne Bedeutung [5]. Nach der Permeation werden die Resorbate in der Binnenstruktur der Mikrovilli
in das Terminalgespinst geleitet, passieren diese Zone, um anschließend
in Vakuolen eingelagert zu werden, die sich aus dem endoplasmatischen
Retikulum abschnüren. In diesen Vakuolen erfolgt Anreicherung und
Transport, bis unter mehrfachem Konflux Granula von lichtmikroskopischer Größe entstehen. Ein geringer Anteil der Stoffe gelangt im Grundplasma bis in den Bereich des Golgi-Apparates und wird in dessen Vakuolen eingelagert. Die Abgabe der Resorbate erfolgt vorwiegend in den
Interzellularraum. Abweichend von diesen Beobachtungen an adulten
Mäusen wurde bei neugeborenen [11] an der Epitheloberfläche eine
rege Pinocytosetätigkeit festgestellt. Sie steht unverkennbar mit der
Aufnahme von Darminhalt in Zusammenhang. In der 3. Lebenswoche
erlischt diese Fähigkeit zugunsten des oben genannten Aufnahme-
modus. Als weiteres Beispiel für die resorptive Tätigkeit sei die Nierentubuluszelle des Hauptstücks angeführt. Die Aufnahme und der intrazelluläre Transport makromolekularer Stoffe aus dem Tubuluslumen
erfolgt durch Pinocytosebläschen. Im supranucleären Bereich haben
sie durch mehrfachen Konflux die Größe lichtmikroskopisch sichtbarer
Vakuolen erreicht. Nun wird mit den Vakuolen unterschiedlich verfahren, je nachdem von welcher Natur die eingelagerten Stoffe sind:
Trypanblau enthaltende Vakuolen verharren im supra- und paranucleären
Bereich der Zelle ohne echten Kontakt mit dem Golgi-Apparat unter
dem Bild der granulären Trypanblauspeicherung [30]. Vakuolen mit
physiologischeren Resorbaten passieren dagegen diese Zone, in der
vermutlich die im einzelnen noch nicht bekannten Umbauprozesse
stattfinden, und geben ihren Inhalt in die Räume der basalen Plasmalemmeinfaltungen ab (bezüglich des Transportes bei der Rückresorption
sei auf das Ref. von Herrn THOENES verwiesen).

4. In Zellen mit bevorzugt *sekretorischer Leistung* wird die alleinige
Beteiligung des Golgi-Apparates nur vereinzelt beobachtet [23], wie auch
die Sekretbildung ausschließlich im endoplasmatischen Retikulum nur
in einigen Zelltypen nachgewiesen wurde. Sie erfolgt in sackartig oder
zisternal erweiterten Räumen des Retikulums oder in ergastoplasmati-
schen Bläschen (Mukopolysaccharide in Chondroblasten [19], Vorstufen

der Bindegewebsgrundsubstanz in Fibroblasten [18], Peptidbildung in der Eiweißdrüse des Huhnes [17], Bildung des Prostatasekretes [5]). Für die Bildung proteinhaltiger Produkte ist die Anwesenheit des Ribosomenbesatzes (Ergastoplasma) notwendig. Im agranulären Retikulum erfolgt anscheinend die Produktion der Steroidhormone [9, 13]. Meist sind an der Sekretbildung Golgi-Apparat und endoplasmatisches Retikulum gemeinsam beteiligt. So übernimmt bei der Milchsekretion [1, 37] der Golgi-Apparat die Bereitung des Kaseins. Das Milchfett entsteht im endoplasmatischen Retikulum. Beide Komponenten werden von der Zelle getrennt in das Drüsenlumen abgegeben. Allerdings muß man aus unseren derzeitigen Kenntnissen von der Proteinsynthese folgern, daß auch die Kaseinbildung nicht primär im Golgi-Apparat abläuft, sondern im endoplasmatischen Retikulum vorbereitet und das Zwischenprodukt zur endgültigen Verarbeitung an den Golgi-Apparat weitergeleitet wird. Bei der Bildung des Neurosekrets sind von mehreren Untersuchern zwei in Größe und Dichte deutlich unterscheidbare granuläre Komponenten nachgewiesen worden. Auch hier entsteht die eine im Golgi-Apparat, die andere im endoplasmatischen Retikulum. Nicht eindeutig geklärt ist die Frage, ob die beiden Komponenten innerhalb der Zelle konfluieren, wie dies bei der Sekretbildung in der Harderschen Drüse beobachtet wird [41]. Eine tiefere Einsicht in den Zusammenhang zwischen endoplasmatischem Retikulum und Golgi-Apparat bei der Sekretbereitung vermitteln neuere elektronenmikroskopische Untersuchungen [7, 14]: Im Ribosomen-besetzten endoplasmatischen Retikulum (Ergastoplasma) werden unter der energetischen Leistung der Mitochondrien Vorstufen der Sekrete gebildet, in den Enden der Schläuche angereichert und in Bläschen abgegeben. Diese gelangen zum Golgi-Apparat und verschmelzen mit den Lamellen, während der Bläscheninhalt in die membranbegrenzten Räume übertritt. Am anderen Ende erfolgt erneut die Anreicherung und Aufnahme in Vakuolen. Durch mehrfachen Konflux entstehen in den Golgi-Vakuolen sich stets vergrößernde Prosekrettropfen, die auf dem Weg zur Zelloberfläche zu den fertigen Sekreten heranreifen. Die von Hirsch entwickelte Vorstellung von der Fließbandproduktion der Drüsenzelle findet durch diese zwar noch nicht mehrfach bestätigte und bewiesene Ansicht eine grundlegende Stütze. Außerdem sind an der Bildung bestimmter Sekrete auch Stoffe beteiligt, die aus dem Kernraum stammen und durch Kernporen austreten [15] oder an der äußeren Kernmembran als Bläschen (blebs) [10] abgegeben werden. Wohin sie geleitet und wo sie in den Verarbeitungsprozeß einbezogen werden, entzieht sich noch unserer Kenntnis.

Literatur

[1] Bargmann, W., u. A. Knoop: Z. Zellforsch. 49, 344 (1959).
[2] Beams, H. W., and R. G. Kessel: J. cell. Biol. 18, 621 (1963).
[3] Bennett, H. St.: J. biophys. biochem. Cytol. Suppl. 2, 99 (1956).
[4] Bergener, M.: Z. Zellforsch. 57, 428 (1962).
[5] Brandes, D., and A. Portela: J. biophys. biochem. Cytol. 7, 505 (1960).
[6] Brandt, Ph. W., and G. D. Pappas: J. biophys. biochem. Cytol. 8, 675 (1960).

[7] Caro, L. G., and G. E. Palade: J. cell. Biol. **20**, 473 (1964).

[8] Chapman-Andresen, C., and H. Holter: Exp. Cell Res. Suppl. **3**, 52 (1955).

[9] Christensen, K., and D. W. Fawcett: J. biophys. biochem. Cytol. **9**, 653 (1961).

[10] Clark, W. H.: J. biophys. biochem. Cytol. **7**, 345 (1960).

[11] Clark, S. L.: J. biophys. biochem. Cytol **5**, 41 (1959).

[12] de Duve, Ch: Funktionelle und morphologische Organisation der Zelle. S. 209. Berlin-Göttingen-Heidelberg: Springer 1963.

[13] Endres, A.: J. biophys. biochem. Cytol. **12**, 101 (1962).

[14] Essner, E., and A. B. Novikoff: J. cell. Biol. **15**, 289 (1962).

[15] Feldherr, C. M.: J. cell Biol. **14**, 65 (1962).

[16] Gusek, W.: Veröffentl. a. d. morph. Pathologie. H. 64. Stuttgart: Fischer 1962.

[17] Hendler, R. W., A. J. Dalton, and G. G. Glenner: J. biophys. biochem. Cytol. **3**, 325 (1957).

[18] Karrer, H. E., and J. Cox: J. Ultrastruct. Res. **4**, 420 (1960).

[19] Knese, K. H., u. A. Knoop: Z. Zellforsch. **53**, 201 (1961).

[20] Komnick, H., u. U. Komnick: Z. Zellforsch. **60**, 163 (1963).

[21] Marchesi, V. T., and R. J. Barrnett: J. cell. Biol. **17**, 547 (1963).

[22] Mercker, H. J., u. P. M. Carstensen: Blut **9**, 329 (1963).

[23] Moericke, V., u. K. E. Wohlfarth-Bottermann: Z. Zellforsch. **51**, 157 (1960).

[24] Napolitano, L.: J. cell. Biol. **18**, 663 (1963).

[25] Rollhäuser, H., u. W. Vogell: Z. Zellforsch. **47**, 53 (1957).

[26] Roth, T. F., and K. R. Porter: J. cell. Biol. **20**, 313 (1964).

[27] Ruska, H., D. H. Moore, and J. Weinstock: J. biophys. biochem. Cytol. **3**, 249 (1957).

[28] Sjöstrand, F.: J. Ultrastruct. Res. **8**, 517 (1963).

[29] Schlote, F. W., u. W. Hanneforth: Z. Zellforsch. **60**, 872 (1963).

[30] Schmidt, W.: Z. Zellforsch. **58**, 573 (1962).

[31] Schneider, L.: J. Protozool. **7**, 75 (1960).

[32] Schwarz, W., u. G. K. Suchowsky: Virchows Arch. path. Anat. **337**, 270 (1963).

[33] Staubesand, J.: Z. Zellforsch. **58**, 915 (1963).

[34] Straus, W.: J. cell. Biol. **20**, 497 (1964).

[35] Ward, R. T.: J. cell. Biol. **14**, 309 (1962).

[36] Wassermann, F., and F. McDonald: Z. Zellforsch. **59**, 326 (1963).

[37] Wellings, S. R., and J. R. Philp: Z. Zellforsch. **61**, 871 (1964).

[38] Wessing, A.: Protoplasma (Wien) **55**, 264 (1962).

[39] Wilbrandt, W.: Verh. Ges. Deutsch. Naturf. Ärzte **102**. Vers. (1962).

[40] Wohlfarth-Bottermann, K. E.: Forschungsber. d. Landes Nordrhein-Westfalen, Nr. 1247 (1963).

[41] Woodhouse, M. A., and J. A. G. Rhodin: J. Ultrastruct. Res. **9**, 76 (1963).

[42] Wotton, R.: Int. Rev. Cytol. **15**, 399 (1963).

Summary

Uptake of substances into the cytoplasm of cells is completed only after permeation through the plasmalemma. Hence pinocytosis does not provide a basically novel type of uptake mechanism, since substances taken up must still cross vacuolar membrane formed from the plasmalemma before they reach the cytoplasm. The intracellular fate of reabsorbed materials usually involves intracellular or transcellular transport. Transport takes place in the cytoplasm or in connecting structures (micro-tubules, endoplasmic reticulum, infoldings of plasmalemma) but mainly in vacuoles. They arise as a result of vesicle formation in the plasmalemma or of pinching off from the Golgi complex, the endoplasmic reticulum and its derivatives, or the peri-nuclear cisterna. The discussion next concerns the subsequent changes in the transport vacuoles and their contents, as demonstrated by electron microscopy and histochemistry, and the significance of lysosomes in this regard. Finally, these processes are discussed in more detail for some concrete cases, such as storage, cytopempsis, reabsorption, and secretion.

Diskussion

Thoenes: Ich wollte nur, um Mißverständnissen vorzubeugen, zu der Frage der Definition etwas im Hinblick auf die basalen Einfaltungen vieler Epithelzellen sagen. Ich glaube, man muß doch festhalten, daß die durch Einfaltung der Zellmembran gebildeten Räume nicht nur nicht intraplasmatisch und auch nicht intrazellulär liegen, sondern wirklich echte Extrazellulärräume sind und in der Regel sogar Interzellulärräume. Ich glaube, daß Ihre Darstellung da etwas mißverständlich sein könnte.

Schnepf: Darf ich zu diesen Problemen der Definition vom Gesichtspunkt des Botanikers Stellung nehmen? Wir haben es in der Pflanzenzelle relativ einfach dadurch, daß wir eine massive Zellwand haben. Alles das, was innerhalb der Zellwand liegt, ist intrazellulär. Das Plasmalemma liegt in der Regel der Zellwand dicht an. Es kann sich aber auch durchaus abheben, das wäre dann ein extraplasmatischer Raum in der Pflanzenzelle. Wie ist es jetzt mit den oft behaupteten, aber gar nicht so sehr oft gesehenen Fusionen der Membranen des endoplasmatischen Retikulums mit dem Plasmalemma? Ruska hat ja die ganze Zelle in verschiedene Phasen aufgeteilt. Es ist ein großes Verdienst Ruskas, daß er gezeigt hat, wie verschieden die einzelnen Räume in der Zelle sein können. Es ist aber vielleicht doch dadurch jetzt ein Bild entstanden, das zu kompliziert geworden ist. Ruska unterscheidet bei der tierischen Zelle zwischen mehreren verschiedenen Plasmasorten. Er spricht von Golgi-Plasma, von Grundplasma, von Reticuloplasma, von innerem Mitochondrienplasma, von äußerem Mitochondrienplasma und von Nukleoplasma. In der Botanik kämen ein inneres Plastidenplasma und ein äußeres Plastidenplasma hinzu und — sehr wichtig! —, wenn man konsequent ist, auch ein Vakuolenplasma. Es kann aber eigentlich kein Vakuolenplasma geben. In der Botanik jedenfalls stellt man gerade die Vakuole dem Plasma gegenüber. Die ganze Nomenklatur würde vereinfacht, wenn man sich darauf einigte, das als das eigentliche Plasma zu bezeichnen, was man normalerweise mit Matrix oder Grundplasma bezeichnet, und alle anderen Räume als nicht-plasmatisch zu bezeichnen. Diese anderen Räume unterscheiden sich vom Grundplasma in verschiedener Weise. Erstens finden dort nicht die grundlegenden Stoffwechselreaktionen des Lebens statt, nämlich keine Eiweißsynthese. Die Träger der Information liegen nur im Grundplasma und im Karyoplasma, beide fusionieren durch die Poren in der Kernhülle miteinander. Die anderen Räume können frei nach außen münden. Das kann das Grundplasma, wie es gestern von Herrn Wohlfarth-Bottermann gezeigt worden ist, nicht: Wenn man eine Partie Grundplasma abtrennt, umgibt es sich sofort mit einer Membran. Das Grundplasma ist nicht mit der äußeren Phase — ich möchte sie einmal als wäßrige Phase bezeichnen — mischbar, während das „Plasma" der anderen Kompartimente durchaus damit mischbar ist, ohne daß eine Membran dazwischen auftritt. Diese Betrachtung könnte man auch noch ausdehnen. Ich möchte zwei verschiedene „Phasen" gegenüberstellen, einmal das Grundplasma — wobei ich jetzt keinen Namen prägen möchte, was viel zu kompliziert ist, sondern nur auf die Unterschiede aufmerksam machen — und zum anderen die darin gelegenen Kompartimente. Diese Kompartimente können einmal miteinander fusionieren — sowohl die Membran als auch der Inhalt —, ohne daß Membranen als Trennwände auftreten, und können zum anderen auch mit dem umgebenden wäßrigen Medium frei fusionieren, wenn auch nur gelegentlich. Etwas kompliziert wird die Sache bei Plastiden und Mitochondrien. Vielleicht darf ich dazu auf einen speziellen Fall hinweisen, den eigenartigen symbiontischen Organismus *Geosiphon pyriforme.* Das ist eine Symbiose zwischen einem Pilz, der eine einzellige Blase bildet, und einer Blaualge, einer Nostoc-Art, die in dieser Blase liegt. Die innigeVerquickung von

Pilzzelle und Algenzelle in feinstrukturellem Bereich [s. Schnepf, Arch. Mikrobiol. **49**, 112 (1964)] legt die Vermutung nahe, daß es sich generell bei dem inneren Plastidenplasma und inneren Mitochondrienplasma um etwas ähnliches wie das Grundplasma eines Endosymbionten handelt — ich möchte betonen, das kann Analogie sein. Der äußere Raum dieser beiden Zellorganellen entspräche dann einem ursprünglich extrazellulären Milieu. Für diese Analogie lassen sich auch biochemische Fakten heranziehen, beispielsweise der DNS-Gehalt von Plastiden und Mitochondrien. Auf Einzelheiten möchte ich jetzt — wie gesagt — nicht eingehen, sondern nur einmal diese Zusammenhänge aufzeigen.

P. Sitte: Wir haben in Heidelberg sehr eingehend diese Fragen durchdiskutiert und bei den verschiedensten pflanzlichen Zelltypen geprüft, ob sich das Modell von Herrn Schnepf, das im Prinzip nur zwei Phasen kennt, bewährt. Ich kann die Frage jetzt, ohne ins Detail zu gehen, bejahen. Es ist eine sehr schöne Konzeption.

Cytopempsis

Von

J. Staubesand, Freiburg i. Br.

Mit 21 Abbildungen

A. Zur Begriffsbestimmung

Schnüren sich Anteile des Plasmalemms einer Zelle unter Einschließung von Tröpfchen extrazellulärer Flüssigkeit zelleinwärts ab, um in Form sublichtmikroskopisch kleiner Bläschen zu wandern und an einer anderen Stelle der Zellwand wieder ausgefaltet zu werden, kann Bläscheninhalt ohne Permeation einer Membran durch das Cytoplasma geschleust worden sein. Die Wirkung dieses von Palade [*62*] (auf Grund elektronenmikroskopischer Befunde am Kapillarendothel) konzipierten Vorganges gliche demnach derjenigen offener Poren mit einem den Vesikeln entsprechenden Durchmesser [*82*]. Moore und Ruska [*57*] haben für diese "transmission by cell" den Begriff *Cytopempsis* geprägt, um ein Geschehen zu kennzeichnen, dessen physiologische Bedeutung im Gegensatz zur Pinocytose nicht in der Aufnahme und *intrazellulären Verarbeitung* von Flüssigkeit, sondern in einem *transzellulär gerichteten Passagemechanismus* gesehen wird. Auch Kisch [*39, 40*] hat sich dagegen gewandt, die Aufnahme und Abgabe von Flüssigkeitströpfchen als Pinocytose oder Mikropinocytose zu bezeichnen. Der von ihm vorgeschlagene Terminus „Tröpfchenaustausch" oder „Tröpfchendurchgang" ("dropletexchange", "droplettransfer") hat sich bislang allerdings nicht durchgesetzt.

Cytopemptische Transporte spielen sich in der Größenordnung sublichtmikroskopischer Strukturen ab und beginnen mit einer Mikropinocytose [*61*]. Sie unterscheiden sich demnach zumindest in der Dimension von der lichtmikroskopischen Pinocytose [*44*]. Während morphologische Merkmale der Mikropinocytose bei sehr zahlreichen Zellarten gefunden wurden (Lit. u. a. bei [*90, 96*]), sind zur Pinocytose lichtmikroskopischer Größenordnung fast nur Zellen mit der Fähigkeit zur Pseudopodienbildung imstande (Abb. 1) (Lit. bei [*14, 99*]). Bei lichtmikroskopisch pinocytierenden Phagozyten entwickeln sich in der Zellperipherie Vakuolen eines Durchmesser um $1-2\,\mu$, die in Richtung auf das Zellzentrum wandern, wobei es zu einer Abnahme ihres Umfangs und zu einer Zunahme ihrer Dichte kommt [*99*]. Da durchaus nicht selten Pinocytose-Vakuolen wieder aus der Zelle ausgestoßen werden, kann auch bei der lichtmikroskopischen Pinocytose das entscheidende Merkmal der Cytopempsis, nämlich die Aufnahme *und* Abgabe von Flüssigkeitströpfchen verwirklicht sein, ohne daß diese Erscheinung als Cytopempsis bezeichnet werden darf. Verschiedene Autoren haben sich, z. T. unter Berücksichtigung von Phagocytose und Speicherung darum bemüht,

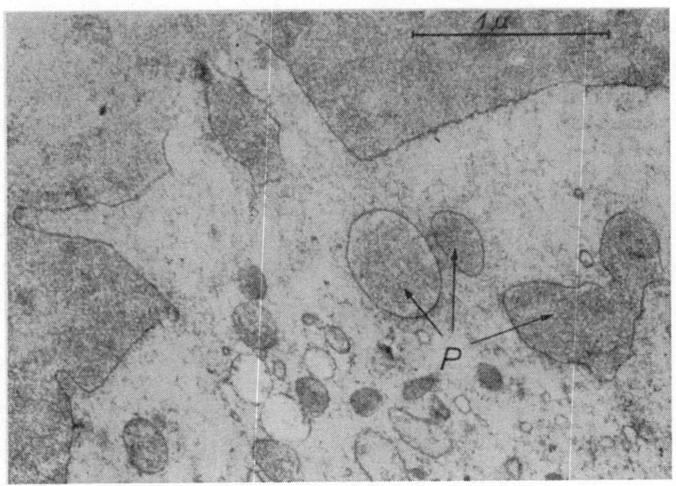

Abb. 1. Neutrophiler Granulozyt (Mensch), aktiviert durch bovines Gammaglobulin 30%. Pseudopodien bilden eine Vakuole. P = rundlich bis polygonal begrenzte Pinocytose-Vakuolen. OsO_4, Vestopal, Uranylacetat-Kontrastierung, EM-9 (Zeiß). Gesamtvergr.: 26000:1 [94]

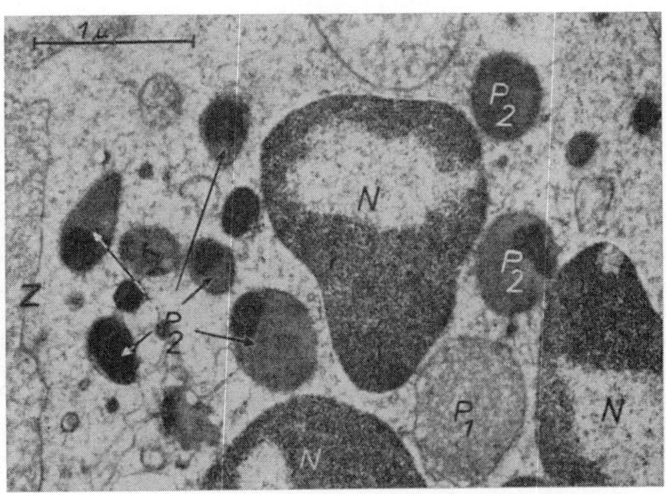

Abb. 2. Neutrophiler Granulozyt (Meerschweinchen), aktiviert durch Human-Gammaglobulin 30%. Z = Zellmembran, N = Kernsegmente; P_1 = Pinocytose-Vakuole mit einem Inhalt, der elektronenmikroskopisch der extrazellulären Immersionsflüssigkeit noch entspricht. In anderen Vakuolen (P_2) ist die pinocytierte Flüssigkeit bereits stärker eingedickt und deshalb kontrastreicher. Stellenweise ist es intravakuolär zur Ausbildung scharf begrenzter Verdichtungszonen gekommen. Technik siehe Abb. 1. Gesamtvergr.: 21000:1 [94]

Pinocytose, Mikropinocytose, Membranfluß bzw. -vesikulation [4] und Cytopempsis entweder als prinzipiell gleichartige Mechanismen anzuerkennen oder als differente Vorgänge gegeneinander abzugrenzen (Lit. u.a.

bei [*16a, 22, 28, 59, 59a, 88, 96, 99*]; vgl. auch die Referate dieser Konferenz von H. Holter und W. Schmidt).

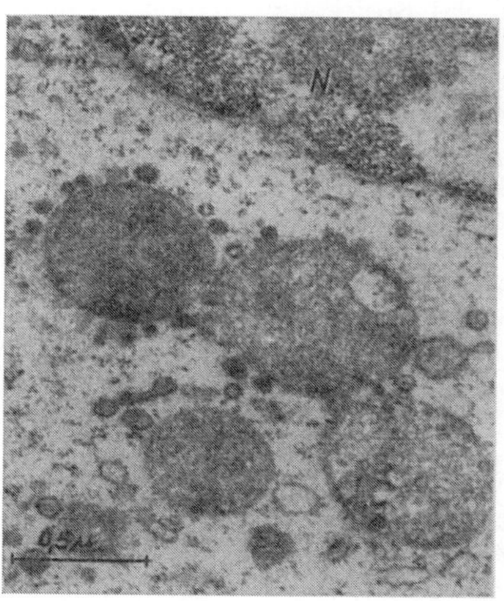

Abb.3. Neutrophiler Granulozyt(Meerschweinchen), aktiviert durch bovines Gammaglobulin 30%. Die linke Vakuole ist kranzartig von Sekundärvesikeln umgeben. *N* = Kern. Technik siehe Abb. 1. Gesamtvergr.: 30000:1 [*94*]

Zusammen mit Wittekind durchgeführte elektronenmikroskopische Untersuchungen an aktivierten, lichtmikroskopisch pinocytierenden neutrophilen Granulozyten zeigen im Cytoplasma Vakuolen unterschiedlicher Größe und Dichte, deren Inhalt entweder dem umgebenden Medium gleicht (Abb. 1) oder als Ausdruck der Eindickung des aufgenommenen Materials bereits konzentriert ist. Mitunter treten innerhalb der Vakuolen elektronendichte, scharf begrenzte Herde auf (Abb. 2), während andere Vakuolen von Sekundär-Bläschen mikropinocytotischer Größenordnung umgeben sind (Abb. 3) [*94*]. Wahrscheinlich liegt hier ein ähnliches Phänomen vor, wie es auch

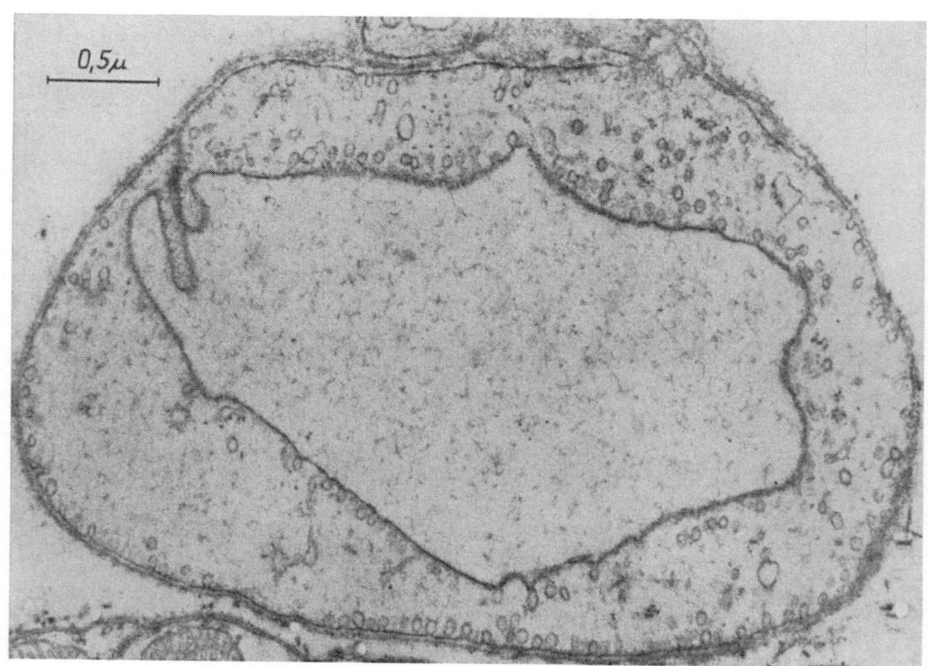

Abb. 4. Muskelkapillare (Ratte). Beachte die vor allem in der unteren Bildhälfte deutliche Anreicherung mikropinocytotischer Vesikel entlang der inneren und äußeren Zellmembran. Technik siehe Abb. 1. Gesamtvergr.: 28700:1. Material: Dr. F. Hammersen, Freiburg i. Br.

Roth [79], Chapman-Andresen und Nilsson [16] sowie Marshall [52a] bei Amöben beschrieben haben, bei denen sich an Nahrungsvakuolen und Pinocytose-Kanälchen sekundär sublichtmikroskopische Vesikel ausbilden können. — Unsere elektronenmikroskopischen Bilder liefern keine Hinweise für eine lokal gesteigerte Mikrovesikulation, die Gropp [28] als Voraussetzung für die Entstehung lichtmikroskopischer Pinocytose-Vakuolen gefordert hat. Im Vordergrund der

Tabelle 1: *Durchmesser mikropinocytotischer Vesikel im Cytoplasma von Endothelzellen*

Autoren	Vesikeldurchmesser in Å
Hager (1961)	200
Casley-Smith u. Florey (1961), Odland (1961)	200— 500
Bennett, Luft u. Hampton (1959)	200— 600
Schulz (1959)	225— 640
Macher u. Vogell (1962)	250— 500
Buck (1958)	300— 500
Palade (1960)	400— 600
Cervos-Navarro (1963)	500
Alksne (1959), Majno u. Palade (1961)	500— 700
Moore (1959)	500— 750
Pappas u. Tennyson (1962)	500—1200
Fernando u. Movat (1964)	570— 630
Parker (1959)	600
Williamson (1964)	600—1000
Policard et al. (1957)	600—1500
Palade (1953), Han u. Avery (1963)	650
Palade (1961)	650— 750
Misotten (1961, 1962)	700
Ruska (1964)	700—1000
Ekholm u. Edlund (1959)	800
Fawcett (1964)	800—1000

lichtmikroskopischen Pinocytose steht die u. U. stunden- und tagelange Auseinandersetzung der Zelle mit den aufgenommenen Flüssigkeitstropfen. Das polymorphe Bild der Pinocytose-Vakuolen ist in hohem Maße abhängig vom Verarbeitungszustand ihres Inhalts. *Demgegenüber sind die mikropinocytotischen Vesikel einheitlicher gestaltet* (Abb. 4). Zwar schwanken die Angaben zum Durchmesser dieser Bläschen (bezogen auf Endothelzellen) immerhin noch zwischen 200 und 1500 Å, die Mehrzahl der Untersucher mißt jedoch Vesikeldurchmesser zwischen 500 und 750 Å (Tab. 1). Vakuolen können nicht nur durch primäre Abfaltung von der Zellmembran (wie bei Phagozyten, Abb. 1), sondern bekanntlich auch in der Tiefe des Cytoplasmas durch Konfluieren mikropinocytotischer Bläschen und ganz andere Mechanismen (z. B. Hypoxydose) entstehen. In Zellen, die nicht zur Pinocytose *lichtmikroskopischer* Dimension imstande sind, scheint die Art der Vakuolenbildung durch Verschmelzung von Vesikeln häufig zu sein (Abb. 12) (vgl. hierzu auch [83, 84]). Bei den durch Mikropinocytose entstandenen *Vesikeln* (bis zu einem Durchmesser von etwa 1000 Å) scheinen *Transportfunktionen* — in die Zelle hinein, aus der Zelle heraus oder durch die

Zelle hindurch — bei den *Vakuolen* (eines Durchmessers über 1000 Å) vor allem *metabolische* und *katabolische Prozesse* im Vordergrund zu stehen. Der Nachweis enzymatischer Aktivitäten in Vesikeln bestimmter Kapillarendothelzellen [*51, 52, 95*] spricht freilich dafür, daß die Bläschen durchaus nicht nur als Transportvorrichtungen aufgefaßt werden dürfen.

B. Zum Ablauf der Cytopempsis

I. Deutungsschwierigkeiten

Da cytopemptische Transportvorgänge der unmittelbaren Beobachtung nicht zugänglich sind, können sie nur durch Kombination verschiedener Zustandsbilder der *fixierten* Zelle in ihrem räumlichen und zeitlichen Ablauf *rekonstruiert* werden. Dabei bleibt der subjektiven Interpretation zwangsläufig ein gewisser Spielraum.

Schon die Diagnose „intrazelluläres Bläschen" ist mitunter problematisch, weil man im Schnitt kein dreidimensionales *Bläschen*, sondern einen *Membranring* sieht, der grundsätzlich Anteil eines *kugel-* oder eines *schlauch*förmigen Gebildes sein kann. Nur durch Auswertung von Serienschnitten — wie es z. B. Wohlfarth-Bottermann [*100*] durchgeführt hat — ließe sich eindeutig klären, ob eine kreisförmig in sich zurückkehrende Membran als abgeschlossenes Bläschen oder als Anteil eines womöglich mit dem Plasmalemm verbundenen Schlauches bzw. verzweigter Röhren aufzufassen ist.

Betrachten wir dazu zwei Kapillaren, die unter vergleichbaren Bedingungen gewonnen wurden: in der einen sind die Endothel-„Vesikel" überwiegend isoliert und annähernd kreisförmig (Abb. 4), in der anderen jedoch bilden sie oft Paare oder Dreiergruppen, stellenweise sind sie auch Y-förmig oder ganz unregelmäßig begrenzt (Abb. 5). Während man im ersten Fall die „intrazellulären Membranringe" ohne weiteres für Anteile kleiner Bläschen zu halten bereit ist, liegt es im zweiten viel näher, sie mit einem dreidimensionalen Röhrensystem in Zusammenhang zu bringen. Auch andere Untersucher [*1, 11, 34, 37*] haben in Endothelzellen Vesikel beschrieben, die sich ineinander öffnen bzw. mit dem endoplasmatischen Retikulum oder mit dem perinukleären Spalt verbinden. Für die Beurteilung der Kinetik transzellulärer Passagevorgänge ist es aber nicht ohne Bedeutung, ob Flüssigkeit oder Substanzpartikel durch abgefaltete, nachfolgend freibewegliche und schließlich wieder ausgestülpte Bläschen oder über vorgebildete, mit dem extrazellulären Raum kommunizierende Röhren im Sinne des endoplasmatischen Retikulum und nach dem Prinzip des "membrane flow" [*4*] erfolgt.

Von dieser Überlegung einmal abgesehen, muß festgehalten werden, daß eine Unterscheidung zwischen Anteilen des endoplasmatischen Retikulum und pinocytotischen Bläschen oft kaum möglich ist [*25*], oder anders ausgedrückt: ein gewisser Prozentsatz der im Schnitt als Bläschen sichtbaren Strukturen könnte nicht zum cytopemptischen System, sondern zum endoplasmatischen Retikulum gehören. Schmidt [*84*], Novikoff [*59*], Komnick [*41*] u. a. haben darauf hingewiesen, daß ein

Teil der Cytoplasma-Vesikel von Lamellen des Golgi-Komplexes abstammt. Als Bläschen, die für eine transzelluläre Passage in Betracht kommen, fallen sie deshalb ebenfalls aus. Die Zahl der eine Zelle durchwandernden Vesikel wird weiterhin durch jene Bläschen vermindert, die der Mikropinocytose, d. h. der Aufnahme und intrazellulären Verarbeitung von Flüssigkeit bzw. Material kolloidaler Größenordnung

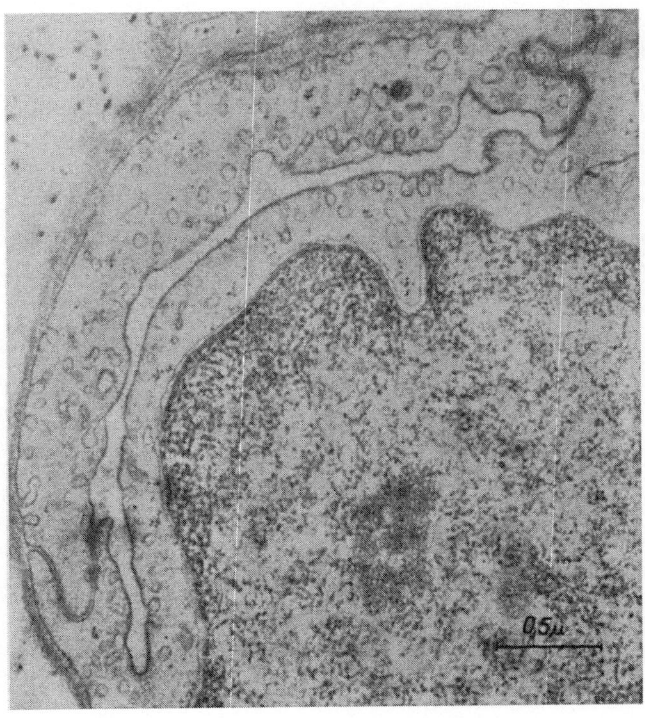

Abb. 5. Kollabierte Muskelkapillare (Ratte). Im Cytoplasma der linken Endothelzelle zahlreiche paarweise miteinander verbundene Vesikel und kurze, z. T. gegabelte oder unregelmäßig begrenzte Röhrchen. Technik siehe Abb. 1. Gesamtvergr.: 28000:1. Material: Dr. F. Hammersen, Freiburg i. Br.

dienen. Diese Bläschen sollen nach der Aufschließung ihres Inhaltes verschwinden können [79] oder mit Zellorganellen bzw. Speicherungsvakuolen fusionieren (Abb. 12).

Auch perlschnurartig aufgereihte Bläschenformationen, die wir in verschiedenen Zelltypen — unzutreffend! — als „Pinocytose-Ketten im Dienst eines transzellulären Flüssigkeitsdurchgangs" beschrieben haben [91, 92, 93], können zwar die Zahl intracytoplasmatischer Vesikel erheblich vermehren, sind jedoch nicht dem cytopemptischen System zuzuordnen.

Die Wahrscheinlichkeit — z. B. an der Stelle eines vorher vorhandenen Pinocytose-Kanals —, eine einzelne *Reihe* von Bläschen im ultradünnen Schnitt zu erfassen, ist minimal. Den Bläschen-„Ketten" müssen also flächenhafte Vesikel-*Felder* entsprechen. Die filamentfreie Randzone der Muskelzellen des Regenwurms enthält auffallend geordnete Bläschenformationen, die im Schnitt entlang gekrümmter oder gewinkelter Achsen mit der Zelloberfläche verbunden sind (Abb. 6). Wir legen diesem Befund heute folgenden Vorgang zugrunde: Zunächst schieben sich zungenförmige Zellfortsätze über das Plasmalemm, dann kommt es zu Kontakten und Verschmelzungen der benachbarten Membranen, und schließlich tritt an deren Stelle ein flächenhafter Verband dicht beieinanderliegender Bläschen (Abb. 7). Auch in Bereichen, in denen eine Zelle einknickt oder in denen sich Zellfortsätze aneinanderschmiegen, kann ein entsprechender vesikulärer Membranzerfall auftreten. Dadurch wird zwar das Wirkungsfeld der nach ihrer Verlagerung in das Zellinnere vesikulierten Zelloberfläche gegen den extrazellulären Raum gesteigert, nicht jedoch — wie wir ursprünglich vermuteten — ein cytopemptischer Flüssigkeitstransport ermöglicht (vgl. hierzu auch [*42, 58, 65, 66, 104*]).

Linear oder gewunden bis verschlungen angeordnete Bläschen-,,Ketten" im Cytoplasma von Megakaryozyten der Ratte (Abb. 8, 9), dürfen ebenfalls nicht für eine pinocytotische oder cytopemptische Aktivität der betreffenden Zelle in Anspruch genommen werden, sondern dienen möglicherweise der Bereitstellung von Membranmaterial, das bei der Abspaltung der Thrombocyten benötigt wird.

Aus den angeführten Beispielen ergibt sich, daß *nur ein*

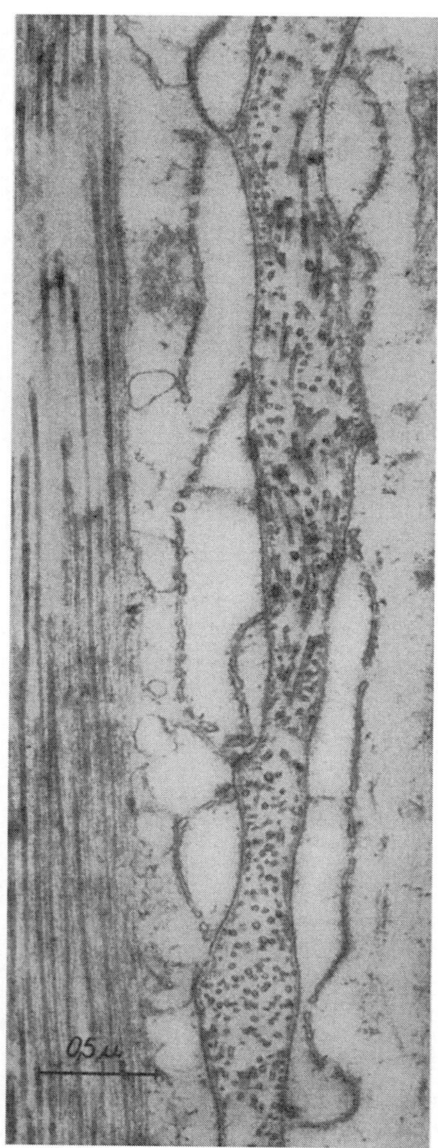

Abb. 6. Mehrere Vesikel-,,Ketten" in der filamentfreien Randzone zweier durch ein (etwa in Bildmitte vertikal verlaufendes) Bindegewebsseptum getrennter Muskelzellen des Regenwurms. Technik siehe Abb. 1. Gesamtvergr.: 30800:1. Orig.

Teil der intracytoplasmatischen Vesikel transzellulären Passagevorgängen zugerechnet werden darf. Dem elektronenmikroskopischen Bild läßt sich meist nicht sicher entnehmen, ob ein Bläschen dieser oder jener Funktion

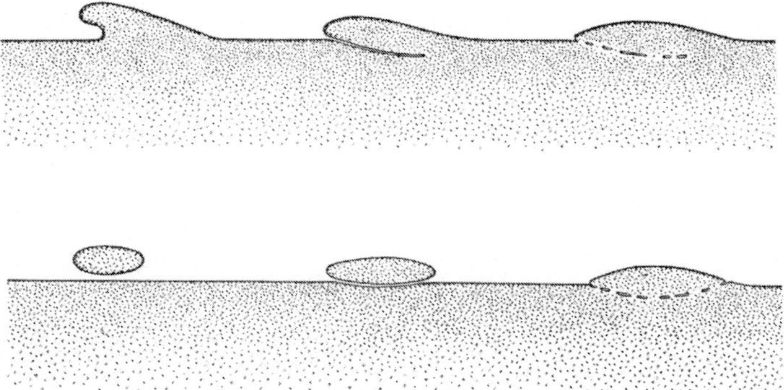

Abb. 7. Schematische Darstellung der Entstehung eines Bläschenfeldes (im Schnitt: einer Bläschen-„Kette" !) durch Anlagerung eines zungenförmigen Fortsatzes an die Oberfläche derselben Zelle. Im Bereich der Berührungsfläche kommt es zu Fusionierungen und Vesikulation beider Membranen. Im oberen Bildteil Längsschnitte, im unteren Querschnitte korrespondierender Stellen. Orig.

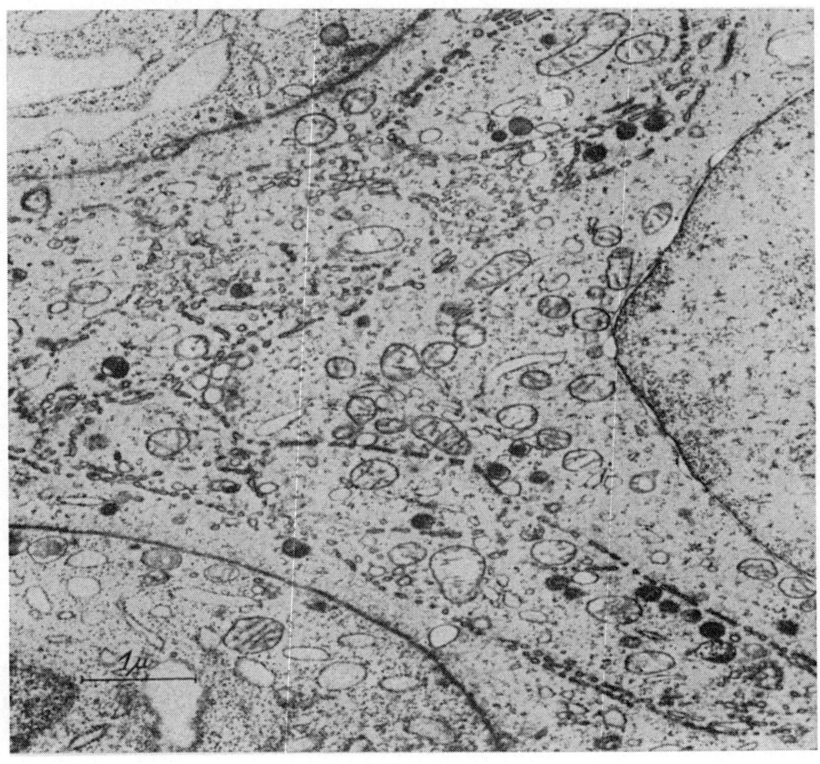

Abb. 8. Megakaryozyt (Ratte). Im Cytoplasma Vesikel-„Ketten", die einzelne Territorien der Zelle zu begrenzen scheinen. Technik siehe Abb. 1. Gesamtvergr.: 17500:1. Orig.

dient. Doch helfen in diesem Zusammenhang rechnerische Überlegungen weiter, auf die ich später zurückkommen werde (s. S. 182).

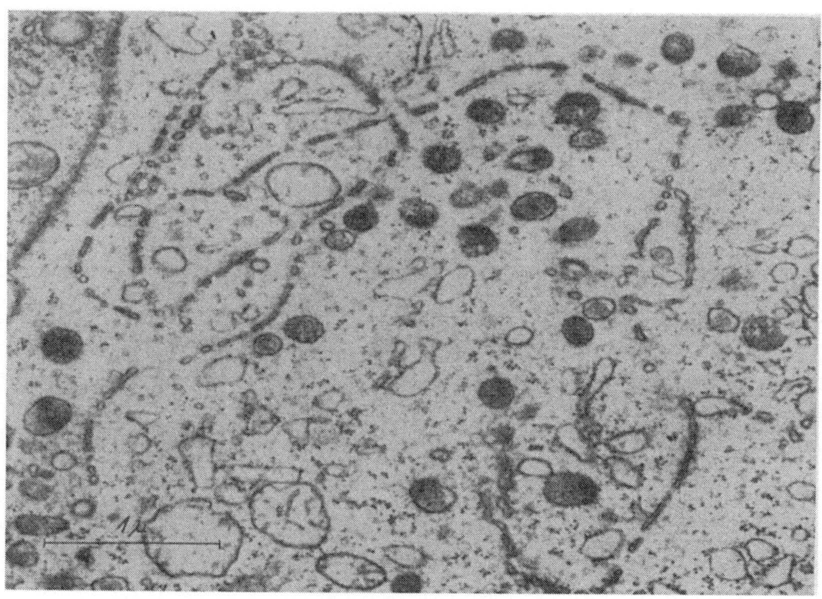

Abb. 9. Megakaryozyt (Ratte). Mehrere verschlungen verlaufende Doppelmembranstrukturen, die stellenweise als „Röhren", stellenweise als Vesikel-„Ketten" in Erscheinung treten. Technik siehe Abb. 1. Gesamtvergr.: 30800:1. Orig.

II. Die Bildung der "inpocketings"

Die Fähigkeit der Zellmembran, auf dem Wege über grübchen-, taschen- oder vielleicht auch rinnenförmige Einsenkungen geschlossene und schließlich frei abgelöste Bläschen zu bilden (Abb. 10), wird nur dadurch ermöglicht, daß das Plasmalemm keine überzugsähnliche, die Zelle stabil abgrenzende Schicht, sondern eine plastisch verformbare, sich Dehnungskräften schnell adaptierende, zur Lösung wie zur Fusionierung befähigte Organisationsform des Plasmas darstellt. Stoßen Membranteile an den Rändern eines sich ausbildenden Bläschens zusammen, wird an der Berührungsstelle wahrscheinlich die molekulare Ordnung vorübergehend verändert, so daß im Augenblick der Bläschenentstehung eine Verschmelzung der Membran erfolgen kann. WOHLFARTH-BOTTERMANN [104] hat kürzlich instruktive Beispiele für die Dynamik cytologischer Membranen zusammengestellt (vgl. hierzu auch [80, 81]). Da ständig Teile des Plasmalemms über Membranfluß und Vesikulationsvorgänge [47] in die Tiefe der Zelle verschoben werden, ist die tatsächliche *Membran*oberfläche im Verhältnis zur lichtmikroskopischen *Zell*oberfläche sehr viel größer. Die Auseinandersetzung zwischen intra- und extrazellulärem Raum vollzieht sich also durchaus nicht nur entlang der äußeren Zellmembran, sondern in einem noch kaum abschätzbaren Ausmaß an den in das Zellinnere verlagerten Membranen. Die konventionellen Begriffe des intra- und extrazellulären Raumes sind demnach in einem physiologischen Sinne recht problematisch geworden [*35a, 88*]. Auch WOHLFARTH-BOTTERMANN [*104*] spricht von „großflächigen extrazellulären Räumen", die durch die Einstülpung der Zellmembran *im Inneren* der Zelle vorhanden sind.

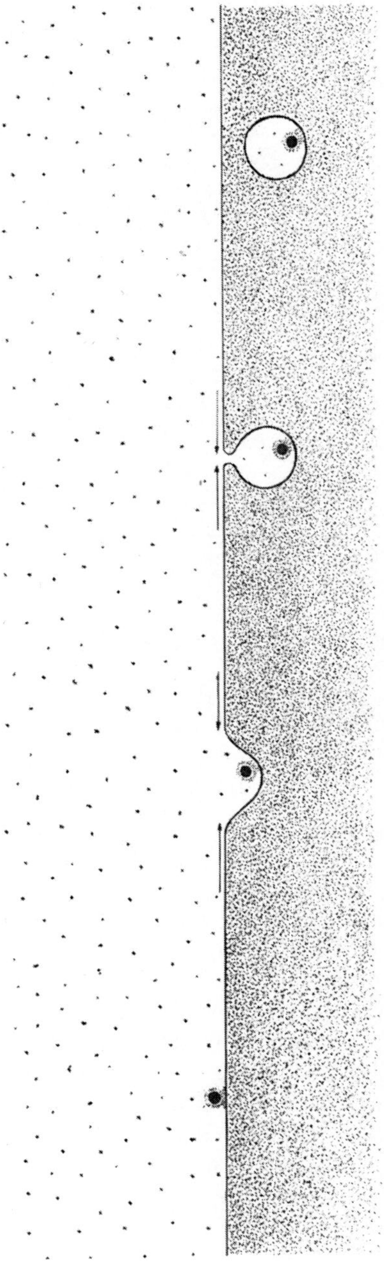

Wodurch die Einfaltung der Zellmembran und dadurch die Bildung eines Bläschens ausgelöst wird, ist ungeklärt. Zur Markierung mikropinocytotischer Stoffaufnahmen benutzte kolloidale Metallpartikel (z. B. Gold, ThO_2 oder HgS) liegen in der Regel nicht „nackt" in Vesikeln, sondern sind von einem mehr oder weniger deutlichen Mantel eines elektronenmikroskopisch sichtbaren Materials — vielleicht handelt es sich dabei um Mucopolysaccharide [*7a, 83*], Fibrin oder ein anderes Protein [*1, 33*] — umgeben. Diese Substanz steht in unseren Bildern meist mit der Bläschenwand in Verbindung (Abb. 13, 15). Mitunter findet man Goldkörnchen, die zwar noch 200 bis 300 mμ von der Zelloberfläche entfernt, aber doch schon innerhalb eines unscharf begrenzten wolkigen Stoffes liegen, während sich das Plasmalemm unter diesem Komplex bereits zu einer kleinen Grube eingesenkt hat (Abb. 11).

Abb. 10. Schema zur Entstehung eines mikropinocytotischen Bläschens, in das ein Partikel kolloidaler Größenordnung eingeschlossen wird. Pfeile: Fließrichtungen der Zellmembran. Orig.

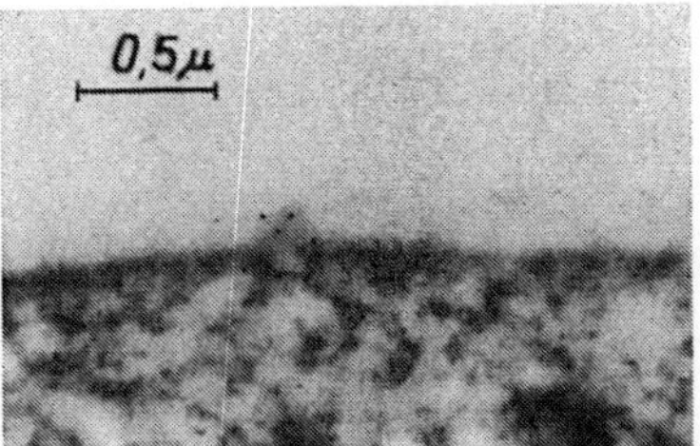

Abb. 11. Oberfläche einer peritonealen Deckzelle (Bufo bufo), deren Plasmalemm sich unter einer Gruppe von Goldpartikeln eingefaltet hat. OsO$_4$, Methacrylat, Elmiskop I (Siemens), Gesamtvergr.: 19000:1. Orig.

III. Das intrazelluläre Schicksal der mikropinocytotischen Vesikel

Bereits Palade [62] hat darauf hingewiesen, daß in Kapillarendothelzellen die Vesikel entlang der dem Lumen und der dem Grundhäutchen zugewandten Zellmembran gehäuft und in der Tiefe der Zellen spärlicher

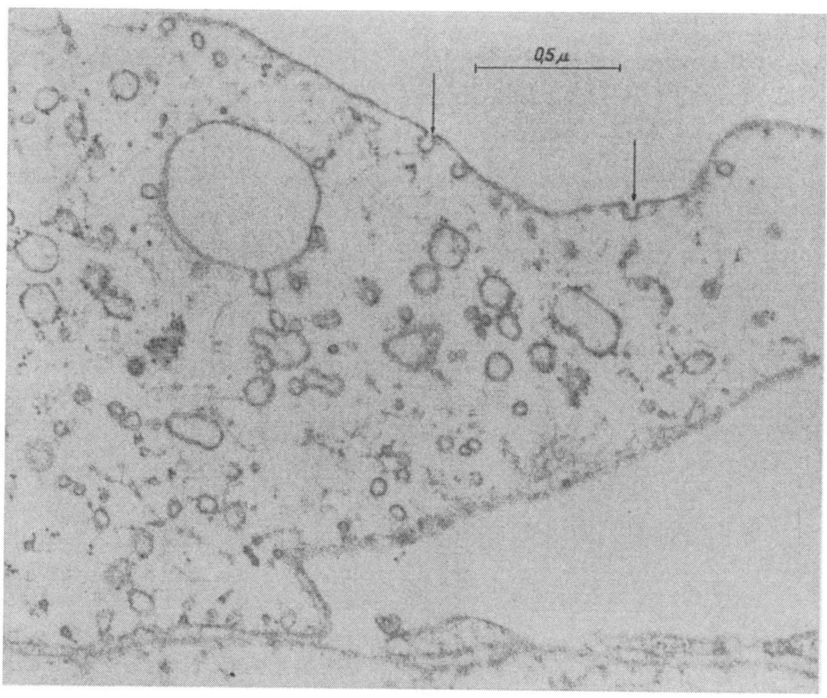

Abb. 12. Peritoneale Deckzelle (Bufo bufo). Pfeile: "inpocketings". Links oben größere Vakuole, die mit mikropinocytotischen Vesikeln kommuniziert. Technik siehe Abb. 11. Gesamtvergr.: 58000:1 [90]

anzutreffen sind (Abb. 3). Dieser auch von anderen Untersuchern bestätigten Erscheinung [5, 21, 23, 24a u. a.] könnte die relativ langsame Ein- und Ausfaltung der Vesikel einerseits und ihre verhältnismäßig rasche transzelluläre Wanderung andererseits zugrundeliegen. Nun wissen wir aus sinnreich kombinierten vital- und elektronenmikroskopischen Untersuchungen Schneiders [85, 86] an Paramecien, daß die Membranen der Tubuli nephridiales in weniger als 10 sec, abhängig von einem zeitlich genau definierbaren Pumprhythmus, mit der Membran des Radialkanals fusionieren und sich wieder zu lösen vermögen. Wenn die Vesikelentstehung am Plasmalemm der Endothelzellen ebenfalls nur so kurze Zeit in Anspruch nehmen sollte, würde die Anreicherung der Vesikel an der Zellober- und -unterfläche nicht mit der Trägheit dieses Vorgangs, sondern vielleicht mit einer Bläschenbereitstellung „auf Vorrat" entlang der Zelloberfläche in Zusammenhang zu bringen sein. Für eine *in Schüben* ablaufende Entstehung und Wanderung der Bläschen durch das Cytoplasma bestehen keine Anhaltspunkte [*90*].

Das Schicksal mikropinocytotischer Vesikel ist nicht einheitlich. Während ROTH [*79*] annimmt, daß sich zahlreiche Mikrovesikel durch Membranruptur wieder auflösen, betont HOLTER [*36*] — und damit stimmen wir überein —, das Verschwinden ihrer Membran sei bisher elektronenmikroskopisch nicht nachgewiesen worden. Ein Teil der Bläschen fusioniert jedenfalls mit Sicherheit zu größeren Vakuolen (Abb. 12), ein anderer mit dem endoplasmatischen Retikulum, das offenbar seinerseits stellenweise an Vakuolen angeschlossen ist (Abb. 13), ein weiterer scheint mit Mitochondrien verschmelzen zu können oder sich in diese Organellen blasenartig einzustülpen (Abb. 14a, b).

Nach GROPP [*28*] beeinflussen die Mitochondrien weniger die initiale Phase

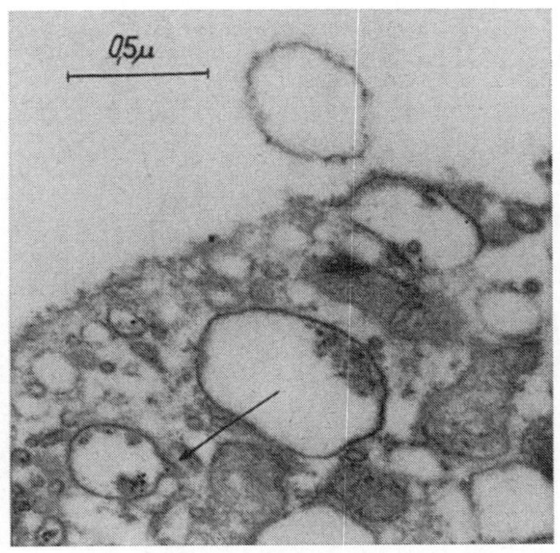

Abb. 13. v. Kupffersche Sternzelle (Bufo bufo) 5 min nach intraperitonealer Injektion von Goldsol. Der Pfeil weist auf ein Röhrchen hin, das an seiner Mündung in eine Vakuole ein Goldpartikel enthält. Andere Gold-Teilchen in Vesikeln, Vakuolen und auf der Zelloberfläche, über der ein Membranring (ausgestoßene Vakuole?) zu erkennen ist. Technik siehe Abb. 11. Gesamtvergr.: 30 800:1. Orig.

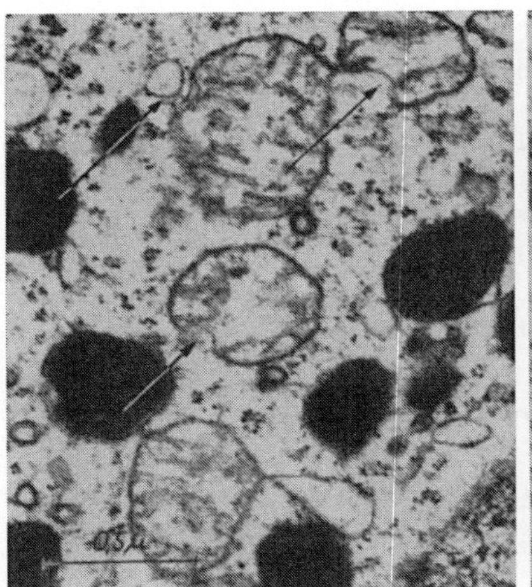

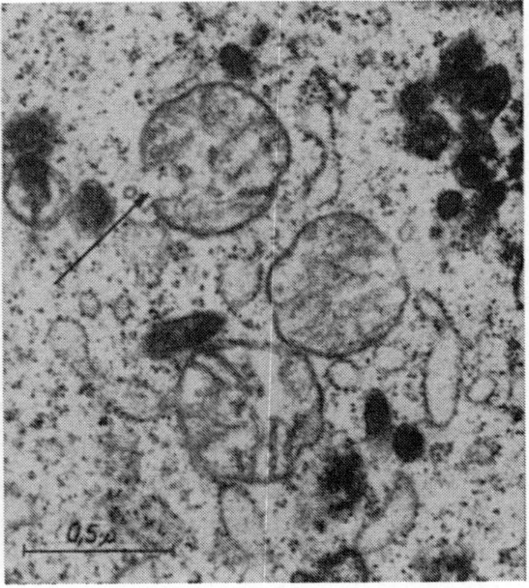

a   b

Abb. 14. Taschenförmige Einstülpungen in Mitochondrien (Leukozyten aus dem Knochenmark, Ratte). *a* = eosinophiler, *b* = neutrophiler Granulozyt. Technik siehe Abb. 1. Gesamtvergr.: 30 800:1. Orig.

der Pinocytose am Zellrand als die weitere Aggregation der Tröpfchen, ihren Transport in das Innere der Zelle und schließlich ihre katabolische Auflösung durch Bereitstellung der notwendigen Energie für diese Prozesse durch Abspaltung von ATP. Vitalmikroskopisch stellte er fest, daß Aggregate pinocytotischer Vakuolen im Cytoplasma oft von zahlreichen Mitochondrien umgeben sind, die mit den Tröpfchen in Kontakt kommen und sich dann wieder zurückziehen. Unsere Befunde

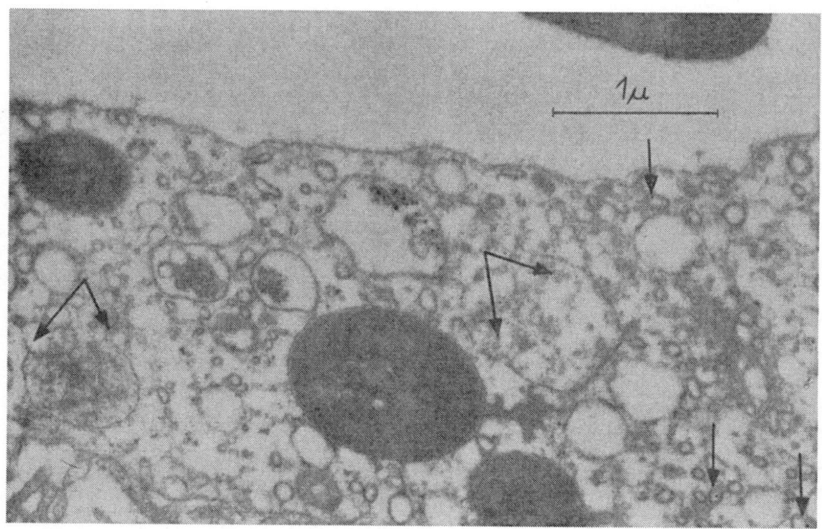

Abb. 15. v. Kupffersche Sternzelle (Bufo bufo) 5 min nach intraperitonealer (!) Injektion von Goldsol. Im Cytoplasma mehrere Vakuolen mit recht unterschiedlichem Inhalt. Eine Vakuole (etwa in Bildmitte am oberen Zellrand) und mehrere Vesikel (Pfeile) enthalten Goldkörnchen. Die Membranen zweier Vakuolen sind stellenweise fragmentiert, stellenweise bereits vollständig verschwunden (Pfeile). Technik siehe Abb. 11. Gesamtvergr.: 28000 : 1. Orig.

sprechen durchaus nicht für eine Transformation pinocytotischer Bläschen zu Mitochondrien [27], sondern nur für die Einbeziehung von Bläschen mikropinocytotischer Größenordnung in die Substanz dieser Organellen.

Die Anreicherung von Goldpartikeln in Vakuolen (Abb. 13, 15) dürfte so gut wie ausschließlich durch Vermittlung mikropinocytotischer Vesikel erfolgt sein (vgl. hierzu auch [11, 18, 33, 59]). 24 und 48 Std nach der Applikation der Markierungsflüssigkeit hat sich die Zahl der „Goldkörnchen-Vakuolen" vielleicht durch Ausschleusung „in toto" (Abb. 13) deutlich vermindert. Während die Membranen anderer Vakuolen gleicher Größenordnung und unterschiedlichen Inhalts häufig Zeichen der Fragmentation bzw. der Auflösung aufweisen (Abb. 15), so daß ihr Inhalt unmittelbar in das Grundplasma inkorporiert werden kann — entsprechende Befunde hat Wohlfarth-Bottermann [100, 104] an der Amöbe *Hyalodiscus simplex* erhoben—, sehen wir niemals vergleichbare Veränderungen an der Membran *jener* Vakuolen, in denen

Tabelle 2: *Verteilung von Goldpartikeln in v. Kupfferschen Sternzellen und in peritonealen Deckzellen bei Kröte (Bufo bufo) und Schildkröte (Emys orbicularis)*

Lokalisation	Sternzellen[1] in %	Deckzellen[2] in %
Auf der Oberflächenmembran	8,7	22,4
In taschenförmigen Einsenkungen der Zell*ober*fläche	3,9	23,5
In Vesikeln	9,4	15,3
In Vakuolen	*76,4*	3,5
„Frei" im Cytoplasma	0	0
In taschenförmigen Einsenkungen der Zell*unter*fläche	0,4	3,1
Unterhalb der Zellen	1,2	*32,2*

[1] Fixierung 5, 10 und 30 min nach intraperitonealer Injektion des Goldsols.
[2] Fixierung 1, 2, 3, 5 und 10 min nach intraperitonealer Injektion des Goldsols.

Goldpartikel abgelagert sind. Eine Vakuolenmembran wird also offenbar nur und erst dann aufgegeben, wenn der von ihr umschlossene Inhalt so weit „assimiliert" worden ist, daß die Barriere gegen das Grundplasma, d. h. gegen das *Milieu interne* der Zelle fallen kann (vgl. hierzu auch [*83, 84*]). Bläschen und Vakuolenmembran sind zwar Abkömmlinge der Zellmembran, können sich jedoch von dieser offenbar in wesentlichen Eigenschaften unterscheiden. Schon 1955 haben CHAPMAN-ANDRESEN und HOLTER [*15*] geglaubt, zeigen zu können, daß sich radioaktiv markierte Glukose nach ihrer Einbringung in Pinocytose - Vakuolen von *Amoeba proteus* rasch über die gesamte Zelle verteilt, obgleich

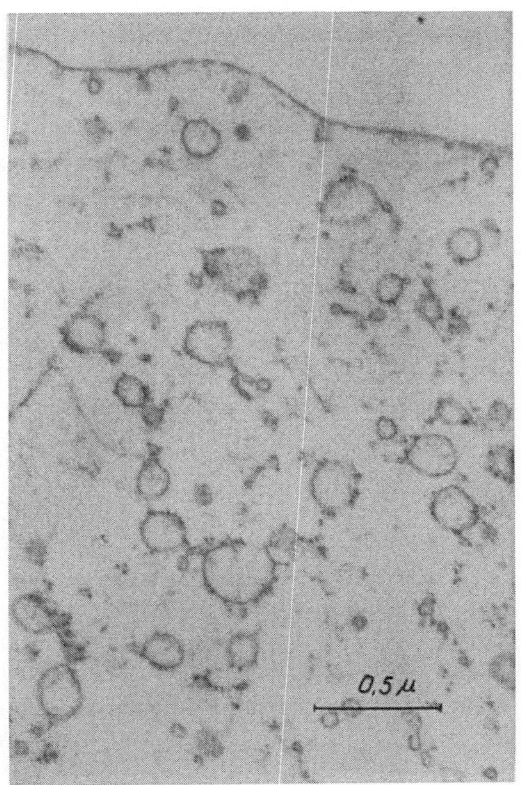

Abb. 16. Peritoneale Deckzelle (Rana temporaria). Zwischen den Vesikeln ist im Grundcytoplasma ein Gespinst feinster Filamente zu erkennen. Technik siehe Abb. 11. Gesamtvergr.: 35000:1 [*90*]

deren Oberflächenmembran für Glukose undurchlässig sein soll. Nach MARCHESI und BARRNETT [*50, 51*] differiert die Fermentaktivität mikropinocytotischer Vesikel des Kapillarendothels gegenüber ATP, ADP und AMP qualitativ von der des Plasmalemms derselben Zellen.

Beim Vergleich speichernder Endothelien (v. Kupffersche Sternzellen) mit peritonealen Deckzellen zeigt sich eine recht unterschiedliche Verteilung der angebotenen Markierungssubstanz (Tab. 2). Während in den Sternzellen die Partikel meistens (nämlich 1364 von 1785 insgesamt ausgezählten Körnchen) in größeren Vakuolen deponiert sind, und sich nur eine relativ kleine Zahl unterhalb der Zellen findet, haben 35,3% der ausgezählten Partikel die mesothelialen Peritonealzellen bereits

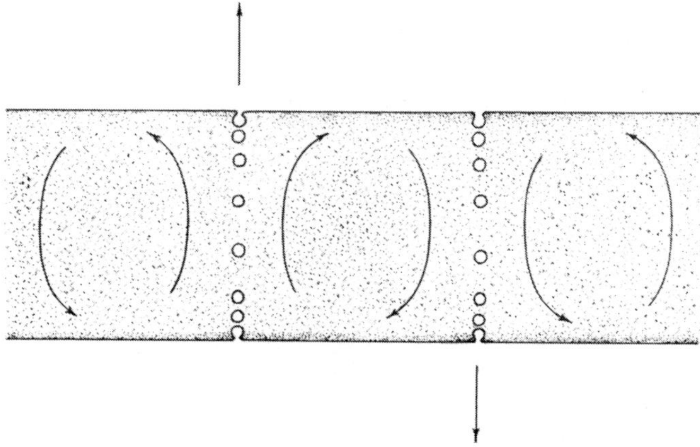

Abb. 17. Schema der gegensinnigen Cytopempsis. Die Vesikulationsrichtungen sind durch gerade Pfeile, vermutete Plasmaströmungen, die die Bläschenbewegung beeinflussen könnten, durch gekrümmte Pfeile markiert. Orig.

1 bis 10 min nach den intraabdominalen Injektionen passiert und liegen entweder in taschenförmigen Einsenkungen der Zell*unter*fläche oder in Spalträumen des subperitonealen Bindegewebes. *Die überwiegende Zahl mikropinocytotischer Vesikel dient demnach in den Sternzellen dem Transport in Richtung auf die Speicherungs-Vakuolen, während in den Mesothelzellen (und entsprechend in Kapillarendothelien) die Vesikel vor allem in die transzelluläre Stoffpassage im Sinne der Cytopempsis eingeschaltet sind.* „Wahrscheinlich bereitet jeder Zelltyp dem pinocytierten Material ein spezifisches Schicksal, in Abhängigkeit von der typischen Ausrüstung mit inneren Strukturen und Enzymen" [99].

Das elektronenmikroskopische Bild liefert bislang keine überzeugenden Hinweise auf die Kräfte, die die *gerichtete Bewegung* der mikropinocytotischen Vesikel bewirken. Ob zarte fädige Strukturen, die man gelegentlich zwischen den Vesikeln antrifft (Abb. 16) „als agierendes Substrat" [104] in Frage kommen, muß weiteren Untersuchungen vorbehalten bleiben.

Wohlfarth-Bottermann [102, 103] hat die Bedeutung von „Plasmafilamenten des Grundcytoplasmas . . ., die bei verschiedenen Objekten offenbar sehr verschieden ausgeprägt sein können und die Fähigkeit besitzen, zu fibrillären Strukturen ‚zusammenzutreten' ", für Bewegungsvorgänge von Amöben und Schleimpilzen überzeugend nachgewiesen. Vielleicht spielen auch Protoplasmaströmungen im Sinne der „Cyclose" [*38*] vor allem bei der sog. gegensinnigen Cytopempsis [*89, 90*] für den Bläschentransport eine Rolle (Abb. 17).

IV. Die Ausschleusung cytopemptisch transportierten Materials

Die letzte Phase des cytopemptischen Transports ist die Ausschleusung des Inhalts mikropinocytotischer Vesikel. Sie kann offenbar auf zweierlei Weise vor sich gehen: entweder wird die Vesikelmembran in

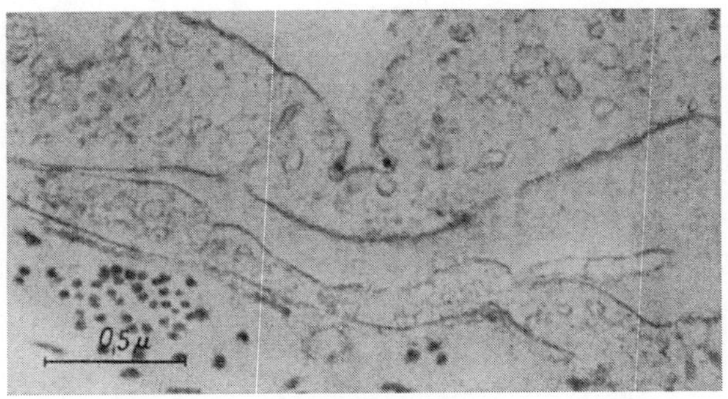

Abb. 18. Peritoneale Deckzelle (Harnblase, Bufo bufo) wenige min nach intravesikaler (!) Injektion von Goldsol. Zwei Goldkörnchen, die nach Passage der Harnblasenschleimhaut auch durch die Peritonealzellen geschleust worden sind, liegen in den Nischen einer grubenförmigen Einsenkung der peritonealen Zelloberfläche. Technik siehe Abb. 1. Gesamtvergr.: 32000:1. Orig.

einer Art Umkehrung ihrer Entstehung in das Plasmalemm einbezogen, während sich die Kuppe der fusionierten Bläschen- und Zellmembran öffnet (Abb. 18, 20), oder das Bläschen wölbt die Zellmembran vor,

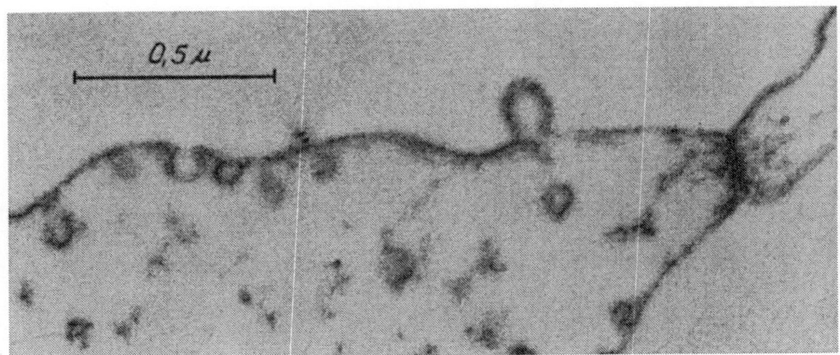

Abb. 19. Oberfläche einer peritonealen Deckzelle (Bufo bufo) nach intraperitonealer Injektion von Goldsol. Links neben einer bläschenförmigen Ausstülpung der Plasmahaut zwei an der Zellmembran haftende Goldkörnchen. Technik siehe Abb. 11. Gesamtvergr.: 51000:1 [*90*]

verschmilzt mit ihr und löst sich anschließend als Ganzes ab (Abb. 19, 20). Der erste Fall würde für die Zelle die Erhaltung eines zuvor in die Tiefe des Cytoplasmas verlagerten Membranteils, der zweite jedoch einen dem Bläschenumfang entsprechenden Membranverlust bedeuten. Beide Möglichkeiten sind bei der Extrusion von Sekretkörnchen aus Drüsenzellen beschrieben worden [2, 3, 19, 76, 77, 78]. Bei der Cytopempsis ist offenbar die Abgabe des Bläscheninhalts *ohne* Membranverlust der Regelfall, die Ausschleusung vollständiger Bläschen — zumindest in unserem Material — seltener.

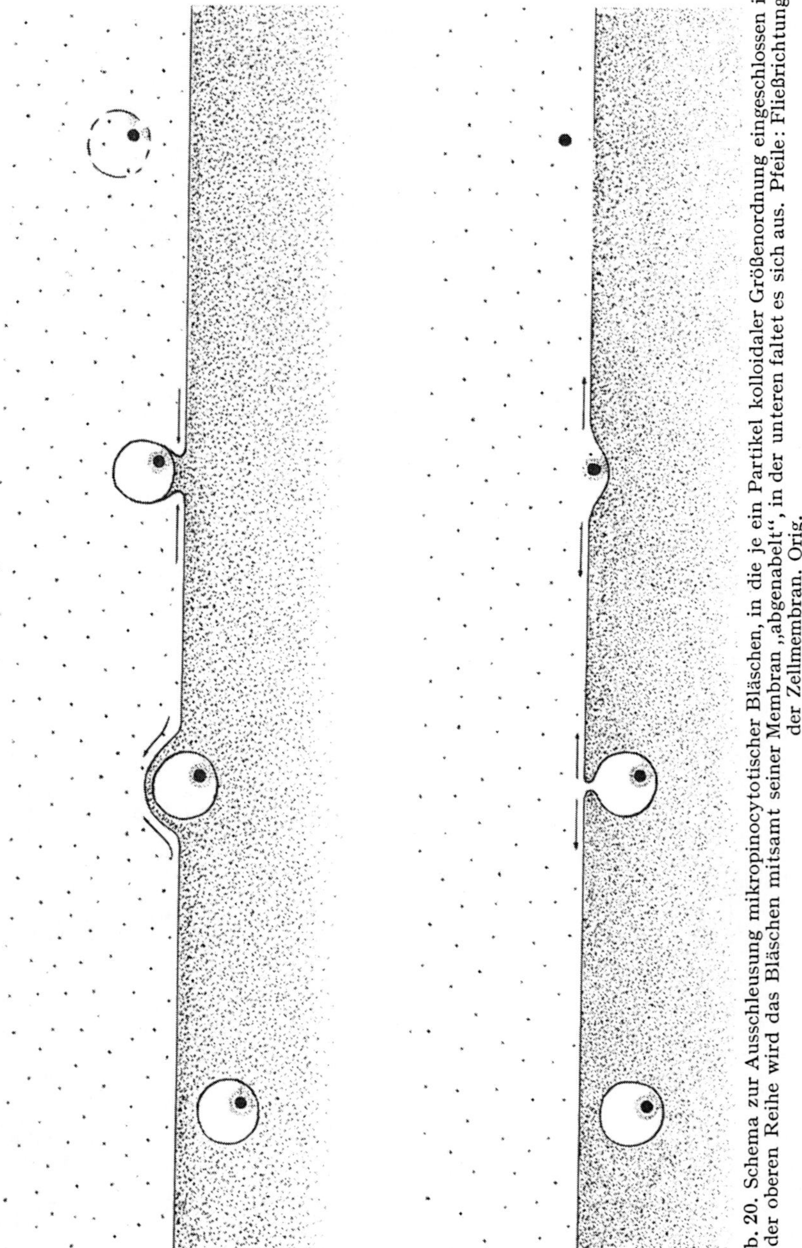

Abb. 20. Schema zur Ausschleusung mikropinocytotischer Bläschen, in die je ein Partikel kolloidaler Größenordnung eingeschlossen ist. In der oberen Reihe wird das Bläschen mitsamt seiner Membran „abgenabelt", in der unteren faltet es sich aus. Pfeile: Fließrichtungen der Zellmembran. Orig.

Ein besonderes Problem hängt damit zusammen, daß dasselbe Partikelchen (z. B. Goldkörnchen), das ursprünglich an der Zellmembran adhärierte und auf diese Weise in ein sich abfaltendes Bläschen eingeschlossen werden konnte, nun auf der anderen Zellseite wieder entlassen wird. Wodurch wird die primäre „Bindung" gelöst? Welche Kräfte bewirken die Trennung des Partikels von der Membran und seinen Abtransport?

C. Die mögliche Bedeutung der Cytopempsis für Transportvorgänge, erläutert am Beispiel endothelialer und mesothelialer Zellen

I. Allgemeines

Ob und in welchem Ausmaß sowie für welche Stoffe die Cytopempsis als Möglichkeit transendothelialer Passagen herangezogen werden darf, ist nach wie vor umstritten. Die beiden sich gegenüber stehenden Extremauffassungen spiegeln sich in folgenden Zitaten: „Die große Permeabilität des Kapillarendothels ist . . . auf sublichtmikroskopische Pinocytose-Mechanismen zurückzuführen" [101]. — "Subsequent experiments using electron-dense particulate markers have, thus far, failed to produce compelling evidence that these vesicles are vehicles of transport in a preferred direction across endothelial cells" [22].

Entscheidende Voraussetzung für den Vollzug einer transendothelialen Cytopempsis ist die Bildung mikropinocytotischer Vesikel (Lit. s. Tab. 1). Es scheinen gewisse quantitative Unterschiede in ihrer Verteilung zu bestehen: z. B. sollen sie in Endothelzellen normaler Kapillaren des Zentralnervensystems [95] und der Retina [52] verhältnismäßig spärlich, in Hautkapillaren häufiger und in Kapillaren der Skelett- und Herzmuskulatur sehr reichlich vorhanden sein [82]. Angaben zur Menge der Vesikel müssen allerdings mit Vorsicht beurteilt werden, weil ihre Zahl bereits unter normalen Bedingungen offenbar von Endothelzelle zu Endothelzelle oder von Kapillarstrecke zu Kapillarstrecke erheblich streuen kann [1, 31, 32]. Deshalb ist auch — solange keine statistisch gesicherten Untersuchungen vorliegen — Zurückhaltung gegenüber jenen, z. T. widerspruchsvollen Angaben geboten, die sich auf eine Vermehrung bzw. Verminderung der Vesikelzahl unter experimentellen Bedingungen beziehen [1, 8, 9, 12, 24a, 35, 46, 56, 72, 75].

Zur Prüfung des cytopemptischen Transports durch ein Endothel (bzw. durch vergleichbare mesotheliale Zellen) werden Stoffe benutzt, die im elektronenmikroskopischen Bild gute Kontraste geben. Bevorzugt kamen zur Anwendung: *Ferritin* [6, 6a, 7, 11, 18, 20, 37, 45, 72], *Thoriumdioxyd* [8, 24, 33, 37, 61, 66, 67, 98], *Quecksilbersulfid* [1, 24a, 33, 48, 61, 97], *Goldsol* [31, 32, 63, 64, 67, 89, 91, 98], *kolloidales Eisen* [24, 37, 67, 89, 91], *Tusche* bzw. *kolloidale Kohle* [24, 43, 49, 105], *metallisches Tellur* [29, 30]. Zur Größe der benutzten Teilchen finden sich u. a. folgende Angaben: ODOR [61] und ALKSNE [1] geben die Durchmesser der intramesothelial (bzw. intraendothelial) gelegenen HgS-Partikel mit 50—220 Å (bzw. 50—250 Å) an. Nach MAJNO und PALADE [48] variieren die Teilchen der verwendeten kolloidalen HgS-Suspension zwischen 70 und 350 Å (über 50% hatten eine Größe von 70–160 Å); in vivo könne der tatsächliche Umfang der Partikel jedoch durch ein adsorbiertes Material eines niederen elektronenmikroskopischen Kontrastes angestiegen sein. Der Durchmesser der Gold-Teilchen beträgt nach Palade [*63, 64*] 40–150 Å.

Die mikropinocytotische *Aufnahme* von Teilchen der genannten Markierungssubstanzen durch Endothel- und Mesothelzellen ist so oft nachgewiesen worden, daß zumindest die erste Phase des cytopemptischen Transports als völlig gesichert gelten kann. Offenbar vermögen Ferritin und Eisenverbindungen das Plasmalemm von Endothel- und Mesothelzellen direkt, d. h. ohne mikropinocytotische Vesikulationsvorgänge zu permeieren [*11, 37, 72, 89, 90, 98*]; das schließt freilich die sekundäre Ablagerung dieser Partikel in Vakuolen nicht aus. ThO_2-, HgS-, Gold- und Tusche-Teilchen werden jedoch in der Regel von Anfang an nicht „frei" im Cytoplasma, sondern in Vesikeln oder Vakuolen angetroffen. Da verschiedene Beobachter [*33, 37, 63, 64, 89, 90, 98*] Markierungspartikel nicht nur in "inpocketings" der Zelloberfläche und in intrazellulären Bläschen, sondern auch in Grübchen der Zell*unter*fläche sowie in perikapillären (bzw. submesothelialen) Bindegewebsspalten (vgl. auch Tab. 2) gefunden haben, kann meines Erachtens *auch der cytopemptische Transport kolloidaler Teilchen durch Endothel- und Mesothelzellen als erwiesen* gelten. Das ermöglicht selbstverständlich noch keine Rückschlüsse auf die physiologische Bedeutung dieses Vorgangs. Der transendotheliale Durchgang von Wasser, hydratisierten Ionen, kleinen organischen Molekülen, überhaupt niedermolekularen Substanzen vollzieht sich viel zu rasch, als daß diesem Phänomen dabei eine wesentliche Rolle zufallen könnte (abgesehen von der u. U. enormen Oberflächenvergrößerung der Zelloberfläche durch Mikropinocytose und der dadurch entsprechend heraufgesetzten Permeationsmöglichkeit). Jedoch wäre es möglich, daß der Austausch hochmolekularer Plasmaproteine und anderer Großmoleküle durch Cytopempsis erfolgt [*53*]. „Wenn nur 0,1% des Plasmavolumens durch Pinocytose transportiert würde, so würde es beinahe den gesamten beobachteten Capillartransport der Plasmaeiweiße erklären" [*73*].

Auch die mikropinocytotische Aufnahme von Lipiden und Lipoiden sowie deren cytopemptische Passage durch Endothelzellen (Atherosklerose-Problem!) ist beschrieben worden [*9, 10, 26, 35*, u. a.], und neuerdings hat Williamson [*97*] auf Möglichkeiten und Bedeutung des transendothelialen cytopemptischen Durchgangs freier Fettsäuren hingewiesen.

II. Gegensinnige Cytopempsis

Wenn der Cytopempsis eine physiologische Bedeutung für die transendotheliale (bzw. mesotheliale) Durchschleusung bestimmter Stoffe zukommt, muß sie *gegensinnig*, d. h. von der Blutseite in Richtung auf die Gewebsflüssigkeit und umgekehrt erfolgen. An unserem Modell (Herzbeutel- und Peritonealmesothel) haben wir die gegensinnige Passage dadurch experimentell geprüft, daß die Markierungssubstanz einmal unmittelbar in die serösen Höhlen (Herzbeutel, Bauchhöhle) und zum anderen intravasal appliziert wurde. Die Wahrscheinlichkeit, Partikel nach ihrer Injektion *in das Gefäßsystem* im Rahmen allgemeiner Austauschvorgänge in Deckzellen der Bauch- oder Herzbeutelhöhle wiederzufinden, war von vornherein nur sehr gering, da angenommen werden

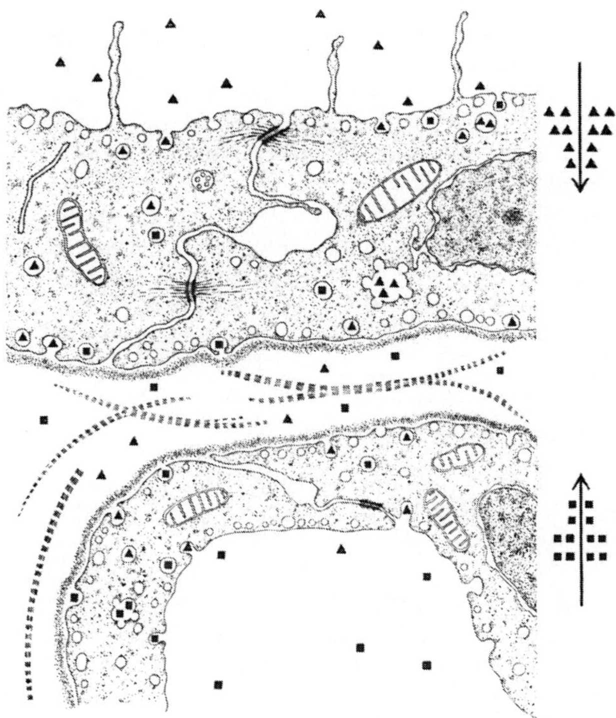

Abb. 21. Schema des gegensinnigen Stoffverkehrs durch Cytopempsis zwischen einer serösen Höhle (oben) und einer Blutkapillare (unten). Die Pfeile rechts, von Symbolen für transportierte kolloidale Substanzen umgeben, weisen auf die entgegengesetzten Richtungen cytopemptischer Passagen hin. Jede Position des Markierungsstoffes konnte experimentell nachgewiesen werden [*90*]

mußte, daß die Masse eines auf diese Weise verabreichten Fremdstoffs von den endothelialen Uferzellen abgefangen wird, noch ehe die Teilchen im Zuge des ständigen Flüssigkeits- und Stoffwechsels zwischen Blut und Peritoneal- bzw. Pericardialserum in den Deckzellen erscheinen konnten. Trotz dieser ungünstigen Ausgangssituation, also fast wider Erwarten, ließen sich die zur Markierung benützten Goldkörnchen auch nach intravasaler Injektion in Vesikeln peritonealer Deckzellen nachweisen [*89, 90*]. Daraus ergibt sich zwingend, daß cytopemptische Transporte eines partikulierten Materials kolloidaler Größenordnung in Mesothel- wie Endothelzellen in entgegengesetzten Richtungen erfolgen (Abb. 21). Das bestätigt und erweitert Befunde HAMPTONs [*33*], der ThO$_2$- und HgS-Teilchen orthograd über die V. portae und retrograd über den Ductus choledochus leberwärts injizierte. Diese — wenn auch etwas unphysiologische — Versuchsanordnung erlaubte Aussagen über entgegengesetzte cytopemptische Transporte durch das Cytoplasma der Leberzellen.

III. Cytopempsis-Zeiten und andere quantitative Angaben

Für die mikropinocytotische *Aufnahme* kolloidaler Partikel durch Endothel- und Mesothelzellen sind *weniger als 60 sec*, für die transzelluläre Cytopempsis einige Minuten zu veranschlagen [*33, 63, 64, 89, 90*]. Den z. T. wesentlich längeren Zeitangaben anderer Untersucher [*1, 6a, 8, 61*, u. a.] könnten Befunde zugrundeliegen, die sich nicht mehr auf die Durchschleusung, sondern schon auf die Ablagerung der benutzten Markierungssubstanzen beziehen.

In Anlehnung an quantitative Überlegungen Parsons' [*70*] hat mein Mitarbeiter Hammersen [*32*] folgende Überschlagsrechnung zur Abschätzung der die Kapillarwand cytopemptisch passierenden Flüssigkeitsmenge angestellt:

In 1000 g Skelettmuskulatur beträgt die Kapillaroberfläche $7 \times 10^{12} \mu^2$ [*74*]. Setzt man für die Oberfläche einer Endothelzelle rund 120 μ^2 ein, entfallen auf die gesamte Kapillaroberfläche (pro 1000 g Muskulatur) $5,8 \times 10^{10}$ Endothelzellen. Da für die Aufnahme von 0,1 ml

Tabelle 3 (nach Parsons 1963): *Für die Aufnahme von 0,1 ml Flüssigkeit benötigte Bläschenzahl*

Bläschendurchmesser in Å	Bläschenzahl
200	$2,4 \times 10^{16}$
500	$1,5 \times 10^{15}$
600	$9,1 \times 10^{14}$
1000	$1,8 \times 10^{14}$

Flüssigkeit bei einem Vesikeldurchmesser von 500 Å $1,5 \times 10^{15}$ Bläschen benötigt werden (Tab. 3), kommen auf jede Zelle $2,6 \times 10^4$ Bläschen mit einem Gesamtvolumen von $1,71 \times 10^{-12}$ ml. Unter der Voraussetzung einer gleichmäßigen Verteilung der Vesikel und einer durchschnittlichen Länge der Zellen von 12 μ ist im 300 Å dicken Schnitt mit 65 Vesikeln pro Endothelzelle zu rechnen. Das entspricht in der Größenordnung den Befunden (Abb. 4, 5). Geht man weiter davon aus, daß der cytopemptische Durchgang der Bläschen 2—3 min dauert, würden bereits in dieser kurzen Zeit durch die Endothelzellen der in 1000 g Muskulatur verlaufenden Kapillaren 0,1 ml Flüssigkeit transportiert. Pro Stunde könnten demnach auf diese Weise 2—3 ml und pro Tag und 1000 g Muskulatur 48—72 ml ausgetauscht werden.

D. Zusammenfassung

Transzelluläre cytopemptische Passagen beginnen mit Bläschenbildungen nach dem Prinzip der Mikropinocytose. Mikropinocytose und Pinocytose der lichtmikroskopischen Dimension unterscheiden sich u. a. dadurch, daß die *Pinocytose-Vakuolen* in der Regel durch Vermittlung von Pseudopodien, *Mikropinocytose-Vesikel* hingegen durch Invagination der Zellmembran entstehen. Während in den je nach ihrem Zustand unterschiedlich großen und auch hinsichtlich ihres Inhalts polymorphen Pinocytose-Vakuolen das aufgenommene Material in erster Linie eingedickt bzw. abgelagert und/oder aufgearbeitet wird, dienen die meist recht uniformen Vesikel offenbar vor allem (wenn auch sicher nicht ausschließlich) Transportvorgängen. Die räumliche und funktionelle Verknüpfung beider Phänomene, die prinzipiell unterschieden werden sollten, wird jedoch besonders dort deutlich, wo lichtmikroskopisch entstandene Pinocytose-Vakuolen von dicht stehenden „Sekundär-Vesikeln" umgeben sind.

Cytopemptische Vesikel unterscheiden sich weder von mikropinocytotischen Vesikeln noch von Bläschen, die sich von den Lamellen des Golgi-Apparates abgeschnürt haben. Auch Anschnitte des endoplasmatischen Retikulum und Vesikel-„Ketten", die durch Membranfusionierungen und anschließenden Zerfall in flächenhafte Bläschenkomplexe entstanden sind, können u. U. mit cytopemptischen Transportbläschen verwechselt werden. Die Zahl der in cytopemptische Vorgänge eingeschalteten Vesikel wird also kleiner sein als die Zahl der im Schnitt sichtbaren Membranringe der in Frage kommenden Abmessungen.

Die Ursachen für Bläschenentstehung, -wanderung und -wiederausfaltung sind noch weitgehend undurchsichtig. Dagegen besteht Klarheit in bezug auf die für den transendothelialen cytopemptischen Durchgang bestimmter kolloidaler Markierungssubstanzen erforderliche Mindestzeit, die in der Größenordnung weniger Minuten liegt. Am Kapillarendothel und an den mesothelialen Deckzellen der serösen Höhlen konnte die gegenläufige Richtung cytopemptischer Transporte experimentell erwiesen werden. Der transendotheliale (und -mesotheliale) Durchgang von Wasser, hydratisierten Ionen und kleineren organischen Molekülen dürfte nur insofern durch Cytopempsis beeinflußt werden, als dieser Vorgang zu einer beträchtlichen Vergrößerung zwischen Extra- und Intrazellularraum führt. Dagegen müssen die Bläschen als Vehikel für den Austausch hochmolekularer Plasmaeiweiße und anderer Großmoleküle in Betracht gezogen werden. Überschlagsmäßig ergibt sich, daß mit Hilfe des Vesikeltransportes durch die Endothelzellen der auf 1000 g Skelettmuskulatur kommenden Kapillaren in 2—3 min 0,1, pro Stunde 2—3 und pro Tag 48—72 ml Flüssigkeit ausgetauscht werden könnten.

Literatur

[1] ALKSNE, J. F.: Quart. J. exp. Physiol. **44**, 51 (1959).
[2] BARGMANN, W., K. FLEISCHHAUER u. A. KNOOP: Z. Zellforsch. **53**, 545 (1961).
[3] —, u. A. KNOOP: Z. Zellforsch. **49**, 344 (1959).
[4] BENNETT, H. ST.: J. biophys. biochem. Cytol. (Suppl. 2) **2**, 99 (1956).
[5] —, J. H. LUFT, and J. C. HAMPTON: Amer. J. Physiol. **196**, 381 (1959).
[6] BESSIS, M., et J. BRETON-GORIUS: Sem. Hôp. Path. et Biol. **1957a**, 411.
[6a] BRUNS, R. R.: Anat. Rec. **145**, 360 (1963) Abstr.
[7] — Sem. Hôp. Path. et Biol. **1957b**, 2173.
[7a] BRANDT, PH. O., and G. D. PAPPAS: J. biophys. biochem. Cytol. **8**, 675 (1960).
[8] BUCK, R. C.: J. biophys. biochem. Cytol. **4**, 187 (1958a).
[9] — Amer. J. Pathol. **34**, 897 (1958b).
[10] — Histogenesis and morphology of arterial tissue. In: M. SANDLER and G. H. BOURNE: Artherosclerosis and its origin. New York and London: Academic Press 1963.

[11] Casley-Smith, J. R., and H. W. Florey: Quart. J. exp. Physiol. **46**, 101 (1961)
[12] Cervós-Navarro, J.: Arch. Psychiat. Nervenkr. **204**, 484 (1963).
[13] — Arch. Psychiat. Nervenkr. **205**, 204 (1964).
[14] Chapman-Andresen, C.: Studies on pinocytosis im amoebae. Lab. Carlsberg, Copenhagen: Danish Sci.Press 73—264 (1962).
[15] —, and H. Holter: Exp. Cell Res. Suppl. **3**, 52 (1955).
[16] —, and J. R. Nilsson: Exp. Cell Res. **19**, 631 (1960).
[16a] Duve, Chr. de: Ciba Foundation Symposion, London: J. & A. Churchill Ltd., 1963.
[17] Ekholm, R., and Y. Edlund: J. Ultrastruct. Res. **2**, 453 (1959).
[18] Farquhar, M. G.: Anat. Rec. **136**, 191 (1960) Abstr.
[19] — Angiology **12**, 270 (1961).
[20] —, S. L. Wissig and G. E. Palade: J. exp. Med. **113**, 47 (1961).
[21] Fawcett, D.W.: The fine structure of capillaries, arterioles and small arteries. In: S. R. M. Reynold and B. W. Zweifach: The microcirculation, Symposion of factors influencing exchange of substances across capillary wall. The University of Illinois Press: Urbana, 1959.
[22] — Selected applications of the electron microscope in histology and cytology. In: B. M. Siegel: Modern Developments in Electron Microscopy. New York and London: Academic Press 1964.
[23] Fernando, N. V. P., and H. Z. Movat: Exp. molecul. Pathol. **3**, 87 (1964).
[24] French, J. E., H. W. Florey, and B. Morris: Quart. J. exp. Physiol. **45**, 88 (1960).
[24a] Friederici, H. H. R., and C. L. Pirani: Lab. Invest. **13**, 250 (1964).
[25] Gansler, H.: Z. Zellforsch. **55**, 724 (1961).
[26] Geer, J. C., H. C. McGill, and J. P. Strong: Amer. J. Pathol. **38**, 263 (1961).
[27] Gey, G. O., P. Shapras, and E. Borysko: Ann. N. Y. Acad. Sci. **58**, 1089 (1954).
[28] Gropp, A.: Phagocytosis and Pinocytosis. In: G. G. Rose: Cinemicrography in Cell Biology. New York and London: Academic Press 1963.
[29] Hager, H.: Arch. Psychiat. Nervenkr. **201**, 53 (1960) Abstr.
[30] — Acta neuropath. (Berlin) **1**, 9 (1961).
[31] Hammersen, F.: Verh. Anat. Ges. 60. Vers. Wien 1964a, Anat. Anz. (Erg.-H.) im Druck.
[32] — 1964b. In Vorbereitung.
[33] Hampton, J.: Acta Anat. (Basel) **32**, 262 (1958).
[34] Han, S. S., and J. K. Avery: Anat. Rec. **145**, 549 (1963).
[35] Hess, R., u. W. Stäubli: Verh. dtsch. Ges. Pathol. **47**, 369 (1963).
[35a] Holter, H.: Intern. Rev. Cytol. **8**, 481 (1959).
[36] — Pinocytosis. Biological structure and function, Vol. I. London and New York: Academic Press Inc. 1960.
[37] Jennings, M. A., V. T. Marchesi, and H. W. Florey: Proc. roy. Soc. B, **156**, 14 (1962).
[38] Kamiya, N.: Protoplasmic streaming. Protoplasmatologia VIII/3a. Wien: Springer 1959.
[39] Kisch, B.: Exp. Med. Surg. **18**, 182 (1960).
[40] — Arch. Kreisl.-Forsch. **42**, 1 (1963).
[41] Komnick, H.: Protoplasma (Wien) **56**, 385 (1963a).
[42] — Protoplasma (Wien) **56**, 605 (1963b).
[43] Kotani, M., M. Rai, and S. Nakao: Okajimas Folia anat. jap. **38**, 149 (1962).
[44] Lewis, W. H.: Bull. Johns Hopk. Hosp. **49**, 17 (1931).
[45] Linss, W., u. G. Geyer: Anat. Anz. **114**, 225 (1964).
[46] Luse, S. A.: J. Histochem. Cytochem. **8**, 398 (1960).
[47] Macher, E., u. W. Vogell: Dermatologie **124**, 110 (1962).
[48] Majno, G., and G. E. Palade: J. biophys. biochem. Cytol. **11**, 571 (1961).
[49] Marchesi, V. T.: Proc. roy. Soc. B. **156**, 550 (1962).
[50] —, and R. J. Barrnett: J. Cell Biol. **17**, 547 (1963).
[51] — — J. Ultrastruct. Res. **10**, 103 (1964).

[52] MARCHESI, V. I., M. L. SEARS, and R. J. BARRNETT: Invest. Ophthal. 3, 1 (1964).
[52a] MARSHALL, J. M.: XIth Intern. Congr. Cell Biol. Providence, Rhode Island, 1964.
[53] MAYERSON, H. J., C. G. WOLFRAM, H. H. SHIRLEY JR., and K. WASSERMAN: Amer. J. Physiol. 198, 155 (1960).
[54] MISOTTEN, M. L.: Bull. Soc. belge Ophthal. 129, 382 (1961).
[55] — Ophthalmologica (Basel) 144, 1 (1962).
[56] MOORE, D. M.: In: S. R. M. REYNOLDS and B. W. ZWEIFACH: The Microcirculation. Symposion on Factors influencing Exchange of Substances across Capillary Wall. The University of Illinois Press: Urbana 1959.
[57] MOORE, D. H., and H. RUSKA: J. biophys. biochem. Cytol. 3, 457 (1957).
[58] NOIROT, CH., et C. NOIROT-TIMOTHÉE: C. R. Acad. Sci. (Paris) 251, 2779 (1960).
[59] NOVIKOFF, A. B.: Biol. Bull. 119, 287 (1960).
[59a] — Lysosomes and related particles. In, J. BRACHE and A. E. MIRSKY: The Cell. Vol. II. New York and London, Academic Press 1961
[60] ODLAND, G. F.: The fine structure of cutaneous capillaries. In: W. MONTAGNE und R. A. ELLIS: Advances in Biology of Skin. Vol. II. Blood Vessels and Circulation 57—70 (1961).
[61] ODOR, D. L.: J. biophys. biochem. Cytol. 2, Suppl. 105 (1956).
[62] PALADE, G. E.: J .Appl. Physics 24, 1424 (1953).
[63] — Anat. Rec. 136, 254 (1960) Abstr.
[64] — Circulation 24, 368 (1961).
[65] PAPPAS, G. D.: Ultrastructure of the Ciliary Epithelium and its Relationship to aqueous Secretion. In: F. W. NEWELL: Glaucoma. Transaction IVth Conf. Josiah Macy jr. Foundation, New York 1940.
[66] —, and G. K. SMELSER: The fine Structure of the Ciliary Epithelium in Relation to aquenous Humor Secretion. In: G. K. SMELSER: The Structure of the Eye. New York and London: Academic Press 1961
[67] —, and V. M. TENNYSON: J. Cell Biol. 2, 227 (1962).
[68] PARKER, F.: Amer. J. Anat. 103, 247 (1959).
[69] — Amer. J. Pathol. 36, 19 (1960).
[70] PARSONS, D. S.: Nature (Lond.) 199, 1192 (1963).
[71] POLICARD, A., A. COLLET et S. PREGERMAN: Acta anat. (Basel) 30, 624 (1957).
[72] RAIMONDI, A. J., J. P. EVANS, and S. MULLAN: Acta Neuropath. 2, 177 (1962).
[73] RENKIN, E. M.: Klin. Wschr. 41, 147 (1963).
[74] —, u. J. R. PAPPENHEIMER: Erg. Physiol. 49, 59 (1957).
[75] RINEHART, J. F., and M. G. FARQUHAR: Anat. Rec. 121, 207 (1955).
[76] ROBERTIS, E. DE, and A. V. FERREIRA: Exp. Cell Res. 12, 568 (1957a).
[77] — — Exp. Cell Res. 12, 575 (1957b).
[78] —, and D. D. SABATINI: Fed. Proc. 19, 70 (1960).
[79] ROTH, L. E.: J. Protozool. 7, 176 (1960).
[80] RUSKA, C.: Z. Zellforsch. 53, 867 (1961).
[81] RUSKA, H.: Marburger Sitz.-Ber. 82, 3 (1960).
[82] — Der Feinbau von Capillaren. In: L. DELIUS u. E. WITZLEB: Probleme der Haut- und Muskeldurchblutung. Bad Oeynhausener Gespräche VI. Berlin-Göttingen-Heidelberg: Springer 1964.
[83] SCHMIDT, W.: Z. Zellforsch. 58, 573 (1962).
[84] — Verh. anat. Ges. 58. Vers. 1962, Anat. Anz. 112, (1963), Erg.-H. 178 (1964).
[85] SCHNEIDER, L.: Verh. dtsch. Zool. Ges. Münster i. W., Zool. Anz. (Suppl.) 457 (1959).
[86] — J. Protozool. 7, 75 (1960).
[87] SCHULZ, H.: Die submikroskopische Anatomie und Pathologie der Lunge. Berlin-Göttingen-Heidelberg: Springer-Verlag 1959.
[88] SIEKEVITZ, PH.: On the Meaning of intracellular Structure for metabolic Regulation. In: G. E. W. WOLSTENHOLME and C. M. O'CONNOR: Ciba Foundation Symposium on the Regulation of Cell Metabolism. London: J. & A. Churchill Ltd. 1959.
[89] STAUBESAND, J.: Klin. Wschr. 38, 1248 (1960).
[90] — Z. Zellforsch. 58, 915 (1963).
[91] — Pharm. Ztg. (Frankfurt) 109, 1130 (1964).

[92] STAUBESAND, J., u. K.-H. KERSTING: Z. Zellforsch. 62, 416 (1964).
[93] —, B. KUHLO u. K.-H. KERSTING: Z. Zellforsch. 61, 401 (1963).
[94] —, u. D. WITTEKIND: Dtsch. med. Forsch. (1964) im Druck.
[95] TORACK, R. M., and R. J. BARRNETT: J. Neuropath. exp. Neurol. 23, 46 (1964).
[96] WEILING, F.: Protoplasma (Wien) 55, 452 (1962).
[97] WILLIAMSON, J. R.: J. Cell Biol. 20, 57 (1964).
[98] WISSIG, S. L.: Anat. Rec. 130, 467 (1958) Abstr.
[99] WITTEKIND, D.: Naturwissenschaften 7, 270 (1963).
[100] WOHLFARTH-BOTTERMANN, K. E.: Protoplasma (Wien) 52, 58 (1960a).
[101] — Zool. Anz. 23, Suppl. 393 (1960b).
[102] — Protoplasma (Wien) 54, 514 (1962).
[103] — Protoplasma (Wien) 57, 747 (1963a).
[104] — Zellstrukturen und ihre Bedeutung für die amöboide Bewegung. Köln und Opladen: Westdtsch. Verl. 1963b.
[105] YAMAGISHI, T.: Nagoya med. J. 7, 1 (1961).

Summary

Transcellular passage by cytopempsis begins with the formation of vesicles according to the principle of micropinocytosis. *Micro*pinocytosis and pinocytosis (of light-microscopic dimensions) differ from each other, among other respects, in that pinocytotic vacuoles are formed by fusing pseudopods, while micropinocytotic vesicles originate in invaginations of the cell membrane itself. Besides their mechanism of formation, the main function of these membrane-bounded intracellular cavities is also different. The pinocytotic vacuoles, which are heterogeneous in their size and contents, store, concentrate, and/or elaborate the material taken up; while the more uniform *vesicles* obviously serve transport purposes, though not exclusively. The functional and local interconnection of both these phenomena, between which a clear distinction should be drawn, becomes especially evident in those areas where pinocytotic vacuoles of light-microscopic dimensions are surrounded by crowded "secondary" vesicles on an ultramicroscopic scale.

Cytopemptic vesicles differ neither from micropinocytotic vesicles nor from vesicles which have been pinched off from the lamellae of the Golgi apparatus. In addition, certain profiles of the endoplasmic reticulum, and rows of vesicles which originate by fusion of membranes and subsequent break-down into planar vesicle complexes, might be confused with cytopemptic transport vesicles. Thus, the number of vesicles actually taking part in the process of cytopempsis is less than the number of membrane-bounded annular profiles of the appropriate dimensions visible in a given section.

The causes of vesicle formation, migration, and subsequent outfolding are still largely obscure. However, the minimal time necessary for the transendothelial passage of certain colloidal tracer particles is well-known and lies in the range of a few minutes. The fact that cytopempsis is a transport mechanism working in both directions has been proven experimentally for capillary endothelium and the mesothelial cells of the serous cavities. The transendothelial (and transmesothelial) passage of water, hydrated ions, and smaller organic molecules is influenced by cytopempsis only insofar as the contact area between the intracellular and extracellular space is thereby considerably enlarged. However, for the exchange of plasmaproteins and other substances with higher molecular weight one has to take the cytopemptic vesicles into account. A rough calculation based on number, surface area, etc., of vesicles and capillaries in skeletal muscle showed that the fluid exchange by means of vesicles through the capillary endothelial cells contained within 1000 g muscle is approximately 0.1 ml in 2—3 min., 2—3 ml per hour, or 48—72 ml per day.

Diskussion

Sievers: Die Perlschnurketten von Vesikeln, die Sie früher als Cytopempsisketten gedeutet haben und heute anders deuten, legen eigentlich die Vermutung nahe, daß sie, wenn sie nicht durch verschiedene Fixierungen belegt sind, Fixierungsartefakte sind. Die Gefahr ist immer groß, daß sich, wenn man solche Reihen von Vesikeln nur nach *einer* Fixierung beobachtet, durch einen Effekt, den man nicht genau kennt, diese Vesikel während der Fixation gebildet haben.

Staubesand: Ich stimme Ihnen vollkommen zu. Bei den Bildern, die Sie jetzt im Sinne haben, die sich auf die Muskulatur des Regenwurms bezogen und auf die Megakaryocyten, haben wir nur mit Osmium fixiert. Allerdings muß ich sagen, für das Problem, um das es hier geht, scheint mir das keine so große Rolle zu spielen. Ich wollte mit diesen Beispielen lediglich darauf hinweisen, daß nicht jedes Bläschen, das man in einer Zelle findet, in Zusammenhang gebracht werden darf mit einem cytopemptischen Transport.

Wohlfarth-Bottermann: Ich möchte nur eine kurze Bemerkung zu der Frage von Herrn STAUBESAND machen, welche Kräfte für die Bewegung dieser Bläschen, ganz gleich, welcher Art, im Cytoplasma verantwortlich sein können. Sie wiesen auf die Filamente des Grundcytoplasmas hin. Ich sehe da keine Alternative wie Sie; denn Sie sagten: Entweder werden sie durch diese Plasmafilamente bewegt oder durch eine Protoplasmaströmung. Wir glauben, daß die Protoplasmaströmung auch in diesen Zellen letztlich durch diese Filamente, durch dieses sog. kontraktile Gelretikulum in Bewegung gesetzt wird. Das würde beides sehr gut zusammen stimmen und wird auch sehr schön bestätigt durch die elektronenmikroskopischen Aufnahmen, die Herr HOLTER zeigte, wo diese Verdichtungen des Ektoplasmas direkt und genau an den Stellen ansetzen, wo die Kräfte benötigt werden, um die Membran in die Zelle hereinzuholen.

Wessing: Wie ist es mit der Begrenzung der Bläschengröße nach oben hin? Wo liegen die Grenzwerte?

Staubesand: Ich würde vorschlagen, daß man membranumgrenzte Räume oberhalb 1000 Å Durchmesser als Vakuolen bezeichnen sollte und unterhalb 1000 Å als Vesikel. Es ist mir immer wieder aufgefallen, daß diese beiden Begriffe Vesikel und Vakuolen synonym gebraucht werden, und das erscheint mir sehr problematisch.

Wessing: Dann gibt es nach Ihrer Definition Cytopempsisvesikel und Cytopempsisvakuolen?

Staubesand: Nein, die Cytopempsis spielt sich an Vesikeln ab. Sie setzt eine Mikropinocytose voraus, d. h. eine Pinocytose elektronenmikroskopischer Dimension, nicht eine Pinocytose lichtmikroskopischer Größenordnung. Die Angaben in der Literatur über die Größe der mikropinocytotischen Vesikel schwanken sehr, zwischen 200 und 1000 Å. Aber die Mehrzahl der Untersucher gibt Größen an — das stimmt auch mit unseren Messungen überein — zwischen 500 und 750 Å.

Wessing: Beim Primärharntransport von Drosophila liegen manchmal Bläschen vor, die sogar über 1 μ Durchmesser erreichen können. Es handelt sich meiner Ansicht nach um Cytopempsisvesikel.

Gersch: Das ist eine sehr entscheidende Frage der Charakteristik, ob man das nach der Größe bestimmen kann oder ob andere Faktoren beteiligt sind.

Holter: Ich würde empfehlen, recht vorsichtig zu sein, von ganz gelegentlichen elektronenmikroskopischen Befunden sich überzeugen zu lassen, daß man mit einer systematischen Fusion z. B. zwischen Mitochondrien und derartigen Cytopempsisvesikeln rechnen kann. Bei Leukocyten haben wir die Cytoplasmaströmungen, die ein zufälliges Treffen von Vakuolen begünstigen würden.

Staubesand: Es ist behauptet worden, daß Pinocytosevakuolen in Mitochondrien übergehen könnten und Mitochondrien aus Pinocytosevakuolen entstehen sollen. Für solche Vorgänge habe ich in unserem Material niemals irgendeinen Anhalt gefunden. Ich wollte nur diese Bilder hier zur Diskussion stellen, um auf mögliche Fehlinterpretationen auf diesem Gebiet hinzuweisen.

Diamond: Da heute soviel von Vakuolen und Bläschen die Rede ist, möchte ich auf ein wichtiges Phänomen hinweisen, wo Bläschen und Vakuolen keine Rolle spielen, nämlich auf den Transport von Wasser in großen Mengen durch Epithelzellen, wie es uns im Magen, im Darm, in der Leber, in der Gallenblase und in Drüsen bekannt ist. Der Mechanismus in diesen Epithelzellen ist ein ganz anderer. Dafür liegen viele Beweise vor. Ich möchte nur zwei anführen. Zuerst ist die Geschwindigkeit, mit der Wasser durch diese Zellen transportiert wird, viel zu groß. Beispielsweise in der Gallenblase und im Darm wird durch die Zellen in ein paar Minuten mehr Wasser durchtransportiert, als das Gesamtvolumen der Zellen ausmacht.

Zweitens werden oftmals nicht nur Goldkörner und dgl. von diesen Zellen nicht aufgenommen, sondern auch feinere Moleküle von etwa 7 Å Durchmesser nicht. Wenn man in die Gallenblase Moleküle verschiedener Größe einführt, wird normalerweise nur Natriumchlorid heraustransportiert. Die anderen Moleküle bleiben, sogar wenn der Durchmesser nur 6 oder 7 Å beträgt, nach 6 Std bis zu 99,9% noch in der Gallenblase. Von einer Aufnahme von Flüssigkeitströpfchen kann in diesen Fällen nicht die Rede sein.

Staubesand: Daß Wasser und Ionen ohne Pinocytose die Zellen passieren, ist vollkommen klar. Es würde mir leid tun, wenn Sie gemeint haben, ich wollte alle Transportprozesse auf Cytopempsis beziehen. Ich habe gerade an dem von Ihnen genannten Beispiel des Gallenblasenepithels Beobachtungen anstellen können. Es gelingt nicht, wenn man eine Goldsollösung in die Gallenblase bringt, im Epithel der Gallenblase Goldsolkörnchen in Vesikeln nachzuweisen oder nur in einem so kleinen Prozentsatz, daß man es praktisch nicht zu berücksichtigen braucht. Für das Gallenblasenepithel spielt offenbar die Cytopempsis nicht die Rolle wie für das Endothel und für das Mesothel. Daß die Cytopempsis als Transportphänomen existiert, darüber gibt es heute keinen Zweifel mehr. Ihre quantitative Bedeutung, sowohl bei den verschiedenen Zellen als auch bei den verschiedenen Stoffen, d. h. ihr quantitatives Verhältnis zum Transmembran-Transport, läßt sich jedoch heute noch nicht abschätzen.

Solomon: I should like to talk about some observations that we have made in white cells that were induced to phagocytose. We were concerned about the general problem of the relationship of pinocytosis to transport processes of the kind that those of us in the "Transport Workers Union" seem to specialise in, and we thought — at least it occurred to us — that one important problem that does not seem to me to be soluble with the electron-microscope was the relative importance of ion transport during phagocytosis as compared with the total flux of ions across the membrane. So we devised some experiments, Dr. Karnovsky and myself, in our laboratory. Karnovsky had been working with white cells and was able to turn on the phagocytotic process at will by feeding them polystyrene particles. The experiment that we devised was to measure the uptake of radioactive potassium as a function of time. These experiments were done some years ago and we have never published them because the problems of the kinetics in making flux measurements when the cell is taking in water as well as ions are difficult. But we did, however, observe the rate of disappearance of radioactive potassium from the medium in two aliquots of white cells; one set was phagocytotic and the other set was not, and as far as we could tell the uptake of potassium fell nearly on the same curve; there was no difference that we could determine between the two curves, whether the cells were phagocytosing or not. This led us to conclude that the amount of ion transport that took place was large compared with the amount of ions that might have been transported during the phagocytotic process. It appears from our preliminary experiments that phagocytosis is not likely to be playing an important role with regard to potassium uptake.

L. Hokin: I would like to add some observations we have made in our laboratory to those that Dr. Solomon mentioned. We have also been studying the induction of phagocytosis in leucocytes with polystyrene particles and have used P^{32} in the medium and have followed the uptake of P^{32} into the cells. Now, you cannot measure the inorganic phosphate inside the cell very accurately because of adherent inorganic phosphate and so on, but the ATP which is in equilibrium isotopically with the intracellular inorganic phosphate is a good measure of the specific activity of the intracellular inorganic phosphate, and we found that when we stimulated phagocytosis with polystyrene particles there was no increase in the uptake of inorganic phosphate into the cells, at least inorganic phosphate that would exchange with the ATP. I think it would be very likely that if there were any increased uptake of inorganic phosphate we would have seen it by higher specific activities of the ATP. So I think that perhaps we can add phosphate to potassium, as two ionic species which do not show increased transport under conditions where phagocytosis had been markedly stimulated.

Tosteson: I wanted to make a general comment about this problem of the reationship between pinocytosis and the transport of small molecules, water and small ions, across the so-called unit membrane. Physiologists who have been interested in the transport of these small substances have not found much value in the process of pinocytosis in explaining the phenomena that are observed in the movement of these small substances across the unit membrane. We have recently made some calculations concerning the active transport of potassium across the red blood cell membrane which, I think, lead to some conclusions that might bring the two kinds of processes closer together. These are kinetic considerations. There are approximately 3,000 sites, places on the red cell membrane, for active transport of potassium — 3,000 sites per cell. There are approximately 3,000 potassium ions actively transported per second. This means that the average turnover time of each site is a second. The membrane is only approximately 100 Å thick. This is true of the plasmalemma and all the intracellular membranes, all of the unit membranes that we have been discussing. A second is an eternity at dimensions of this size. The time required for a sodium ion, for example, to diffuse across 100 Å distance is 10^{-8} seconds. The relaxation time for rotary diffusion of a tobacco mosaic virus, which is several thousand Å long is a milli-second — very fast compared to the average transit time.

Thus, although we have no knowledge of the size of membrane unit, what the physiologists have called "the carrier", that moves or is moved in the passage of the potassium or sodium ion across the membrane, we must admit that it could be a very large piece. If one imagines that the membrane structure is micellar, it could be that one micelle moves in relation to a neighbouring micelle in a form of picapinocytosis, if you wanted to coin a new word.

Stoffausscheidung bei Protozoen

Von

L. Schneider, Bonn

Mit 2 Abbildungen

A. Einleitung

I. Die Sonderstellung der Protozoen

Die Protisten nehmen als einzellige Organismen im Tier- und Pflanzenreich eine Sonderstellung ein. Das Fehlen *echter* Gewebe und Organe bedingt, daß alle für die Erhaltung und Fortpflanzung des Organismus notwendigen Lebensfunktionen in einer einzelnen Zelle ablaufen. Von besonderer Bedeutung für die Stoffausscheidung sind verschiedene Faktoren:

1. Die Protozoenzelle befindet sich nicht wie die Metazoenzelle in einem physiologisch gleichartigen Medium, vielmehr können die chemischen und physikalischen Faktoren dieses umgebenden Mediums ständigen mehr oder weniger großen Schwankungen unterworfen sein.

2. Die Protozoenzellen besitzen im allgemeinen an ihrer ganzen Oberfläche die gleichen physiologischen Potenzen, d. h. sie weisen zumeist keine physiologische oder funktionelle Polarität auf (z. B. eine Unterteilung in Zellbasis und Zellapex), die bei sekretorisch- und exkretorisch-tätigen Metazoenzellen in physiologischer Hinsicht von größter Bedeutung sind.

3. Die Stoffausscheidung (vor allem der Exkrete) findet in das gleiche Medium statt, aus dem die Zelle auch wieder die nötigen Baustoffe aufnimmt. Da bei den meisten Protozoen die ganze Zelloberfläche der Stoffaufnahme und -ausscheidung dient, muß der Zellmembran ein Wahlvermögen für die aufzunehmenden und auszuscheidenden Stoffe zugeschrieben werden.

II. Exkretion und Sekretion

Die Art der Stoffe, die von den Protozoen ausgeschieden werden, läßt eine Sekretion und Exkretion unterscheiden, jedoch lassen sich zwischen beiden keine scharfen Grenzen ziehen. Als Sekrete werden im allgemeinen Stoffausscheidungen bezeichnet, die durch ihren chemischen Charakter irgendeinen Nutzeffekt für die Zelle haben, z. B. Schleimausscheidungen für den Beutefang und bei Bewegungsvorgängen, Ausscheidungen von organischen oder anorganischen Stoffen bei der Schalen-, Gehäuse- und Cystenbildung. Die Exkretion umfaßt dagegen alle aus dem Stoffwechsel abfallenden und für die Zelle unbrauchbaren Endprodukte, die lebensnotwendig aus der Zelle entfernt werden müssen. Wie schon erwähnt, kann die Stoffausscheidung auf die ganze Zelloberfläche (Zellmembran) verteilt sein. Es können aber auch spezielle Zelldifferenzierungen (Organellen) für die Exkretion oder Sekretion vorhanden sein. Schließlich kann eine „intrazelluläre" Stoffausscheidung (meist in Form von Kristallen) in durch Membranen vom übrigen Cytoplasma abgegrenzte Räume („Vakuolen") erfolgen. Die qualitative und quantitative Untersuchung der ausgeschiedenen Substanzen stößt verständlicherweise auf große präparative Schwierigkeiten, jedoch sind in jüngster Zeit mit besonderen Mikromethoden erste Erfolge erzielt worden [3, 8].

B. Exkretion

I. Orte der Stoffausscheidung

1. Zellmembran

Eine Stoffausscheidung durch die Zellmembran scheint bei fast allen Protozoen, vor allem aber bei denjenigen Formen verbreitet zu sein, die als Zellgrenzstruktur (Zellmembran) nur eine „einfache" Elementarmembran besitzen, z. B. viele Flagellaten, Amöben und intrazelluläre parasitische Protozoen. Diese Annahme basiert darauf, daß molekular-disperse Stoffe die als Elementarmembran (unit membrane) ausgebildete Zellmembranen permeieren können, wenngleich der Mechanismus dieser Membranpermeabilität weitgehend ungeklärt ist. Mit Sicherheit können außerdem kolloidale Stoffe mit Hilfe von Membranflußmechanismen die Zellmembran „umgehen". Dieser elektronenoptisch nachweisbare Vorgang, der bei der Stoffausscheidung nach dem umgekehrten Modus der Pinocytose abläuft, spielt vermutlich ebenfalls eine große Rolle bei Protozoen mit „einfacher" Zellmembran. Dabei werden die Exkrete oder Sekrete während oder unmittelbar nach ihrer Bildung in der Zelle von einer Elementarmembran umhüllt und als „Bläschen" zur Zellmembran transportiert, wo sie sich entweder nach Verschmelzung mit der Zellmembran nach außen öffnen und ihren Inhalt entleeren (wobei die „Bläschenmembran" zu Zellmembran konvertiert), oder die Bläschen werden zusätzlich von Zellmembranmaterial umhüllt und nach Abnabelung in das umgebende Medium abgeschieden.

2. Zellorganellen

a) Kontraktile Vakuolen

Es ist unumstritten, daß vor allem den pulsierenden bzw. kontraktilen Vakuolen eine mehr oder weniger große Bedeutung bei der Stoffausscheidung zukommt. Diese vermutlich hauptsächlich der Osmoregulation dienenden Organellen finden sich bei vielen Protozoen, vor allem bei den Arten, die in anisotonischem Medium (z. B. Süßwasser) leben. In ihrer Ausbildung lassen sich beträchtliche Unterschiede erkennen. Im einfachsten Fall stellen sie „pulsierende Vakuolen" dar, die von einer Elementarmembran umgeben sind. Sie können im strömenden Cytoplasma passiv Ortsveränderungen erfahren und somit auch an jeder beliebigen Stelle der Zelloberfläche entleert werden. Die Füllung (Diastole) dieser einfachen pulsierenden Vakuolen geschieht durch Fusion vieler kleiner mit Flüssigkeit gefüllter Bläschen (Abb. 1) mit der Vakuolenwand, wobei der Inhalt dieser Bläschen an die stetig anschwellende

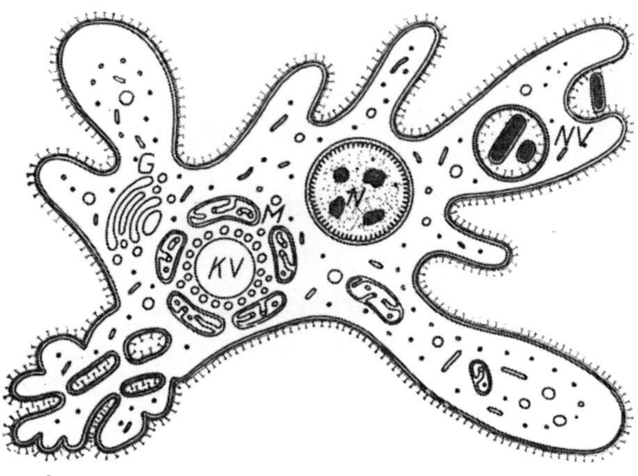

Abb. 1. Schematische Darstellung der Amöbenzelle (*A. proteus*) mit pulsierender Vakuole [26]. Die pulsierende Vakuole (*KV*) wird von „Nephridialplasma" umgeben, das aus zahlreichen kleinen flüssigkeitsgefüllten Bläschen und Tubuli aufgebaut ist. Durch Fusion dieser Vesikel mit der Zentralvakuole wird Flüssigkeit an die dabei anschwellende Vakuole abgegeben. Um die pulsierende Vakuole und das umgebende Nephridialplasma herum liegen zahlreiche Mitochondrien (*M*), die als Energiequelle für die Exkretionstätigkeit der pulsierenden Vakuole angesehen werden müssen. *N* = Nukleus, *NV* = Nahrungsvakuole, *G* = Golgikomplex

Vakuole abgegeben wird. Zum Teil liegen aber in der Randzone um die Vakuole herum zahlreiche feine Tubuli, die ebenfalls ihren Inhalt in die pulsierende Vakuole abgeben. Auf Grund elektronenmikroskopischer Befunde [21, 23, 24] muß angenommen werden, daß diese Tubuli mit dem Kanalsystem des endoplasmatischen Retikulums in Verbindung stehen (vgl. Nephridialtubuli von *Paramecium* Abb. 2). Nach Erreichen eines maximalen Füllungszustandes tritt die Vakuole in Konnex mit der Zellmembran, wobei Vakuolen- und Zellmembran zunächst verschmelzen. Die Vakuole öffnet und entleert sich dann an dieser Stelle nach außen. Ob die Entleerung der Vakuole infolge des Druckes in der Blase oder/und durch den Druck erfolgt, den das umgebende zurückgedrängte Plasma ausübt, ist noch ungeklärt. Es hat den Anschein, daß bei einigen Organismen (z. B. Euglenoiden) die Vakuolenwand nach der Entleerung zu Zellmembran konvertiert, und daß dann anschließend eine völlig neue Vakuole gebildet wird [10]. Bei anderen Protisten dagegen (z. B. *A. proteus*) scheint nach ihrer Entleerung die kollabierte Vakuole wieder von der Zellmembran losgetrennt zu werden und damit als ständige Pulsationsvakuole erhalten zu bleiben.

Daß bei der Stoffausscheidung Energie verbraucht wird, erscheint plausibel. Mit diesem Energieaufwand wird jedenfalls die oft zu beobachtende Ansammlung von Mitochondrien in einem „Hof" um die pulsierende Vakuole herum in Zusammenhang gebracht (Abb. 1). Ein Energieverbrauch geht auch daraus hervor, daß bei Verminderung der O_2-Menge die Pulsation bis zum Stillstand verlangsamt wird [10]. Da bei diesem „einfachen" Typ der Exkretionsvakuolen elektronenoptisch

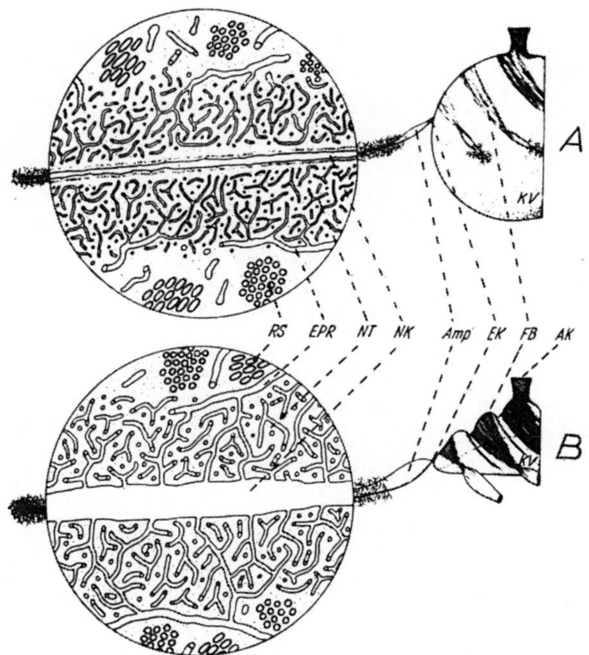

Abb. 2. Schematische Darstellung des Nephridialapparates bei *Paramecium* [*23, 24*]. A = Radialkanäle in Systole, kontraktile Vakuole in Diastole. B = Radialkanäle in Diastole, kontraktile Vakuole in Systole. Das Nephridialplasma um die zur Zentralvakuole hinführenden Radialkanäle wird von feinen verzweigten Tubuli (NT) gebildet. Peripher führen schlauchförmige Strukturen des endoplasmatischen Retikulums (EPR) an das Nephridialplasma heran und haben direkten Anschluß an die Nephridialtubuli. Diese münden im Stadium der Diastole (B) offen in die Nephridialkanäle (NK) ein, im Stadium der Systole (A) ist diese Verbindung jedoch unterbrochen (Verhinderung des Rückflusses aus dem Lumen der Nephridialkanäle in die Tubuli). Die Entleerung (Systole) der Radialkanäle und der kontraktilen Vakuole wird durch kontraktile Elemente bewirkt. Sie verlaufen entlang den Radialkanälen als einzelne Fibrillen und sind an der kontraktilen Vakuole zu flachen Fibrillenbändern (FB) zusammengefaßt. Um das Nephridialplasma herum liegen röhrenförmige Strukturen (RS), deren Funktion bisher ungeklärt ist. Amp = Ampullenförmige Anschwellung der Radialkanäle vor dem in die Zentralvakuole führenden Einspritzkanal (EK). AK = Ausführungskanal, der ebenfalls von kontraktilen Elementen (spiralig) umgeben wird. Weitere Erläuterungen im Text

keine kontraktilen Elemente nachgewiesen werden können, die in unmittelbarem Zusammenhang mit den Vakuolen stehen und für die Pulsation verantwortlich gemacht werden könnten, muß angenommen werden, daß hier nur „elastische" Kräfte, nämlich die Kompression des umgebenden Plasmas und evtl. die elastische Dehnung der Vakuolenmembran, die Entleerung bewirken. Es besteht allerdings auch die

Möglichkeit, daß das (umgebende) Cytoplasma eine Kontraktilität besitzt, die die Systole der Vakuole bewirkt.

Die pulsierende Vakuole der Euglenoiden mündet in das Geisselsäckchen, das eine Einsenkung der Zelloberfläche darstellt. Sie wird von recht großen Sekundärvakuolen umgeben. Bei der Entleerung der Vakuole in das Geißelsäckchen geht die Vakuolenwand in der Wand des Geißelsäckchens auf, worauf eine neue Exkretionsvakuole jeweils durch Zusammenfließen der Sekundärvakuolen gebildet wird [18].

Weitaus höher differenziert sind die kontraktilen Vakuolen der Ciliaten, die mit zuführenden Kanälen ausgestattet sind. Als eines der höchst differenzierten Exkretionssysteme bei Protozoen ist das von *Paramecium* zu betrachten (Abb. 2). Elektronenmikroskopische Untersuchungen [23, 24] ergaben, daß die zur zentralen Vakuole hinführenden Radialkanäle von feinen röhrenförmigen Strukturen (Nephridialplasma) umsponnen werden, die im Stadium der Diastole (der Radialkanäle) offen in die Kanäle einmünden, während sie an ihrer Peripherie in ständiger direkter Verbindung mit dem Kanalsystem des endoplasmatischen Retikulums stehen. In der Systole wird die offene Verbindung zwischen Nephridialtubuli und Radialkanälen durch Abtrennen der Tubuli unterbrochen, vermutlich um einen Rückfluß von Exkretionsflüssigkeit in die Nephridialtubuli zu verhindern. Die Radialkanäle, die vor ihrer Einmündung in die kontraktile Vakuole durch blasenförmige Anschwellungen, die Ampullen, verdickt sind und dann in die Einspritzkanäle übergehen, werden von kontraktilen Elementen begleitet, die sich an der Vakuole zu flachen Fibrillenbändern anordnen. Diese Fibrillenbänder überziehen die nach außen gerichtete Wand der Vakuole, führen weiter zum Exkretionskanal, den sie spiralig umgeben und enden am Exkretionsporus unter der Zelloberfläche. In der Endphase der Diastole der Radialkanäle läuft vermutlich eine Kontraktionswelle dieser kontraktilen Strukturen entlang den Radialkanälen, verengt und entleert diese (Systole) zunächst in die anschwellenden Ampullen, die sich darauf durch die Einspritzkanäle in die dabei anschwellende Zentralvakuole entleeren. Die sich nunmehr in Diastole befindliche Exkretionsvakuole wird durch die fortschreitende Kontraktionswelle ebenfalls entleert, und danach schließlich der geöffnete Exkretionskanal wieder verschlossen. Der Kontraktionswelle folgt direkt die Erschlaffung der kontraktilen Elemente, so daß sich die Radialkanäle bereits wieder im Stadium der Diastole befinden, wenn die Zentralvakuole noch im Stadium der Systole ist. Die Pulsationsfrequenz bei *Paramecium* beträgt durchschnittlich etwa 10 sec, kann aber auch in Abhängigkeit verschiedener Faktoren (Temperatur, Konzentration und Zusammensetzung des umgebenden Mediums sowie experimentelle Einflüsse z. B. Druck u. a.) wesentlich kürzere oder längere Intervalle aufweisen.

Zu den Exkretionsorganellen müssen wahrscheinlich auch die Pusulen der Dinoflagellaten gezählt werden [10], obwohl eine Pulsation dieser Vakuolen nicht mit Sicherheit beobachtet werden konnte. Sie sind meistens in Zweizahl vorhanden und bestehen in der Regel aus einer Vakuole, die sich durch einen Kanal in eine der Geisselporen öffnet. Beide Pusulen können auch durch einen feinen Kanal verbunden sein [5]. Zum Teil besitzen sie bei den gepanzerten Dinoflagellaten „sekundäre Aussackungen" ihrer Vakuolenwand.

b) Trichocysten

Nach WOHLFARTH-BOTTERMANN [35] stellen möglicherweise auch die Trichocysten Organellen der Osmoregulation dar. Übereinstimmend mit KRUSZYNSKI [15] konnte nachgewiesen werden, daß ausgeschleuderte Trichocysten mehr oder weniger große Mengen von Salzen (u. a. Calcium in feinen Gipskristallen) enthalten können. „Die fortlaufende spontane Ausscheidung von Trichocysten sowie ihr Gehalt an Salzen sind starke Hinweise für eine Exkretionsfunktion dieser Organelle. Eine mengenmäßige Regulation der Exkretion von Salzen durch eine verschieden starke Salzeinlagerung und durch eine unterschiedlich große Ausscheidung von Trichocysten beweisen eine osmoregulatorische Funktion, da der Regulationsmechanismus in biologisch sinnvoller Weise den gestellten experimentellen Bedingungen entsprach: bei plötzlicher Hypotonisierung des Mediums werden mit Hilfe der Trichocysten in verstärktem Maße Salze ausgeschieden" [35].

3. Intrazelluläre Speicherung von Exkreten

Eine besondere Art der Stoffausscheidung aus dem Stoffwechselgeschehen der Zelle stellt die bei vielen Protozoen zu beobachtende Speicherung von Exkretstoffen in Form von Kristallen oder anderen polymorphen Zelleinschlüssen dar. Diese Einschlüsse können mehr oder weniger intensiv gefärbt sein und werden dann auch als Pigmente bezeichnet, z. B. das blaugrüne Pigment bei *Stentor coeruleus*, die verschiedenfarbigen Pigmente bei *Nassula*-Arten und das rote Haematochrom bei manchen Flagellaten (vor allem *Haematococcus pluvialis*). Zu dieser Kategorie werden aber auch die braunen Pigmentkörner der Malariaparasiten gerechnet. Es handelt sich dabei um Hämatin, das aus der Umwandlung des Hämoglobins resultiert. Die Bildung der bei den Protozoen weit verbreiteten Exkretkristalle scheint u. a. von der Ernährungsweise abhängig zu sein. Form und Anzahl der Exkretkristalle wechseln nämlich bei *Amoeba proteus* und *A. dubia* mit der Nahrung [19, 20]. Bei *Paramecium* treten bei ausschließlicher Bakterienfütterung keine Exkretkörner, dagegen bei Zusatz von faulem Fleisch oder Eiweißlösungen große Exkretkristalle auf [27]. Elektronenmikroskopische Aufnahmen der Exkretkristalle von *Paramecium* [25] zeigen, daß sie von einer Membran als Permeabilitätsschranke umgeben werden, die sie vom stoffwechselaktiven Grundplasma isoliert. Die derart isoliert gespeicherten Exkrete können auf verschiedene Art aus der Zelle ausgeschieden werden. Eine der Möglichkeiten besteht darin, daß sie in Exkretvakuolen eingeschlossen werden, die dann an der Zelloberfläche entleert werden z. B. bei *Trichosphaerium* [22] und bei *Cochliopodium bilimbosum* [1]. Bei den Foraminiferen werden die Exkretkristalle zusammen mit anderen Exkretstoffen (Xanthosomen) als Sterkome aus den Zellen ausgeschieden [5]. Bei den Ciliaten schließlich scheinen sie sich in unmittelbarer Nähe der kontraktilen Vakuole (durch Umwandlung des schwerlöslichen neutralen Salzes des phosphorsauren Kalkes in das lösliche saure Salz) aufzulösen, und mit der Exkretionsflüssigkeit aus der Zelle entfernt werden zu können [27].

II. Mechanismus der Stoffausscheidung

Die Ausscheidung der Stoffwechselendprodukte aus dem Organismus stellt auch bei den Protozoen einen lebenswichtigen Prozeß dar. Diese Exkretion, aber auch die Sekretion, scheint bei den meisten Protisten hauptsächlich durch die Zellmembran zu erfolgen. Dabei muß sicher zwischen einer Ausscheidung molekular-disperser Stoffe, die die Zellmembran offensichtlich direkt permeieren können (auch die aus mehreren Membranen aufgebaute Pellikula) und Substanzen unterschieden werden, die in kolloidaler oder wasserunlöslicher „fester" Form vorliegen und die Zellmembran nur mittels „Membranflußmechanismen" passieren können. Während bei Metazoen-Zellen darüber schon zahlreiche Untersuchungen vorliegen, existieren solche für die Protozoen kaum. Zum anderen erscheint es sehr fraglich, inwieweit es möglich ist, die an Metazoen-Zellen erhobenen Befunde, Theorien und Hypothesen über die Ausscheidungsprozesse auf die Protozoen anzuwenden oder zu übertragen, weil hier ja andere physiologische Bedingungen herrschen als im mehrzelligen Gewebe. Vor allem, und das ist für alle diese Vorgänge von besonderer Bedeutung, liegt bei den meisten Protozoen (Süßwasserformen) ein erheblicher osmotischer Druckunterschied zwischen Zelle und umgebenden Medium vor, der einen ständigen Wasserstrom in die Zelle bewirkt.

Es muß deshalb bei diesen Formen als zusätzliches „Exkret" das osmotisch eingedrungene Wasser, außer dem aus dem Stoffwechsel stammenden Wasser (das nur etwa 1% der Gesamtmenge des ausgeschiedenen Wassers ausmacht [13]), ausgeschieden werden. Offenbar ist dies den Zellen nur mit Hilfe der pulsierenden bzw. kontraktilen Vakuolen möglich. Nach wie vor müssen deshalb diese Organellen als Osmoregulatoren angesehen werden. Es kann aber kein Zweifel daran bestehen, daß mit dem Wasser zusammen, meistens in diesem gelöst, andere Exkretstoffe die Zelle verlassen. Derartige Stoffe sind vermutlich u. a. das Kohlendioxyd und Endprodukte des Stickstoff-Stoffwechsels (Ammoniak, Harnstoff oder Harnsäure) wie im folgenden noch im einzelnen angeführt werden wird. Es ist unwahrscheinlich, daß diese Stoffe direkt in die (als Sammelblase fungierenden) pulsierenden oder kontraktilen Vakuolen abgeschieden werden. Die Mikromorphologie der Nephridialorganellen legt vielmehr den Verdacht nahe, daß diese Abscheidung schon einige Schritte vorher erfolgt. Bei den „einfachen" Organellen (z. B. bei den Amöben), bei denen kleine flüssigkeitsgefüllte Bläschen als „Nephridial, plasma" die pulsierende Vakuole umgeben, kann angenommen werden, daß die Exkretabscheidung in diese Bläschen erfolgt, zumal sie in ihrer Gesamtheit eine wesentlich größere „aktive" Oberfläche bereitstellen als die Sammelblase. Es wäre aber auch denkbar, daß die Exkrete bereits bei der Absonderung („Entmischung") der *„Exkretionsflüssigkeit"* vom Grundplasma, also bei der Bildung der kleinen Bläschen, in dieser gelöst sind. Für die komplizierter gebauten Nephridialorganellen der Ciliaten ist von besonderer Bedeutung, daß das Nephridialplasma eine direkte Verbindung mit dem Kanalsystem des endoplasmatischen Retikulums (Abb. 2) besitzt [*21, 23, 24*]. Dies läßt, zumindestens bei *Paramecium* und den anderen Protozoen mit tubulärem, vernetztem Nephridialplasma, die Vermutung zu, daß das endoplasmatische Retikulum in diesen Fällen eine Transportfunktion für verschiedene in Wasser gelöste Stoffe ausübt, und daß in dem Nephridialplasma eine Rückresorption von solchen Stoffen stattfindet, die dem Stoffwechsel wieder nutzbar gemacht werden. In diesem Zusammenhang könnten auch die langen röhrenförmigen Strukturen, die das Nephridialplasma entlang den Radialkanälen umgeben und begleiten, gedeutet werden. Der schon von v. GELEI [*7*] gezogene Vergleich mit den „exogen arbeitenden Nephridien höherer Metazoen" [*6*] erscheint danach gar nicht so abwegig.

III. Qualitative und quantitative Bestimmungen

Wie schon einleitend erwähnt, stehen genauen Bestimmungen der qualitativen und quantitativen Stoffausscheidung große präparative Schwierigkeiten entgegen. Zahlreiche Untersuchungen über die Exkretstoffe des Stickstoff-Stoffwechsels liegen für *Paramecium* vor [*34*], und man geht in der Annahme nicht fehl, daß sie hauptsächlich durch die kontraktilen Vakuolen ausgeschieden werden. Nachgewiesen wurden Ammoniak, Harnstoff und Harnsäure, und zwar hauptsächlich im Kulturmedium. Es ist allerdings bis heute umstritten, welche der 3 Substanzen tatsächlich ausgeschieden werden. WEATHERBY [*31—33*] postulierte, daß von dem im Kulturmedium nachgewiesenen Ammoniak und Harnstoff nur der Harnstoff von *Paramecium* ausgeschieden wird, während das Ammoniak durch Hydrolyse des Harnstoffs im Kulturmedium entstehen soll. Nach Vorstellung dieses Autors wird auch nur ein Teil des Harnstoffs von den kontraktilen Vakuolen ausgeschieden, während der größere Teil die Zellmembran passieren soll. Auch DARBY [*4*] glaubt, daß nur Harnstoff ausgeschieden wird, der im Kulturmedium von den Bakterien zu Ammoniak abgebaut wird. Dagegen fand JONES [*11*] in Kulturen von *P. multimicronucleatum* größere Mengen Harnsäure. Nach LUDWIG [*17*] wird bei *Paramecium* außerdem die bei der Respiration entstehende Kohlensäure durch die kontraktilen Vakuolen ausgeschieden. Nach seinen Berechnungen reicht nämlich die von den Vakuolen ausgeschiedene Wassermenge gerade zur Lösung des entstehenden CO_2 aus.

Neuerdings wurden bei *Tetrahymena* Phosphate, Purine und Pyrimidine als Exkretprodukte nachgewiesen [*16*], während bei *Paramecium aurelia* eine Ausscheidung von Hypoxanthin und in kleineren Mengen Adenin und Guanin beschrieben wurde [*30*].

Die ins umgebende Medium ausgeschiedenen Exkrete können für andere Organismen des gleichen Lebensmilieus giftig sein. Unter

Umständen, vor allem bei zu kleinem Lebensraum, kann aber auch eine Selbstvergiftung (Autointoxikation) eintreten. Eine besondere Bedeutung kommt den Stoffausscheidungen bei den parasitisch lebenden Protozoen zu. Ihre pathogene Wirkung kann, außer der Entziehung von Körpersubstanzen und mechanischer Schädigung des Wirtes, eine Giftwirkung sein, die entweder auf der Ausscheidung der *Exkrete* beruht, oder durch Fermente bzw. Zerfallstoffe beim Tode der Protozoen hervorgerufen wird [*5*].

Die qualitative Analyse der „Exkret"-Kristalle stößt ebenfalls auf erhebliche Schwierigkeiten, ganz abgesehen davon, daß die chemischen Bestandteile dieser Kristalle und Kristalloide bei den verschiedenen Protozoen sicher unterschiedlicher Natur sind. Die zahlreichen Untersuchungen ergaben Zusammensetzungen aus Harnsäure, Calciumoxalat, Calciumcarbonat, Leucin, sekundäres oder tertiäres Calciumphosphat, Aminosäuren, Calcium-chloro-phosphat, Natrium-urat, Quarz und aus einem mit Magnesiumsalz substituierten Glycin. Bei *Paramecium*-Zellen sollen die Kristalle aus Calciumphosphat, das an einen organischen Rest gebunden ist, bestehen [*27*]. Bernheimer [*1*] postuliert jedoch auf Grund seiner Untersuchungen, daß die Kristalle bei *Paramecium* möglicherweise in Beziehung zu Melamin (Cyanursäureamid) stehen. An isolierten und rekristallisierten Kristallen von *Amoeba proteus, A. dubia* und *Chaos chaos* konnte mit mikroanalytischen Methoden, Röntgenbeugung, Infrarot-Spektralanalyse und petrographischer Analyse nachgewiesen werden, daß sie aus Triuret (Carbonyl-biuret) bestehen [*3, 8*].

Direkte quantitative Messungen der ausgeschiedenen Stoffe sind kaum möglich, dagegen läßt sich bei der Stoffausscheidung mittels der kontraktilen Vakuolen zumindest die ausgeschiedene Flüssigkeitsmenge an Hand der Vakuolengröße (im maximalen Füllungszustand) und der Pulsationsfrequenz bestimmen. Um bei der außerordentlich differierenden Zellgröße der Protozoen in etwa vergleichbare Werte zu erhalten, erweist es sich als nützlich, die Zeit anzugeben, in der eine dem Körpervolumen entsprechende Flüssigkeitsmenge ausgeschieden wird. Diese Zeit beträgt z. B. bei *Amoeba proteus* (bei $19-27°$ C in Heu-Infusion) $3,9-13,2$ Std, bei *Paramecium caudatum* (bei $15-23°$ C in Kulturlösung) $15-49$ min, bei *Carchesium aselli* (bei $14,5-16°$ C in Londoner Leitungswasser) 25 min und bei *Cothurnia curvula* (bei $14,5-16°$ C in Seewasser) $4^1/_2$ Std [*12*]. Vergleichsweise beträgt die Ausscheidungsmenge von Primärharn in der menschlichen Niere [*9*] etwa 150 l/Tag, die adäquate Menge zum Körpervolumen wird also in etwa 12 Std bzw., wenn man den tatsächlich aus dem Körper ausgeschiedenen Harn (etwa $1,5$ l/Tag) berücksichtigt, in etwa 50 Tagen ausgeschieden.

Die quantitative Flüssigkeitsausscheidung, und damit vermutlich auch die quantitative exkretorische Stoffausscheidung, ist, wie z. T. schon erwähnt, abhängig von chemischen und physikalischen Faktoren. Zu den chemischen Faktoren zählen u. a. die Konzentration der im Kulturmedium gelösten Stoffe, aber auch das Vorhandensein spezieller, die Ausscheidung hemmender oder fördernder Substanzen (z. B. Sauerstoffangebot), die aber im einzelnen nicht aufgeführt werden sollen.

C. Sekretion

Die Sekretion umfaßt hauptsächlich die Ausscheidung von Schleimen, die für den Beutefang und die Bewegung von Bedeutung sein können, und die Ausscheidung von Hüllsubstanzen für die Ausbildung von Schalen und Gehäusen sowie für die Cystenbildung. Eine Schleimausscheidung im Dienste der Bewegung ist allerdings nach wie vor umstritten. Sie scheint auch nicht aktiv zur Lokomotion beizutragen. Eine Fortbewegung durch Schleimausscheidung wurde vor allem bisher für die Gleitbewegung der Gregarinen angenommen (Theorie des Vorwärtsschiebens [28], Theorie des Quellungsdruckes [29]). Auf Grund von elektronenmikroskopischen Erkenntnissen [14] ist der Sitz der Gleitbewegung jedoch die Pellikula bzw. nur deren innere Membran, in der feine kontraktile Elemente liegen. Eine Schleim-(Sekret-)Ausscheidung im Dienste des Beutefanges findet sich vor allem bei festsitzenden Protozoen und ist am ausgeprägtesten bei den Fangtentakeln der Suctorien entwickelt. Der „Fangschleim" kann gleichzeitig eine lähmende oder verdauende Wirkung besitzen.

Die zu Hüllbildungen (Schalen und Gehäuse) führenden Ausscheidungen sind nur bei bestimmten Organismen anzutreffen. Die Sekrete können dabei sehr verschiedene Substanzen darstellen, die auch recht unterschiedliche Härtegrade aufweisen, nämlich Gallerten, Pektin (Pseudochitin), Cellulose und echtes Chitin als organische Stoffe. Eine weitere Verfestigung kann durch Imprägnierung oder Einlagerung von ebenfalls aus der Zelle ausgeschiedenen anorganischen Substanzen (Calcium, Silicium, Eisensalze u. a.) oder durch Einlagerung von Fremdkörpern (Sand, Fremdschalen u. a.) bewirkt werden.

Die Cystenbildung durch Ausscheidung von Stoffen, auf äußere oder innere Reize hin, ist bei den Protozoen, vor allem bei den Süßwasser- und parasitischen Formen weit verbreitet, ihr Chemismus ist aber noch weitgehend ungeklärt. Die Ausscheidung der Sekrete findet hauptsächlich durch die ganze Zelloberfläche bzw. in bestimmten Arealen derselben statt. Die Stoffe können dabei anscheinend die Zellmembran entweder in niedermolekularer Form permeieren oder durch Membranflußmechanismen (Ausschleusung durch die Zellmembran) ausgeschieden werden. Zum Teil können aber auch bei der Hüllbildung vorgefertigte „Bauelemente" (z. B. die aus Tektin bestehenden „Protrichocysten" oder Tektinstäbchen [2], die bei verschiedenen Ciliaten die Cystenhülle bilden) aus der Zelle ausgestoßen werden.

D. Zusammenfassung und Schlußbetrachtung

Die Sonderstellung der Protozoen als einzellige Organismen ist auch von Bedeutung für die Sekretion und Exkretion, weil die Zellen im allgemeinen keine strukturell ausgebildete Polarität im Dienste des Stoffaustausches und des Stofftransportes besitzen. Weiterhin leben sie im allgemeinen in einem unphysiologischen Medium, das oftmals außerdem ein osmotisches Druckgefälle zur Zelle aufweist. Schließlich findet die

Exkretion in das gleiche Medium statt, aus dem gleichzeitig die Aufnahme lebensnotwendiger Substanzen erfolgt. Die Exkretion und Sekretion kann entweder durch die Zellmembran hindurch oder mit Hilfe besonderer Organellen (kontraktile Vakuolen, Pusulen (?) und Trichocysten) erfolgen. Eine besondere Bedeutung kommt dabei sicher den kontraktilen Vakuolen zu, die auch als Organellen der Osmoregulation angesehen werden. Als Beispiele werden der Feinbau der relativ einfachen „pulsierenden" Vakuole von *Amoeba proteus* und der Feinbau des hoch differenzierten Nephridialapparates von *Paramecium* beschrieben. Einige Protozoen weisen die Fähigkeit auf, Exkrete „intrazellulär" durch Bildung von Exkretkristallen auszuscheiden. Der Mechanismus der Stoffausscheidung bei Protozoen ist noch weitgehend unbekannt. Die Befunde, Theorien und Hypothesen an Metazoenzellen lassen sich nicht ohne weiteres zur Erklärung der Stoffausscheidungsmechanismen bei den Protozoen übertragen. Außerdem bereitet die qualitative und quantitative Untersuchung der Exkretprodukte erhebliche Schwierigkeiten. Einige dieser nachgewiesenen Exkretprodukte werden aufgezählt, aber auch hier bestehen Schwierigkeiten bei der Interpretation, welche Stoffe tatsächlich ausgeschieden werden, und welche Stoffe im Kulturmedium sekundär durch Aufspaltung oder Zerfall entstehen. Zum Beispiel findet man im Kulturmedium von *Paramecium* Ammoniak, Harnstoff und Harnsäure, aber es wird wahrscheinlich nur der Harnstoff ausgeschieden. Es ist aber zu erwarten, daß mit Hilfe elektronenoptisch-cytochemischer Nachweismethoden mehr und neue Einzelheiten über die Sekretion und Exkretion bei Protozoen erhalten werden.

Literatur

[1] Bernheimer, A. W.: Arch. Protistenk. **90**, 365 (1938).
[2] Bresslau, E.: Naturwissenschaften **9**, 57 (1921).
[3] Carlström, D., and K. M. Møller: Exp. Cell Res. **24**, 393 (1961).
[4] Darby, H. H.: Arch. Protistenk. **65**, 1 (1929).
[5] Doflein, F., u. E. Reichenow: Lehrbuch der Protozoenkunde. 6. Aufl. Jena: VEB G. Fischer 1953.
[6] Fortner, H.: Arch. Protistenk. **56**, 295 (1926).
[7] Gelei, J. v.: Biol. Zbl. **45**, 676 (1925).
[8] Griffin, J. L.: J. biophys. biochem. Cytol. **7**, 229 (1960).
[9] Hargitay, B., u. W. Kuhn: Z. Elektrochemie **55**, 539 (1951).
[10] Haye, A.: Arch. Protistenk. **70**, 1 (1930).
[11] Jones, E. P.: Univ. Pitt. Bull. **29**, 1 (1933).
[12] Kitching, J. A.: Biol. Rev. **13**, 403 (1938).
[13] — Protoplasmatologia, Bd. III, D 3a (1956).
[14] Kümmel, G.: Arch. Protistenk. **102**, 501 (1958).
[15] Kruszynski, J.: Arch. Protistenk. **92**, 1 (1939).
[16] Leboy, Ph. S., S. G. Cline, and R. L. Conner: Protozool. **11**, 217 (1964).
[17] Ludwig, W.: Arch. Protistenk. **62**, 12 (1928).
[18] Mainx, F.: Arch. Protistenk. **60**, 305 (1928).
[19] Mast, S. O.: Biol. Bull. **75**, 389 (1938).
[20] — Biol. Bull. **77**, 391 (1939).
[21] Pitelka, D. R.: Electron-microscopic structure of protozoa. Oxford-London New York-Paris: Pergamon Press 1963.
[22] Schaudinn, F.: Abh. K. Preuß. Akad. Wiss. Berlin 1899, Anh. 1.

[23] Schneider, L.: Verh. dtsch. Zool. Ges. in Münster, Zool. Anz. Suppl. 23, 457 (1959).

[24] — J. Protozool. 7, 75 (1960).

[25] — Unveröffentlicht.

[26] —, u. K. E. Wohlfarth-Bottermann: Protozoen, In: G. C. Hirsch, H. Ruska u. P. Sitte: Grundlagen der Cytologie (in Vorbereitung). Jena: VEB Gustav Fischer Verlag.

[27] Schewiakoff, T.: Z. wiss. Zool. 57, 32 (1893).

[28] Schewiakoff, W.: Z. wiss. Zool. 58, 340 (1894).

[29] Sokolow, B.: Arch. Protistenk. 27, 260 (1912).

[30] Soldo, A. T., and W. J. van Wagtendonk: J. Protozool. 8, 41 (1961).

[31] Weatherby, J. H.: Biol. Bull. 52, 208 (1927).

[32] — Physiol. Zool. 2, 375 (1929).

[33] — The contractile vacuole. In: G. H. Calkins, and F. M. Summers: Protozoa in Biological Research. New York: Columbia Univ. Press 1941.

[34] Wichterman, R.: The biology of Paramecium. New York, Toronto: The Blakiston Comp. Inc. 1953.

[35] Wohlfarth-Bottermann, K. E.: Arch. Protistenk. 98, 169 (1953).

Summary

The unique position of the protozoa as unicellular organisms is of significance for secretion and excretion, because the cells generally have no structural polarity for exchange and transport of substances. Furthermore, they generally live in an unphysiological medium, whose osmotic pressure is often different from that of the cell. Finally, excretion takes place into the same medium out of which the uptake of essential substances simultaneously takes place. Excretion and secretion are either mediated by the cell membrane or by special organelles (contractile vacuoles, pusules (?), and trichocysts). Certainly the contractile vacuoles, which are assumed to be osmoregulatory organelles, are of special importance for excretion. As examples, the fine structure of the relatively simple pulsating vacuoles of *Amoeba proteus* and of the highly differentiated nephridial apparatus of *Paramecium* is described. Some protozoa can dispose of excretory products "intracellulary" by forming excretory crystals. The mechanism of excretion in protozoa is still largely unknown. Observations, theories, and hypotheses derived from metazoan cells cannot be used without modification to explain excretory mechanisms in protozoa. Furthermore, the qualitative and quantitative analysis of excretory products is difficult. Some excretory products of protozoa are listed, but there are difficulties in interpretation, such as in deciding what are the primary excretory products and what substances arise secondarily in the culture medium through splitting or breakdown. For example, one finds ammonia, urea, and uric acid in culture media of *Paramecium*, but possibly only urea is excreted. Cytochemical techniques and electron microscopy may be expected to provide new information about secretion and excretion in protozoa.

Diskussion

Ullrich: Die pulsierenden Vakuolen sind bereits so groß, daß man neben den morphologischen auch andere Methoden zur Aufklärung der Funktion benutzen kann. B. Schmidt-Nielsen hat z. B. die pulsierende Vakuole von Amöben punktiert und durch Gefrierpunktsmessung des Punktates gefunden, daß der Vakuoleninhalt gegenüber dem Zellplasma hypoton ist [Science 139, 606 (1963)]. Dieser Befund, daß ein hypotones Sekret von der Zelle in einem Arbeitsgang möglicherweise sogar durch aktiven Wassertransport gebildet wird, ist um so bedeutungsvoller, als andere hypotone Sekrete, z. B. der Niere, Speichel- und Schweißdrüsen zunächst als isotones Primärsekret vorliegen und dann in einem zweiten Arbeitsgang durch Rückresorption von gelösten Stoffen, hauptsächlich von NaCl hypoton werden. Ist es möglich, daß im Nephridialsystem *von Paramecium* auch zunächst ein isotones Primärsekret vorliegt, das durch Rückresorption von gelösten Stoffen hypoton wird?

Schneider: Bei *Paramecium* könnte man die Spekulation anstellen — wir wissen gar nichts darüber —, daß in den Strukturen des endoplasmatischen Retikulums, die ganz sicher hier eine Transportfunktion besitzen, sozusagen der „Primärharn" angeliefert wird und daß dann unter Umständen die Tubuli nephridiales mit der Rückresorption in Zusammenhang zu bringen sind. Inwieweit das stimmt, das entzieht sich zur Zeit unserer Kenntnis.

Diamond: Herr Ullrich, es ist doch wohl möglich, daß in einem Gang eine hypotonische Lösung hergestellt wird, ohne daß es sich dabei um aktiven Wassertransport handelt. Es sind mir mindestens zwei Mechanismen bekannt, die das schaffen könnten. Der erste ist der sog. Kodiffusionsmechanismus. Wenn NaCl aktiv transportiert wird, kann es durch Reibung Wasser mit sich nehmen, und die durch Reibung hergestellte Lösung kann hypotonisch, isotonisch oder hypertonisch sein. Eine zweite Möglichkeit wäre der sog. Doppelmembranmechanismus. Da handelt es sich um zwei Membranen, die Poren von verschiedenem Durchmesser haben. Es ist gezeigt worden, daß man mit so einem System entweder hypotonische, isotonische oder hypertonische Lösungen erhalten kann. Eine hypotonische Sekretion beweist also nicht, daß der Wassertransport aktiv sein muß.

Wessing: Ich möchte zu diesem Thema noch bemerken, daß Ramsay bei solchen Insekten, die nicht mit Wasser zu sparen brauchen, im Lumen der Malpighischen Gefäße stets einen hypotonischen Primärharn findet. Es gibt allerdings auch Insekten, die mit Wasser sparen müssen, z. B. die landbewohnenden Fliegen. Bei ihnen kann der Primärharn bis zur Kristallisation vieler Substanzen entwässert werden. Bei vermutlich gleichem Transportmechanismus ist der resultierende osmotische Wert also ganz unterschiedlich.

Thoenes: Haben die Protisten, die keine pulsierenden Vakuolen besitzen, Golgi-Apparate? Es gibt doch Hypothesen, daß die pulsierende Vakuole irgendwie ein Äquivalent des Golgi-Apparates sein soll.

Schneider: Diese Frage kann ich nicht generell beantworten. Bei *Paramecium* habe ich trotz einiger Tausend Aufnahmen bisher noch keinen Golgi-Apparat einwandfrei identifizieren können. Aber ich kann mit Sicherheit sagen, daß andere Protisten mit pulsierender Vakuole, wie z. B. *Amoeba proteus*, gleichzeitig Golgi-Apparate besitzen.

Druckfiltration als ein Mechanismus der Stoffausscheidung bei Wirbellosen

Von

Georg Kümmel, Berlin

Mit 11 Abbildungen

A. Einleitung

Als Geburtsjahr der Erkenntnis, daß die Druckfiltration einen wichtigen Ausscheidungsmechanismus darstellen kann, müssen wir das Jahr 1844 ansehen. Damals stellte Ludwig seine berühmt gewordene Hypothese der Harnbildung der Wirbeltiere auf. Er nahm an, daß in den Glomeruli der Wirbeltierniere durch den Blutdruck ein Filtrat abgepreßt wird (Abb. 10b), das Wasser und alle kleinmolekularen Substanzen enthält, während alle höhermolekularen Stoffe, besonders Eiweiße, durch die Dichtigkeit des Glomerulusfilters zurückgehalten werden. Nach Ludwigs Meinung sollte die Bildung des Endharnes keine aktive Zelltätigkeit benötigen, sondern sich ausschließlich durch passive Vorgänge, wie Osmose, einstellen.

Dieser letzte Teil der Ludwigschen Hypothese war zu einfach und mußte bekanntlich stark modifiziert werden. Doch ist die Meinung Ludwigs über die Bildung des Primärharnes für die Mehrzahl der Wirbeltiere auch heute noch gültig. Nur in den aglomerulären Nieren einiger mariner Teleostier liegen andere Verhältnisse vor.

Man hat den ersten Teil der Ludwigschen Hypothese auch auf andere Tiergruppen übertragen. Es ist meine Absicht, über das Vorkommen eines solchen Filtrationsmechanismus gerade bei diesen Tiergruppen zu berichten. Die Verhältnisse bei den Wirbeltieren sind ja wohl allgemein bekannt, sie sollen daher nur an manchen Stellen zum Vergleich herangezogen werden.

Da in diesem Bericht die Druckfiltration im Mittelpunkt steht, müssen wir uns zuerst über die folgenden vier Begriffe klarwerden: Filter, Filtration, Druckfiltration und Filtrat. Ein *Filter* können wir immer als eine Art Membran betrachten, die Stoffe nach ihrer Teilchengröße auswählt. Als *Filtration* im weitesten Sinne (Filtration s. l.) werden wir jede Bewegung von Flüssigkeit durch einen Filter aus einem Raum I in einen Raum II ansehen, bei der Stoffe oberhalb einer kritischen Teilchengröße von dem Filter im Raum I zurückgehalten werden. Ist die Filtermembran nur für Wasser, Ionen und kleinere Moleküle durchlässig, werden dagegen Makromoleküle (insbesondere die meisten Proteine) zurückgehalten, sprechen wir von Ultrafiltration. Die Bewegung der Flüssigkeit beruht insofern auf rein physikalischen Phänomenen, als dabei keine aktiven Transportvorgänge von Zellen eine Rolle spielen. Es muß aber irgendein aktiver Vorgang eingreifen, der die Ursache der Flüssigkeitsbewegung ist. Dieser Vorgang kann einmal in der Ausbildung eines

Konzentrationsgradienten zwischen Raum I und Raum II bestehen. Ich will den Ausdruck Filtration aber auf die Fälle beschränken, in denen der aktive Vorgang in der Ausbildung eines hydrostatischen Druckgradienten besteht. Dann liegt eine *Druckfiltration* vor. Infolge des Druckgefälles wird ein *Filtrat*, das nur Stoffe unter einer durch die Art der Membran bedingten kritischen Teilchengröße enthält, aus Raum I in den Raum II abgepreßt. Die Konzentration kleinmolekularer Stoffe im Filtrat ist etwa gleich der Konzentration dieser Stoffe in der ursprünglichen Flüssigkeit. Gewisse Konzentrationsunterschiede können allerdings etwa auf Grund der Donnanverteilung auftreten. Ein hydrostatischer Druck ist für eine Ultrafiltration erst dann ausreichend, wenn er den kolloidosmotischen Druck der Flüssigkeit im Raum I und einen möglichen Gegendruck im Raum II übersteigt.

Für eine so charakterisierte Druckfiltration müssen demnach zwei Bedingungen erfüllt sein:

a) Das Vorhandensein eines Filters, das einen Raum I von einem Raum II trennt.

b) Das Vorliegen eines ausreichenden hydrostatischen Druckgradienten zwischen Raum I und Raum II.

Zu a). Die Natur biologischer Filtermembranen, die in Ausscheidungsorganen auftreten, ist sehr verschieden. In einem Ausscheidungsorgan können zudem mehrere Filtermembranen hintereinandergeschaltet sein. Die Feinstruktur der Membranen ist nicht genügend bekannt (s. unten). Um das Verhalten der Membran bei der Filtration verständlich zu machen, ist es am besten, vereinfachend anzunehmen, es handle sich um typische Porenmembranen mit einem Siebeffekt. Dann ist die entscheidende Größe für die Passage von Stoffen der Moleküldurchmesser bzw. sein Verhältnis zur Porenweite [*48*]. Bei Annäherung der Moleküldurchmesser an die Porenweite wird ein Siebeffekt deutlich, und zwar schon, wenn der Porendurchmesser noch größer als der Moleküldurchmesser ist. Alle Substanzen, die größer als die Poren sind, können überhaupt nicht passieren. Substanzen mit einem Moleküldurchmesser, der ungefähr dem Porendurchmesser entspricht, unterliegen einem Siebeffekt, ihre Filtration ist eingeschränkt. Die Verhältnisse sind dadurch noch komplizierter, daß die meisten biologischen Membranen keine einheitliche Porenweite besitzen. Im Grundsätzlichen ändert sich aber dadurch nichts.

Die Kennzeichnung der biologischen Filtermembranen als Porenmembranen ist, wie oben betont, sicher eine vereinfachende Annahme. In allen Fällen, in denen die Feinstruktur solcher Membranen gut bekannt ist, hat man nie wirklich Poren nachweisen können [*15, 21, 46*]. Diese Membranen haben eine feinfibrilläre Struktur (s. auch Abb. 8b). Doch könnten sich die Lücken zwischen den sich kreuzenden Fibrillen wie Poren verhalten.

Zu b). Wir müssen drei Möglichkeiten in Betracht ziehen, die zu einem Druckgradienten zwischen Raum I und Raum II führen können: Die Erzeugung eines Überdruckes in Raum I, eines Unterdruckes in Raum II oder die Addierung von Über- und Unterdruck.

Bei der Erzeugung eines Unterdruckes wirkt wohl immer der Schlag von Geißeln oder Wimpern, der zu einer Flüssigkeitsbewegung von der Filtermembran weg und so zu einem gewissen Unterdruck an der Filtermembran führen kann. Ein Überdruck ist dagegen immer auf den hydrostatischen Druck des Blutes oder allgemein der Körperhöhlenflüssigkeit zurückzuführen. Dieser Druck wird vorzüglich durch die Pulsationen des Herzens oder anderer kontraktiler Abschnitte des Gefäßsystems erzeugt, aber auch z. B. durch die Spannung von Körpermuskeln oder bei Organverschiebungen infolge von Bewegungen usw.

Die Filtration kann wohl immer nur ein Ausscheidungsmechanismus unter mehreren anderen sein. Ich erinnere an die bekannte Dreiheit der Vorgänge: Filtration, Sekretion und Reabsorption. Die Frage nach der Reihenfolge der drei Vorgänge liegt auf der Hand. Man wird der Ansicht MARTINs [25] zustimmen, nach der die Filtration, wo sie überhaupt vorkommt, der initiale Vorgang sein wird. Dadurch wird ein primärer Harn gebildet, der durch die anderen Vorgänge verändert werden kann, bevor er als Endharn ausgeschieden wird.

Die eindeutigste Prüfungsmethode, ob eine Filtration vorliegt, ist die Gewinnung und Analyse des Primärharns. Entspricht dessen Zusammensetzung einem Filtrat, d. h. sind die Konzentrationen aller das Filter passierenden Substanzen im Primärharn (U_p) und in der Ausgangsflüssigkeit (hier allgemein mit A, wenn es sich um Blut handelt mit B bezeichnet) etwa gleich $\left(\dfrac{U_p}{A} = 1 \text{ oder } \dfrac{U_p}{B} = 1 \right)$, so ist ein Druckfiltrationsvorgang sehr wahrscheinlich.

Doch stehen dieser Art von Prüfung nicht selten unüberwindliche Schwierigkeiten entgegen. Außerdem ist ein negatives Resultat kein eindeutiger Beweis gegen eine Filtration, da durch Sekretion und Reabsorption das Filtrat sogleich verändert werden kann. Man muß daher andere Methoden zur Prüfung anwenden: Da beim Filtrationsprozeß aktive Leistungen der Zellen keine Rolle spielen (s. oben), wird er durch Abkühlung oder durch Vergiftung des Organs mit geeigneten Substanzen (z. B. Phlorrhizin) nicht oder zumindest nur geringfügig verändert werden[1]. Die aktiven Vorgänge Sekretion und Reabsorption werden aber gehemmt. Der Endharn (U) müßte sich daher in seiner Zusammensetzung einem Filtrat nähern, wenn ein Filtrationsvorgang vorliegt $\left(U \to U_p \text{ bzw. } \dfrac{U}{A} \text{ oder } \dfrac{U}{B} \to 1 \right)$. Ein vergleichbares Ergebnis kann man mit Stoffen erreichen, die im Normalfall nach der Filtration völlig reabsorbiert werden. Erhöht man die Konzentration solcher Stoffe in der Ausgangsflüssigkeit über die Schwellenkonzentration, die gerade noch vollständig reabsorbiert wird, erscheint der Stoff im Endharn. Das Phänomen einer solchen Schwelle ist an sich schon ein Hinweis auf eine Filtration.

Eine weitere sehr einfache Methode wäre die Verwendung einer Testsubstanz, die nur filtriert, aber nicht sezerniert oder reabsorbiert wird. Setzt man eine solche Substanz der Ausgangsflüssigkeit zu und erscheint die Substanz dann im Harn, so wäre eine Filtration (s. l.) bewiesen. Leider ist die Beweiskraft dieser Methode von ziemlich umständlichen Prüfungsmethoden abhängig [43]. Denn ich kann einer Substanz nicht ohne weiteres ansehen, ob sie die geforderten Eigenschaften besitzt oder nicht. Und wenn ich für eine Substanz festgestellt habe, daß sie bei einem bestimmten Tier die geforderten Eigenschaften besitzt, ist es keineswegs sicher, daß das für alle anderen Tiere gilt. Für Wirbeltiere erfüllt das Polysaccharid

---

[1] Das gilt natürlich nur, wenn nicht auch Vorgänge, die die Ursache der Flüssigkeitsbewegung sind, beeinflußt werden!

Inulin diese Bedingungen. Ob das auch für alle Wirbellosen gilt, ist ungeklärt. Und doch wird Inulin sehr häufig bei physiologischen Untersuchungen an Exkretionsorganen benutzt, allerdings nicht zur eigentlichen Prüfung der Filtration.

Auch Untersuchungen der Stoffwechselaktivität und der Fermentverteilung in den einzelnen Abschnitten der Exkretionsorgane können wichtige Hinweise geben. So ist anzunehmen, daß die Regionen und Strukturen, an denen ein Filtrationsvorgang abläuft, nur einen relativ geringen Energieverbrauch haben.

Stets wird man sich bemühen, diese physiologischen Prüfungen durch Untersuchung der Morphologie zu unterstützen. Mit anderen Worten, man wird nach morphologisch definierten Filtrationsorten suchen. Bei Exkretionsorganen, bei denen eine direkte physiologische Prüfung aussichtslos erscheint, z. B. bei sehr geringer Größe der Organe, ist die Untersuchung der Morphologie natürlich von ganz besonderer Bedeutung.

Als Fazit dieser einleitenden Betrachtungen können wir sagen: Um einen Druckfiltrationsvorgang im Ausscheidungsprozeß sicherzustellen, sind folgende Fragen zu klären:

1a. Die Frage nach dem Organabschnitt, an dem ein Filtrationsvorgang möglicherweise ablaufen kann.

1b. Die Frage nach dem morphologisch definierten Filtrationsort.

2. Die Frage nach dem Druckgradienten.

3. Die Frage nach der Art der Filtermembran.

4. Die Frage nach der Zusammensetzung des Primärharns. Wenn diese Frage nicht direkt zu klären ist, müssen indirekte Methoden angewandt werden (s. oben).

In den folgenden Abschnitten soll anhand von Beispielen das Vorkommen einer Druckfiltration beim Ausscheidungsprozeß illustriert werden. Die Beispiele werden sich, wie ich bereits andeutete, ausschließlich auf Wirbellose beziehen. Ich werde dabei nicht nach dem System vorgehen, sondern nach der Art der Exkretionsorgane.

B. Beispiele

I. Protonephridien

Die Protonephridien sind wohl die am weitesten verbreiteten Ausscheidungsorgane überhaupt. Sie kommen bei Plathelminthen, Nemertinen, Entoprocten, Rotatorien, Gastrotrichen, Kinorhynchen, Acanthocephalen, Priapuliden und einigen Anneliden bei adulten Tieren vor. Außerdem sind sie als larvale Ausscheidungsorgane weit verbreitet.

Ein Protonephridium besteht immer aus einem oder mehreren sich verzweigenden, innen blindgeschlossenen Kanälen, die nach außen führen. Die Kanäle sind durch verschiedenartig gebaute Zellen, sog. Terminalzellen abgeschlossen, die wie eine Kuppe den Kanalanfängen aufsitzen. In das Kanalinnere entsenden die Terminalzellen ein bis mehrere Undulipodien, die einen ganzen Schopf, eine „Wimperflamme", bilden können. Die Undulipodien sind demnach von einer röhrenförmigen Kanalwand umgeben (Abb. 1, 2a).

Abb. 1. Schema einer Terminalzelle (Cyrtocyte) vom Miracidium des großen Leberegels. Rechts: Totalansicht; links: Aufsicht auf einen Querschnitt in der Reusenregion. Nach Kümmel u. Brandenburg [22]

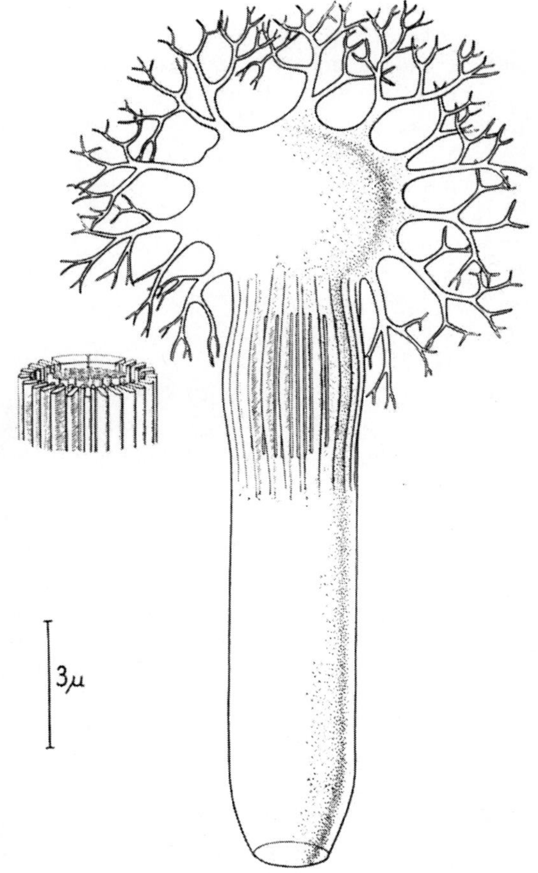

Abb. 2a—c. a Längsschnitt durch eine Terminalzelle (Cyrtocyte). Links ein Reusenstäbchen, rechts die Verbindungs„membran" getroffen. b Halber Querschnitt durch einen Anneliden (Regenwurm). Sehr schematisch. c Schnitt durch ein Cölomsäckchen eines Krebses, sehr schematisch. + = Überdruck, — = Unterdruck, Pfeile deuten den Weg der Exkretionsflüssigkeit an. Bl = Blutlakune, BM = Bauchmark, C = Cölomraum, D = Darm, DG = Dorsalgefäß, GL = Flüssigkeitserfüllte Gewebslücken, LC = Lumen des Cölomsäckchens, N = Nephridium, NT = Trichter des Nephridiums, SG = Subneuralgefäß, VG = Ventralgefäß, WF = Wimperflamme

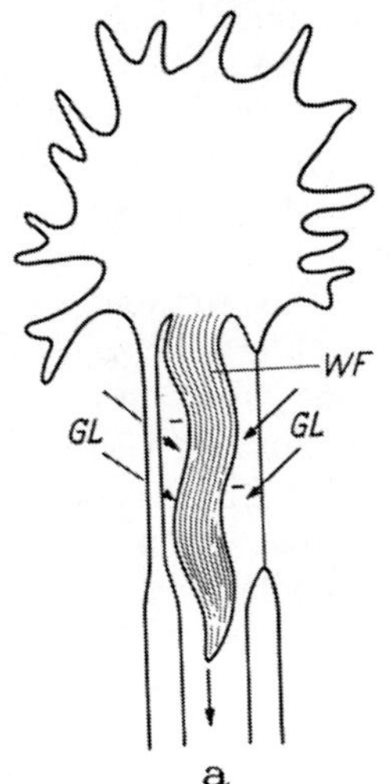

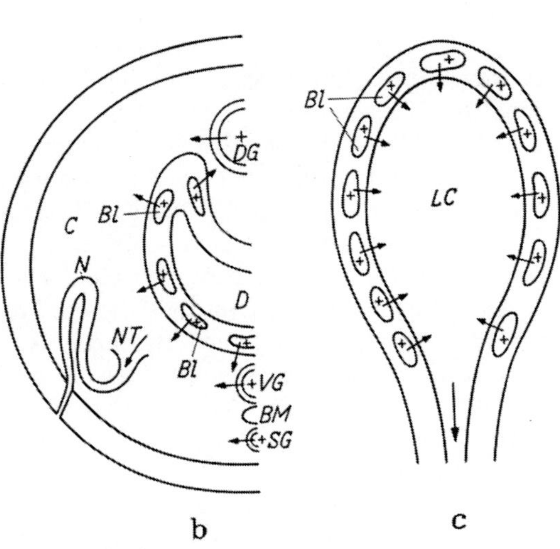

Abb. 2

Wir haben uns der wohlbegründeten Ansicht MARTINs [25] angeschlossen, daß der Filtrationsvorgang der initiale im Prozeß der Harnbildung ist. Danach wird der Abschnitt des Exkretionsorganes, in dem eine Filtration möglicherweise ablaufen kann, der erste, innerste sein, im Falle der Protonephridien also der Abschnitt mit den Terminalzellen. Und an diesen wird man nach entsprechenden Strukturen suchen. Tatsächlich hat man an einer ganzen Reihe von Terminalzellen in den letzten Jahren bei der Untersuchung der Feinstruktur solche Strukturen gefunden, und zwar, wie es zu erwarten war, in der röhrchenförmigen Wand um die Undulipodien (*Stenostomum* [18], *Fasciolamiracidium* [16, 17], *Schistosoma* [42], *Codonocephalus* [36], *Urnatella* [18], *Pedizellina* [4], *Chaetonotus* [3], *Priapuliden* [19], *Glycera* [5], Larve von *Lymnaea* [4]).

Wenn auch der Wandbau bei den einzelnen Formen im Detail nicht geringe Unterschiede aufweist (Abb. 1, 3, 4), so ist das Bauprinzip doch einheitlich: Die Wand des Röhrchens, die eine Bildung bzw. einen Teil der Terminalzelle darstellt, ist nicht in ganzer Ausdehnung kompakt, sondern immer, zumindest stellenweise, in feine Stäbchen oder Leisten aufgelöst. Zwischen den Stäbchen bleiben Lücken bzw. Schlitze von einigen 100 Å Breite frei. Das Röhrchen, das die Undulipodien umschließt, ähnelt so einer Reuse. Wir haben daher für diesen charakteristischen Zelltyp den Namen Reusengeißelzellen oder Cyrtocyten vorgeschlagen [22], zu denen außer den Terminalzellen auch die Choanocyten der Schwämme gehören.

Diese Strukturen könnten auch funktionell als Reusen bzw. sehr grobe Filter angesehen werden. In den meisten, wenn nicht überhaupt in allen Fällen treten jedoch zu den Stäbchenreusen noch Membranen oder membranartige Strukturen hinzu. So werden in den untersuchten Cyrtocyten fast immer die Schlitze zwischen den Stäbchen von membranartigen Strukturen überbrückt (Abb. 2a, 3, 4a, b, d). Nur bei der Larve von *Lymnaea* scheint eine solche Verbindungs„membran" sicher zu fehlen. Dafür hüllt eine andere Membran die ganze Reuse ein bzw. liegt der Reuse auf. Eine solche umhüllende Membran tritt auch bei einer Reihe von Formen zusätzlich neben der Verbindungs„membran" auf [*Urnatella* (Abb. 4b), *Pedizellina*, *Priapuliden* (Abb. 4a)]. Diese umhüllende Membran entspricht wohl einer Basalmembran, sie ist bei den verschiedenen Formen recht ähnlich. Dagegen ist die Natur der Verbindungs„membran" sehr unterschiedlich und praktisch ungeklärt [19]. Manchmal macht sie den Eindruck einer Einheitsmembran, doch reicht die Auflösung meiner Bilder zu sicheren Aussagen nicht aus. Die Verbindungs„membran" ist auch nicht immer im elektronenmikroskopischen Bild deutlich sichtbar. Ihre Deutlichkeit hängt dann wahrscheinlich stark von der Art der Fixierung und besonders der Nachkontrastierung ab.

Wir müßten in diesen Membranen die eigentliche Filtermembran sehen, da die Reuse selbst ein viel gröberes Filter darstellen würde. Allerdings wissen wir nichts über die Permeabilitätseigenschaften der verschiedenen Membranen. Daher ist auch gänzlich unbekannt, welche

Bedeutung den einzelnen Membranen beim Filtrationsprozeß zukommen könnte, wenn zwei Membranen, Basalmembran und Verbindungs-„membran", hintereinander vorhanden sind. Ein ähnliches Problem ist aus dem Glomerulus der Wirbeltiere bekannt [37]. In der Antennendrüse der Krebse wird es uns noch einmal begegnen.

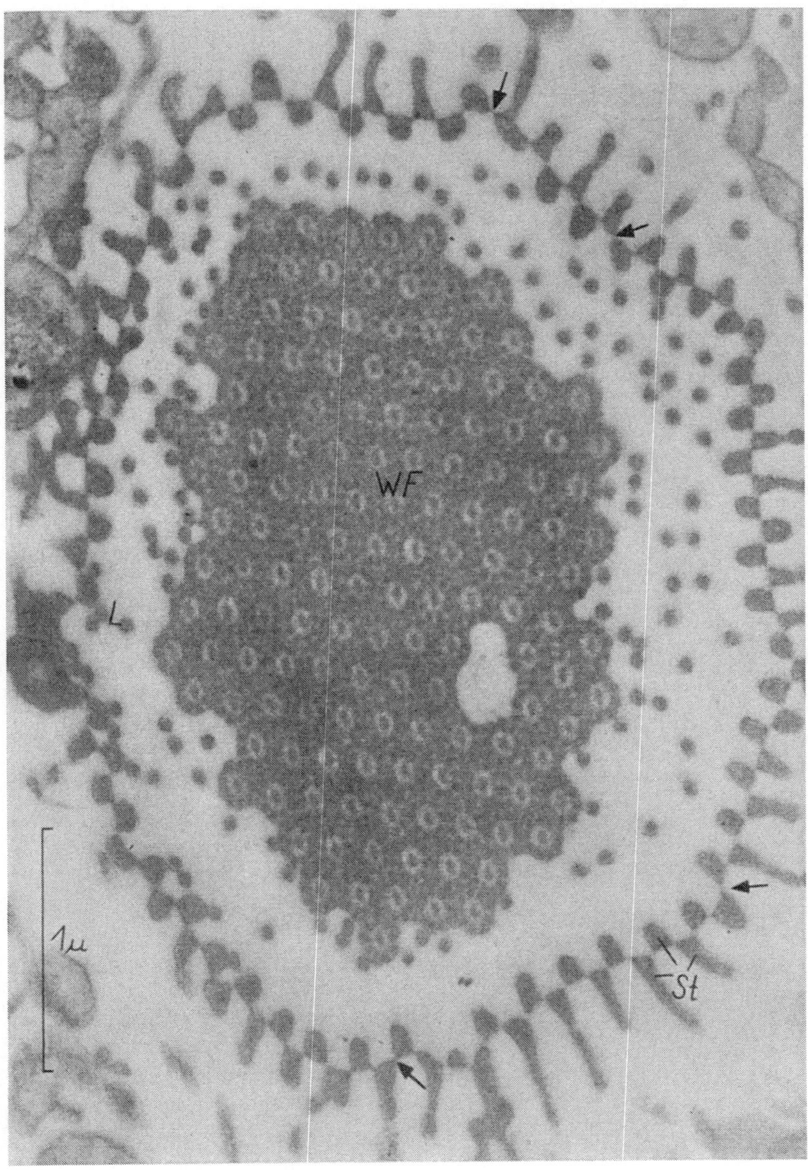

Abb. 3. Querschnitt durch eine Terminalzelle (Cyrtocyte) vom Miracidium (vgl. Abb. 1) in der Reusenregion. *L* = Leptotrichien, *St* = Stäbchen der Reuse, quer, *WF* = Wimperflamme. → = Verbindungs„membran". Nach [17]

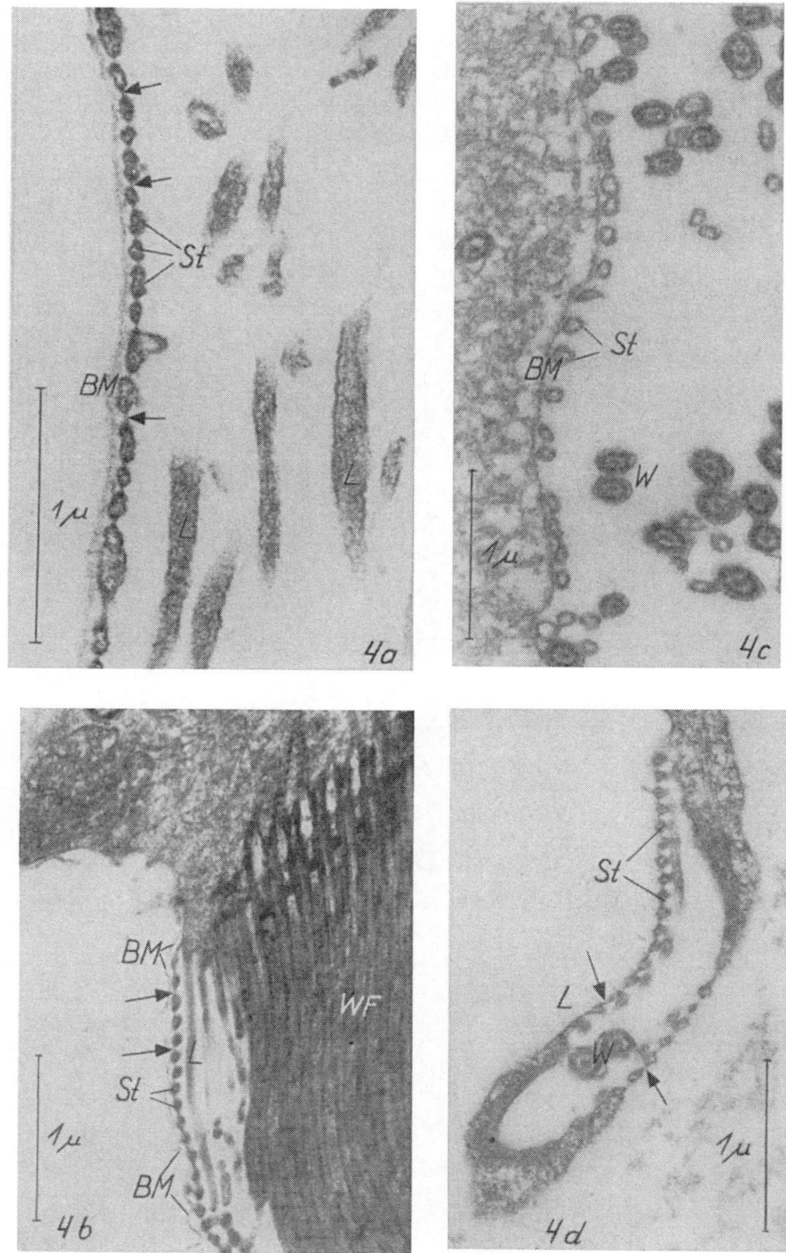

Abb. 4a—d. a Längsschnitt durch die Wand einer Cyrtocyte von *Halicryptus (Priapulida)*.
b Längsschnitt durch eine Cyrtocyte von *Urnatella (Entoprocta)*. c Schnitt durch die
Wand einer Cyrtocyte von *Lymnaea* (Larve). Nach BRANDENBURG [4]. d Schräger Querschnitt durch eine Cyrtocyte von *Stenostomum (Turbellaria)*. → = Verbindungs„membran"
BM = Basalmembran, *L* = Leptotrichien, *St* = Stäbchen der Reuse, quer, *W* = Wimpern
im Reusenlumen, *WF* = Wimperflamme

Es ist heute noch nicht sicher, ob an allen Terminalzellen solche oder ähnliche Strukturen vorkommen, die auf einen Filtrationsvorgang hinweisen [47]. Das Fehlen oder besser der Verlust derartiger Strukturen ist durchaus denkbar: Wie bei einigen Wirbeltieren aus glomerulären Nieren mit Filtration aglomeruläre Nieren hervorgegangen sind, könnten an Protonephridien analoge Umwandlungen eingetreten sein.

Die Reusen sind in allen untersuchten Fällen von flüssigkeitserfüllten Räumen umgeben, seien es Gewebslücken der primären Leibeshöhle oder Cölomräume [19]. Aus diesen Räumen könnte demnach Flüssigkeit in einem Filtrationsvorgang in das Reusenlumen, weiterhin in die abführenden Kanäle und dann nach außen gelangen (Abb. 2a).

Der bis jetzt bekannte Bau der Terminalzellen spricht also durchaus zugunsten eines hier ablaufenden Filtrationsvorganges. Ungeklärt ist die Frage, ob ein für eine Filtration ausreichender Druckgradient verfügbar ist.

Es war wohl zuerst REISINGER [35], der, fußend auf GOODRICH [10], die Meinung vertrat, daß der Schlag der Undulipodien in der Terminalzelle zu einem Druckgradienten führen könnte. Ihm sind viele Autoren gefolgt [1, 2, 16, 22, 25]. Allerdings sind auch starke Einwände geäußert worden, besonders von CARTER [7]. Dieser vertrat die Auffassung, daß der Wimperschlag nie zu einem ausreichenden Filtrationsdruck führen könnte. Er gibt als höchsten Wert, den der Unterdruck in der Terminalzelle erreichen kann, einige mm Wassersäule an. Dieser Wert ist aber an einem Schwamm gemessen worden, und zwar an einer ganzen Geißelkammer. Man kann gegen die Übertragung dieses Wertes auf die Terminalzelle starke Bedenken erheben, wie auch PANTIN [27] betont. Der Bau einer Terminalzelle und einer Geißelkammer ist so unterschiedlich, daß auch die erzeugten Drucke bzw. Unterdrucke kaum übereinstimmen werden. Ich halte es daher für durchaus möglich, daß der Schlag der Undulipodien im Reusenröhrchen zu einem ausreichenden Filtrationsdruck führt, also einen Unterdruck erzeugt, der den kolloidosmotischen Druck der Körperhöhlenflüssigkeit übersteigt (Abb. 2a).

Die Untersuchung der Morphologie ist bei den Protonephridien für die Klärung der Frage, Filtration oder nicht, besonders bedeutsam, da physiologische Experimente wegen der Kleinheit der Organsysteme schwierig und auch kaum durchgeführt worden sind. So ist es vor allen Dingen bisher nicht gelungen, die Exkretionsflüssigkeit zu gewinnen und zu analysieren. Die wenigen funktionellen Beobachtungen, die man angestellt hat, sind aber mit der Ansicht eines Filtrationsvorganges zu vereinen.

So hat PANTIN [27] an der Landnemertine *Geonemertes* zuweilen Oscillationen an der Röhrchenwand der Terminalzelle beobachtet. Man kann annehmen, daß diese Wellen Druckschwankungen innerhalb des Röhrchenlumens anzeigen, hervorgerufen durch den Schlag der Wimperflamme, da diese frei im Röhrchen schwingt. Der gleiche Autor hat an *Asplanchna*, einem Rotator, festgestellt, daß hier die Frequenz des Wimperschlages proportional der gebildeten Harnmenge ist (zit. nach [32]). Diese Beziehung ist am einfachsten so deutbar, daß der Wimperschlag einen Filtrationsvorgang in Gang setzt.

14*

Auch die Verteilung der alkalischen Phosphatase in den Protonephridien von *Geonemertes* und *Rhynchodemus* (Turbellar) läßt sich gut
mit der hier vertretenen Ansicht eines Filtrationsvorganges in den Terminalzellen vereinbaren [8]. Dieses Ferment fehlt nämlich praktisch
vollständig in den Terminalzellen, während es in manchen Abschnitten
der ausführenden Kanäle reichhaltig vorkommt. Die alkalische Phosphatase spielt wahrscheinlich bei vielen aktiven Transportvorgängen eine
Rolle, für den passiven Vorgang einer Filtration ist sie dagegen nicht
notwendig.

Wenn wir jetzt noch einmal alle aufgeführten Ergebnisse Revue
passieren lassen, wird, glaube ich, folgender Eindruck deutlich: Im
Prozeß der Harnbildung in den Protonephridien spielt ein Filtrationsvorgang eine Rolle. Doch ist ein Eindruck noch kein Beweis. Wir müssen
weitere Untersuchungen abwarten.

II „Offene" Exkretionsorgane

Diese Exkretionsorgane bestehen wieder aus ausführenden exkretorisch tätigen Kanälen, die aber innen nicht durch Terminalzellen abgeschlossen sind, sondern sich in einen Cölomraum öffnen. Sie sind also
stets an einen Cölomraum gebunden. Die Cölomräume sind sehr unterschiedlich ausgebildet. Auch die Öffnungen in die Cölomräume und die
Kanäle können bei den einzelnen Tierformen sehr verschiedenartig
gebaut sein und sind keineswegs immer miteinander homologisierbar [11].

Die so charakterisierten Exkretionsorgane sind auch weit verbreitet,
z. B. bei Anneliden, Onychophoren, Crustaceen, Arachnomorphen und
Mollusken. Etwas eingehender untersucht sind nur die Exkretionsorgane von Anneliden, Crustaceen und Mollusken, ich werde mich daher
bei der Besprechung auf diese Tierformen beschränken.

1. Anneliden

Häufig, aber keineswegs immer, ist die Öffnung zum Cölomraum
ein ziemlich weiter Trichter (Abb. 2b). Das Lumen der Exkretionsorgane steht dann mit dem Cölomraum in freier Verbindung. Daher muß
der Primärharn mit der Cölomflüssigkeit identisch sein [32, 34].

Suchen wir nach der Stelle, an der ein Filtrationsvorgang ablaufen
kann, müssen wir uns außerhalb der eigentlichen Exkretionsorgane umse-
hen, und zwar an der Grenze des Cölomraumes: Es wäre möglich, daß hier
eine Filtration aus dem Blut in den Cölomraum stattfinden kann
(Abb. 2b)[1]. Meines Wissens ist die Feinstruktur dieser Region bisher nur
ungenügend bekannt, so daß wir aus der Morphologie keine Stütze
für diese Ansicht besitzen. Ebenso ist das Problem des für eine Filtration
notwendigen Druckgradienten offen: Man kennt den Druck in den Gefäßen und Kapillaren nicht.

---

[1] Bahl [1, 2] hat für einige Oligochaeten die Vermutung geäußert, daß auch
eine Filtration aus dem Blut direkt in das Lumen des Exkretionsorganes statthaben
könnte. Die Anordnung der Kapillaren, die das Nephridium versorgen, und deren
Feinbau machen aber wenigstens für *Lumbricus* diese Möglichkeit sehr unwahrscheinlich [12].

Besser ist man über die Osmolarität und die Zusammensetzung des primären Harnes (Cölomflüssigkeit) einerseits und des Blutes andererseits bei einigen Regenwurmarten unterrichtet [1, 2, 14, 33]. Leider lassen sich aber die Angaben bei den verschiedenen Formen in einigen Punkten nicht miteinander zur Deckung bringen. In einem Punkt stimmen aber die beiden eingehenden Untersuchungen von BAHL [1, 2] und KAMEMOTO [14] überein: Es bestehen wesentliche Unterschiede in den Konzentrationen auch kleinmolekularer Substanzen und anorganischer Ionen zwischen der Cölom- und Blutflüssigkeit. Sicher ist also die Cölomflüssigkeit kein einfaches unverändertes Filtrat. Das schließt aber die Möglichkeit einer Druckfiltration als ersten Vorgang in der Harnbildung nicht aus. Nur müßten sich an diesen Vorgang sekretorische oder/und resorbierende Prozesse anschließen, die das Filtrat verändern.

Möglich wäre auch die Bildung eines Primärharnes in einem Prozeß nach Art einer Dialyse, wie es GRASZYNSKI [12] vorschlägt. Doch sind die Verhältnisse auch bei den am besten untersuchten Formen, den Oligochaeten, noch zu undurchsichtig, um gesicherte Aussagen zuzulassen. Es wäre zudem sehr gut möglich, daß bei den verschiedenen Formen auch prinzipielle Unterschiede bestehen.

2. Crustaceen

Wie bei manchen Arachnoiden ist bei den Krebsen der mit dem Cölomraum verbundene Cölomraum, das sog. Cölomsäckchen (Sacculus), klein und praktisch ein Teil des Exkretionsorganes. In das Säckchen mündet ein sehr unterschiedlich gebauter exkretorisch tätiger Kanal, der nach außen führt (Abb. 5a).

Morphologisch und physiologisch gut untersucht sind nur die Antennendrüsen einiger Dekapodenarten, und zwar besonders die des Hummers und die verschiedener Flußkrebsarten. Ich werde mich daher in diesem Bericht auf die letztgenannten Formen beschränken.

Wenn wir nach dem Organabschnitt suchen, an dem eine Filtration ablaufen kann, werden wir unsere Aufmerksamkeit wieder auf den ersten Abschnitt, das Cölomsäckchen, richten müssen. Dieses ist innen bei den hier betrachteten Formen vielfach gekammert (Abb. 5b), indem das eigentliche Epithel sich ins Innere einfaltet [23, 28]. Dabei gelangen offensichtlich auch bindegewebige Zellen der Säckchenhülle in die Falten, die somit aus solchen „Bindegewebszellen" und den Epithelzellen bestehen (Abb. 5c) [21]. Zwischen „Bindegewebszellen" und Epithelzellen liegen in großer Zahl endothellose Blutlakunen (Abb. 5c), die aus einem Seitenast der Antennenarterie mit Blut versorgt werden [32].

Die anatomischen und lichtmikroskopischen Untersuchungen haben somit keinen direkten Hinweis auf eine Struktur ergeben, an der eine Filtration ablaufen könnte. Erst die Untersuchung der Feinstruktur hat das vermocht [20, 21]. Wie oben dargetan, begrenzen Epithel- und „Bindegewebszellen" die Blutlakunen. Beide Zellarten kleiden die Blutlakunen mit einer Art Basalmembran von mehreren 1000 Å Dicke aus (Abb. 6—8). Für unseren Zusammenhang ist nur die Feinstruktur der Epithelzellen von Interesse. Diese entsenden zur Basalmembran hin

viele Ausläufer, die sich der Basalmembran anlegen (Abb. 6). Seitlich
gehen von den Ausläufern in großer Zahl verschiedenartig gestaltete,
häufig zungen- oder fingerförmige, Fortsätze aus, die sich mit den Fortsätzen anderer Ausläufer verzahnen (Abb. 6, 8). So überziehen Ausläufer

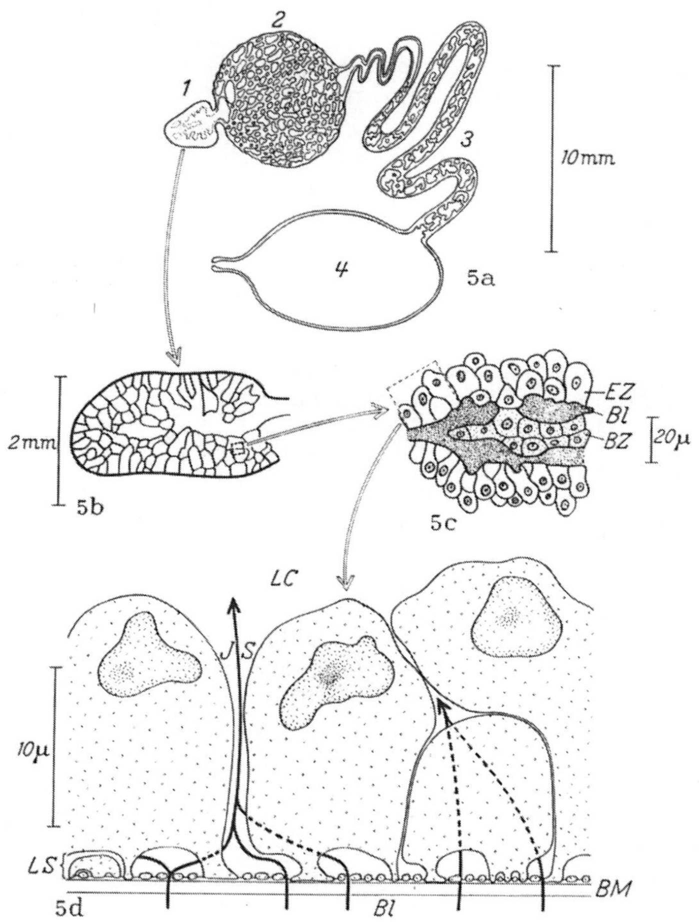

Abb. 5 a—d. a Auseinandergelegte Antennendrüse vom Flußkrebs. Flächenschnitt. Schema
nach PETERS [28]. 1 = Cölomsäckchen, 2 = Labyrinth, 3 = Nephridialkanal, 4 = Harnblase. b Schnitt durch das Cölomsäckchen, nach MALUF [23], vereinfacht. c Eingerahmter
Bereich von b, vergrößert. Schnitt durch eine Einfaltung (Kammerwand) n. mikroskop.
Präparat, schematisch. EZ = Epithelzellen, BZ = „Bindegewebszellen", Bl = Blutlakunen. d Eingerahmter Bereich von c, vergrößert. Schnitt durch einige Epithelzellen,
stark, schematisiert. Die Pfeile deuten den Weg der Exkretionsflüssigkeit während und nach
der Filtration an. Bl = Blutlakunen, BM = Basalmembran, IS = Interzellularspalt,
LS = Lückensystem, LC = Lumen des Cölomsäckchens. Nach [21]

und Fortsätze als ein dichtes Netz die Basalmembran; es bleiben nur
zwischen den Fortsätzen mäanderförmige Schlitze von 300—400 Å
Breite frei (Abb. 6, 8). Diese eben geschilderten Differenzierungen er-

geben eine frappierende Ähnlichkeit der Epithelzellen mit den Podozyten der Wirbeltierniere.

Zwischen den Ausläufern mit ihren Fortsätzen und den eigentlichen
Zellkörpern bildet sich ein in sich geschlossenes Lückensystem, das

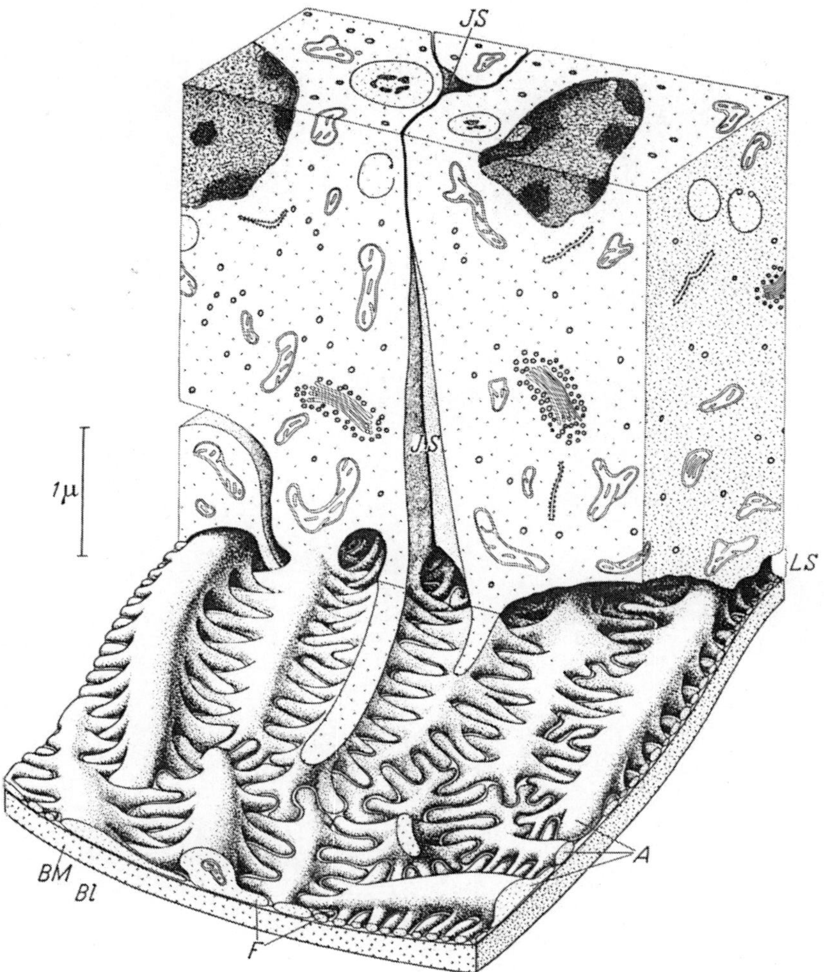

Abb. 6. Blockdiagramm von drei Epithelzellen des Cölomsäckchens von *Cambarus* (Flußkrebs), schematisiert. Ein Teil des Lückensystems ist freigelegt, um die Ausläufer und Fortsätze sichtbar zu machen. *A* = Ausläufer, *Bl* = Blutlakune, *BM* = Basalmembran,
F = Fortsätze, *IS* = Interzellularspalt, *LS* = Lückensystem. Nach [21]

sich im Basisbereich der Epithelzellen ausbreitet (Abb. 6, 7). Zur Basalmembran hin wird das Lückensystem durch eine feine Membran (weniger
als 100 Å Dicke) abgeschlossen, die die Schlitze zwischen den Fortsätzen überspannt (Abb. 6, 7). Sie erinnert an analoge Membranen der
Wirbeltierpodozyten [37, 49]. Diese Membran ist genau wie bei Cyrtocyten nicht immer deutlich sichtbar.

Mit dem Cölomsäckchenlumen steht das basale Lückensystem durch Interzellularspalten in freier offener Verbindung (Abb. 5d, 6). Stoffe, die aus dem Lumen der Blutlakunen in das Lumen des Cölomsäckchens

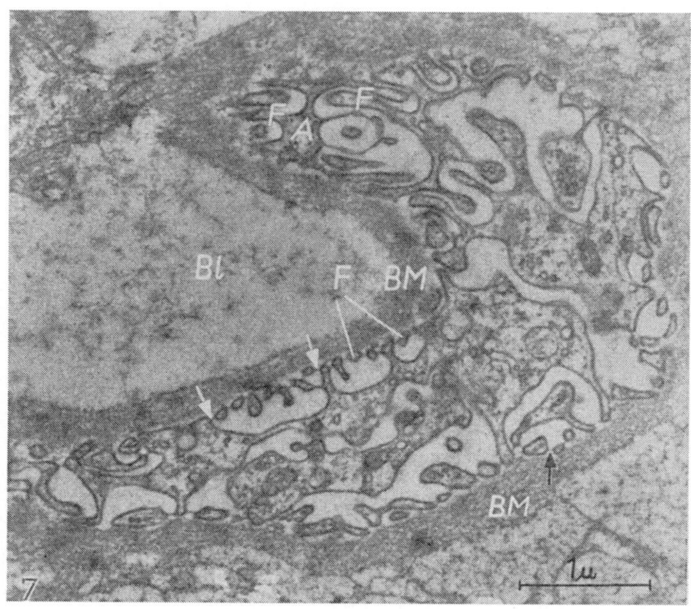

Abb. 7. Flächenschnitt in der Region des Lückensystems von Epithelzellen des Cölomsäckchens von *Cambarus* (Flußkrebs). *A* = Ausläufer, *Bl* = Blutlakune, *BM* = Basalmembran, *F* = Fortsätze, → = Verbindungs,,membran". Nach [21]

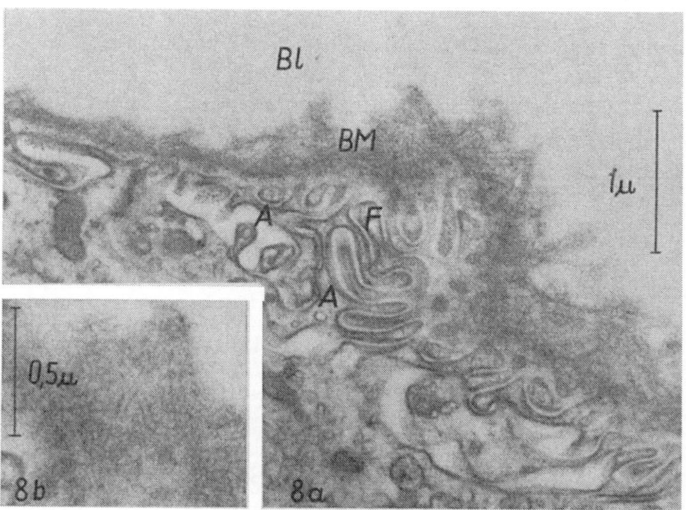

Abb. 8a u. b. a Tangentialschnitt durch den basalen Bereich einer Epithelzelle. *A* = Ausläufer, *Bl* = Blutlakune, *BM* = Basalmembran, *F* = Fortsätze. b Ausschnitt aus a, vergrößert. Nach [21]

gelangen sollen, brauchen demnach keine zelligen Elemente, sondern nur die beiden Membranen, Basalmembran und feine Membran zwischen den Fortsätzen, zu passieren (Abb. 5d). Wenn ein Filtrationsvorgang vorkommt, sind diese Membranen die einzigen Barrieren, sie würden als Filter wirken.

Die Frage, ob ein ausreichender Druckgradient für eine Filtration vorliegt, hat PICKEN [29] durch Messung des kolloidosmotischen Druckes des Blutes und des hydrostatischen Druckes in der Leibeshöhle (gemessen im sog. Sternalsinus) zu klären versucht. Er stellte eine Differenz der beiden Drucke von mehreren cm Wassersäule fest. Der hydrostatische Druck in den Blutlakunen ist zwar unbekannt, aber wahrscheinlich auf Grund der ziemlich direkten Blutversorgung [32] des Organabschnittes höher als im Sternalsinus. Daher werden im Cölomsäckchen wahrscheinlich Druckverhältnisse herrschen, die eine Filtration zulassen (Diskussion, s. [39]).

Eine Analyse der Flüssigkeit im Cölomsäckchen, die als primärer Harn gelten muß, ist ebenfalls wenigstens am Flußkrebs von zwei Seiten in Angriff genommen worden. Schon 1936 stellte PETERS [28] an *Potamobius fluviatilis* (wohl = *Astacus astacus*) fest, daß die Cl-Konzentration der Flüssigkeit im Cölomsäckchen der im Blut entsprach. An zwei weiteren Flußkrebsarten (*Austropotamobius pallipes* und *Orconectes virilis*) zeigte RIEGEL [39] in jüngster Zeit, daß die Flüssigkeit im Cölomsäckchen zwar blutisotonisch ist, ihre Cl-Konzentration aber im Gegensatz zu der Angabe PETERS kleiner als im Blute ist, allerdings nur geringfügig. Wenn eine Filtration aus dem Blut stattfindet, muß daher durch Sekretion und Reabsorption der Primärharn sogleich verändert werden.

Viele weitere Versuche sind angestellt worden, um den Vorgang der Harnbildung genau kennenzulernen. Nur wenige kann ich davon erwähnen. So wurde die Ausscheidung von Inulin getestet. Beim Hummer, der einen angenähert blutisotonischen Endharn ausscheidet, war das $\frac{U}{B}$ Verhältnis für Inulin ungefähr 1 [6]. Beim Flußkrebs dagegen war dieses Verhältnis stets über 1 und unterhalb einer Blutkonzentration von 1 mg-% Inulin nahezu konstant [40]. Diese Werte sprechen für eine Wasserreabsorption nach einer Filtration.

Glucose erscheint bei normalen Blutkonzentrationen nicht im Endharn. Vergiftet man aber die Niere mit Phlorrhizin, so tritt Glucose beim Hummer [6] und Flußkrebs [40] auch im Endharn auf (Glucosurie). Der Endharn nähert sich also in seiner Zusammensetzung dem Primärharn. Das gleiche gilt, wenn man die Konzentration an Glucose im Blut stark erhöht [6, 40]. Auch dann kommt es zu einer Glucosurie, Glucose wirkt also als Schwellensubstanz. Auch diese Ergebnisse sind am leichtesten durch einen Filtrationsvorgang deutbar.

Auch die anderen Experimente (Wasserinjektionen in den Krebs [38], Atmungsmessungen [28], Abkühlung [38]) sprechen alle für eine Filtration als initialen Vorgang. Wenn es auch heute noch keinen endgültigen

Beweis für diese Ansicht gibt, so ist doch die Fülle der Hinweise morphologischer und physiologischer Art überzeugend (Abb. 2c,5d). Die Annahme von Maluf [24], daß der gesamte Harn in das Exkretionsorgan hinein sezerniert wird, ist sicher nicht haltbar. Doch ist es sehr gut möglich, daß der Primärharn schon innerhalb des Cölomsäckchens durch Sekretion und Reabsorption verändert wird [21, 39].

3. Mollusken

Die Exkretionsorgane der Mollusken öffnen sich ebenfalls fast immer in einen Cölomraum, das das Herz umgebende Perikard, und zwar durch den sog. Renoperikardialgang. Das Aussehen der Exkretionsorgane bei den einzelnen Formen ist sehr unterschiedlich. Doch handelt es sich im Grunde stets wieder um Kanäle, die nach außen führen, d. h. meist in die Mantelhöhle münden. Häufig ist der erste Abschnitt zu dem sog. Nierensack angeschwollen.

Wie bei den vorangegangenen Tierformen können wir vermuten, daß durch Filtration in die Cölomhöhle, d. h. in das Perikard hinein, ein primärer Harn abgesondert wird, der dann durch den oder die Renoperikardialgänge in die eigentlichen Exkretionsorgane gelangen könnte

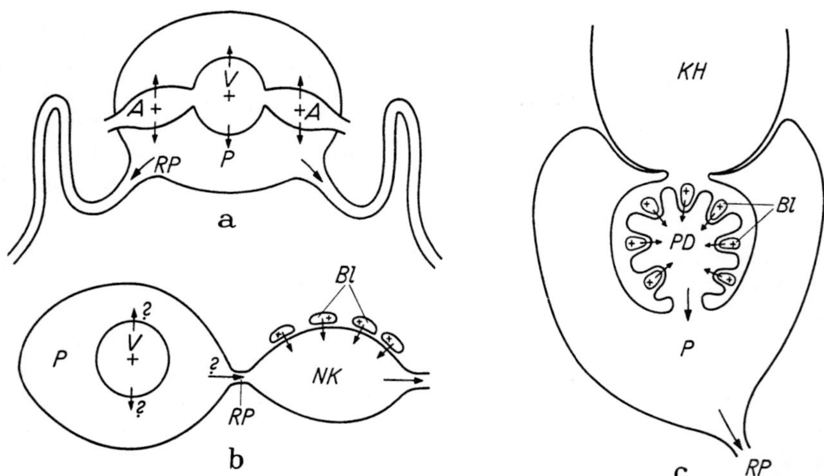

Abb. 9a—c. a Querschnitt durch das Perikard mit Herz und Exkretionsorganen einer Muschel oder eines primitiven Gastropoden, sehr schematisch. Die Pfeile durch die Wand des Ventrikels gelten für eine Muschel, durch die Atrien für einen Gastropoden *(Haliotis)*, siehe Text. b Querschnitt durch das Perikard mit Herz und Exkretionsorgan eines Landpulmonaten, sehr schematisch. c Schnitt durch ein Kiemenherz mit Anhang (Perikardialdrüse) und durch das Perikard eines Cephalopoden, sehr schematisch. A = Atrien, Bl = Blutgefäß bzw. Blutlakunen, KH = Kiemenherz, NK = Nierenkammer, P = Perikardialhöhle, PD = Perikardialdrüse, RP = Renoperikardialgang, V = Ventrikel. Die Pfeile deuten den Weg der Exkretionsflüssigkeit an, + = Überdruck

(Abb. 9a, c). Als näherer Ort der Filtration können besonders die pulsierenden Abschnitte des Blutgefäßsystems in Frage kommen, die vom Perikard umhüllt werden. Die Feinstruktur dieser Stellen ist allerdings in keinem Fall bekannt.

Der Beweis einer Filtration in die Perikardialhöhle hinein ist zuerst von PICKEN [30] an der Süßwassermuschel *Anodonta* geführt worden. Er zeigte zunächst, daß der hydrostatische Druck im Ventrikel (etwa 6 cm Wasser) den kolloidosmotischen Druck des Blutes (etwa 4 mm Wasser) weit übersteigt. Später haben FLORKIN und DUCHATEAU [9] tatsächlich den Ventrikel als den eigentlichen Filtrationsort identifizieren können.

PICKEN und FLORKIN u. DUCHATEAU haben auch Perikardialflüssigkeit gewonnen und vergleichend mit dem Blut analysiert. Es stellte sich heraus, daß die Perikardialflüssigkeit weitgehend einem Filtrat entspricht. Es scheint also wenigstens für *Anodonta* sicherzustehen, daß die Primärharnbildung durch Filtration in das Perikard hinein erfolgt.

PICKEN [30] war auch der erste, der bei Gastropoden, und zwar bei dem Süßwasserpulmonaten *Lymnaea*, eine Filtration in das Perikard hinein nachgewiesen hat. Auch hier stellte er einen ausreichenden Filtrationsdruck fest. Er konnte ferner zeigen, daß die Perikardialflüssigkeit blutisotonisch ist und außerdem wesentlich weniger Eiweiß als das Blut enthält. Allerdings ist die genaue Zusammensetzung der Perikardialflüssigkeit ebensowenig bekannt wie der genaue Filtrationsort.

Sehr viel weiter führt eine Arbeit von HARRISON [13] an dem primitiven Gastropoden *Haliotis*. Die Autorin führte Testsubstanzen (Inulin u. a.) in die vordere Arterie ein und maß die Konzentrationen gleichzeitig im Blut und in der Perikardialflüssigkeit. Sie stellte fest, daß die Konzentrationen dieser Substanzen im Blut und Perikard nahezu gleich waren. Sie schloß daher auf eine Filtration, die sich nach ihrer Meinung wahrscheinlich durch die Wände der Atrien in die Perikardialhöhle hinein vollzieht (Abb. 9a).

Es scheint also zumindest bei manchen Gastropoden ähnlich zu sein wie bei *Anodonta*. Bei anderen Gastropoden, so bei den bisher untersuchten Landpulmonaten *Achatina* und *Helix*, liegen anscheinend andere Verhältnisse vor. Hier findet in die Perikardialhöhle entweder nur eine sehr geringe oder gar keine Filtration statt. Dafür spricht auch die Enge der Renoperikardialgänge, die offensichtlich nicht zur Ausleitung des Primärharnes ausreichen [45]. Daß aber doch der primäre Vorgang in einer Filtration besteht, scheint sicher zu sein. So ist nach VORWOHL [45] die Nierenkammerflüssigkeit bei *Helix* und *Achatina* blutisotonisch und außerdem eiweißfrei, sie wird als Primärharn angesehen. Auch die Experimente mit Testsubstanzen, so mit Inulin und verschiedenen Farbstoffen, die MARTIN et al. (zit nach [25]) an *Achatina* unternommen haben, sprechen für eine Filtration. Sie bestimmten die Konzentration dieser den Tieren zugeführten Substanzen im Blut und Harn, gleichzeitig analysierten sie auch den Glucosegehalt. Das $\frac{U}{B}$ Verhältnis für Inulin war bei freier Ableitung des Harnes durch Katheter, also gemessen unter einem niedrigen Druck, etwa gleich eins, dagegen für die Farbstoffe und Glucose ungleich eins. Höhere Werte als eins erhielten sie für das $\frac{U}{B}$ Verhältnis auch des Inulins, wenn der freie Abfluß gehemmt ist, also ein Stau eintritt. Dieser erhöhte Wert wird wahrscheinlich durch die Reabsorption von Wasser bedingt, die erst bei einem Stau auftritt, wie er nativ wohl immer in den Exkretionsorganen vorkommt. Vergiftet man das Organ, werden die $\frac{U}{B}$ Verhältnisse für alle Substanzen angenähert eins[1]. Alle diese Versuche lassen sich am einfachsten durch einen Filtrationsvorgang deuten.

Wo der Ort des Filtrationsvorganges ist, scheint nicht völlig geklärt zu sein. Wahrscheinlich vollzieht er sich aber durch die Zellagen der Nierenkammer (Abb. 9b) [45]. Die Niere wird unter anderem durch Blutgefäße, die direkt vom Ventrikel des Herzens ausgehen, versorgt, so daß auch hier für einen ausreichenden Filtrationsdruck gesorgt ist [25].

Bei dem einzigen gut untersuchten Cephalopoden, *Octopus*, ist wiederum eine Filtration in die Perikardialhöhle hinein nachgewiesen worden [31]. Die Filtration findet aber hier nicht an den Wänden des Herzens oder der Kiemenherzen statt, sondern in den Kiemenherzanhängen (Abb. 9c). Diese Kiemenherzanhänge, auch Perikardialdrüsen genannt, sind Wucherungen des Perikardepithels, das sich zu einer Kammer schließt, deren Inneres nur an einer Stelle mit dem eigentlichen Perikardlumen kommuniziert. Die Anhänge stehen mit dem Kiemenherzen durch einen kurzen Stiel in Verbindung, durch den Blut in die Wandung der Anhänge tritt [44]. Dadurch werden die Wandepithelien reichlich mit Blut bespült. Man muß sich wohl vorstellen, daß die Filtration durch die Epithellage hindurch in das Lumen der Anhänge hinein abläuft, und der primäre Harn dann durch die Öffnung in das eigentliche Perikard und von da aus in die Exkretionsorgane gelangt (Abb. 9c). Der notwendige Blutdruck ist durch die direkte Blutzufuhr aus dem Kiemenherzen, das kontraktil ist, gewährleistet.

Auch bei den Mollusken besteht also wohl allgemein der initiale Vorgang des Harnbildungsprozesses in einer Filtration. Der Ort der Filtration ist aber bei den einzelnen Formen unterschiedlich. Das ist eigentlich nicht verwunderlich: Immer dort, wo sich ein genügender Druckgradient ausbildet, kann es an geeigneten Strukturen zu einer Filtration kommen.

III. Die Exkretionsorgane (Cyrtopodozyten) von Branchiostoma

Die metamer angeordneten Exkretionsorgane des Lanzettfischchens *Branchiostoma* werden meist unter die Protonephridien eingereiht, da sie Solenozyten besitzen sollen. Die Untersuchung der Feinstruktur berechtigt aber zu einer gesonderten Besprechung [5].

An diesen Exkretionsorganen lassen sich zwei Teile unterscheiden (Abb. 10a): 1. Ein Nephridialkanal, der in den Peribranchialraum mündet. 2. Viele langgestreckte Zellen, die wie Stecknadeln im Kanalanfang stecken. Nur die letztgenannten Zellen brauchen uns hier näher zu interessieren.

---

[1] Das gilt nur bei freier Ableitung des Harnes, da die Wasserresorption wenigstens bei Vergiftung mit Phlorrhizin nicht gehemmt wird.

Die eigentlichen Körper dieser Zellen, die dem Stecknadelkopf entsprechen, liegen den Wänden von Bluträumen an (Abb. 10a, 11). Von jedem Zellkörper geht ein langes Röhrchen aus, das einen Cölomraum, das subchordale Cölom, durchzieht und schließlich die Wand des Nephridialkanals durchbohrt [11]. Im Inneren jedes Röhrchens schlägt eine Geißel. Die erwähnten Bluträume sind ohne Endothel, sie

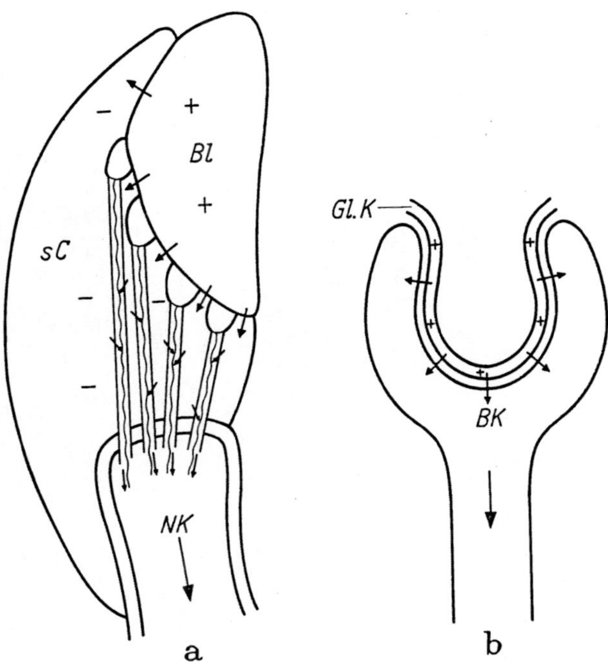

Abb. 10a u. b. a Schnitt durch ein Exkretionsorgan von *Branchiostoma*, sehr schematisch. b Schnitt durch einen Glomerulus eines Wirbeltiers, schematisch. Die Pfeile deuten den Weg der Exkretionsflüssigkeit an, + = Überdruck, — = Unterdruck, *BK* = Raum der Bowmanschen Kapsel, *Bl* = Blutraum, *GlK* = Glomeruluskapillare, *NK* = Nephridialkanal, *sC* = Subchordales Cölom

sind nur von einer Basalmembran ausgekleidet. Diese ist offensichtlich das Produkt der den Bluträumen anliegenden Zellkörper und außerdem von gefingerten Ausläufern, die von den Zellkörpern ausgehen. Die Zellkörper und ihre ineinander verzahnten Ausläufer bilden auf der Oberfläche der Bluträume einen fast lückenlosen Belag. Das erinnert wieder sehr an die Podozyten aus der Wirbeltierniere [5]. Allerdings konnten wir beim Lanzettfischchen keine Membran sehen, die die freibleibenden Schlitze überbrückt. Ein weiterer wesentlicher Unterschied ist der Besitz des Röhrchens, das bei den Podozyten der Wirbeltiere nie vorkommt[1].

Die Wand des Röhrchens ähnelt in ihrem Bau dem der Reusenröhrchen von Cyrtocyten. Sie ist aus 10 dreikantigen Längsstäbchen aufgebaut. Doch ist im Gegensatz zu allen bekannten Cyrtocyten das

---

[1] In diesen Zellen sind also Merkmale von Podozyten und Cyrtocyten vereinigt. Wir haben sie deshalb Cyrtopodocyten genannt [22].

Röhrchen durch keine Membran abgeschlossen, sondern das Lumen des Röhrchens steht in freier Verbindung mit dem Lumen des Cölomraumes (Abb. 11).

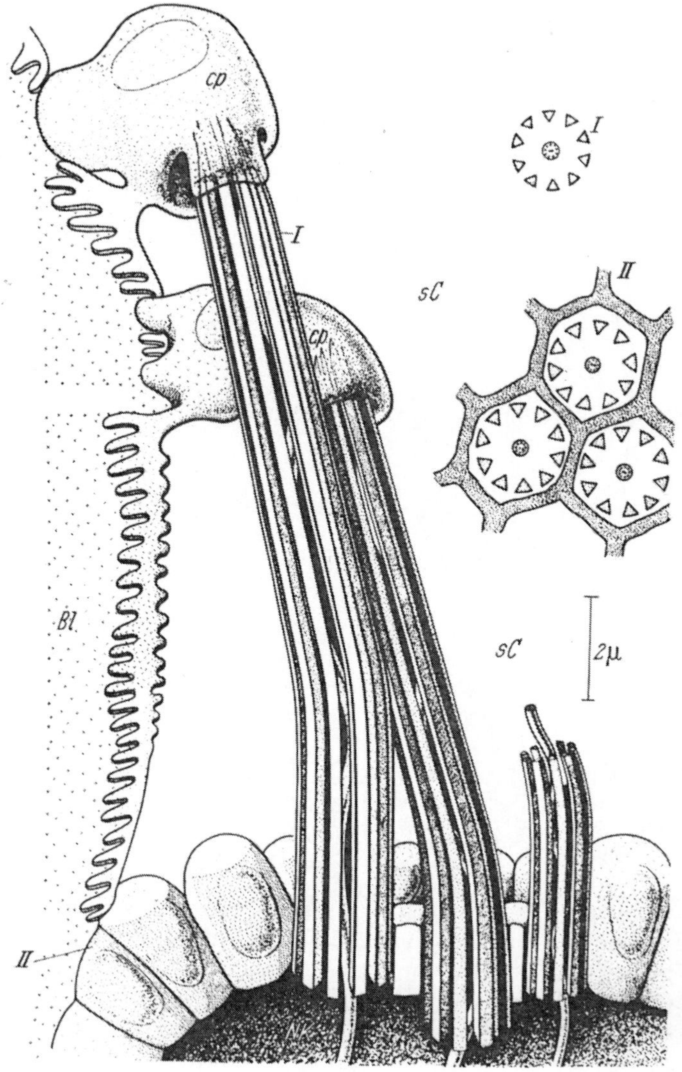

Abb. 11. Cyrtopodocyten aus einem Exkretionsorgan von *Branchiostoma*. *Bl* = Blutraum, *CP* = Cyrtopodocyten, *NK* = Nierenkanal, *sC* = Subchordales Cölom, I bzw. II = Querschnitte durch die gekennzeichneten Bereiche. Nach Brandenburg und Kümmel [5]

Die eben beschriebenen Differenzierungen sprechen durchaus für einen Filtrationsvorgang (Abb. 10a): Der Blutdruck im Zusammenwirken mit dem Sog im Cölomraum, hervorgerufen durch den Geißelschlag in den Reusenröhrchen, ergibt einen Druckgradienten, der zu einer Filtration aus dem Blut in den Cölomraum hineinführt. Als Filtermembran kann nur die Basalmembran an der Oberfläche der Bluträume dienen, da die Röhrchen selbst ja keine derartige Membran besitzen. Der so gebildete primäre Harn wird durch die seitlich offene Röhrchenwand weiter durch den Geißelschlag in das Röhrchenlumen und dann in den Nephridialkanal getrieben.

Diese Vorstellung einer Filtration wird durch den Bau sehr nahegelegt, sie kann sich aber auf keinerlei Experimente stützen, sondern nur auf den gesunden Menschenverstand. Sie muß also noch bewiesen werden.

C. Schlußbetrachtungen

Wir haben in den vorangegangenen Abschnitten eine Reihe von Tiergruppen kennengelernt, bei denen ein Druckfiltrationsvorgang im Harnbildungsprozeß sichergestellt, eine Reihe anderer, bei denen ein derartiger Vorgang mit einiger oder großer Wahrscheinlichkeit zu vermuten ist.

Die Frage liegt auf der Hand, ob dieser Vorgang allgemein verbreitet ist. Bei der Beantwortung dieser Frage müssen wir natürlich alle Tiergruppen ohne eigentliche Exkretionsorgane ausscheiden. Dann würde ich die Frage bejahen, allerdings mit gewissen Einschränkungen. Es gibt ja offensichtlich einzelne Tierformen oder auch Tiergruppen mit Exkretionsorganen, an denen ein derartiger Vorgang nicht abläuft, so etwa die Wirbeltiere mit aglomerulären Nieren, möglicherweise die Planarien (Turbellarien) mit Protonephridien ohne Cyrtocyten [47] und die große Gruppe der Insekten mit Malpighischen Gefäßen[1]. Man kann aber annehmen, daß diese Verhältnisse sekundär erworben sind. Daher sollte man die obige Aussage vielleicht etwas vorsichtiger so formulieren: Im Harnbildungsprozeß war ursprünglich immer ein Druckfiltrationsvorgang wesentlich, der auch bei den meisten Tierformen erhalten geblieben ist. Bei einigen Formen ist er sekundär verlorengegangen.

Sehr auffällig ist eine gewisse Gleichartigkeit im Bau des eigentlichen Filtrationsortes. Es kommt hier zu einer „Auflösung" des Zelleibes in Stäbchen oder Leisten, zwischen denen schmale Schlitze freibleiben. Diese Schlitze werden fast immer von Membranen abgeschlossen, und diese sind das eigentliche Filter. Die Frage drängt sich auf, was der Vorteil oder der Sinn einer solchen Anordnung ist. Ideal wäre eine möglichst große Filterfläche, diese müßte aber einem ausreichenden Druckgradienten standhalten. Die in der Natur verwirklichte Anordnung kann man als Kompromißlösung zwischen dem Ideal einer möglichst großen Filterfläche und der Notwendigkeit einer gewissen Stabilität betrachten. Das gleiche wäre mit einer Porenwand zu erreichen. Tatsächlich ist eine derartige Anordnung bei vielen Kapillarwandungen der Wirbeltiere verwirklicht.

Die Stäbchen oder Leisten könnten neben dieser stützenden Funktion noch andere Aufgaben erfüllen. Sie könnten nämlich die Fähigkeit zu Formänderungen haben und so die Schlitze verbreitern bzw. verengen.

---

[1] Eine Filtration s. l. kann vorkommen. Bei Insekten sind wohl die Mechanismen der Primärharnbildung nicht gleichartig.

Damit wäre die Möglichkeit gegeben, die Filtrationsrate zu regulieren. An Wirbeltiernieren sind solche Formänderungen beschrieben worden, doch kennt man Analoges bisher nicht aus Exkretionsorganen von Wirbellosen.

Als Filter dienen die erwähnten Membranen, die die Schlitze zwischen den Stäbchen abschließen. Damit ist die wichtige Feststellung gemacht, daß die Exkretionsflüssigkeit keine eigentlichen Zellen oder Zellteile bei der Filtration zu passieren braucht. Nicht selten sind, wie wir sahen, zwei verschiedene Membranen ausgebildet. Über die Bedeutung der einzelnen Membranen beim Filtrationsprozeß lassen sich in diesen Fällen nur Vermutungen äußern [37]. Auch die Natur der Verbindungs-„membranen" bei Cyrtocyten und Epithelzellen im Cölomsäckchen der Krebse ist ungeklärt. Das sind nur einige der vielen Probleme, die noch ihrer Lösung harren.

Literatur

[1] Bahl, K. N.: Quart. J. micr. Sci. **85**, 343 (1945).
[2] — Biol. Rev. **22**, 109 (1947).
[3] Brandenburg, J.: Z. Zellforsch. **57**, 136 (1962).
[4] — Persönl. Mitteilung.
[5] —, and G. Kümmel: J. Ultrastruct. Res. **5**, 437 (1961).
[6] Burger, J. W.: Biol. Bull. **113**, 207 (1957).
[7] Carter, G. S.: A General Zoology of the Invertebrates, 3rd. ed. London: Sidgwick & Jackson 1948.
[8] Danielli, J. F., and C. F. A. Pantin: Quart. J. micr. Sci. **91**, 209 (1950).
[9] Florkin, M., et G. Duchateau: Physiol. comp. **1**, 29 (1948).
[10] Goodrich, E. S.: Quart. J. micr. Sci. **41**, 439 (1899).
[11] — Quart. J. micr. Sci. **86**, 115 (1945).
[12] Graszynski, K.: Zool. Beitr. N. F. **8**, 189 (1963).
[13] Harrison, F. M.: J. exp. Biol. **39**, 179 (1962).
[14] Kamemoto, F. I., A. E. Spalding, and S. M. Keister: Biol. Bull. **122**, 228 (1962).
[15] Kurtz, St. M., and J. F. A. McManus: J. Ultrastruct. Res. **4**, 81 (1960).
[16] Kümmel, G.: Z. Naturforsch. **13b**, 677 (1958).
[17] — Protoplasma **51**, 371 (1959).
[18] — Z. Zellforsch. **57**, 172 (1962).
[19] — Z. Zellforsch. **62**, 468 (1964).
[20] — Naturwissenschaften **51**, 200 (1964).
[21] — Zool. Beiträge N. F. (im Druck).
[22] —, u. J. Brandenburg: Z. Naturforsch. **16b**, 692 (1961).
[23] Maluf, N. S. R.: Zool. Jb. 3. Abt. **59**, 515 (1939).
[24] — Biol. Bull. **81**, 235 (1941).
[25] Martin, A. W.: In: Recent Advances in Invertebrate Physiology, p. 247. Oregon: University of Oregon Publications 1957.
[26] — Ann. Rev. Physiol. **20**, 225 (1958).
[27] Pantin, C. F. A.: Quart. J. micr. Sci. **88**, 15 (1947).
[28] Peters, H.: Z. Morph. Ökol. Tiere **30**, 355 (1935).
[29] Picken, L. E. R.: J. exp. Biol. **13**, 309 (1936).
[30] — J. exp. Biol. **14**, 20 (1937).
[31] Potts, W. T. W., and A. W. Martin: Proc. Internat. Congr. of Zoology 1963, Vol. I, p. 78 (1963).
[32] —, and G. Parry: Osmotic and Ionic Regulation in Animals. Oxford-London-New York-Paris: Pergamon Press 1964.
[33] Ramsay, J. A.: J. exp. Biol. **26**, 46 (1949).
[34] — In: Active Transport and Secretion, Symposia of the Society for Experimental Biology, VIII, p. 1. Cambridge: University Press 1954.

[35] REISINGER, E.: Zool. Anz. **54**, 200 (1922).
[36] — Zool. Anz. **172**, 16 (1964).
[37] RHODIN, J. A. G.: J. Ultrastruct. Res. **6**, 171 (1962).
[38] RIEGEL, J. A.: J. exp. Biol. **38**, 291 (1961).
[39] — J. exp. Biol. **40**, 487 (1963).
[40] —, and L. B. KIRSCHNER: Biol. Bull. **118**, 296 (1960).
[41] SCHMIDT-NIELSEN, B.: Ann. Rev. Physiol. **25**, 631 (1963).
[42] SENFT, A. W., E. D. PHILPOTT, and A. H. PELOFSKI: J. Parasit. **47**, 217 (1961).
[43] SMITH, H. W.: Principles of Renal Physiology. New York: Oxford University Press 1956.
[44] STROHL, J.: In: Wintersteins Handbuch der vergleich. Physiol., Bd. II/2, S. 443. Jena: Gustav Fischer 1924.
[45] VORWOHL, G.: Z. vergl. Physiol. **45**, 12 (1962).
[46] WESSING, A.: Protoplasma (Wien) **55**, 433 (1963).
[47] WETZEL, B. K.: Fifth Internat. Congr. for electron Microscopy Q10 (1962).
[48] WILBRANDT, W.: Fortschr. Zool. **12**, 28 (1960).
[49] YAMADA, E.: J. biophys. biochem. Cytol. **1**, 551 (1955).

Summary

The introduction deals first with the definition of filter, filtrate, filtration, and filtration by pressure. For filtration by pressure two conditions must be fulfilled:
a) the presence of a filter separating a compartment I from a compartment II;
b) the presence of a sufficient gradient of hydrostatic pressure between compartments I and II.

These two conditions are then discussed. The main methods for proving the presence of filtration by pressure are described.

In the succeeding sections some examples of the presence of filtration by pressure as an excretory mechanism are discussed.

I. Protonephridia: the weir of the flame bulbs indicates filtration. It is uncertain whether a sufficient pressure gradient exists. Further investigations are necessary.

II. "Open" organs of excretion: in annelids filtration by pressure is possible but quite unproved. Our knowledge is too scanty.

In some crustaceans we find in the first part of the excretory organs cells reminiscent of the podocytes in vertebrate glomeruli. These morphological structures and numerous physiological investigations indicate that filtration by pressure is rather probable in this case.

In three different classes of molluscs physiological experiments have largely proved the existence of filtration by pressure. The site of filtration varies, and the filtrate usually reaches the pericardium first.

III. In the excretory organs of *Branchiostoma* the morphological structures point to filtration. There have been no physiological investigations.

As a final observation, the general structural principles of the filtration site are pointed out, and the functional significance of these structures is discussed.

Diskussion

Hirsch: Es erscheint mir fraglich, ob man nach Ihren morphologischen Befunden überhaupt noch von einer Filtration sprechen kann; denn alle diese Schlitze, die Sie zeigten und die als Filterporen in Frage kämen, sind durch die darunterliegende Basalmembran abgeschirmt oder durch Membrandiaphragmen — die auch bei den Podozyten des Glomerulum nachgewiesen wurden — verschlossen.

H. Sitte: Die Basalmembran kann sehr wohl als Filter in Betracht gezogen werden, nur dürfen wir ihre Struktur nicht mit einem mechanischen Sieb mit Löchern konstanter Größe vergleichen, sondern mit einem Filterpapier. Bei einem Filterpapier werden auch bestimmte Teilchen einer bestimmten Größe zurückgehalten. Man könnte aber kaum sagen, daß eine gewisse Porengröße da ist. Wohl aber kann man sagen, daß es sich verhält, als wären Poren eines bestimmten Durchmessers vorhanden; denn auch da passieren Partikel einer gewissen Größe nicht mehr, und was darunter liegt, passiert. Dabei ist natürlich eine statistische Streuung vorhanden. Ein solches Netzwerk müssen wir nach all dem, was uns von der Feinstruktur bekannt ist, in der Basalmembran annehmen. Deshalb sucht man dort vergebens nach Poren im Sinne von Löchern, wird aber sehr wohl eine Abgrenzung von hochmolekularen und niedermolekularen Bestandteilen im Rahmen einer Druckfiltration bekommen.

Kümmel: Wie Herr SITTE zeigte, kann das Gelgerüst der Basalmembran durchaus als Filter fungieren. Schwierig ist natürlich die Deutung der Diaphragmen. Aber auch hier könnte ich mir vorstellen, daß diese irgendwie als Filter fungieren, daß dort ,,Poren" sind, die im Elektronenmikroskop nicht sichtbar gemacht werden können.

Solomon: For some years my colleagues and I have been interested in trying to measure pores that are too small to be seen by the electron microscope, pores such as we believe exist, for example, in the red cells in man. Now, the methods of measuring these pores are largely physical, and one such measurement has to do with the comparison of the rate of the entrance of water under osmotic pressure with that under diffusion, and this leads to an estimate of an equivalent pore radius in the red cells of 3.5 Å. Another measurement that seems to me to be independent is derived from a determination of the ability of substances which can pass through the pores to create osmotic pressure and by a complex argument that I do not want to go into here, we have come to the conclusion by this measurement that the pores are 4.2 Å in radius. These numbers are, in our view, in very good agreement indeed. Thirdly, GIEBEL and PASSOW in Hamburg made measurements of the size of molecules which would go through beef red cells and they came out with a pore radius of approximately 4 Å. So all three of these measurements lead us to believe that there is a submicroscopic structure which is able to discriminate between molecules. We do not believe that we know what this pore should look like; we do not even know whether the pore is open all the time. We have introduced a term which I would like to recommend to you which is "equivalent pore radius". This is the size that the hole would have if, indeed, it was a series of right circular holes bored through the membrane, each with ideal physical properties, but we suspect, indeed, that we are dealing with irregular apertures that may be open or may be closed.

Diamond: Ich möchte noch etwas zu der Frage der sog. osmotischen Filtration bemerken. Die osmotische Filtration unterscheidet sich von der Druckfiltration dadurch, daß sie stattfinden kann, wenn kein hydrostatischer Druck vorhanden ist, und zwar kommt das auf folgende Weise zustande: Wenn eine gelöste Substanz durch eine Membran aktiv transportiert wird, dann können diese Moleküle Wasser mit sich reißen, und dieser Wasserstrom reißt dann die kleineren Moleküle mit sich. Das Musterbeispiel für osmotische Filtration bei Wirbeltieren ist wohl die Abscheidung der Primärgalle in der Leber. Der Grund, warum mir die osmotische Filtration besonders wahrscheinlich erscheint, ist die Beziehung zwischen hydrostatischem Druck und osmotischem Druck. Eine Atmosphäre hydrostatischer Druck hat dieselbe Wirkung wie 40 mmol osmotischer Druck. Eine Atmosphäre hydrostatischer Druck kommt selten in der Biologie vor. Aber 40 mmol sind schon ein kleiner Unterschied. Es ist möglich, daß in vielen anderen Fällen bei Wirbellosen, wie beispielsweise RAMSAY schon bei den Malpighischen Gefäßen gezeigt hat, der Wasserstrom und das Mitreißen der kleinen Moleküle durch aktiven Transport einer gelösten Substanz erfolgt und nicht unter hydrostatischem Druck.

Tosteson: One of the problems in understanding filtration through biological membranes is that we know very little about the physics of flow through very small pores. Recently artificial membranes with true right-angled cylindrical pores have been made by BEAN. He makes these membranes by peeling off a sheet of mica, approximately 7 microns thick, and mounts it in vacuo about an inch or an inch and a half away from a fission source. The nuclear fission fragments penetrate the mica making approximately 10^5 holes per square centimetre, and each little hole is about between 20 and 30 Å in diameter. He can then place these pieces of mica in hydrogen fluoride (the hydrogen fluoride etches only the walls of the pores). By regulating the length of time of exposure, he can prepare membranes with homogeneous pores ranging in size from 20 to 30 Å in diameter up to several thousand Å in diameter. The study of flow through these membranes will be of great interest in relation to biological filtration processes.

Gersch: Wenn man die Terminalzelle des Protonephridiums des Leberegels in Abb. 1 mit den Solenocyten von *Branchiostoma* in Abb. 11 vergleicht, so fällt doch auf, daß das Endkölbchen im ersten Fall — entsprechend seiner Lage in einem parenchymatösen Gewebe — lange und verzweigte Fortsätze besitzt, während es im zweiten Fall, wo es an ein Cölom bzw. an einen Blutraum grenzt, ziemlich glatt ist. Nach meinen bisherigen Kenntnissen ist es gerade das Wesen dieses Protonephridiums, daß es sich in ein Parenchym hineinsenkt, so daß man annehmen müßte, daß sich der Filtrationsmechanismus in erster Linie an den Stellen abspielt, an denen die Fortsätze in das Parenchym hineingreifen. Haben Sie dazu eine Meinung?

Kümmel: Wir haben in diesem Endkölbchen strukturell keinerlei Anhaltspunkte für die Möglichkeit einer Filtration gefunden. Nun muß ich erwähnen, daß es in diesem Parenchym feine Gewebslücken gibt. Um den Reusenabschnitt herum sind solche Gewebslücken immer deutlich ausgeprägt. Welches die Bedeutung dieser vielen und verzweigten Fortsätze ist, ist schwer zu sagen. Es ist fraglich, ob sie Ausdruck einer Transportfunktion oder Verankerung sind, denn man findet auch bei parenchymatösen Würmern Protonephridien mit Endkölbchen, die kuppelförmig gestaltet sind und eine ziemlich glatte Oberfläche besitzen.

Die Funktion der Malpighischen Gefäße

von

ARMIN WESSING, Bonn

Mit 15 Abbildungen

Die Nieren der Insekten, nach ihrem Entdecker Malpighische Gefäße[1] genannt, sind schlauchförmige Organe, die an der Stelle, wo der Mittel- in den Enddarm übergeht, in den Verdauungskanal einmünden. Wie alle Exkretionsorgane sorgen sie für die Erhaltung eines gleichbleibenden, optimalen Milieus im Körper.

Ihre Funktion ist keineswegs einheitlich, sondern wird vor allem von ökologischen Faktoren bestimmt, die auf Art und Ablauf interner physiologischer Prozesse wesentlichen Einfluß haben. So zeigen aquatil und terrestrisch lebende Tiere, Wasservergeuder, Wassersparer, Nahrungsspezialisten usw. deutliche Unterschiede. Schließlich haben die Gefäße im Laufe der Stammesgeschichte nicht nur morphologische, sondern auch funktionelle Veränderungen erfahren.

In dieser Darstellung sollen vor allem die neueren, elektronenmikroskopischen Ergebnisse der Zellfunktionsforschung vorgetragen werden, die das Bild über die Arbeit der Malpighi-Gefäße bedeutend erhellen. Lichtmikroskopische und physiologische Untersuchungen werden nur insoweit kurz erwähnt, als sie zum Verständnis und zur Diskussion wesentlich beitragen. Die wertvollen Zusammenfassungen der letzten Jahre [*20, 48, 73, 98*] sollen also hierdurch keineswegs ersetzt werden.

Die Arbeit der Malpighi-Gefäße ist eng verknüpft mit der Funktion anderer Organe, wie der des Mitteldarmes und des Fettkörpers, die sogar als konkurrierende Exkretionsorgane auftreten können, und mit der Funktion des Enddarmes und des Rektums, die wertvolle Stoffe reabsorbieren können, die mit dem Harn aus den Gefäßen in den Enddarm gelangen.

A. Morphologie der Malpighi-Gefäße
[s. *48, 98*]

Die schlauchförmigen Malpighischen Gefäße (M.T.) liegen in unterschiedlicher Länge und Anzahl (zwei bis über hundert) frei in der Leibeshöhle der Insekten oder sind zwischen anderen Organen (Fettkörper, Darm) eingebettet. Sie werden allseitig vom Blut[2], das die Leibeshöhle erfüllt, umspült.

---

[1] Auch Myriapoden, Chilopoden und viele Arachniden besitzen Malpighische Gefäße.

[2] Blut ohne Zellen = „Hämolymphe".

Das Lumen der Schläuche wird von zwei bis mehreren, einschichtig gelagerten Zellen umhüllt. Einen abweichenden sog. kryptonephridialen Bau zeigen die Gefäße der Käfer, vieler terrestrisch lebender Schmetterlingslarven, die des Ameisenlöwen (*Myrmeleon*) und der Zikaden. Hier liegen die Anfangsstücke oder mittlere Gefäßabschnitte der Rektalblase oder anderen Teilen des Darmtraktes an, oft von einer besonderen Membran gegen das Blut abgegrenzt. Diese räumliche Beziehung hat wahrscheinlich stoffwechselphysiologische Gründe, die aber noch nicht ganz geklärt sind. Dünnwandige Stellen in den Darmzellen, denen die Gefäße anliegen, lassen auf einen lebhaften Stofftransport (Wasser, bestimmte Ionen o. ä.) vom Darm zum Malpighi-Gefäß schließen [*1, 30, 34, 35, 47, 52—55, 76*].

Die Gefäße sind gelegentlich durch gespeicherte oder ausgeschiedene Farbstoffe gefärbt. Dabei handelt es sich oft um das gelb-orangefarbige *Riboflavin*, bei vielen Dipteren (Brachyceren) darüber hinaus um das gelbe *3-Hydroxykynurenin* und das rote *Drosopterin* [*84, 91, 92*], bei anderen Tierarten oft um Farbstoffe aus der Nahrung (*Chlorophyll*, Blutabbbauprodukte o. ä.).

Zwischen den Gefäßen eines Tieres können deutliche Unterschiede in Form und Färbung auftreten, die auf eine verschiedene Funktion der Schläuche hinweisen [*71*], die auch gelegentlich durch eine unterschiedliche Ausscheidung von injizierten Farbstoffen nachgewiesen wurde [*21, 33*]. Die einzelnen Tubuli aber lassen fast immer eine deutliche regionale Gliederung erkennen, oft besitzt ein Gefäß 2—4 lichtmikroskopisch gut voneinander zu unterscheidende Abschnitte, die bei der Harnbildung meist verschiedene Funktionen erfüllen. So besitzt *Rhodnius* einen distalen und einen proximalen Gefäßteil [*95, 99*], *Drosophila* einen erweiterten distalen, einen proximalen und dazwischen einen kurzen Übergangsabschnitt [*85*]; bei *Dacus* kann man auf Grund von Enzymreaktionen vom proximalen Abschnitt noch eine mittlere Region abgliedern [*5, 42*]; bei *Acheta* setzen sich 3 [*13*], bei Zikaden 4 Regionen deutlich voneinander ab [*74*]. Alle Regionen zeigen funktionelle Unterschiede.

Meist münden mehrere Gefäße in einen Sammelkanal (Ureter), der am Ende zu einer Ampulle erweitert sein kann und stets aus einer Schicht vieler Zellen zusammengesetzt ist. Letztere besitzen meist die typische Struktur von Darmzellen. Das Lumen kann sogar, wie der Enddarm, von einer Chitinschicht ausgekleidet sein [*13*]. Den Ureter umfassen Muskelzellen, die durch peristaltische Pumpbewegungen den Gefäßinhalt dem Enddarm zuführen. Muskelelemente können aber auch das Gefäß in seiner ganzen Länge umgeben oder aber völlig fehlen[1]. Der Eintritt des Harnes in den Enddarm kann durch einen Sphinkter geregelt werden (PALM, siehe [*48*]).

Die Konsistenz des abströmenden Urins hängt weitgehend von den Wasserverhältnissen ab. Je nach Tierart und je nach ökologischen und stoffwechselphysiologischen Gegebenheiten ist er klar, breiig oder fester kristallin. Sein Wassergehalt kann sich aber im Laufe einer kurzen Zeit verändern [*94*]. Neben den exkretorisch arbeitenden Zellen kommen bei einzelnen Tierarten auch Zellen mit einer anderen physiologischen Bedeutung vor; so treten z. B. bei *Locusta* und *Acheta* schleimbildende Mucozyten auf [*12, 26, 38*].

I. Der Aufbau und die Funktionsstrukturen der Zellen

Über die Feinstruktur der Malpighi-Gefäßzellen liegen bereits aus 7 verschiedenen Insektengruppen Untersuchungen vor: *Orthoptera*: *Melanoplus differentialis* [*6*], *Carausius morosus* [*45*], *Blatta orientalis* [*45*], *Acheta domestica* [*7, 8, 11—13*], *Dissosteira carolina* [*81*], *Periplaneta americana* [*78*], *Oedaleus* [*80*], 13 verschiedene Orthopterenarten [*13*]. *Libellula: Aeschna spec.* [*13*]. *Heteroptera: Rhodnius prolixus* [*99*]. *Homoptera*: 5 Zikadenarten [*23*], *Macrosteles fascifrons* [*74*].

---

[1] Ob die bei *Drosophila* innerhalb der Zellen der Malpighi-Gefäße auftretenden Filamentbündel kontraktil sind oder die Zellen an ihrer Basis nur verfestigen, ist noch unklar [*85*].

Diptera: Drosophila melanogaster [84–90, 92, 93], *Dacus oleae* [5, 42].
Lepidoptera: Galleria mellonella [16]. *Hymenoptera: Apis mellifica* [45],
Bombus lapidarius [45].

Die Ergebnisse dieser Untersuchungen lassen folgende allgemeine
Aussage zu: Die Zellen zeigen ungefähr den gleichen polaren Aufbau
wie die Zellen des proximalen Tubulus im Nephron der Vertebratenniere.
Die basale Region bildet unter einer aus feineren [87] oder dickeren,
periodisch gestreiften [81] Fibrillen zusammengesetzten Basalmembran
mehr oder weniger tief in den Zelleib eindringende Zellmembraneinfaltungen oder Zellverzahnungen, die, wie bei *Drosophila* [85], innerhalb der
Zelle ein großflächiges, dicht-vernetztes, extrazelluläres System von Spalträumen bilden können („basales Labyrinth").

Die gegenüberliegende apikale Region besitzt den aus lichtmikroskopischen Untersuchungen bereits bekannten Bürstensaum, der aus
Zellausstülpungen besteht, die je nach Intensität der Zellarbeit mehr
oder weniger lang sein [87, 98] und evtl. seitlich anastomosieren können
[13, 42, 81]. Sowohl das basale Labyrinth als auch der Bürstensaum vergrößern auf beiden Seiten die physiologisch wirksame Oberfläche der
Zelle um ein Vielfaches.

Zwischen den polaren Regionen liegt der intermediäre Abschnitt
der Zelle mit ein bis zwei Zellkernen, zahlreichen Golgi-Apparaten und
der Hauptmasse des weitverzweigten Kanalsystems des endoplasmatischen Retikulums (ER), das ohne Ribosomen bleiben (glattwandig)
oder mit RNS-Grana (rauhwandig) besetzt sein kann. Oft bildet es
ein Ergastoplasma (ERG), d. h. geschichtete Membrankomplexe mit
angelagerten Ribosomen [5, 42, 99]. Dieses Kanalsystem des ER erstreckt sich auch in die basale und apikale Region. In jedem Mikrovillus des Bürstensaumes kann ein Zentralkanal auftreten [11, 13, 81],
der mit dem ER der tiefer gelegenen Zellschichten Verbindung hat.
Solche Kanäle, die in den Mikrovilli aller resorbierenden Zellen von
Drosophila vorkommen [85] und in letzter Zeit auch aus den Mikrovilli
anderer Zellarten bekannt wurden, sind wahrscheinlich in Malpighi-
Gefäßen häufiger anzutreffen; darauf lassen die Bilder von *Rhodnius* [99]
und *Dissosteira* [81] schließen. Sie haben meist Rosenkranzform, so daß
im Schnitt fast immer eine Reihe von Vesikeln erscheint, und die Ver-
bindungskanäle seltener sichtbar sind.

Die Bildung kristalliner Exkrete erfolgt entweder intrazellulär
(*Acheta* [7, 8, 13], *Periplaneta* [78], *Bombus* [45], verschiedene Zikadenarten [23, 74, 99] und *Rhodnius* [99]) oder extrazellulär im Gefäßlumen
(*Carausius* [45], *Rhodnius* [99], *Dacus oleae* [5, 42]), wobei die Kristalle
oft zwischen den Mikrovilli liegen [5, 99]. Häufig aber bleiben die Exkrete,
wenn das Malpighi-Gefäß gar kein oder wenig Wasser reabsorbiert, im
Lumen gelöst. Die geformten Exkrete sind rundlich, konzentrisch geschichtet und bestehen aus Harnsäure, Uraten, Calcium-, Magnesium-
phosphaten und -carbonaten oder bei Zikaden, wo sie eine dodekaeder-
förmige Gitterstruktur bilden (sog. Brochosomen), aus einem basischen
Protein. Sie liegen einzeln oder in größeren Mengen in Vesikeln [23, 74]
und entstehen an der Zellbasis zwischen den Zellmembranfalten, wan-

dern durch die Zelle und werden am Zellapex ins Lumen abgegeben
[23, 74]. Für *Acheta* [7, 13] ließ sich diese Abgabe nicht beobachten;
möglicherweise werden hier die Grana vor ihrer Ausstoßung nochmal
gelöst, um dann im Lumen wieder auszukristallisieren.

Vesikel, die in bestimmten Gefäßregionen vermehrt auftreten,
sind im übrigen schon mehrmals beschrieben worden [5, 6, 13, 42, 99].
Solche Vesikel sollen, ganz gleich, ob sie Exkretkristalle oder gelöstes
Exkret transportieren, aus Golgi-Vesikeln gebildet werden [5, 74].
Bei der Wanderung der Vesikel von der Zellbasis zum Lumen werden die
Inhaltsstoffe immer dichter. Die Vesikel können auch zusammenfließen und
zeigen gelegentlich eine Verbindung mit endoplasmatischen Kanälen.
Bei *Acheta* [11, 13] liegen parallel neben den Zellmembraneinfaltungen
größere Mengen von langgestreckten oder gekrümmten Tubuli, die vor
allem nach dem Trinken oder nach Injektion von wäßrigen Lösungen
in Erscheinung treten („hydrophile Strukturen"). Diese Tubuli können
dann auch gruppenweise von einer gesonderten Membran umhüllt
sein und gehören entweder zum ER oder stellen besondere Vesikelformen
dar, die sich, wie BERKALOFF meint, aus dem Membranmaterial per-
forierter Zellmembraneinfaltungen entwickeln.

Die Mitochondrien zeigen eine charakteristische Verteilung: Die
meisten konzentrieren sich in den Cytoplasmakompartimenten des
basalen Labyrinthes und im apikalen Zellbereich vor oder in den Mikrovilli des Bürstensaumes. Bemerkenswert ist eine Transformation von
Mitochondrien der Intermediärregion, die schon bei mehreren Tierarten beschrieben wurde [5, 9, 13, 42, 74, 80, 85, 99]. Dabei kugeln sich die
Mitochondrien ab und schmelzen ihre Innenstrukturen ein. So entstehen
homogene Körper, in denen dann eine Schichtung auftritt und die schließlich völlig abgebaut werden können [85]. Ihre Bildung kann durch Injektion von Fremdeiweiß stimuliert werden [9]. Bei *Acheta* und *Drosophila* lagert sich in diesen Körpern *Riboflavin* ab. Ob diese Transformation stoffwechselphysiologisch von Bedeutung ist, bedarf noch der Klärung[1]. Da bei *Drosophila* diese Erscheinung vor allem kurz nach der Verpuppung anzutreffen ist und eine große Anzahl von Mitochondrien
erfaßt, handelt es sich wahrscheinlich um den Abbau von überzähligen
Mitochondrien. Gelegentlich treten in den Zellen Glykogenansammlungen
[12], multivesikuläre Körper und Cytosomen auf [42].

Bei *Acheta* [13], *Dissosteira* [81] und *Rhodnius* [99] ist die Interzellular-
substanz in Form von „septierten Desmosomen" [100] organisiert.

Besondere Beachtung verdienen Unterschiede der Zellstrukturen
zwischen den Gefäßabschnitten in bezug auf ihre Funktion. Am ausgeprägtesten sind diese Differenzierungen offenbar bei den phylogenetisch
hoch entwickelten Zikaden: Bei *Macrosteles fascifrons* [74] zeigt fast
jede der vier lichtmikroskopisch gut voneinander abgesetzten Regionen
einen anderen Zellaufbau: Besonders bemerkenswert ist, daß der

---

[1] WIGGLESWORTH u. SALPETER [99] nehmen an, daß aus ihnen die ebenfalls
zirkulär geschichteten Exkretgrana hervorgehen, ein Befund, den man noch einmal prüfen sollte.

Region III basale Zelleinfaltungen und ein Bürstensaum fast völlig fehlen, dieser Gefäßabschnitt aber trotzdem große Mengen kristalliner Exkrete ausscheidet. Diese beiden Zelldifferenzierungen sind demnach nicht unbedingt für einen Exkrettransport von der Hämolymphe zum Lumen notwendig. Die proximale Resorptionsregion (I) besitzt diese Strukturen dagegen in ausgesprochenem Maße. Bei anderen Arten (*Acheta, Rhodnius, Dacus*) treten intrazelluläre Exkretgrana bzw. vesikuläre Elemente in distaleren (oberen) Gefäßregionen auf, wohingegen im proximalen (unteren) Abschnitt allenfalls im Lumen Exkretgrana liegen können. Im distalen Gefäßabschnitt geht demnach offenbar der Transport von Exkreten ins Lumen vor sich und im proximalen Teil und Ureter erfolgt eine Reabsorption von wertvollen Stoffen. Diese Funktionsverteilung innerhalb der Malpighi-Gefäße hatte WIGGLES-WORTH bereits 1931 [*96*] bei *Rhodnius* nachgewiesen.

Bei *Acheta, Rhodnius* und *Dacus* wurden zwischen den oberen und unteren Gefäßregionen noch andere Unterschiede beschrieben, die die Ausbildung der Zellmembraneinfaltungen und Mikrovilli und die Anzahl und Lagerung der Mitochondrien betreffen.

Außer den für *Acheta* bereits erläuterten Veränderungen der Zellstruktur durch funktionelle Belastung [*11, 13*], wäre zu erwähnen, daß bei *Rhodnius* 1 Std nach Fütterung die Mikrovillispitzen und die darin liegenden Zisternen des ER in der proximalen Region durch den Wasserfluß anschwellen und nach 48 Std die schon besprochenen intra- und extrazellulären Exkretgrana erscheinen [*99*].

II. Die Funktion der Malpighischen Gefäße
von *Drosophila melanogaster*

[*84—93*]

Durch eingehende Untersuchungen der Malpighischen Gefäße (M.T.) von *Drosophila melanogaster*, der Taufliege, konnte ich in den letzten Jahren einen besseren Einblick in die Funktion dieser Exkretionsorgane erhalten.

Drosophila besitzt, wie alle Dipteren, 4 Malpighi-Gefäße: 2 vordere, deren distale Anfangsstücke bei Larve und Puppe verdickt sind, und 2 hintere Schläuche, die in ihrer ganzen Länge gleichförmig gestaltet sind, denen also die distale Anschwellung fehlt. Sowohl die beiden vorderen als auch die beiden hinteren Gefäße vereinigen sich zu je einem von Muskeln umgebenen kurzen Ureter, der die Exkrete in den Enddarm leitet (Abb. 1). Bei dem vorderen Gefäßapparat wird die Erweiterung während der Puppenzeit allmählich eingeschmolzen, so daß die Imago 4 gleichförmige M.T. besitzt [*85*]. Spritzt man Farbstoffe, elektronenmikroskopische Kontrastmittel oder andere Markierungsstoffe in die Hämolymphe, dann zeigt sich der erste und der intensivste Transport vom Blut zum Lumen in diesen erweiterten Anfangsstücken [*88*]. Hier erfolgt demnach ein kräftiger, transzellulärer Strom von Stoffen aus dem Blut zum Lumen. Der hier eingeschleuste Primärharn[1] wird dann in den weiter proximal gelegenen Gefäßabschnitten durch Reabsorption von Wasser und von wertvollen Substanzen wieder verändert. Die Reabsorption von Wasser läßt sich durch Injektion von Farbstoffen in die Hämolymphe nachweisen [87], die nach distalem Transport ins Lumen erst im proximalen Gefäßabschnitt so stark entwässert werden, daß sie hier auskristallisieren (Abb. 2a, b). Im übrigen tritt bei *Drosophila* in reabsorbierend tätigen Zellen immer eine Vorstufe der braunen Augenpigmente (Ommochrome), das gelbe *3-Hydroxykynurenin*, auf [92]. Diese Aminosäure bildet sich aus einer in der larvalen Hämolymphe und im Fettkörper [66–69] auftretenden weiteren Vorstufe, dem *Kynurenin*. Dieses wird durch ein in den Zisternen des ER der proximalen Zellen auftretendes Enzym, der *Kynureninoxydase*, oxydiert, wonach das gebildete *Oxykynurenin* in ampullenartigen Erweiterungen des ER gespeichert wird [84, 85]. Während der Puppenzeit leeren sich diese „Ampullen", das Speicherprodukt wird dann vermutlich zur Pigmentbildung der Augen verwendet. Das *Oxykynurenin* reduziert Osmiumtetroxyd schnell zu metallischem Osmium und ist dadurch und durch andere chemische und physikalische Nachweismethoden leicht zu identifizieren [84, 85, 92].

Vom erweiterten Gefäßanfang bis zum Ureter nimmt der Stofftransport vom Blut zum Lumen immer mehr ab, die Reabsorption dagegen zu. Das Fehlen des distalen Gefäßabschnittes bei der Imago und bei den hinteren M.T. der Larve und Puppe ist insofern verständlich, als im imaginalen

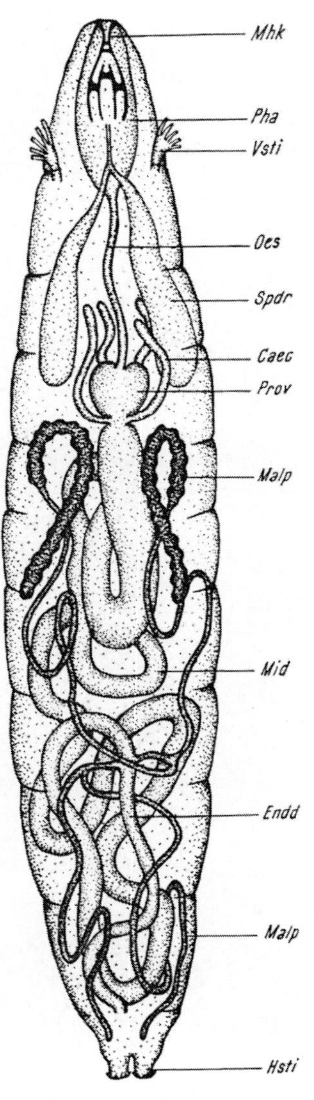

Abb. 1. *Drosophila melanogaster*, Larve, Lage des Darmtraktes und seiner Anhangsorgane. *Malp* = Malpighische Gefäße. Die beiden oberen und die beiden unteren Gefäße münden über einen kurzen Ureter in den Darmkanal. Die vorderen M.T. sind distal erweitert. (*Caec* = Caecum, *Endd* = Enddarm, *Hsti* = Hinterstigmen, *Oes* = Oesophagus, *Mhk* = Mundhaken, *Pha* = Pharynx, *Prov* = Proventrikulus, *Spdr* = Speicheldrüse, *Vsti* = Vorderstigmen)

---

[1] Als Primärharn möchte ich nur das im Lumen befindliche Exkret bezeichnen, das durch die Reabsorption der unteren Gefäßabschnitte verändert wird, nicht dagegen die Harnflüssigkeit, die durch die Zellen geschleust und bei dieser Passage noch verändert wird. (Herrn Dr. KÜMMEL, Berlin, danke ich für seine diesbezügliche Diskussion.)

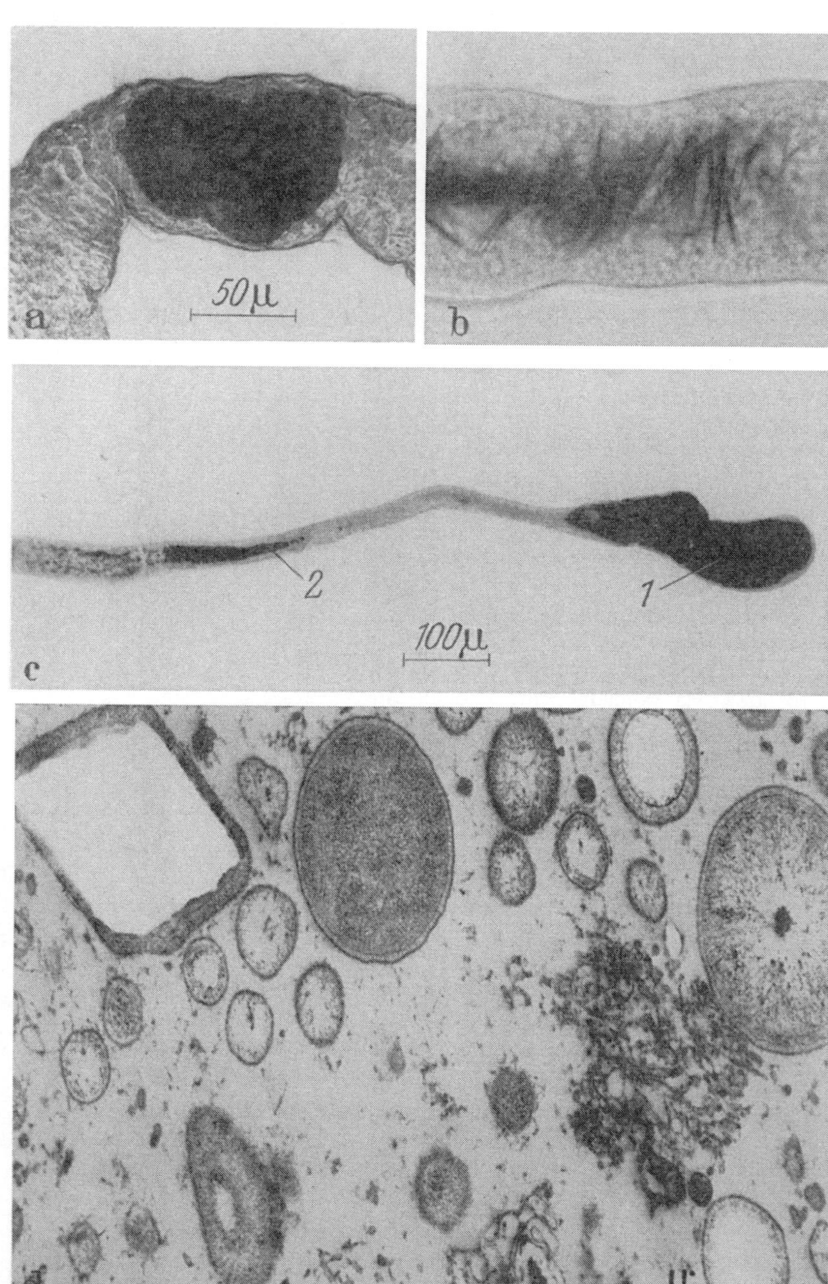

Abb. 2a—d. Auskristallisieren von Farbstoffen durch Reabsorption von Wasser im Lumen des proximalen Hauptstückes der larvalen M.T. von *Drosophila* nach Injektion in das Blut. a Großer Hämoglobinkristall. b Indigokarmin. c Darstellung der Harnsäure mit einem Silbersalz in den M.T. einer älteren Puppe von *Drosophila*. Die Harnsäure wird im distalen, erweiterten Abschnitt abgeschieden, sammelt sich hier an (*1*) und wird allmählich durch das Lumen zum Enddarm abgeführt (*2*). d Kristalline Exkrete und Membranhaufen im Gefäßlumen einer älteren Puppe nach Fixierung mit OsO_4—$K_2Cr_2O_7$. Bei den rundlichen Kristallen handelt es sich um Harnsäure oder Urate. E.O. 5900:1, E.V. 16000:1 (E.O. = Elektronenoptische Vergrößerung, E.V. = Endvergrößerung)

Körper bzw. in den hinteren Körperabschnitten vermutlich die Stoffwechselaktivität und damit der Anfall von Exkretstoffen geringer sind.

Die vorderen larvalen M.T. besitzen also 4 Abschnitte: einen distalen (*Anfangsstück*) und einen langen proximalen Abschnitt (*Hauptstück*), dazwischen ein *Übergangsstück* und schließlich den *Ureter* (Abb. 14). Die Zellen aller Regionen sind zwar nach dem für Malpighi-Gefäße typischen polaren Muster (basale Zellmembraneinfaltungen und gegenüberliegend ein Bürstensaum) aufgebaut, zeigen aber in den einzelnen Gefäßabschnitten ganz charakteristische Abwandlungen.

Das gesamte Lumen ist in seiner ganzen Länge von kristallinen Exkreten erfüllt, die bei der Larve alle leicht wasserlöslich sind. Man muß demnach einen *starken osmotischen Gradienten von der Hämolymphe zum Lumen* annehmen.

Die *Harnsäure* (oder ihre Salze) bzw. *Hypoxanthin* und *Xanthin*, die Abbauprodukte des N-Stoffwechsels, werden während der Larvenzeit nur im Fettkörper gespeichert, die M.T. aber exzernieren während dieser Zeit nur chromatographisch nachweisbare Spuren dieser Purine. Erst bei der Histolyse des Fettkörpers während der Puppenzeit werden dann große Mengen Harnsäure (oder Urate) durch die M.T. und den Mitteldarm ausgeschieden (Abb. 2c, d).

Das Exkret im Lumen der Gefäße ist, wie schon für *Rhodnius* [96] nachgewiesen wurde, starken Schwankungen des pH-Wertes unterworfen. Bei der Larve liegt der Wert nach Messungen mit verschiedenen Farbindikatoren im distalen Gefäßabschnitt etwa bei pH 8–8,5, in der proximalen Region dagegen etwa bei pH 6, dazwischen findet man Übergangswerte. Im Ureter kann das pH sogar auf etwa 5,5 sinken. Ähnliche Verhältnisse zeigen die hinteren Malpighi-Gefäße. Der „Primärharn" ist demnach im Gegensatz zur Hämolymphe, die sauer ist (pH 6,3–6,8)[1], alkalisch, nach Reabsorption aber sauer. Der durch den Ureter abströmende Sekundärharn senkt den pH-Wert des Darminhaltes, der im Vorderdarm ungefähr einen pH-Wert von 6,5 und im Mitteldarm von 8,2–8,5 hat, im Enddarm wieder auf etwa 6,8. Der Wert steigt aber bis zur Ausmündung des Enddarmes, vermutlich durch weitere Reabsorption, noch einmal leicht an. Die *saure Reaktion des proximalen Harnes wird bei der Larve von Drosophila nicht wie bei Rhodnius durch die Anwesenheit von Harnsäure* herbeigeführt. Die gleichen pH-Werte wurden auch für die Gefäße der Puppe von *Drosophila* ermittelt, deren M.T. große Mengen Harnsäure ausscheiden.

In den folgenden Ausführungen werden nur die vorderen M.T., die bei Larve und Puppe distal erweitert sind, besprochen.

1. Der distale Gefäßabschnitt (Anfangsstück)

Während das Gefäß in der proximalen Region im Querschnitt nur aus 2 Zellen zusammengesetzt ist, besteht es im distalen (oberen) Anfangsstück aus mehreren Zellen, die mehr oder weniger stark abgeplattet sind und dadurch ein großes Lumen umschließen. Den Zellen fehlt ein typischer Bürstensaum; allenfalls treten schlecht entwickelte

---

[1] Nach Boche u. Buck [*14*] 6,29—7,15.

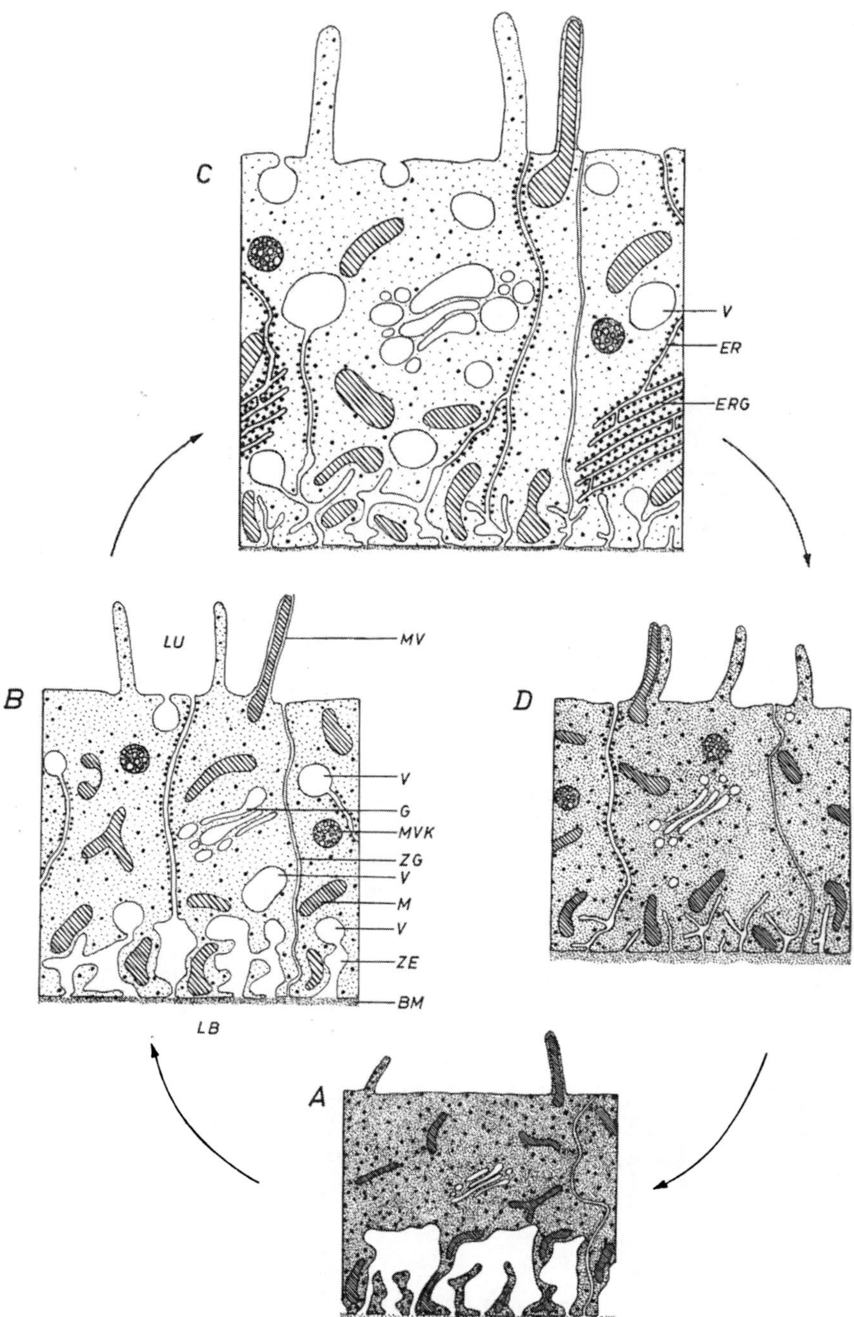

Abb. 3. Schematische Darstellung des Zellfunktionszyklus, durch den der Primärharn im erweiterten, distalen Anfangsstück der M.T. von *Drosophila* im teilentwässerten Zustand zum Lumen befördert wird. A. Stärkste Kondensation der Zelle mit erweiterten, mit Harn gefüllten, basalen Zellmembraneinfaltungen. B. Abschnürung der gefüllten Vesikel von

kurze oder sehr dünne Mikrovilli auf, die nie einen zentralen Kanal besitzen, der, wie schon gesagt, für alle Mikrovilli reabsorbierender Zellen der M.T. charakteristisch ist.

Die Zellen zeigen im elektronenoptischen Bild ganz verschiedene Aspekte. Eine sinnvolle Aneinanderreihung von Aufnahmen ließ erkennen (Abb. 3), daß hier ein Zellfunktionsrhythmus vorliegt, durch den das Exkret zum Lumen befördert wird [88]. Dabei ist die Zelle mit allen Organellen zunächst stark kondensiert, die basalen Zellmembraneinfaltungen sind erweitert und vermutlich durch den Blutdruck mit Hämolymphe-ähnlicher Flüssigkeit gefüllt, die die als Ultrafilter wirkende Basalmembran passiert hat, die als Durchflußregler möglicherweise eine wichtige Rolle spielt (Abb. 3A). Die Erweiterungen schnüren größere und kleinere Vesikel ab, die zum Lumen wandern und hier ihren Inhalt abgeben (Abb. 3, B, C). Die Vesikel sind fast immer, wie auch für *Dacus oleae* nachgewiesen wurde [5], mit Kanälen des ER verbunden. Während der Vesikelwanderung (Cytopempsis [46]) nimmt die Zelle immer mehr Wasser auf, wodurch sich ihre Strukturen auflockern. Dabei schwellen vor allem die langen Vesikel der zahlreichen Golgi-Apparate an, die dann in kleinere Vesikel zerteilt ins Cytoplasma abgegeben werden (Abb. 3C). Ihr weiteres Schicksal ist noch unklar. Vielleicht wird durch sie die Zelle wieder entwässert und das Wasser zum Blut zurücktransportiert. Anschließend nimmt der Vesikelfluß fast völlig ab, die Zelle geht wieder in den kondensierten Zustand über (Abb. 3, D).

Während der Cytopempsis werden die zunächst stark erweiterten Zellmembraneinfaltungen immer englumiger und kleiner und verschwinden schließlich bis auf geringe Reste völlig. Besondere Bedeutung kommt sicherlich auch dem endoplasmatischen Retikulum zu, denn seine Kanäle schließen nicht nur an die Zellmembraneinfaltungen (Abb. 4a, 6a, Pfeil!), sondern auch an die Lumen-begrenzende Zellmembran an, so daß sie ebenfalls eine Stofftransportfunktion ausüben könnten. Die Endoplasmakanäle zeigen im übrigen in allen Gefäßregionen eine Beziehung zu den Golgi-Apparaten; sie können sowohl durch sie hindurchlaufen, als sich auch mit ihren seitlichen, runden Vesikeln verbinden (Abb. 4b, Pfeil!). Die endoplasmatischen Zisternen nehmen außerdem auch mit den Interzellularen Verbindung auf, die demnach ebenfalls Stoffe weiterleiten könnten.

Meist befinden sich benachbarte Zellen im gleichen Funktionszustand, arbeiten aber oft auch völlig asynchron. Dieser Funktionszyklus wurde auch für den distalen, erweiterten Anfangsteil der Gefäße der Puppe

---

den Einfaltungen und Abwandern zum Lumen. Verbindung der Vesikel mit ER-Kanälen, die mit den Einfaltungen, und den Interzellularen in Verbindung stehen können (*BM* = Basalmembran, *ER* = Kanäle des endoplasmatischen Retikulums, *G* = Golgiapparat, *M* = Mitochondrium, *MVK* = multivesikulärer Körper, *MV* = Mikrovillus des Bürstensaumes, *V* = Vesikel, *ZE* = basale Zellmembraneinfaltungen). C. Stärkste Hydratation der Zelle. Basale Zellmembraneinfaltungen und Vesikelfluß nehmen ab. Entwicklung von typischem Ergastoplasma (*ERG*), das in die Kanäle des *ER* übergeht. D. Beginnende Kondensation der Zelle, Vesikelfluß fast völlig verschwunden, basale Zellmembraneinfaltungen stark reduziert (nach [88] verändert!)

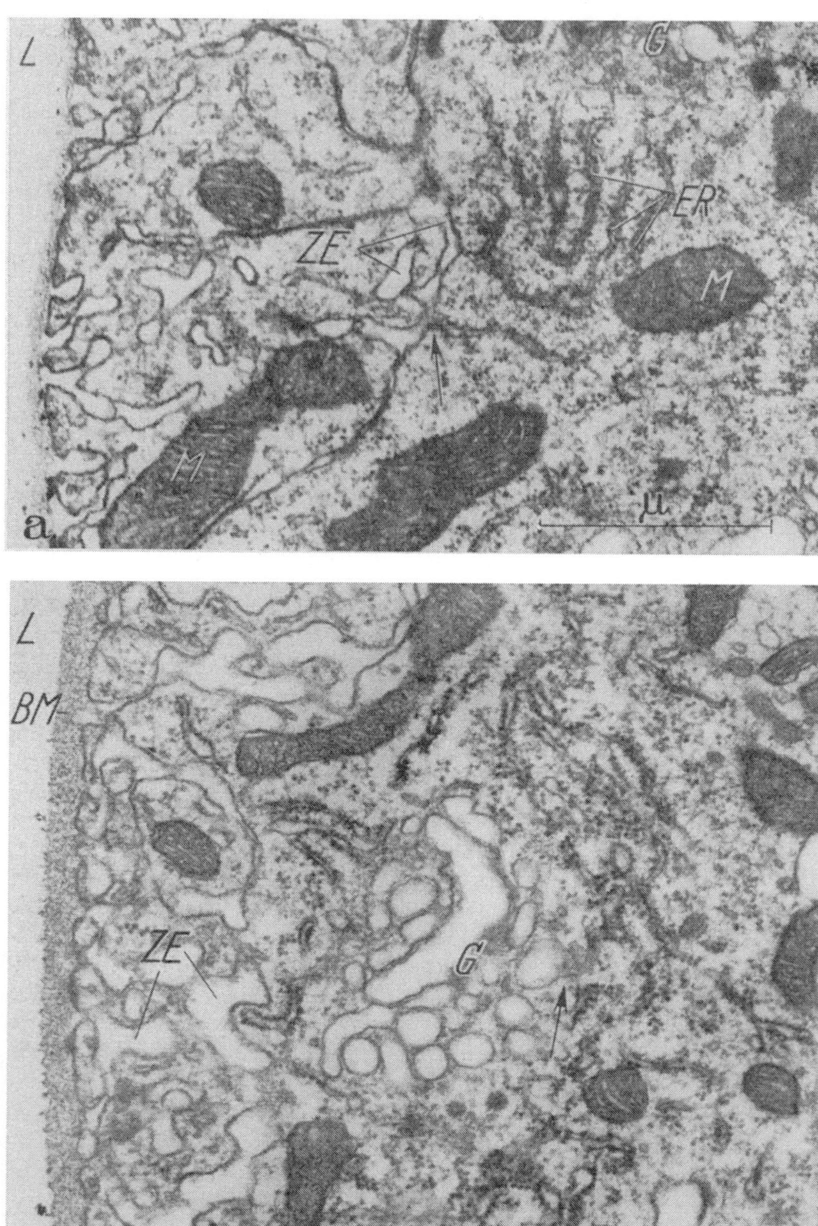

Abb. 4a u. b. Verbindungen der Kanäle des endoplasmatischen Retikulums (*ER*) in den Zellen des distalen, erweiterten Anfangstückes der M.T. von *Drosophila*. Verbindungen: a mit der der Leibeshöhle (*L*) zugewendeten Zellmembran (Pfeil, *ZE* = basaleZellmembraneinfaltungen), b mit dem Vesikel eines Golgiapparates (Pfeil!) (*BM* = Basalmembran, *M* = Mitochondrien). a E.O. 6250:1, E.V. 30500:1, b E.O. 6250:1, E.V. 29000:1.

nachgewiesen, wo allerdings, entsprechend der herabgesetzten Stoffwechsellage, der Vesikelfluß geringer ist. Trotzdem läßt sich bei alten Puppen die Entstehung der Vesikel aus den Einfaltungen gut verfolgen, da jetzt eine lebhafte Passage von *Harnsäure* (oder Uraten) ins Lumen

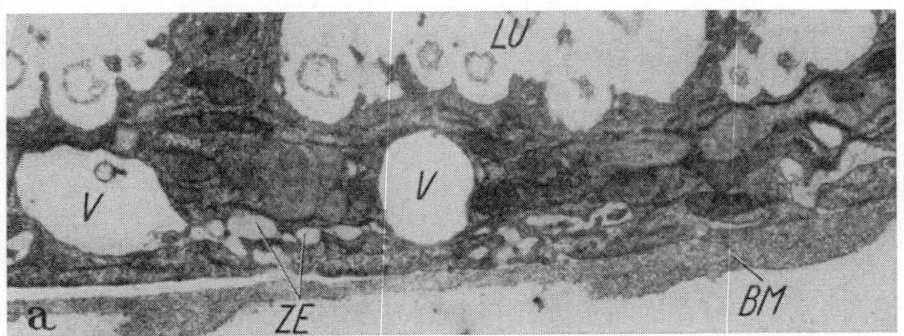

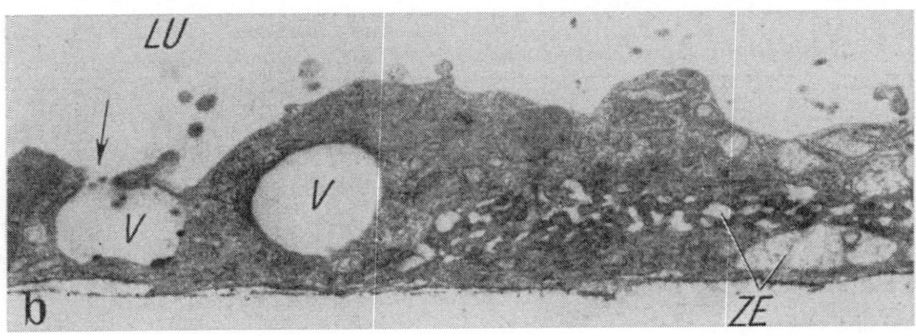

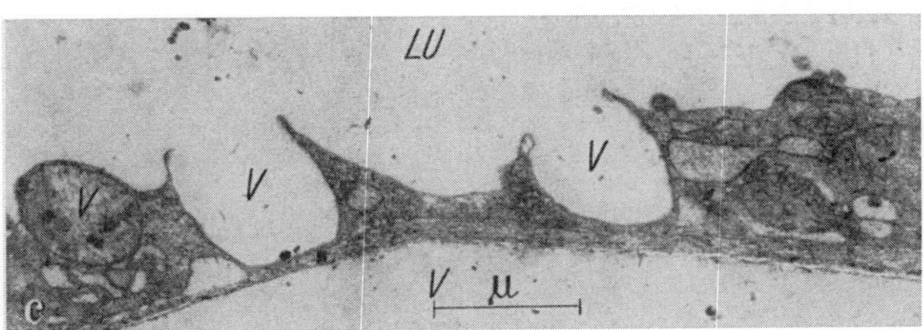

Abb. 5a—c. Primärharntransport im distalen, erweiterten Anfangsstück der M.T. der alten Puppe von *Drosophila*, 60 min nach Injektion von *Myofer*, *Goldsol* und *Goldchlorid*lösung in das Blut. Verstärkter Vesikelfluß vom Blut zum Lumen. a Bildung eines großen Vesikels von den erweiterten Einfaltungen und Abgabe von „halbkristallinen" Exkreten ins Lumen. b Öffnung eines Vesikels zum Lumen, der Konzentrationen von Myofer zeigt (Pfeil!). c Öffnung der Vesikel zum Lumen; die schwarzen Partikel sind wahrscheinlich Zusammenballungen von Goldsol-Teilchen. E.O. 6250:1, E.V. 17000:1

einsetzt, die vorher im Fettkörper gespeichert worden waren. Die Vesikel durchwandern die schmale Zelle, ihre Membran nimmt Kontakt mit der apikalen Zellmembran auf, verschmilzt damit, die Kontaktstelle öffnet sich, und der Vesikelinhalt ergießt sich ins Lumen (Abb. 5a—c).

Eine Bestätigung für das Vorhandensein des beschriebenen Funktionsrhythmus der distalen Zellen ließ sich dadurch erbringen, daß

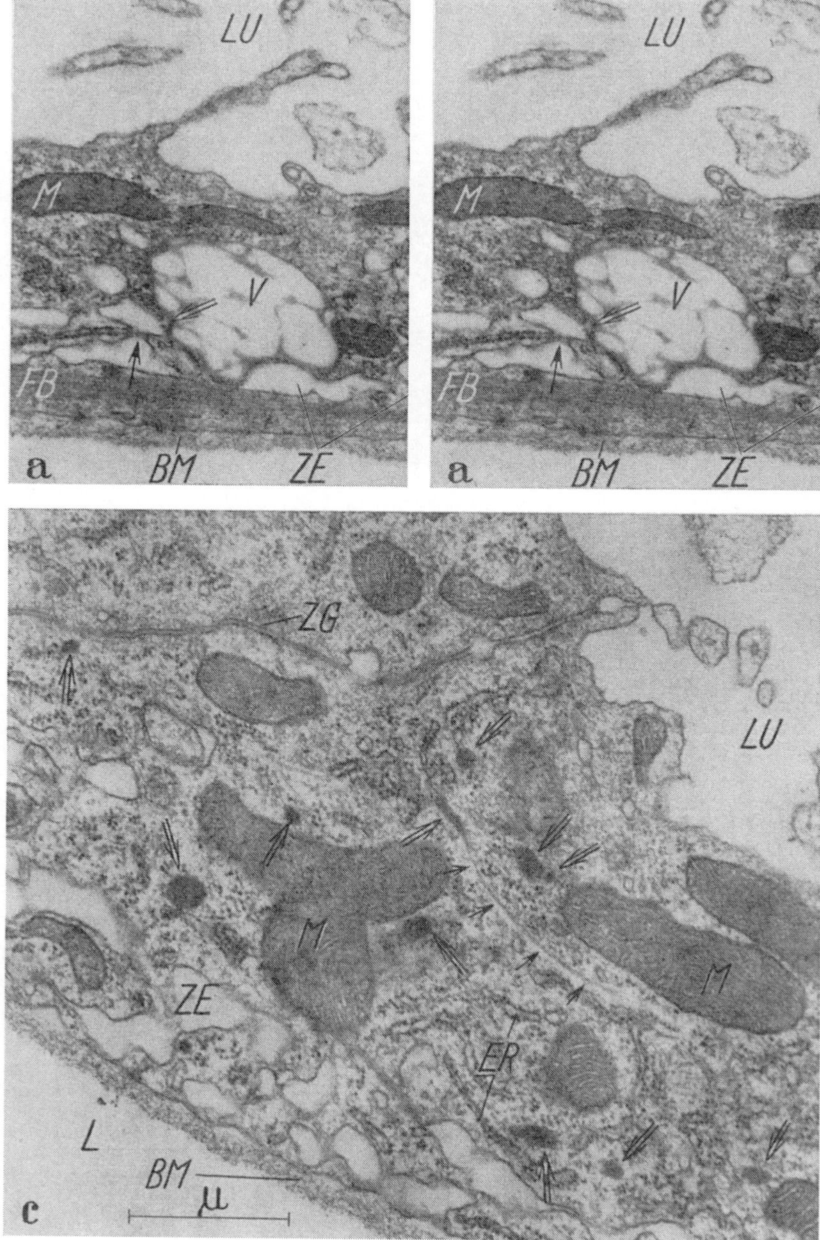

Abb. 6a—c. Transport von *kolloidalem Silber (Silberprotein)* im distalen, erweiterten Gefäßanfang der larvalen M.T. von *Drosophila* (c 5 min, a und b 10 min nach Injektion). a Markierung eines aus den Einfaltungen (Doppelpfeil!) entstandenen Vesikels (*V*), der Pfeil
weist auf eine Verbindung eines *ER*-Kanales mit den Zellmembraneinfaltungen (*ZE*).
FB = Fibrillenbündel, *BM* = Basalmembran. b Kondensation des intravesikulären, kolloidalen Silbers, wahrscheinlich durch den benachbarten Golgiapparat (*G*). *LU* = Gefäß-

1. die Vesikel mit verschiedenen Stoffen elektronenoptisch markiert werden konnten und 2. eine Markierung während der stärksten Hydratisierung und der Kondensationsphase der Zelle immer ergebnislos verlief.

Verfütterung von *3-Hydroxykynurenin* an Larven[1] zeigte schon nach 15 min eine deutliche Ablagerung dieser Aminosäure in den Vesikeln, die sich nach 30 min noch verstärkt hatte. Da das *Oxykynurenin* in der verabreichten Dosierung keinen deutlichen Elektronenkontrast ergibt, *werden transportierte Stoffe demnach in den Vesikeln durch Entwässerung konzentriert. Das ist sicherlich der Grund für die fortschreitende Hydratisierung der Zelle während des Vesikelflusses.*

Auch nach Injektion von 2 Silberproteinaten (*Argentum colloidale* und *Albumosesilber*, Firma Merck, Darmstadt) ließen sich nach 5 min große Vesikel markieren, die offensichtlich von den Zelleinfaltungen abgetrennt worden waren; deren Inhalt wurde entwässert und dadurch elektronenoptisch sichtbar (Abb. 6b). Die häufige Verbindung dieser Vesikel mit endoplasmatischen Kanälen, Ergastoplasma und die immer wieder auftretende, benachbarte Lage des Golgi-Apparates (Abb. 6b, G) ließ eine Beteiligung dieser Membransysteme am Materialtransport bzw. der Entwässerung der intravesikulär transportierten Exkretstoffe vermuten. Durch diese Entwässerung erfolgt natürlich eine Zunahme des Elektronenkontrastes, der vor der Entleerung der Vesikel seinen Höhepunkt erreicht. Diese Ergebnisse zeigen, daß die Zelle bei der Cytopempsis möglichst wenig Wasser zum Lumen transportiert, und daß das Wasser noch innerhalb der Zelle durch die Vesikelmembran in das Grundcytoplasma befördert wird und wahrscheinlich auf noch unbekannte Weise zum Blut zurückgelangt.

Die Transportfunktion der Vesikel ließ sich auch durch längeres Verfüttern von *Eisen-III-chlorid* und Injektion von *Hämoglobin* (Firma Schuchardt, München) nachweisen (Abb. 8b). Neben den durch Elektronenkontrastmittel markierten Vesikeln treten immer solche auf, die nicht markiert sind. Wahrscheinlich transportieren sie osmotisch aktive Stoffe, vermutlich anorganische Salze, da ihr Inhalt nach Entwässerung sich gegen das Fixierungsmittel hypertonisch verhält und die Vesikelmembranen durch den bei der Fixierung im Inneren entstehenden, erhöhten hydrostatischen Druck zerreißen (Abb. 8b, V, Doppelpfeile!). Bei den Vesikeln, die mit kolloidalen Substanzen gefüllt sind, tritt diese Erscheinung verständlicherweise nicht auf, ist aber nach Markierung mit $FeCl_3$ häufiger anzutreffen.

Eingehende Versuche mit „*Myofer*" (Eisen-Dextrankomplex, Farbwerke Hoechst, Frankfurt/Main[2]) ließen erkennen, daß der Transport

---

[1] Diese Aminosäure wurde natürlich nur an *Drosophila*-Mutanten verfüttert, die nicht zu deren Bildung befähigt sind.

[2] Molekulargewicht 10000—20000.

---

lumen, L = Blut der Leibeshöhle. c Kleinere Silberkondensationen im Cytoplasma (Doppelpfeile!), die mit einer Membran umgeben werden, die mit den Kanälen des *ER* in Verbindung treten. Die Pfeile weisen auf einen in Bildung begriffenen *ER*-Kanal hin, der zu einem Silberkondensat hinführt. *BM* = Basalmembran, *ZE* = basale Zellmembraneinfaltungen, *M* = Mitochondrium, *ZG* = Zellgrenze, an 3 Stellen erweitert (Artefakt?), L = Leibeshöhle, *LU* = Gefäßlumen. E.O. 12500:1, E.V. 21500:1

von Stoffen von der Hämolymphe zum Lumen auch auf andere Weise vor sich gehen kann. Schon 1 min nach der Injektion ist dieses Kontrastmittel durch die Basal- und die Zellmembran in die Zelle eingedrungen und erfüllt das gesamte Grundcytoplasma. Aber bereits zu diesem Zeitpunkt, und verstärkt 1–2 min später, treten überall kleinere Ansammlungen von Myofer auf, um die eine Membran gebildet wird, die häufig Anschlüsse an das Zisternensystem des ER und des Ergastoplasmas erkennen lassen (Abb. 7a). Schon 5 min p.i. ist das Grundcytoplasma wieder fast vollkommen klar (Abb. 7b) und das eingedrungene Myofer in vesikulären Erweiterungen des ER dicht konzentriert; dieser Eindruck verstärkt sich noch 10 min p.i. (Abb. 8a), 50 min später nimmt der Transport wieder ab.

Die Abgabe des im ER konzentrierten Myofers in das Lumen konnte bisher nie beobachtet werden, vielleicht wird es kurz zuvor wieder „gelöst" und über ER-Kanäle, die an die apikale Zellmembran anschließen, aus der Zelle geschleust. Im Lumen haftet es auf der Oberfläche der Exkretkristalle, deren Umriß im elektronenmikroskopischen Bild dadurch deutlich wird (Abb. 8a).

Nach diesen Befunden erfolgt die transzelluläre *Passage von Myofer* nicht durch Cytopempsis, sondern *nur über das Grundcytoplasma und die Kanäle des ER*; die in der Cytoplasma-Matrix auftretenden Ansammlungen von Myofer werden dabei von einer neuen Membran umgeben, an die Kanäle des ER Anschluß nehmen.

Diese zweite Art des Exkrettransportes ließ sich *auch* für *die Passage von Silberprotein* nachweisen (Abb. 6c), so daß dieser Weg nicht nur von Polysacchariden, sondern offenbar auch von anderen großmolekularen Stoffen gewählt wird. Silberprotein wird offenbar durch beide Passagearten zum Gefäßlumen geschafft.

Alle diese Ergebnisse sprechen dem distalen, erweiterten *Anfangsstück* der M.T. von *Drosophila nur einen Stofftransport vom Blut zum Lumen*, nicht dagegen einen *trans*zellulären Stoffstrom in umgekehrter Richtung zu, da kein Hinweis für eine Rückresorption von Stoffen aus dem Lumen besteht. Sie zeigen außerdem, daß der Primärharn nicht, wie z. B. im Glomerulum des Nephrons der Vertebratenniere, durch einen Filtrationsmechanismus ins Lumen gelangt, sondern daß dessen *transzelluläre Passage nach Art einer Sekretion mit Hilfe von Vesikeln abläuft, deren Inhalt schon während ihrer Wanderung durch eine Entwässerung den osmotischen Verhältnissen des Lumens angepaßt wird*, wodurch dem unteren Gefäßabschnitt ein Teil der Reabsorptionsarbeit erspart bleibt. *Die Cytopempsis erscheint hier also als ein ausgezeichneter Mechanismus, um zwei durch eine Zellage getrennte Flüssigkeiten unterschiedlichen osmotischen Druckes miteinander in Beziehung treten zu lassen.* Die natürlichen Permeabilitätsschranken zwischen den beiden Flüssigkeitsphasen, die Zellmembranen, können durch einen Membranflußmechanismus (BENNETT) am leichtesten überwunden werden. Ohne Zweifel ist dieser bei *Drosophila* „bergauf", also entgegen dem Konzentrationsgefälle laufende *Transportvorgang ein aktiver*. Die größte aktive Arbeit aber wird vor allem *bei der Konzentration des intravesikulären Primär-*

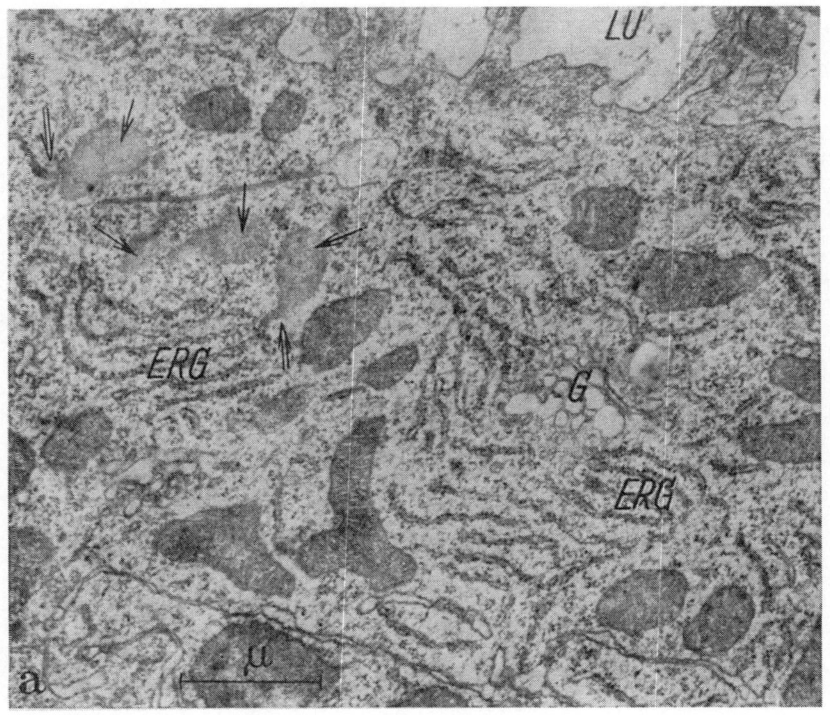

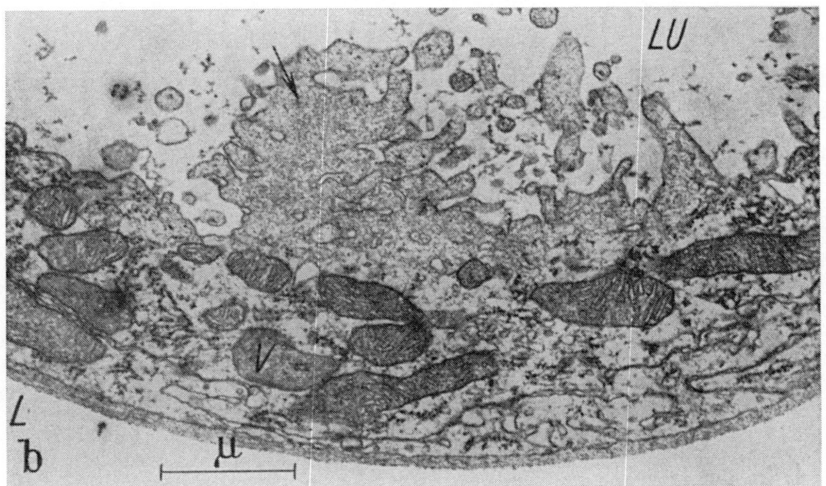

Abb. 7a u. b. Transport von „*Myofer*" (Eisen-Dextran-Komplex) in den Zellen des distalen, erweiterten Gefäßanfanges der larvalen M.T. von *Drosophila* nach Injektion in das Blut. Die Passage erfolgt über das Grundcytoplasma und das *ER*. a Nach 2 min ist das *Myofer* im Grundcytoplasma der ganzen Zelle verteilt und in zahlreichen Verdichtungen im Grundcytoplasma konzentriert (Pfeile!) mit beginnender Membranumhüllung. Daran schließen Kanäle des *ER* an (Doppelpfeile!). E.O. 6250:1, E.V. 18500:1. b Nach 5 min ist das Cytoplasma wieder klar geworden, nur die apikale Region ist noch gelegentlich mit *Myofer* besetzt (Pfeil!). E.O. 6250:1, E.V. 17500:1

16*

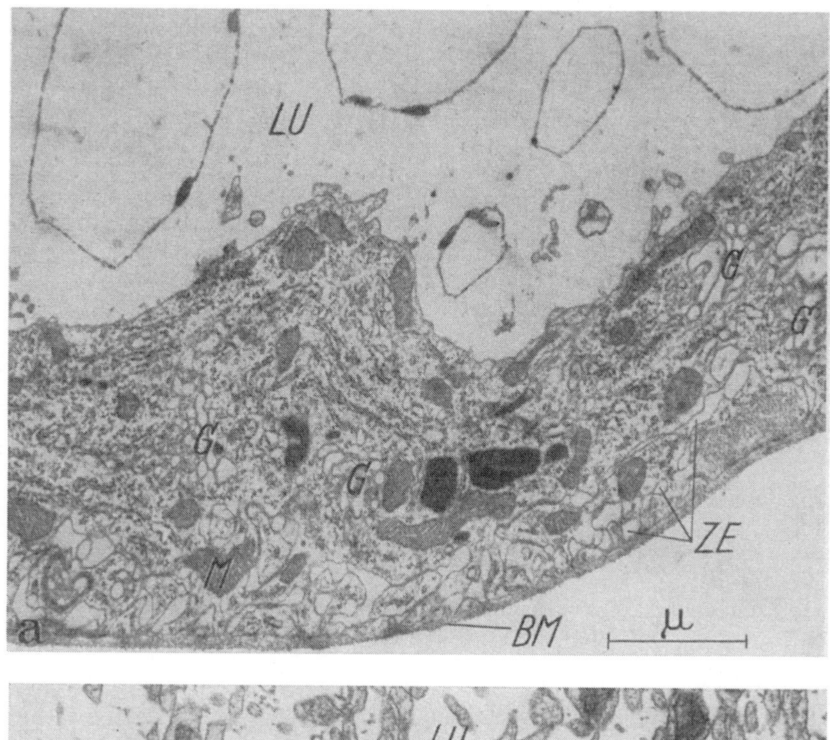

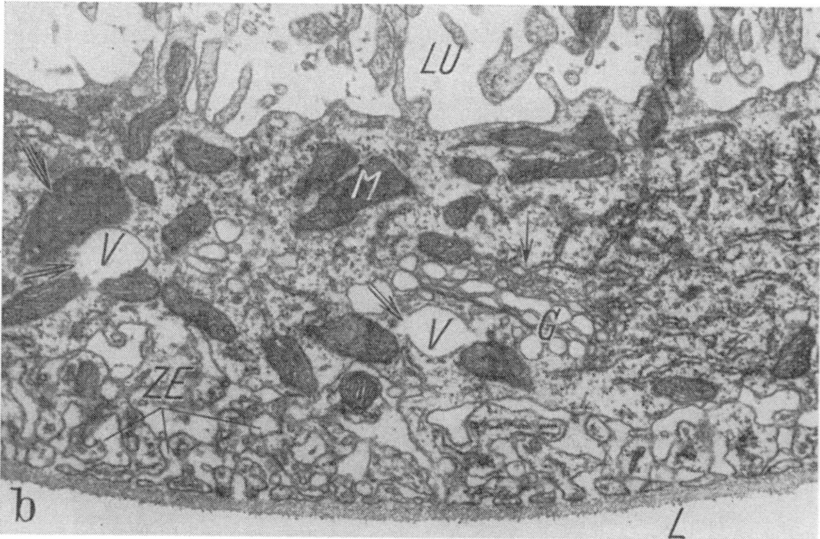

Abb. 8 a u. b. Querschnitte durch Zellen des distalen, erweiterten Anfangsstückes der larvalen M.T. von *Drosophila*. a 10 min p.i. von *Myofer* in das Blut. Dichte Ansammlungen des Markierungsmittels im *ER*. Im Gefäßlumen (*LU*) sind die wasserlöslichen Exkretkristalle außen von *Myofer* umgeben worden und kennzeichnen noch nach dem Auflösen der Kristalle deren äußere Form. b 5 min p.i. von *Hämoglobin*. Gute Markierung eines Vesikels links (dreifacher Pfeil!). Zwei unmarkierte (hyperosmotische) Vesikel (*V*) sind nach der Fixierung geplatzt (Doppelpfeil!). Der Golgiapparat (*G*) steht mit Kanälen des *ER* in Verbindung (einfacher Pfeil!). *BM* = Basalmembran, *ZE* = basale Zellmembraneinfaltungen, *L* = Leibeshöhle, *M* = Mitochondrien. E.O 6250:1, E.V. 18 000:1

harnes und weiterhin beim Rücktransport des Wassers aus der Zelle zum Blut geleistet. Dieser aktive Transport läuft demnach nicht an den polaren Grenzmembranen der Zelle, sondern an den intrazellulären Membranen der Vesikel ab. *Durch angeschlossene Kanäle des ER an die Vesikel wird* entweder noch *während der Vesikulation Flüssigkeit nachgeliefert*, oder das Wasser zum Blut zurückgebracht. Die mit Hilfe eines anorganischen Salzes, einer Aminosäure, mit Proteinen und einem Polysaccharid erzielten Markierungsergebnisse lassen vermuten, daß das gesamte Exkret über Vesikel, das Grundcytoplasma, die Kanäle des endoplasmatischen Retikulums und möglicherweise über die Interzellularspalten zum Gefäßlumen geleitet wird. Natürlich erfordert die intravesikuläre Konzentration der Exkretflüssigkeit bei der Vesikelwanderung einen gewissen Zeitraum, der nur dann erreicht werden kann, wenn der Wanderweg des Vesikels lang oder dessen Verweildauer in der Zelle vergrößert ist. Daher ist es verständlich, daß in Zellbereichen, in denen die Zelle stark abgeplattet ist (dabei sind die polaren Zellgrenzen oft nur von einem bei stärkerer elektronenoptischer Vergrößerung sichtbaren Cytoplasmasaum getrennt), nur selten Vesikel auftreten.

Der im Lumen des distalen Gefäßbereiches herrschende verhältnismäßig hohe pH-Wert von 8–8,5 (im Vergleich zur Hämolymphe mit einem Wert von 6,3–6,8) macht es wahrscheinlich, daß neben der Vesikulation und der damit gekoppelten Harnkonzentrierung noch ein spezieller, aktiver Transport bestimmter Stoffe oder Ionen in das Lumen vorhanden sein muß (z. B. aktive Ausscheidung von Kalium-Ionen, vgl. [56–62]), der hier für die Erhöhung des pH-Wertes verantwortlich zu machen ist.

2. Das Übergangsstück

Das elektronenoptische Bild der arbeitenden Zelle des *Übergangsstückes* (Abb. 9) ist im Vergleich zu dem des distalen Abschnittes völlig anders. Hier ist die Zelle mit einem wohlentwickelten Bürstensaum und tieferen Zellmembraneinfaltungen ausgestattet, deren Lumina zwischen verschiedenen Zellen nicht so starke Differenzen aufweisen wie im Anfangsstück. Daß die Zellen u. a. reabsorbierend tätig sind, zeigt die Speicherung von *Oxykynurenin* in Erweiterungen des ER und das Vorhandensein von Zentralkanälen in den Mikrovilli. Endoplasmatisches Retikulum, Ergastoplasma und zahlreiche Golgi-Apparate sind, wie im Anfangsstück, gut entwickelt. Die ER-Kanäle können auch hier an die Lumina der Zellmembraneinfaltungen anschließen. Ob die Zentralkanäle der Mikrovilli an deren Ende ausmünden, ist noch nicht geklärt. Die Zellen dieses Abschnittes sind offenbar nicht in einen Funktionsrhythmus mit einer ständigen Veränderung der Hydratationsverhältnisse des Cytoplasmas und seiner Organellen eingeschaltet, da der Transport von Exkreten hier offensichtlich geringer, und die Reabsorption von Wasser apikalwärts in das Lumen verschoben ist. Hier lassen nämlich benachbarte Gruppen von Mikrovilli des Bürstensaumes zwischen sich einen freien Raum, den sie durch Zusammenlegen ihrer Enden und einen innigen Membrankontakt gegen das Lumen allseitig dicht abschließen. In diesem intermikrovillären Raum bildet sich durch

Wasserrückresorption allmählich ein Exkretkristall (Abb. 10a). Ähnliche Bilder wurden schon für *Rhodnius* [99] und *Dacus* [5] nachgewiesen. Die so entstandenen Kristalle werden später ins Lumen entlassen, indem die Mikrovilli sich voneinander lösen. Zu den intermikrovillären Räumen bringen Vesikel die Exkrete (Abb. 10b). Daneben münden

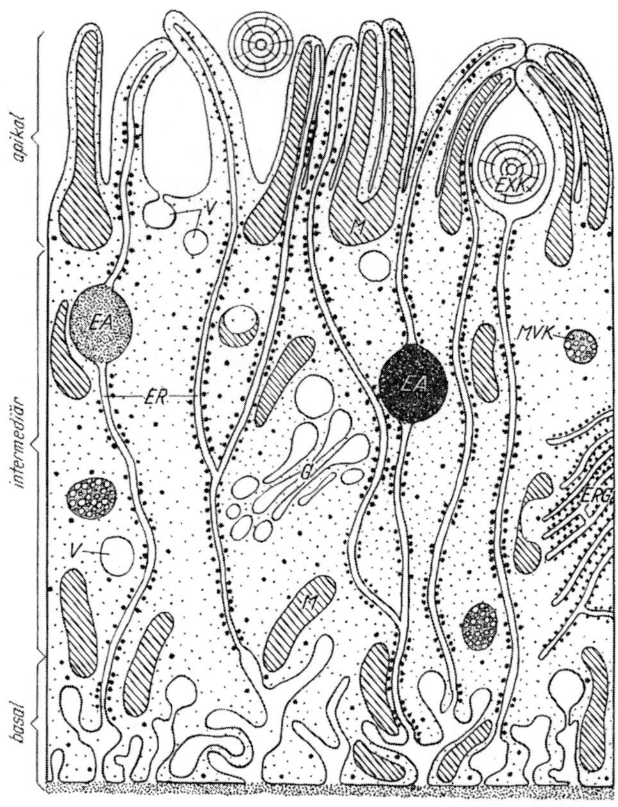

Abb. 9. Schematische Darstellung des Aufbaues einer Zelle des Übergangsstückes der M.T. von *Drosophila*. *Basal:* Zellmembraneinfaltungen, an die *ER*-Kanäle anschließen. *Intermediär:* Speicherung von *3-Hydroxykynurenin* und Fett in Erweiterungen der Endoplasmakanäle (*EA*), Golgiapparate (*G*), Ergastoplasma (*ERG*), multivesikuläre Körper (*MVK*). *Apikal:* Mikrovilli des Bürstensaumes mit Mitochondrien (*M*) bilden zwischen sich abgeschlossene Räume, zu denen Vesikel (*V*) wandern und an die *ER*-Kanäle anschließen und in denen sich Exkretkristalle (*EXK*) bilden

hier auch Kanäle des ER (Abb. 10c), die sicherlich ebenfalls eine Transportfunktion haben, doch ist die Richtung der Stoffpassage innerhalb ihrer Zisternen bisher ungeklärt.

Während die Mitochondrien im distalen Gefäßanfang entsprechend dem Energiebedarf in der Zelle gleichmäßig verteilt sind und nur selten Mitochondrien in die schmalen Mikrovilli eindringen, zeigen sie im Übergangsstück eine Tendenz zur polaren Lagerung im Bürstensaum und zwischen den basalen Zellmembraneinfaltungen. Offensichtlich wird hier durch die Reabsorption von Wasser und anderen wertvollen Stoffen hauptsächlich an zwei Orten Energie benötigt, nämlich beim Übertritt der Stoffe in die Zelle und bei ihrer Ausscheidung in die extrazellulären Lumina der Zellmembraneinfaltungen. Vor allem wird hier Wasser

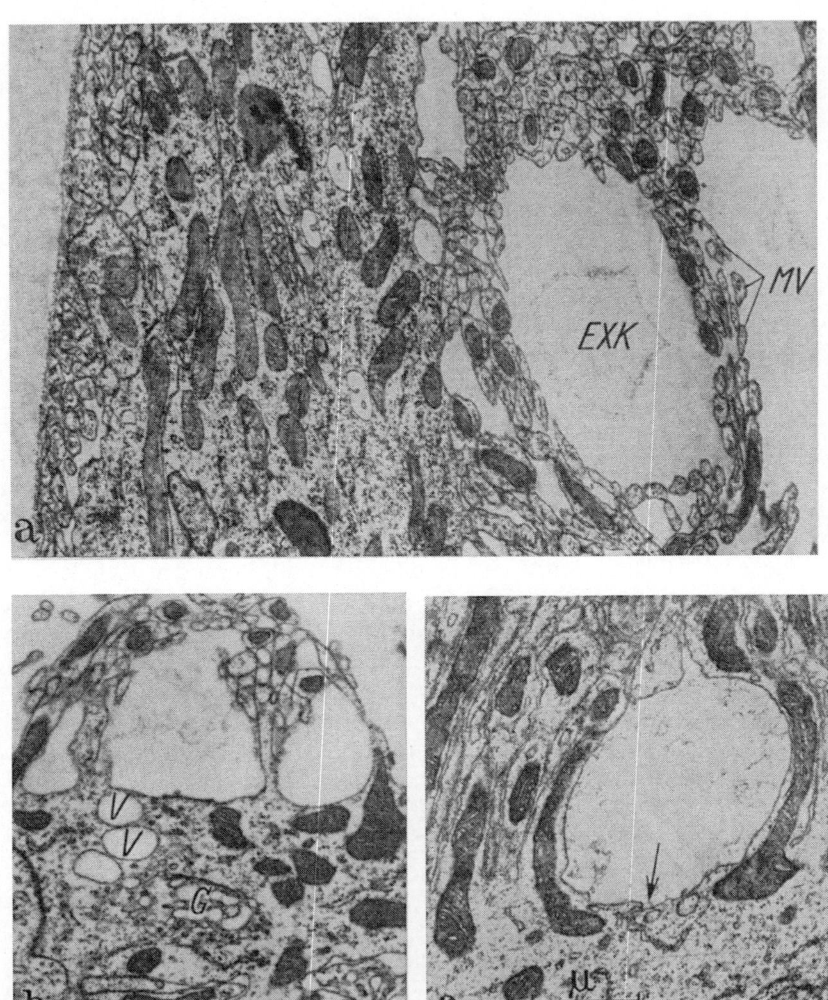

Abb. 10a—c. Übergangsstück der larvalen M.T. von *Drosophila*. a Bildung eines (wasser-löslichen) Exkretkristalls (*EXK*) zwischen den Mikrovilli (*MV*) des Bürstensaumes, von denen jeder einen Zentralkanal des *ER* besitzt. b Zu den intermikrovillären Räumen wandern Vesikel (*V*). *G* = Golgiapparat. c Anschluß eines *ER*-Kanals an einen inter-mikrovillären Kristallbildungsraum (Pfeil!). E.O. 6250:1, E.V. 14000:1

reabsorbiert, was sich dadurch nachweisen läßt, daß normalerweise eine intermikrovilläre Exkretkristallbildung nur in der Übergangzone, nicht dagegen im proximalen Gefäßabschnitt vorkommt, aber 1—2 Tage nach völligem Wassermangel, der eine intensivere Wasserreabsorption not-wendig macht, auch in dieser Region zu finden ist.

Das Übergangsstück fehlt den Gefäßen der Puppe, die infolgedessen während dieser Entwicklungsperiode den Primärharn des Lumens nicht in dem bei Larven üblichen Maße entwässern können [87]. Zudem wird die Exkretflüssigkeit bei ihrer transzellulären Passage im distalen Abschnitt bei Puppen ebenfalls nicht so stark konzentriert wie bei der Larve.

Der Übergangsabschnitt bildet bei imaginalen und den hinteren larvalen Gefäßen den Gefäßanfang; allerdings können hier bei der Imago in einzelnen Zellen den Mikrovilli die Zentralkanäle fehlen.

3. Der proximale Gefäßabschnitt (Hauptstück)

Dieser Abschnitt nimmt den längsten Anteil der Gefäße ein. Er ist durch *Riboflavin, Drosopterin* und granulär gespeichertes *Hydroxykynurenin* rötlichgelb gefärbt. Letzteres wird, wie schon erwähnt, nur in Erweiterungen des ER deponiert; daneben tritt in gleicher Form auch Fett auf. Die Speicherung beider Stoffe nimmt von der Übergangsregion bis zum Beginn des Ureters zu.

Die Zellen sind (Abb. 11) mit tiefen, stark vernetzten Zellmembranfalten und langen Mikrovilli ausgestattet. Jeder Mikrovillus besitzt einen zentralen Kanal, der an dessen Spitze häufig ins Lumen mündet, andererseits Verbindung mit den endoplasmatischen Kanälen der intermediären Zellregion hat, die die Zelle vor allem in polarer Richtung durchziehen und die auch oft an die Lumina der basalen Zellmembraneinfaltungen anschließen [85]. Die Lumina der in den Mikrovilli liegenden Kanäle sind keineswegs in ihrer ganzen Länge geöffnet, sondern an vielen Stellen kollabiert, so daß ein transzellulärer Stofftransport nicht alle diese Bahnen ungehindert passieren kann (s. Abb. 11). In den Zisternen des ER wird durch *Kynureninoxydase Kynurenin* zu *3-Hydroxykynurenin* verwandelt und hier in Erweiterungen dieses Kanalsystems deponiert. Das in gleichen „Ampullen" akkumulierte Fett kann durch Kanäle direkt mit dem Oxykynureninkonkrementen verbunden sein. Das zeigt, daß *das ER die Fähigkeit hat, Stoffe, die im gleichen Kanal dicht nebeneinander liegen, einwandfrei voneinander zu scheiden.*

Die endoplasmatischen Kanäle lassen auch hier die gleichen Beziehungen zu den Golgi-Apparaten erkennen, wie sie für die anderen Regionen beschrieben wurden. Ein typisches, organisiertes Ergastoplasma tritt im proximalen Abschnitt nie auf. Die Golgi-Apparate sind oft den Oxykynurenin-Depots benachbart und wirken vielleicht bei der Akkumulation der Aminosäure in den ER-Zisternen mit. Die endoplasmatischen Kanäle können auch die intermediäre Region der Zelle durchlaufen, ohne zu einem Oxykynurenin- oder Fettdepot zu gelangen, so daß eine direkte Verbindung zwischen Zellmembraneinfaltung und Gefäßlumen entstehen kann.

Multivesikuläre Körper, die in allen Zellen des M.T. in großer Anzahl zu finden sind, entstehen in der proximalen Region, wahrscheinlich an den tiefsten Zellmembraneinfaltungen, wandern apikalwärts und entleeren ihre Bläschen in der Nähe des Gefäßlumens. Offensichtlich bringen sie Stoffe von der Zellbasis zum Lumen. Dagegen ist ihre Bedeutung im Übergangs- und Anfangsstück unklar.

Die Mitochondrien dieser Region liegen fast nur in den Mikrovilli und zwischen den basalen Zellmembraneinfaltungen, also an den Stellen, wo für die Reabsorptionsarbeit Energie benötigt wird. Diese Organellen

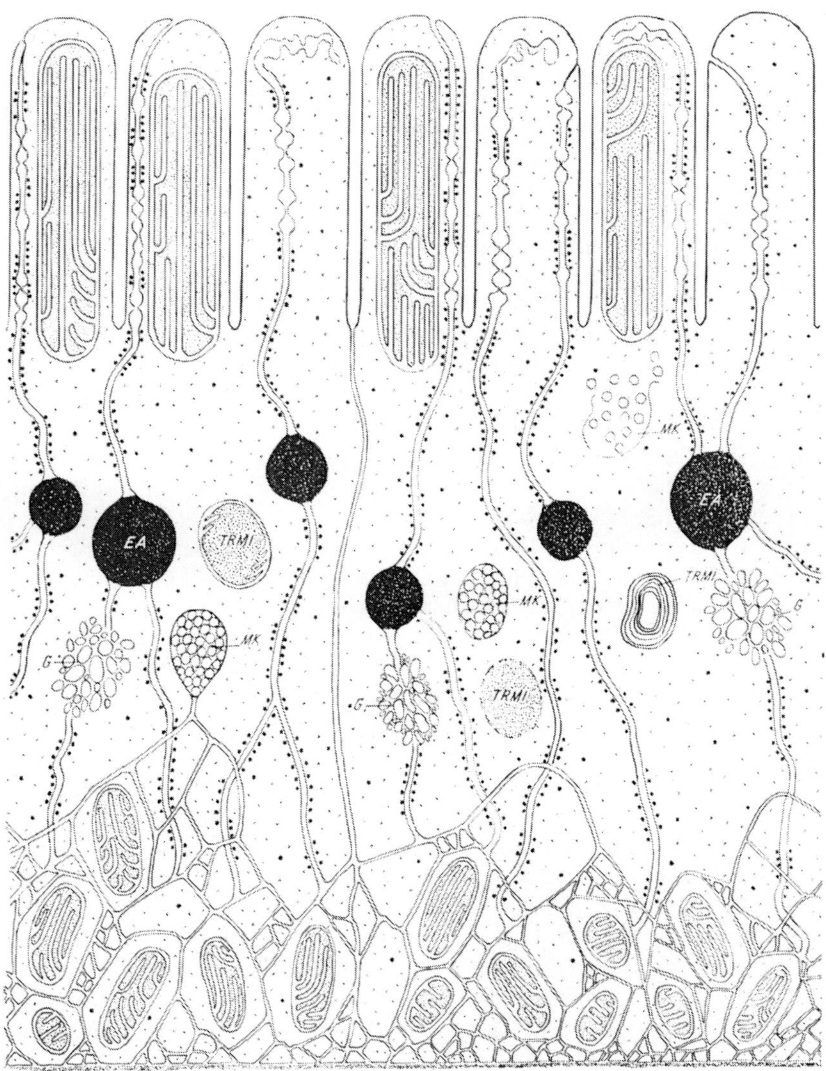

Abb. 11. Schematische Darstellung einer Zelle des proximalen Hauptstückes der larvalen M.T. von *Drosophila* [*85*, verändert!]. *Basal:* Viele anastomosierende Zellmembraneinfaltungen bilden ein „basales Labyrinth". Daran schließen *ER*-Kanäle an. *Intermediär:* In Erweiterungen von *ER*-Kanälen (*EA* = Endoplasmaampullen) wird *3-Hydroxykynurenin* und Fett gespeichert. Multivesikuläre Körper (*MV*) entstehen an den Zelleinfaltungen, wandern ab und lösen sich am Lumen auf. Transformation von Mitochondrien (*TRMI*), *G* = Golgiapparate, durch die Endoplasmakanäle laufen. *Apikal:* Mikrovilli des Bürstensaumes mit Zentralkanälen des *ER*, die teilweise Anschluß an die Zellmembran haben und eine offene Verbindung zwischen Lumen und Blut herstellen. Mitochondrien und deren Cristae in den Mikrovilli sind polar ausgerichtet

und ihre Cristae sind vornehmlich parallel zur polaren Achse der Zelle ausgerichtet. Nur einige intermediäre Mitochondrien werden transformiert.

Besonders auffallend ist, daß der Bau sämtlicher Strukturen der Zellen der proximalen Region von der Stoffwechselaktivität gesteuert wird [87]: Die larvale Zelle, an die verhältnismäßig hohe stoffwechselphysiologische Anforderungen gestellt werden, — denn hier werden wertvolle Substanzen des Primärharnes reabsorbiert —, ist groß und· hat tiefe, zahlreiche, stark vernetzte Zellmembraneinfaltungen, lange Mikrovilli, in die viele Mitochondrien eindringen, ein reich entwickeltes ER, zahlreiche Golgi-Apparate, die fast nur aus Vesikeln bestehen, und ein lockeres Grundcytoplasma. Die Mitochondrien sind groß mit dunkler Matrix und besitzen viele Cristae mitochondriales.

Läßt man Larven hungern oder senkt man den Stoffwechsel durch längeren Aufenthalt bei tiefen Temperaturen (4—6° C), dann ziehen sich die Mitochondrien aus den Mikrovilli, die dabei kleiner werden, zurück; das endoplasmatische Retikulum wird bis auf wenige Kanäle abgebaut, vor allem verschwinden die Zentralkanäle der Mikrovilli. Auch die basalen Einfaltungen werden größtenteils eingeschmolzen (s. auch [5]), und die zwischen ihnen liegenden Mitochondrien verteilen sich gleichmäßig in der Zelle.

Noch deutlicher wird der Stoffwechseleinfluß auf die Zellarchitektur nach der Verpuppung (Abb. 12), denn während dieser Zeit wird ja keine Nahrung aufgenommen, und der Anfall von Exkreten während der Puppenzeit ist gering. Lediglich die im Fettkörper gespeicherte Harnsäure wird ausgeschieden. Innerhalb der ersten 10 Std nach der Puppenbildung sind der Bürstensaum, die basalen Zellmembraneinfaltungen und das endoplasmatische Retikulum bis auf geringe Reste völlig abgebaut [87]. Oxykynurenin- und Fettdepots werden fast völlig geleert. Die Golgi-Apparate bilden jetzt wieder Doppelmembrankomplexe, zeigen also den typischen Bau dieser Zellstrukturen. Die Mitochondrien verteilen sich gleichmäßig in der Zelle, sie werden kleiner, die Matrix verliert allmählich ihren starken Elektronenkontrast, die Cristae werden kurz, und ihre polar orientierte, parallele Lagerung geht verloren. Viele Mitochondrien werden durch Transformation abgebaut. Die Puppenzelle besitzt eine dicke Basalmembran, ist klein und das Cytoplasma dicht.

Zwingt man aber die M.T. der Puppe durch Injektion von Fremdstoffen (*Myofer* und *Goldchlorid*) zu einer Aktivierung der Zellarbeit, dann entsteht schon nach 30 min p.i. ein Zellbild, das dem der larvalen Zelle wieder ähnlicher wird. Allerdings erscheint das Cytoplasma wesentlich dunkler durch seinen Reichtum an Ribosomen, die, wie eingehende Untersuchungen gezeigt haben [89, 90], nach der Verpuppung durch Abgabe von Nucleolarsubstanz aus dem Kern ins Cytoplasma gelangt sind (Abb. 13).

Der Aufbau der imaginalen Zelle liegt entsprechend der Stoffwechsellage zwischen dem der Larve und dem der Puppe [85, 87]. Einfaltungen, Bürstensaum, ER und andere Funktionssysteme sind nur mäßig entwickelt, die Zelle erreicht nicht das Volumen einer larvalen Zelle.

Auch auf den Kern, seine elektronenoptisch sichtbaren Chromo-
somenanteile und den Nucleolus hat die Stoffwechselaktivität der Zelle
einen wesentlichen Einfluß [88]. Der larvale Kern ist groß, und alle
Strukturen sind stark aufgelockert. Nach der Verpuppung schrumpft

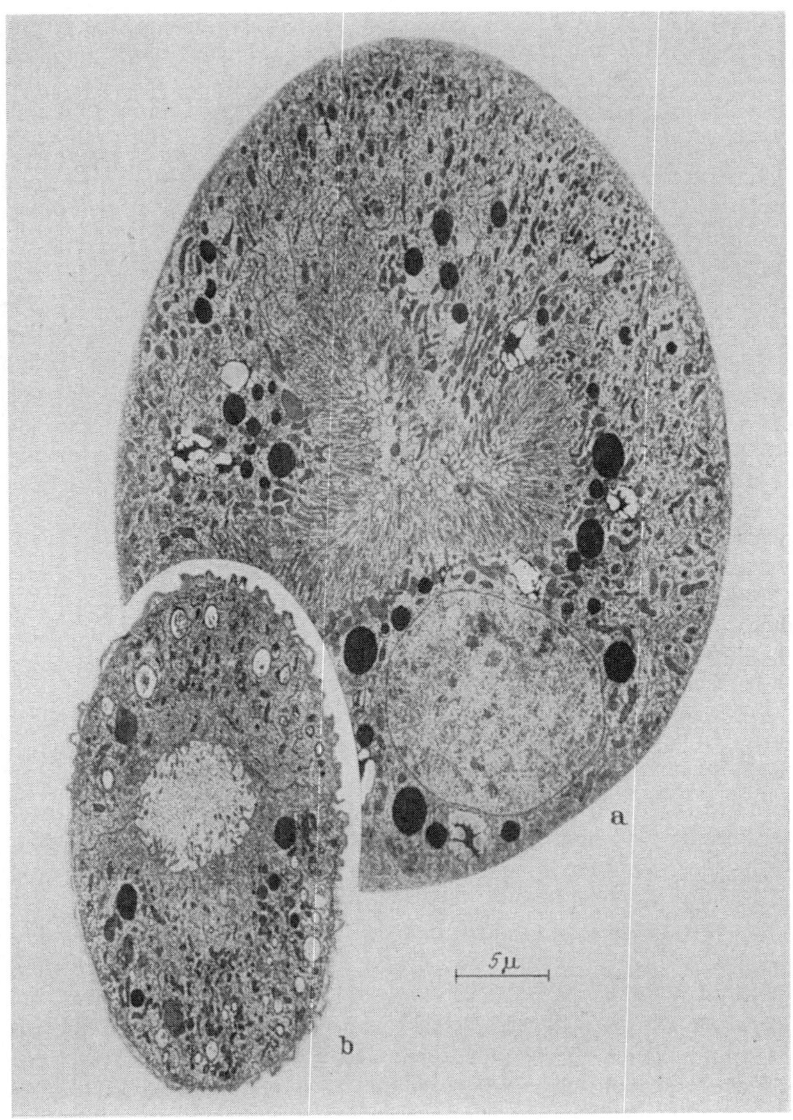

Abb. 12a u. b. Querschnitt durch den proximalen Abschnitt eines M.T. von *Drosophila*.
a Larve. b Junge Puppe. Nach der Verpuppung werden die Zellen kleiner, die Basalmembran
schrumpft, das gespeicherte *Oxykynurenin* und das Fett verschwinden, basale Zellmembraneinfaltungen und Bürstensaum werden fast ganz eingeschmolzen, viele Mitochondrien nach
Transformation abgebaut, die anderen geben ihre polare Lage auf. E.O. 2000:1, E.V.
2500:1

er, und seine internen Strukturen gehen in einen kondensierten Zustand über. Einen ähnlichen Aspekt zeigen die Innenstrukturen des larvalen Kernes unter dem Einfluß von Hunger oder nach längerer Senkung der Zuchttemperatur auf 4—6° C. Nach dem Schlüpfen wird der Kern wieder größer und seine Strukturen lockern sich wieder auf.

Der Aufbau der *Zellen des proximalen Gefäßabschnittes*, die zum Ureter hin zunehmende Speicherung von Oxykynurenin und Fett, das Auskristallisieren von injizierten Farbstoffen im Lumen dieses Abschnittes und das Fehlen eines Vesikelflusses, wie er für den distalen Abschnitt und das Übergangsstück charakteristisch ist, sprechen dafür, daß hier *vorwiegend reabsorbiert* wird. Durch Verfüttern und durch Injektion von elektronenmikroskopischen Markierungsstoffen (*Hydroxykynurenin, Goldchlorid* und *Myofer*) ließ sich aber nachweisen, daß auch dieser Gefäßabschnitt die Fähigkeit hat, Stoffe zum Lumen zu leiten[1].

Schon 5 min nach einer Injektion (15 min nach Fütterung) von *Oxykynurenin* in das Blut einer Larve liegen im Cytoplasma große Vesikel. Diese entstehen, wie bei der transzellulären Passage von Stoffen im distalen Abschnitt, aus den Zellmembraneinfaltungen, in denen die aus dem Blut durch die Basalmembran eingedrungenen Stoffe konzentriert werden. Hier lösen sich allerdings größere Komplexe von Einfaltungen vom „basalen Labyrinth" ab und wandern apikalwärts. Die Konzentration der Stoffe setzt schon dann ein, wenn diese abwandernden Komplexe noch mit dem basalen Faltensystem Verbindung haben. Der genaue Mechanismus der Ablösung ist noch nicht geklärt. Vermutlich werden die zwangsläufig mit abgetrennten, zwischen den Einfaltungen liegenden Cytoplasmabezirke nicht mitbefördert, sondern von den umgebenden Membranen gelöst. Letztere werden später meist von einem gemeinsamen Vesikel umfaßt, dessen Wandung wahrscheinlich durch Zusammenfügen von abwandernden Membranen oder aber durch Neubildung entsteht. Auch an diese großen Vesikel schließen immer Kanäle des ER an; in der Übergangszone, wo gleiche Bilder auftreten, kann auch das Ergastoplasma Verbindung mit ihnen aufnehmen. Da die Vesikel $3-4\mu$ dick sein können, wird vermutlich über diese Kanäle weiteres Exkret zu ihnen hingeleitet und hier entwässert. Dadurch entsteht in den Vesikeln nach der Fixierung ein höherer hydrostatischer Druck, durch den die Membranen gelegentlich zerreißen können.

Die mit *Oxykynurenin* und mit Membranmaterial gefüllten Vesikel vermehren sich später nicht mehr. Diese für den Organismus wertvolle Aminosäure, die bei Larven normalerweise nicht im Blut enthalten ist, nehmen nämlich die Zellen des proximalen Abschnittes nur anfänglich direkt aus dem Blut auf. Danach (nach 1 Std) gelangt das *Oxykynurenin* dann vor allem mit dem Primärharn im distalen Abschnitt ins Lumen und wird schließlich proximal von den Zellen reabsorbiert. Dieses reabsorbierte *Oxykynurenin* wird auf andere Weise, nämlich nur in Form von vielen, kleinen, dichten Konzentrationen zunächst im apikalen Be-

---

[1] Dabei ist allerdings unklar, ob diese Versuche die normalen Verhältnisse wiedergeben. Die folgend beschriebenen Bilder treten nämlich nur im Experiment, nicht dagegen bei normalen Verhältnissen auf.

reich der Zelle in den Zisternen des endoplasmatischen Retikulums gespeichert. Die Konkremente wandern später allmählich zusammen und bilden immer größere Konglomerate von *Oxykynurenin*. Nach 6 Std ist es

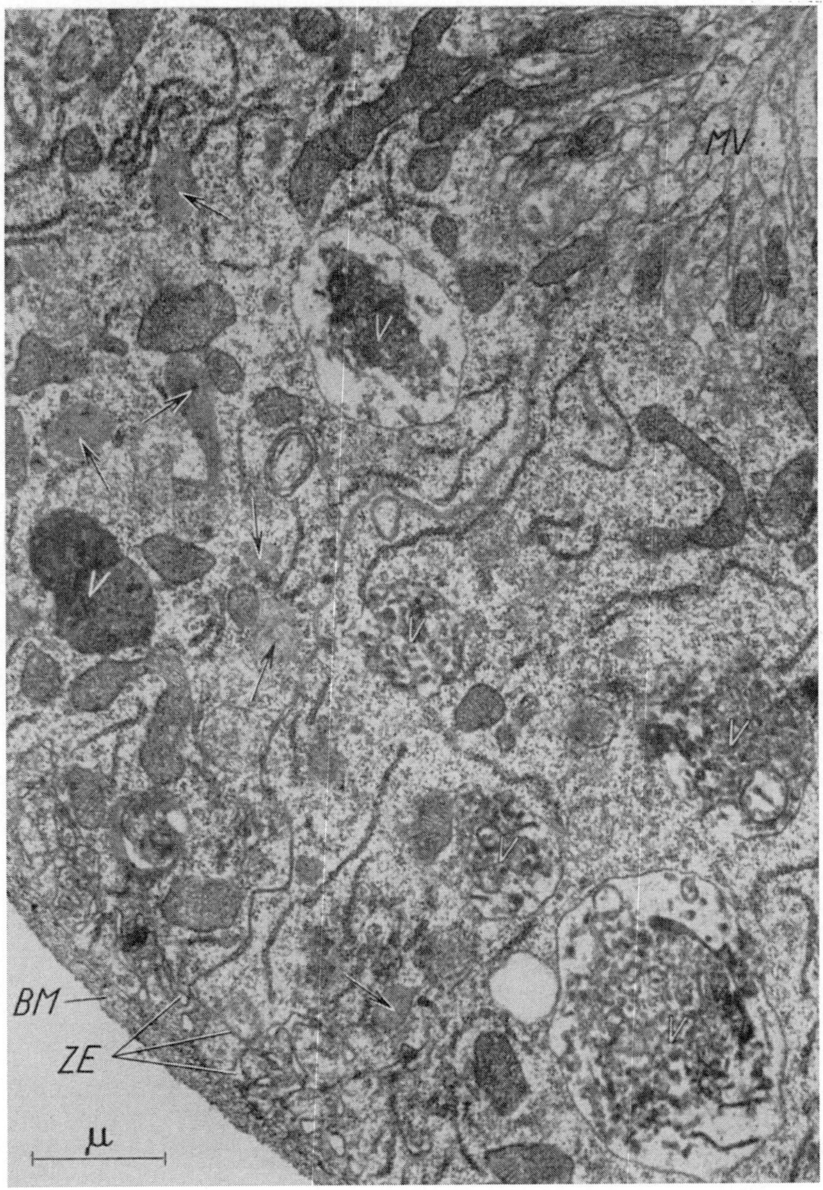

Abb. 13. Transport von *Goldchlorid* und *Myofer* durch die Zellen des proximalen Hauptstückes der M.T. einer alten Puppe von *Drosophila*, 60 min p.i. der Kontrastmittel. Speicherung der Kontrastmittel in großen Vesikeln (*V*), die mit viel Membranmaterial und darin kondensiertem Markierungsmittel gefüllt sind. Ansammlungen von *Myofer* im Grundcytoplasma (Pfeile!). E.O. 6250:1, E.V. 17 500:1

dann nicht mehr möglich, solche im Experiment mit *Oxykynurenin* versorgte Zellen einer *Drosophila*-Augenfarbmutante von denjenigen der Wildform zu unterscheiden, die ja das *Oxykynurenin* ebenfalls im ER deponieren [*84*]. Die anfänglich entstandenen großen Vesikel sind nach 3 Std nicht mehr zu erkennen, da ihr *Oxykynurenin* stärker kondensiert, und diese Akkumulate dann nicht mehr von den Konzentrationen des aus dem Lumen reabsorbierten *Oxykynurenins* zu unterscheiden sind.

Injiziertes *Goldchlorid* oder *Myofer* wird ebenfalls durch die proximalen Zellen exzerniert. Dabei treten zunächst gleiche Konzentrate in den basalen Zellmembraneinfaltungen auf. Auch hier werden größere Teile der Einfaltungen abgelöst, bilden membrangefüllte Vesikel, die unter ständigem Kontakt mit den ER-Kanälen zum Lumen wandern.

Auch die Gefäße der *Puppe* sind in der Lage, injiziertes *Myofer* und *Goldchlorid* über kleinere Einfaltungen, die sich nach der Injektion neu bilden, und über Vesikel zum Lumen zu leiten (Abb. 13). Die in den Vesikeln liegenden Membranen werden dabei mit ausgestoßen. Man findet nämlich im Gefäßlumen relativ oft Membranhaufen zwischen den Exkretkristallen (Abb. 2d). *Myofer* wird, wie im distalen Abschnitt, auch in den proximalen Zellen darüber hinaus über das Grundcytoplasma und ER-Kanäle lumenwärts geleitet.

Die Untersuchungen lassen erkennen, daß die *Zellen der proximalen Zone* neben der Reabsorption, wenn es die Stoffwechsellage unter experimentellen Bedingungen erfordert, *auch exzernieren* können. Diese Fähigkeit ist allem Anschein nach im ganzen Gefäßabschnitt vorhanden. Außerdem lassen die Ergebnisse den Schluß zu, daß das *„basale Labyrinth" in den Stoffstrom sowohl zum Blut als auch zum Lumen eingeschaltet sein kann*. Auch hier entstehen die transportierenden Vesikel in einer modifizierten Form aus der basalen Zellmembran, auch hier handelt es sich also um „Cytopempsis".

Die Unfähigkeit der Puppengefäße, deren proximale Zellen fast keine Einfaltungen und Mikrovilli besitzen, injizierte Farbstoffe, wie das larvale Gefäß, im Lumen so stark zu konzentrieren, daß sie auskristallisieren, zeigt aber, daß diese Strukturen vor allem der Reabsorption dienen [*87*].

4. Der Ureter

Die oberen und die unteren Gefäße vereinigen sich auf beiden Körperseiten zu je einem Ureter, der durch peristaltische Kontraktionen von Muskelzellen, die ihn umgeben, den stark konzentrierten, sauren Harn in den Enddarm abführt. Zahlreiche Tracheolen versorgen diesen Abschnitt mit Sauerstoff. Sie liegen teilweise zwischen den Epithel- und den Muskelzellen, so daß ihr Lumen durch den Druck der Muskeln ständig rhythmisch komprimiert wird, wodurch ein besserer Gasaustausch gewährleistet ist. Das Sarkoplasma der Muskelzellen kann an einigen Stellen gefenstert sein, so daß durch solche Aussparungen der transzelluläre Stoffstrom hindurchtreten kann. Der Feinstrukturaspekt der Ureterzellen ist charakteristisch und gleicht dem der Mitteldarmzellen. Die Mikrovilli sind lang, besitzen aber keinen Zentralkanal und die Mitochondrien dringen nicht in sie ein, halten sich aber massiert basalwärts von ihnen auf. Durch die unterschiedlichen Funktionszustände der Zelle wechselt die Ausbildung der basalen Einfaltungen: Meist dringen sie tief fast bis zum Bürstensaum in die Zelle ein, erreichen aber nie das Gefäßlumen; dabei sind sie weitlumig. Seltener treten sie in geringer Anzahl auf und sind dann kurz und nur mit einem engen Lumen ausgestattet. Zwischen ihnen konzentrieren sich viele Mitochondrien.

Die Interzellularen gehen auch hier, wie in den Zellen der anderen Gefäßabschnitte, in die Einfaltungen über, wo ihr weiterer Verlauf oft nicht mehr zu verfolgen ist. Im distalen Anfangs- und im Übergangsstück waren die Interzellularen häufig mit dem ER oder dem Ergastoplasma verbunden. Diese Anordnung sprach für eine transzelluläre Stoffpassage durch die Interzellularräume. Im Ureter sind diese oft eine Strecke weit stark verengt, so daß in solchen Interzellularen kein Stofftransport ablaufen kann.

Das ER ist gut entwickelt, speichert aber weder Fett noch Oxykynurenin. Die dem Darm näher liegenden unteren Ureterabschnitte besitzen Zellen, die den Mitteldarmzellen immer ähnlicher werden. Sie speichern im Grundcytoplasma große Glykogenmengen.

Die Zellen des Ureter sind, wie die des Mitteldarmes, in der Lage,

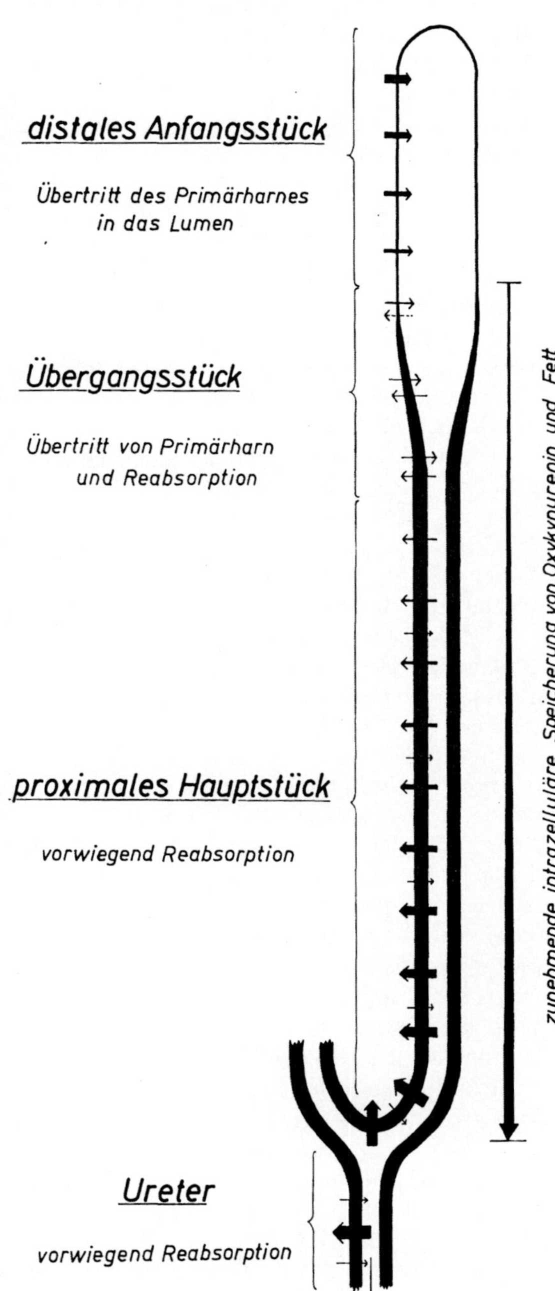

Abb. 14. Richtungen der transzellulären Stoffpassagen im vorderen larvalen M.T. von *Drosophila*

injizierte Stoffe (*Myofer, Goldchlorid*) durch Cytopempsis in das Gefäßlumen abzuführen, ihr Zellaufbau spricht allerdings für eine *vorwiegend
reabsorbierende Funktion* dieser Gefäßregion.

*Die M.T. von Drosophila können also im Experiment in ihrer ganzen
Länge Exkrete zum Lumen befördern.* Dabei ist die Stoffpassage zum
Lumen im distalen erweiterten Abschnitt am stärksten. *Eine Reabsorption kann dagegen nur für das Übergangs- und das proximale Hauptstück
des Tubulus nachgewiesen werden, ist aber wahrscheinlich auch im Ureter
vorhanden* (Abb. 14).

B. Enzyme der Malpighischen Gefäße

Die Nierentubuli der Insekten gehören neben der Muskulatur und dem
Fettkörper zu den stoffwechselaktivsten Organen des Körpers. Daher
sind in ihnen eine große Anzahl von Enzymen lokalisiert, die nicht nur
in den Exkretstoffwechsel eingreifen, sondern die Stoffe während ihrer
transzellulären Passagen wie in einem regulatorischen Stoffwechselorgan auch umformen können.

Die meisten der nachgewiesenen Enzyme sind nur auf bestimmte
Gefäßregionen beschränkt. Ein für Insekten allgemein gültiges
Schema läßt sich jedoch nicht aufstellen, da das Enzymmuster von der
Insektenart, deren Ernährung und von der Entwicklung des betreffenden
Tieres abhängig ist. Vergleiche zwischen verschiedenen Nahrungs-
spezialisten ließen nur wenige Übereinstimmungen erkennen [40].

Alkalische und *saure Phosphatase* [22, 24] findet sich vor allem in der
proximalen Region. Bei *Acheta* und *Phormia regina* ist die *alkalische
Phosphatase* auf den Bürstensaum des proximalen Abschnittes konzentriert [10, 79]. Die Enzyme und ihr Verteilungsmuster in den imaginalen Gefäßen von *Dacus oleae* sind in Abb. 15 dargestellt [5]. *ATP-ase,
β-Oxybuttersäuredehydrogenase* und *TPNH-Diaphorase* lassen sich nur
bei der Larve nachweisen. Larve und Imago zeigen im übrigen in der
Anordnung der Enzyme deutliche Unterschiede. *ATP-ase* tritt auch in
den M.T. von *Periplaneta* auf [70]. Eine *Aminosäureoxydase*, die Aminosäuren in α-*Ketosäuren* umwandelt, wurde bei *Periplaneta, Blatella,
Galleria* und bei *Oncopeltus* nachgewiesen [3]. Die Gefäße von *Acridia,
Blaps, Mantis* und *Apis* besitzen eine *Lipase* [40]. In *Schistocerca-* und
Periplaneta-Gefäßen kommen 2 *Transaminasen* vor [31]. Die für *Drosophila melanogaster* nachgewiesene *Kynureninoxydase* ist wahrscheinlich
nur für terrestrisch lebende Fliegen charakteristisch [84]. Gorka [28] hat
im übrigen bei Käfern typische Verdauungsenzyme, wie *Amylase* und
Saccharase, nachweisen können. Im übrigen fand er, daß Gefäßextrakte
die Eiweiß- und Fettverdauung des Mitteldarmes fördern können. Bei
Käfern und Orthopteren können auch *Dipeptidasen* auftreten [72],
bei *Calliphora*-Larven wurde eine auf Peptide wirkende *Desaminase*
gefunden [18].

Die Enzyme des Citronensäurezyklus und der Atmungskettenphosphorylierung deuten darauf hin, daß die Malpighi-Gefäße einen lebhaften Atmungsstoffwechsel besitzen [50], der die für die zahlreichen

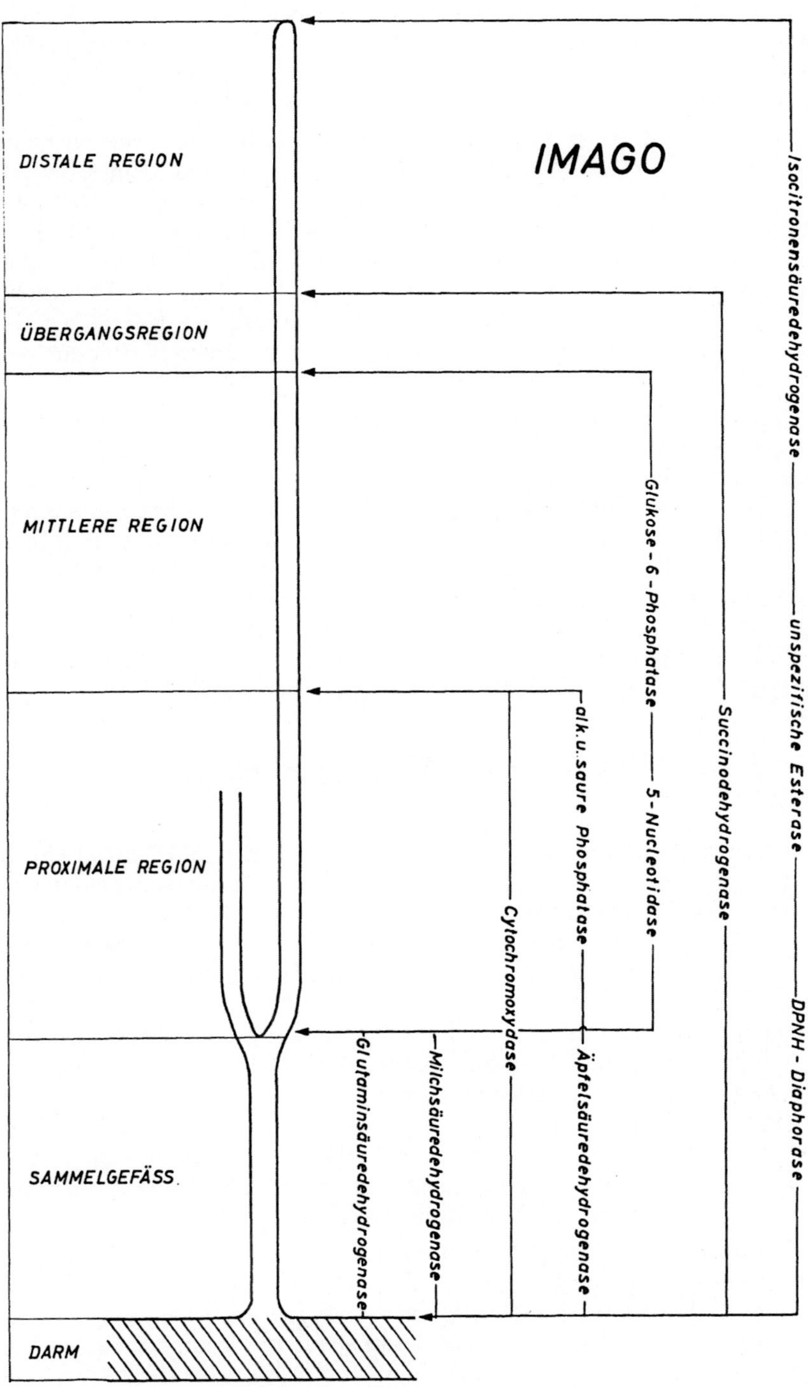

Abb. 15. Verteilung der Enzyme im imaginalen Gefäß von *Dacus oleae*. Zusammengestellt nach Angaben von Baccetti, Mazzi und Massimello [5]

Transportvorgänge notwendigen, energiereichen Verbindungen liefert, für die das spaltende Enzym, die *ATP-ase*, ebenfalls nachgewiesen wurde. Die *alkalische Phosphatase* spielt vermutlich, wie in der Vertebratenniere, eine wichtige Rolle bei der Reabsorption von Sacchariden.

Bemerkenswert ist das Auftreten von Enzymen, die verschiedene Umbau- und Abbauvorgänge katalysieren. Dieser Befund deutet darauf hin, daß dem Malpighi-Gefäß, außer seiner exkretorischen Funktion, neben dem Fettkörper (vgl. [*82*]), eine wichtige Bedeutung als regulatives Stoffwechselorgan zukommt (vgl. [*24*, *28*]). Die durchgeschleusten Stoffe können bei ihrer transzellulären Passage umgebaut und eventuell, wie bei *Drosophila*, in den Zellen gespeichert werden. So bauen die M.T. von *Carausius* C^{14}-markiertes *Valin* in mehrere andere Verbindungen um [*64*].

Kristalline Exkrete können im Lumen auch gespeichert werden [*25*, *36*, *98*]: So entleert sich das distale Anfangsstück der vorderen M.T. von *Drosophila* selbst nach tagelangem Hungern nicht, während das Lumen des proximalen Gefäßteils dagegen bald frei von kristallinen Exkreten ist. Welche Bedeutung diese „Speicherexkretion" hat, wurde bisher noch nicht völlig geklärt. Bei *Carausius* sollen die distalen, erweiterten Anfangsstücke Calciumcarbonat, das für die Eientwicklung gebraucht wird, speichern [*62*]. Die in den M. T. auftretenden Verdauungsenzyme und -aktivatoren lassen die Herkunft dieser Organe aus dem Darmepithel erkennen, dessen Tätigkeit sie unterstützen können.

C. Physiologie der Malpighischen Gefäße

[20, 48, 73, 83, 98]

Die Art der Erhaltung eines relativ konstanten inneren Milieus durch die Exkretionsorgane ist weitgehend von ökologischen Gegebenheiten abhängig [*73*]: Landinsekten gewinnen Wasser und Salze, um die Verluste durch die Verdunstung und die Tätigkeit der Exkretionsorgane auszugleichen. Süßwasserinsekten müssen ebenfalls Salze aus dem umgebenden Milieu erhalten, dabei aber größere Mengen von Wasser ausscheiden; Salzwasserinsekten nehmen Wasser auf und exzernieren Salze. Damit ist ersichtlich, daß je nach Lebensraum die Arbeit eines regulierenden Organs ganz verschieden sein kann. Nicht nur die Malpighi-Gefäße greifen in diesen Stoffhaushalt ausgleichend ein, sondern auch andere Transportsysteme, wie der Darm, der Fettkörper, die Analpapillen und möglicherweise auch die äußere Epidermis. Sie alle sind an der Erhaltung optimaler innerer Verhältnisse aktiv beteiligt.

Die Zusammensetzung des ausgeschiedenen Harnes ist von Art zu Art verschieden und wird von ökologischen und physiologischen Faktoren bestimmt. Neben Wasser wurden u. a. *Natrium*, *Kalium*, *Magnesium*, *Calcium*, *Eisen*, *Chlorid*, *Carbonat*, *Bicarbonat*, *Phosphat*, *Oxalat*, *Sulfat*, *Hypoxanthin*, *Xanthin*, *Harnsäure* und ihre Salze, *Allantoin*, *Allantoinsäure*, *Harnstoff*, gelegentlich auch *Aminosäuren*, *Kreatin* und *Kreatinin* und bei Fleischfressern oft *Ammoniak* nachgewiesen. Wichtigstes

N-Exkret ist die *Harnsäure*. Vielfach wird sie aber, wie bei *Drosophila*, vor allem im Fettkörper gespeichert [*29, 75, 77*], und nur geringe Mengen werden durch die M.T. abgeführt.

Unsere Kenntnisse über die Physiologie der Malpighischen Gefäße verdanken wir vor allem den zahlreichen Arbeiten von RAMSAY, die er an verschiedenen Insektenarten durchführte:

Bei Wasserinsekten (Mückenlarven) verläßt der Primärharn ohne eine erkennbare Reabsorption das Lumen der M.T. Er ist dabei hypoosmotisch zum Blut. Erst im Enddarm und im Rektum erfolgt die Rückgewinnung der für den Organismus wertwollen Stoffe. Bereits bei diesen Versuchen ergab sich ein wichtiger Befund: Die Kalium-Konzentration des Primärharnes ist wesentlich größer als die des Natriums. Das $\frac{C_T}{C_H}$-Verhältnis $= \left(\dfrac{\text{Konzentration des Primärharnes im Tubulus}}{\text{Konzentration in der Hämolymphe}} \right)$ ist für Natrium 0,17–0,28, für Kalium aber 29–32. Daraus wird gefolgert, daß K aktiv in das Gefäßlumen sezerniert wird; es muß dabei gegen einen elektrochemischen Gradienten transportiert werden. Natrium und andere gelöste Stoffe sollen dagegen passiv durch die Gefäßwand ins Lumen diffundieren. Steigt die Na- oder K-Konzentration des Blutes, dann nimmt der Na-Gehalt des Lumens bis zur Isotonie zu. Die M. T. der untersuchten Wasserinsekten können, außer Kalium, keinen anderen Stoff im Lumen in einem höheren Maße konzentrieren, als er im Blut vorhanden ist [*56, 58*].

Die Entdeckung des aktiven Kalium-Transportes sprach für die Hypothese von WIGGLESWORTH [*96*], nach der die Harnsäure bei *Rhodnius* im distalen Abschnitt als leichter lösliches Kaliumurat einwandert und daher den Primärharn alkalisch macht, daß proximal aber das Kalium reabsorbiert wird, wobei die schwerer lösliche Harnsäure ausfällt, und der aus dem Gefäß abströmende Harn sauer wird. Daß im distalen Abschnitt nur Sekretion, im proximalen Reabsorption vorkommt, war durch Anlegen von Ligaturen nachgewiesen worden.

RAMSAY konnte an einer größeren Anzahl von Landinsekten, wie *Rhodnius, Locusta, Carausius, Pieris, Dytiscus, Tenebrio, Tabanus* und *Aëdes* ebenfalls einen aktiven Kaliumtransport und meistens auch eine Akkumulation von positiven Ladungen im Lumen auffinden [*57, 59*]. Bei *Rhodnius* wird das Kalium schon im proximalen Abschnitt teilweise wieder resorbiert, wie schon früher vermutet wurde [*96*]. Inzwischen ist eine aktive Kaliumsekretion auch bei *Schistocerca* gefunden worden [*51*]. Der *Befund der aktiven Kalium-Sekretion* besitzt offensichtlich *eine für Insekten allgemeine Bedeutung und wird von* RAMSAY *für einen fundamentalen Prozeß der Harnbildung* gehalten.

Alle weiteren Versuche wurden an den oberen Gefäßen von *Carausius (Dixippus) morosus*, größtenteils an isolierten Tubuli durchgeführt, die in Hämolymphe (oder in enteiweißtem „Serum"), dagegen nicht in künstlichen Medien, normal funktionieren sollen. Isolierte Gefäße können z. B. Phenolrot aus der umgebenden Flüssigkeit im Lumen konzentrieren [*60*]. Der Primärharn von *Carausius* ist hypo- bis isoosmotisch zum Blut. Pro Stunde werden 6 mm³ Primärharn ins Lumen ausgeschieden, das ist etwa $^1/_{24}$ des Wassergehaltes des Blutes [61]. Der Primärharn verhält sich zur Hämolymphe immer alkalisch, wird aber durch Reabsorption von Kaliumionen im Rektum wieder sauer. Isolierte Gefäße können nicht nur Kalium, sondern auch Natrium [62], Wasser und Phosphat [60] aktiv ins Lumen transportieren.

Die Ausscheidungsrate von Kalium ist aber 10mal größer als die von Natrium. Die Rate der Primärharnbildung steigt durch Verdünnung und fällt durch Konzentration der Hämolymphe, außerdem auch dann, wenn der Druck des Primärharnes zu groß wird; möglicherweise wird durch die stärkere Ausdehnung der Gefäßwand die transzelluläre Stoffpassage unterbrochen. Wird Natrium- oder Kaliumchlorid in das Blut injiziert, dann zeigen die M. T. nur ein relativ schwaches Regulationsvermögen [61]

Alle Abschnitte der Gefäße scheiden in vitro aktiv Natrium, Kalium und Wasser aus, allerdings ist proximal das Na/K-Verhältnis größer als in den anderen Abschnitten. Kalium und Phosphat regen den Harnfluß an, Natrium nur in geringem Maße, hohe Calcium-Konzentrationen setzen dagegen den Harnfluß herab [62, 63]. Alle Stoffe sind im Lumen isolierter Gefäße in geringerer Konzentration vorhanden als im Blut, außer Wasser, Natrium, Kalium und Phosphat. Die Kalium-Konzentration des Mediums bestimmt die Intensität des Harntransportes ins Lumen. Die Menge der aktiv und passiv übertretenden Stoffe ist von ihrer Konzentration im Medium abhängig [62]. Wird dessen Ionenzusammensetzung stark verändert, arbeiten isolierte Gefäße normal weiter [63].

Auch organische Substanzen, wie Aminosäuren, Zucker, Polysaccharide und Harnstoff gelangen passiv ins Gefäßlumen [64, 65]. Je größer ihr Molekulargewicht ist, um so langsamer erfolgt der Übertritt, der *nach* Ramsay *ein reiner Diffusionsvorgang sein und unter dem Einfluß eines Konzentrationsgradienten ablaufen soll.* Außer Harnsäure [61] und bestimmten Farbstoffen (s. auch [36]) ist die Konzentration aller getesteten Verbindungen im Primärharn geringer. Stoffwechselphysiologisch wichtige Substanzen gelangen demnach ins Lumen und werden vom Enddarm und Rektum reabsorbiert. Die $\frac{C_T}{C_H}$-Rate ist auch hier unabhängig von der Konzentration im Medium, d. h. je höher die Konzentration eines Stoffes im Medium wird, um so größer ist die Menge des ins Gefäßlumen übertretenden Stoffes. Dabei ließen sich aber keine Schwellenwerte ermitteln. Für jede Aminosäure wurde eine charakteristische Rate ermittelt. Nur Harnstoff kann den Harnfluß fördern.

Im proximalen Teil isolierter Gefäße können Rohrzucker und Harnstoff auch entlang einem Konzentrationsgefälle aus dem Lumen ins Blut übertreten, nach Ramsay ein weiterer Beweis dafür, daß der umgekehrte Transport vom Blut zum Lumen ein reiner Diffusionsvorgang ist. — Aus den besprochenen Ergebnissen kann man schließen, daß die Malpighi-Gefäße *in vivo* nur Kalium, Harnsäure und bestimmte Farbstoffe konzentrieren, daß aber die M.T. von *Carausius in vitro* darüber

hinaus noch Wasser, Natrium und Phosphat gegen einen Konzentrationsgradienten aktiv zum Lumen befördern können.

Nach Untersuchungen von PATTEN und CRAIG an *Tenebrio molitor*
[49] wird eine hypertonische Lösung von den Gefäßen schneller aufgenommen als eine isotonische. Aminosäuren und Harnstoff wandern aus
dem Medium sofort in das Lumen, Proteine (Gelatine) dagegen nicht.
Wasser, Salze oder Aminosäuren werden nicht selektiv zum Lumen geleitet. Druckerhöhung führt nicht zu einem verstärkten Harnfluß. Bei
Periplaneta ist die Exkretion temperaturabhängig [50].

MADDRELL [37] hat kürzlich nachgewiesen, daß die Harnbildung
einer hormonalen Steuerung unterliegt: Bei *Rhodnius* wird die Diurese
nämlich durch ein Hormon der Hämolymphe stimuliert, das durch ein
Neurosekret der Ganglienzellen des Mesothorax aktiviert wird. KOLLER
[32] hatte schon 1948 nachgewiesen, daß sich die Bewegungen der
Malpighi-Gefäße durch Hirnextrakte beschleunigen lassen.

Bei der Funktion der M.T. könnten offenbar folgende Faktoren eine
Rolle spielen (s. auch [77]).

1. Der osmotische Druck der Hämolymphe.

2. Das Volumen der Hämolymphe.

3. Die Konzentration von anorganischen und organischen Stoffen in
der Hämolymphe (dabei handelt es sich meist um Stoffwechselprodukte).

4. Die Anwesenheit von schädlichen oder nutzlosen Fremdstoffen
in der Hämolymphe.

5. Die Anwesenheit eines Hormons.

Der Übertritt von Stoffen in das Lumen kann entweder über die
ganze Länge des Gefäßes ausgedehnt oder nur auf den distalen oder
mittleren Abschnitt der Gefäße beschränkt sein. Die Reabsorption von
wertvollen Stoffen aus dem abfließenden Primärharn findet entweder
im proximalen Abschnitt der M. T. oder im Enddarm und Rektum statt.
Die Konsistenz des in den Enddarm eintretenden Sekundärharnes und
sein Gehalt an wertvollen Stoffen sind daher ganz verschieden [36,
48, 49, 56—64, 76, 96, 97].

D. Besondere Aufgaben der Malpighischen Gefäße
(siehe [98])

Neben der Exkretion übernehmen die M.T. bei einigen Tierarten
auch noch andere Funktionen. So können sie oft Substanzen speichern,
die im Stoffwechsel eine wichtige Rolle spielen. Neben der schon häufig
beschriebenen Akkumulation von Riboflavin ([19, 43, 44] u. a.) speichern
das Übergangs- und das Hauptstück der M.T. von *Drosophila* Pterine,
Tryptophan und dessen Derivate, die alle für die Bildung der Augenpigmente gebraucht werden [84, 85, 92, 93]. Außerdem spricht die
starke Anreicherung des Cytoplasmas mit Ribonucleoproteiden nach
der Verpuppung für eine Beteiligung der M.T. am Aufbau des imaginalen
Körpers [89, 90].

Außerdem können die Gefäße anderer Insektenarten ein Klebsekret
zur Fortbewegung, Sekrete zur Kokonbildung oder Seidenproduktion

oder Calciumcarbonat zur Verhärtung der Cuticula oder der Eischalen liefern.

E. Schlußwort

Die Nierentubuli der Insekten sind zu vergleichen mit den bei bestimmten Teleostiern vorkommenden aglomerulären Nieren [*15, 42, 64*], da bisher kein Nachweis für eine Filtration des Harnes in das Lumen erbracht werden konnte. Als Mechanismen der Harnbildung kommen *nur Sekretion und Reabsorption* in Betracht.

Die *Sekretion* des Harnmaterials in das Lumen läuft wahrscheinlich hauptsächlich über einen Vesikelfluß („Cytopempsis") ab, wobei die beförderten Stoffe in getrennten Vesikeln oder verschiedenen Gefäßabschnitten transportiert werden können. Der Harn wird entweder nur in einem kleinen, meist distalen Gefäßabschnitt zum Lumen geschafft, oder alle Zellen des Tubulus sind gleichmäßig daran beteiligt.

Die *Reabsorption* der für den Stoffwechsel wertvollen Substanzen läuft entweder noch im proximalen Abschnitt des Tubulus an, oder aber der Enddarm und seine Rektaldrüsen sind dafür verantwortlich.

Bei der Sekretion werden alle Stoffe der Hämolymphe zum Lumen geschafft. Vermutlich hält aber die Basalmembran als Ultrafilter blutspezifische, großmolekulare Proteine zurück. Hämoglobin, Metallproteine, Polysaccharide und deren Abbauprodukte können sie im Experiment aber noch passieren.

Die Untersuchungen an Drosophila ergaben, daß die basale Zellmembran offensichtlich ein Wahlvermögen für die angebotenen Stoffe der Hämolymphe besitzt und demnach bereits an der Zellbasis eine stoffliche Segregation des vesikulär transportierten Materials eintritt. Somit wäre ein aktiver Transport bestimmter Stoffe ins Lumen durch einen Vesikelfluß, bei dem das bewegte Material während seiner transzellulären Passage mehr oder weniger stark entwässert wird, leicht *möglich.* Wenn aber die transportierten Stoffe dabei nicht entwässert werden, *kann ein derartiger Mechanismus eine Diffusion vortäuschen. Die Cytopempsis ist demnach ein Transportmechanismus, der sich ausgezeichnet für Stoffbewegungen gegen ein oder auch entlang einem Konzentrationsgefälle eignet.* Beide Möglichkeiten können in den Malpighischen Gefäßen vorkommen.

Daß neben dem Vesikelfluß noch eine andere Transportart vorkommt, an der das Grundcytoplasma und das endoplasmatische Retikulum beteiligt sind, geht aus den Befunden an *Drosophila* deutlich hervor.

Die exkretorisch-tätigen Zellen der Insektennieren besitzen alle den gleichen Aufbau (basales Labyrinth und Bürstensaum). Aus dieser Zellarchitektur kann man lediglich auf einen polaren Stofftransport schließen, nicht aber auf die Richtung der Stoffpassage. Solche Zellen sind in der Lage, Substanzen sowohl zum Blut als auch zum Gefäßlumen zu schaffen, wobei entgegengesetzte Transporte über die gleichen Funktionsstrukturen laufen können. Bei Zellen, die vorwiegend reabsorbierend tätig sind, besitzen allerdings die basalen Zellmembraneinfaltungen und der Bürstensaum eine relativ große Oberfläche.

Die so zahlreich nachgewiesenen akzessorischen Funktionen der Malpighi-Gefäße lassen erkennen, daß die Insektennieren oft in mannigfacher Art an Stoffwechselvorgängen beteiligt sein können, die zur Exkretion wenig Beziehung haben.

Diese Untersuchungen wurden mit Unterstützung der Deutschen Forschungsgemeinschaft durchgeführt.

Literatur

[1] ALLEGRET, P., et. M. A. MAGUER: C. R. Acad. Sci. (Paris) **256**, 3507 (1963).
[2] AMOURIQ, L.: Bull. Soc. Zool. France **85**, 21 (1960).
[3] AUCLAIR, J. L.: J. Ins. Physiol. **3**, 57 (1959).
[4] BACCETTI, B.: 11. Int. Kongr. Entom., Wien 1960, Verh. **1**, 752 (1961).
[5] —, V. MAZZI e. G. MASSIMELLO: Redia **48**, 47 (1963).
[6] BEAMS, H. W., T. N. TAHMISIAN, and R. L. DEVINE: J. biophys. biochem. Cytol. **1**, 197 (1955).
[7] BERKALOFF, A.: C. R. Acad. Sci. (Paris) **246**, 2807 (1958).
[8] — C. R. Acad. Sci. (Paris) **248**, 466 (1959a).
[9] — C. R. Acad. Sci. (Paris) **249**, 1934 (1959b).
[10] — C. R. Acad. Sci. (Paris) **249**, 2120 (1959c).
[11] — C. R. Acad. Sci. (Paris) **250**, 2609 (1960a).
[12] — C. R. Acad. Sci. (Paris) **250**, 2061 (1960b).
[13] — Ann. Sci. natur. Zool. (Paris), Sér. 12, **2**, 871 (1960c).
[14] BOCHE, R. D., and J. B. BUCK: Physiol. Zool. **15**, 293 (1942).
[15] BONÉ, G. J., et. H. J. KOCH: Ann. Soc. roy. Zool. Belg. **73**, 73 (1942).
[16] BRADFIELD, J. R. G.: Quart. J. micr. Sci. **94**, 351 (1953).
[17] BROWN, A. W.: J. exp. Biol. **14**, 87 (1937).
[18] —, and L. FARBER: Biochem. J. **30**, 1107 (1936).
[19] BUSNEL, R. G., et A. DRILHON: C. R. Soc. Biol. (Paris) **137**, 752 (1943).
[20] CRAIG, R.: Ann. Rev. Ent. **5**, 53 (1960).
[21] CUÉNOT, L.: Arch. Biol. **14**, 293 (1895).
[22] DAY, M. F.: Austr. J. Sci. Res., B, **2**, 31 (1949a).
[23] —, and M. BRIGGS: J. Ultrastruct. Res. **2**, 239 (1958).
[24] DRILHON, A., et R.-G. BUSNEL: Bull. Soc. Chim. biol. (Paris) **27**, 415 (1945).
[25] EASTHAM, M. S.: Quart. J. micr. Sci. **69**, 385 (1925).
[26] GAGNEPAIN, J.: C. R. Acad. Sci. (Paris) **242**, 2777 (1956).
[27] — Bull. Soc. Zool. France **81**, 395 (1957).
[28] GORKA, A.: Zool. Jb. Abt. Allg. Zool. Physiol. **34**, 233 (1914).
[29] GUPTA, P. D., and R. N. SINHA: Ann. Entom. Soc. Amer. **53**, 632 (1960).
[30] HEYMONS, R., u. M. LÜHMANN: Zool. Anz. **102**, 78 (1933).
[31] KILBY, B. A., and E. NEVILLE: J. exp. Biol. **34**, 276 (1957).
[32] KOLLER, G.: Biol. Zbl. **67**, 201 (1948).
[33] KOWALEWSKY, A.: Congr. Intern. Zool. Moskau, **1**, 187 (1892).
[34] LISON, L.: Bull. Acad. roy. Belg. Cl. Sci (5), **23**, 317 (1937).
[35] — Ann. Soc. roy. Zool. Belg. **69**, 195 (1938).
[36] — Mém. Acad. roy. Belg. Cl. Sci. **19**, 1 (1942).
[37] MADDRELL, S. H. P.: Nature (Lond.) **194**, 605 (1962).
[38] MARTOJA, R.: Acta histochem. **6**, 185 (1959).
[39] MAZZI, V., e. B. BACCETTI: Redia **41**, 315 (1956).
[40] — — Redia **42**, 277 (1957a).
[41] — — Redia **42**, 383 (1957b).
[42] — — Z. Zellforsch. **59**, 47 (1963).
[43] METCALF, R. L.: Arch. Biochem. **2**, 55 (1943a).
[44] — Fluorescence-microscopic Studies of the Physiology and Biochemistry of the Malpighian System of Periplaneta americana (L.). Cornell Univ. Libr. Ithaca, N. Y. 401 (1943b).

[45] Meyer, G. F.: Z. Zellforsch. **47**, 18 (1957).
[46] Moore, D. H., and H. Ruska: J. biophys. biochem. Cytol. **3**, 457 (1957).
[47] Patay, R.: C. R. Soc. Biol. (Paris) **129**, 1098 (1938).
[48] Patten, R. L.: In: Insect Physiology (Ed.: K. D. Roeder). New York: J. Wiley & Sons, Inc. London: Chapman & Hall, Ltd. 387 ,1953.
[49] —, and R. Craig: J. exp. Zool. **81**, 437 (1939).
[50] —, J. Gardner, and A. D. Anderson: J. Ins. Physiol. **3**, 256 (1959).
[51] Phillips, J. E.: Rectal absorption of water and salts in the locust and blowfly. Ph. D. Thesis, Univ. Cambridge 1961.
[52] Poll, M.: Bull. Ann. Soc. Ent. Belg. **72**, 103 (1932).
[53] — Rec. Inst. Zool. Torley-Rosseau **5**, 73 (1934).
[54] — Mém. Mus. roy. Hist. Nat. Belg. (2), **3**, 635 (1936).
[55] — Ann. Soc. roy. Zool. Belg. **69**, 9 (1938).
[56] Ramsay, J. A.: J. exp. Biol. **28**, 62 (1951).
[57] — J. exp. Biol. **29**, 110 (1952).
[58] — J. exp. Biol. **30**, 79 (1953a).
[59] — J. exp. Biol. **30**, 358 (1953b).
[60] — J. exp. Biol. **31**, 104 (1954).
[61] — J. exp. Biol. **32**, 183 (1955a).
[62] — J. exp. Biol. **32**, 200 (1955b).
[63] — J. exp. Biol. **33**, 697 (1956).
[64] — J. exp. Biol. **35**, 871 (1958).
[65] —, and J. A. Riegel: Nature (Lond.) **191**, 1115 (1961).
[66] Ritzki, T. M.: J. cell Biol. **16**, 513 (1963).
[67] —, and R. M. Ritzki: J. cell Biol. **12**, 149 (1962).
[68] — — J. cell Biol. **17**, 87 (1963).
[69] — — J. cell Biol. **21**, 27 (1964).
[70] Sacktor, B., G. M. Thomas, J. C. Moser, and D. J. Block: Biol. Bull. **105**, 166 (1953).
[71] Savage, A. A.: Quart. J. micr. Sci. **103**, 417 (1962).
[72] Schlottke, E.: Z. vergl. Physiol. **24**, 210 (1937).
[73] Shaw, J., and R. H. Stobbart: Adv. Insect Physiol. **1**, 315 (1963).
[74] Smith, D. S., and V. C. Littau: J. biophys. biochem. Cytol. **8**, 103 (1960).
[75] Spiegler, D. E.: J. Ins. Physiol. **8**, 127 (1962).
[76] Srivastava, P. N.: J. Ins. Physiol. **8**, 223 (1962).
[77] —, and P. D. Gupta: J. Ins. Physiol. **6**, 163 (1961).
[78] Stadhouders, A. M., u. J. A. M. Jacobs: Naturwissenschaften **48**, 509 (1961).
[79] Stay, B.: J. morphol. Philadelphia **105**, 457 (1959).
[80] Tsubo, I. J.: Nara med. Ass. **12**, 301 (1961).
[81] —, and. P. W. Brandt: J. Ultrastr. Res. **6**, 28 (1962).
[82] Urich, K.: Ergebn. Biol. **24**, 155 (1961).
[83] Waterhouse, D. F., and M. F. Day: In: Insect Physiology (Ed.: K. D. Roeder) New York: J. Wiley & Sons; London: Chapman & Hall, Ltd. 331 (1953).
[84] Wessing, A.: Verh. dtsch. Zool. Ges., Saarbrücken 1961, Zool. Anz., Suppl. **23**, 166 (1962a).
[85] — Protoplasma **55**, 264 (1962b).
[86] — Protoplasma **55**, 294 (1962c).
[87] — Protoplasma **56**, 433 (1963).
[88] — Verh. dtsch. Zool. Ges., München 1963, Zool. Anz., Suppl. **27**, 549 (1964a).
[89] — Z. Zellforsch. (im Druck).
[90] — Verh. dtsch. Zool. Ges., Kiel 1964, Zool. Anz., Suppl. **28** (im Druck).
[91] — Unveröffentlicht.
[92] —, u. R. Danneel: Z. Naturforsch. **16**b, 387 (1961).
[93] —, u. A. Bonse: Z. Naturforsch. **17**b, 620 (1962).
[94] Wigglesworth, V. B.: J. exp. Biol. **8**, 411 (1931a).

[95] Wigglesworth, V. B.: J. exp. Biol. **8**, 428 (1931b).
[96] — J. exp. Biol. **8**, 443 (1931c).
[97] —- Insect Physiology. London: Methuen & Co. 1934.
[98] — Physiologie der Insekten. Basel-Stuttgart: Birkhäuser 1959.
[99] —, and M. M. Salpeter: J. Ins. Physiol. **8**, 299 (1962).
[100] Wood, R. L.: J. biophys. biochem. Cytol. **6**, 343 (1959).

Summary

The malpighian tubules of insects are tube-like excretory organs (Fig. 1) which clear pernicious and useless substances from the blood and divert the excreta into the hind-gut. Their function not only depends upon ecological factors but in addition has changed during the course of phylogeny.

As in the nephron of aglomerular kidneys, the primary urine reaches the tubule lumen only by secretion and diffusion and not by filtration. The primary urine enters the tubule lumen over its whole length or else only in certain segments. In this process, so much water can be absorbed from the urine-forming material that it crystallizes out within the cells. The reabsorption of useful substances from the lumen takes place either in the proximal section of the tubule or in the hind-gut.

Just as in the nephron of the vertebrate kidney, all cells have a typically polar structure with deep basal infoldings or cell interdigitations and an apical brush border.

The cells of the enlarged initial segment of the anterior malpighian tubules in larvae and pupae of *Drosophila melanogaster* transport urine from the blood into the lumen. Fed and injected indicator substances for electron microscopy ($FeCl_3$, 3-hydroxykynurenine, silver protein, haemoglobin) are transported to the tubule lumen by "cytopempsis" (vesicular transport; Figs. 5, 6, 8b). During intracellular passage water is absorbed from the contents of the vesicles so that they approach the osmotic conditions of the concentrated primary urine in the lumen. In the next step the intracellular water is transported back to the blood (Fig. 3). Thus, in this case cytopempsis connects two liquid phases with different osmotic pressures.

Injected iron-dextran ("Myofer") is not transported in vesicles but is taken up in the ground cytoplasm, collected in the endoplasmic reticulum, and brought to the tubule lumen (Fig. 7, 8a). The channels of the endoplasmic reticulum frequently join the polar cell membranes (Fig. 4a) and obviously have a transport function.

The proximal segments following the enlarged initial segment of the tubule reabsorb useful substances from the lumen (Figs. 2a, 2b, 9, 11, 12). These substances are partly accumulated intracellularly in the endoplasmic reticulum and ground cytoplasm (tryptophane, kynurenine, 3-hydroxykynurenine, fat, drosopterin, riboflavin, etc.). The cells are large and have long microvilli and basal infoldings penetrating deeply into the cell. Mitochondria occur only in the polar regions, where they evidently supply energy for reabsorption (Figs. 11, 12a). Cell architecture changes during decrease of cell activity (e. g., after pupation or as results of cold or hunger). Under these circumstances the cells become smaller, the basal infoldings and brush border disappear, and the mitochondria migrate to the center of the cell (Fig. 12b). Nuclear as well as cytoplasmic structures change. The reabsorptive capacity of these cells is experimentally reduced.

Experimentally, the whole reabsorbing proximal segment of the tubule can transport foreign substances injected into the blood (Myofer, gold chloride) by vesicles (cytopempsis) through the ground cytoplasm and endoplasmic reticulum into the lumen. As in the distal segment, water is also reabsorbed from the transported substances here (Fig. 13, 14).

The numerous enzymes demonstrated in the malpighian tubules of various insects indicate that these organs have an active metabolism.

The tubules of many insects actively secrete potassium into the lumen. Ramsay considers this a fundamental process in the formation of insect urine. All other substances are said to move solely by diffusion resulting from concentration gradients. This latter finding is invalid for *Drosophila*, where the primary urine reaches the lumen by an active process, the cytopempsis.

In addition to their excretory function, malpighian tubules may also participate in other physiological metabolic processes.

Diskussion

Komnick: Vielleicht darf ich zunächst eine rein strukturelle Frage anschneiden: Ursprünglich glaubte man, daß das basale Labyrinth durch Einfaltungen der basalen Zellmembran entsteht. Bei verschiedenen Vertebratenzellen mit Transportfunktion hat sich inzwischen immer mehr herausgestellt, daß es ein System von Interzellularen ist. Ich habe bisher geglaubt, die Malpighischen Gefäße von *Drosophila* seien eines der wenigen Beispiele, wo es sich wirklich um Einfaltungen handelt. Leider bestärkt mich nun ein Bild, das Sie heute gezeigt haben, darin, daß es wohl doch Interzellularen sind.

Wessing: Das ganze Gefäß wird in der proximalen Region nur durch zwei Zellen zusammengesetzt. Die Zellen sind hochpolyploid, die Kerne besitzen Polytänchromosomen. Wenn das basale Labyrinth bei *Drosophila* aus Interzellularen hervorgegangen wäre, dann müßte die „Verzahnung" über ungeheuer große Areale verlaufen, und das ist kaum vorstellbar. Man kann zwar in dieser Region keine Funktionsunterschiede zwischen benachbarten Zellen feststellen, aber im distalen Abschnitt kann man das, weil dort der besprochene Funktionszyklus abläuft. Das hier demonstrierte Bild aus dieser Gefäßregion, auf dem zwei Zellen mit ganz verschiedenen Funktionszuständen aneinanderstoßen, die eine Zelle ist wasserreich, die andere wasserarm, beweist, daß im Malpighi-Gefäß von *Drosophila* echte Zellmembraneinfaltungen vorkommen. Die „Verzahnung" der Zellen ist dagegen nur unbedeutend.

Kuyper: Ich möchte nur ganz kurz über Arbeiten berichten, die Stadhouders in Nijmegen über Malpighische Gefäße angestellt hat. Er hat gearbeitet mit *Periplaneta americana*, d. h. einer Orthoptere, die ganz niedrig im System der Insekten steht. Die großen Unterschiede, die ich jetzt aufzeigen werde, werden also vielleicht durch die Stellung im System verursacht.

Bei *Periplaneta* gibt es eine große Zahl Malpighischer Gefäße, die während des Lebens immer mehr zunehmen. Bei älteren Tieren haben wir sogar einige hundert solcher Gefäße. Man kann an den einzelnen Gefäßen zwei Abschnitte unterscheiden, nämlich einen distalen Teil und einen proximalen Teil. Ein Übergangsstück, wie es Herr Wessing beschrieben hat, gibt es bei *Periplaneta* nicht. Eine weitere Besonderheit bei *Periplaneta* ist, daß im Lumen der Malpighischen Gefäße nie Exkremente gefunden werden, weder lichtmikroskopisch noch elektronenmikroskopisch. Das heißt, die Malpighischen Gefäße haben im distalen Teil eine ganz andere Funktion. Das ist eine Speicherniere. Alles, was elektronenmikroskopisch wahrnehmbar aus der Leibeshöhle aufgenommen wird, wird im distalen Teil angehäuft und geht überhaupt nicht in das Lumen hinein, vielleicht nur ganz kleine Moleküle oder Ionen. In den distalen Zellen findet man eine große Menge von konzentrischen Gebilden, die von Berkaloff als Urate gedeutet worden sind. Es ist aber weder mikrochemisch noch histochemisch Urat in den Malpighischen Gefäßen von *Periplaneta* darzustellen. Bewiesen ist, daß es sich um Karbonat und Phosphat handelt. Die ersten Kristalle entstehen in Erweiterungen des endoplasmatischen Retikulums und wachsen dann allmählich an zu schönen konzentrischen Gebilden, die alle im Zellkörper bleiben. Für einen Vesikeltransport im distalen Teil gibt es keine elektronenmikroskopischen Anzeichen. Im proximalen Teil kann man eine Reihe von Vesikeln finden. Aber Stadhouders interpretiert das so, daß es nur einen Vesikeltransport vom Lumen aus gibt. Das ist also eine Reabsorption. Er findet in den Mikrovilli eine ganze Menge von kleinen Vesikeln, die immer weiter zur Basis der Zelle hinwandern und dann in "multi-vesiculate bodies" aufgehen, die eine positive saure Phosphatasereaktion geben, also zu deuten sind als Cytolysosomen. Es gibt auch bei *Periplaneta* dunkle und helle Zellen. Aber es gibt keine Anzeichen dafür, daß es hier um typische Veränderungen im Sinne eines Funktionszyklus gehen sollte, sondern die dunklen Zellen sind nur zu interpretieren als junge Zellen, die auswachsen, bis sie hell sind, und dann aktiv am Prozeß teilnehmen. Bei *Periplaneta* sind Struktur und der funktionelle Prozeß völlig anders. Obwohl selbstverständlich *Periplaneta* und *Drosophila* gar nicht zu vergleichen sind, weder in ökologischer noch in anatomischer Hinsicht, möchte ich nur darauf hinweisen, daß die Malpighischen Gefäße bei verschiedenen Tierarten so verschieden sein können, daß man mit der Interpretation an nur einer Tierart ganz vorsichtig sein muß und nicht verallgemeinern darf.

Füller: Es gibt eine ganze Reihe von Arbeiten, die sich mit den Malpighischen Gefäßen der Insekten beschäftigen. Aber es gibt ja auch andere Tiergruppen, die ebenfalls Malpighische Gefäße besitzen. Von dem Gesichtspunkt ausgehend, habe ich vor einem halben Jahr begonnen, Malpighische Gefäße primitiver tracheater Arthropoden, nämlich von Chilopoden, zu untersuchen. Da sehen die Verhältnisse wieder etwas anders aus. Sie knüpfen aber in gewissem Sinne an die Verhältnisse bei *Periplaneta* an. Die Chilopoden besitzen nur ein Paar Malpighische Gefäße, die in der Grobstruktur relativ undifferenziert sind, eigentlich gar nicht in mehrere Abschnitte differenziert sind. In der gesamten Länge der Malpighischen Gefäße ist ein Bürstensaum ausgebildet, allerdings in etwas unterschiedlicher Weise insofern, als im distalen Teil die Mikrovilli in etwas geringerer Anzahl vorhanden sind als im Hauptabschnitt und im proximalen Teil. Ein grundsätzlicher Unterschied gegenüber allen Befunden bei Insekten liegt darin, daß die Mitochondrien niemals in den Mikrovilli liegen. Sie sind wesentlich größer im Durchmesser als die Mikrovilli. Es ließ sich weiterhin — zunächst wenigstens annähernd — feststellen, daß die Mitochondrien in den einzelnen Abschnitten eine etwas unterschiedliche Verteilung zeigen, und zwar im distalen Abschnitt mehr nach dem Lumen hin gelagert unter den Mikrovilli, im proximalen Abschnitt mehr im Bereich der basalen Einfaltungen.

Wessing: Zunächst ist zu sagen, daß die Längsdifferenzierung der Malpighischen Gefäße innerhalb des gesamten Insektenreichs sehr unterschiedlich ist. Je höher die systematische Stellung, desto größer auch die Differenzierung. Wir finden bei den Heteropteren vier verschiedene Regionen, von den Hymenopteren liegen noch keine Untersuchungsergebnisse vor. Bei den Dipteren, die im System auch relativ hochstehen, finden wir ebenfalls vier, manchmal sogar fünf Regionen. Bei den Orthopteren liegen die Verhältnisse so, daß das gesamte Gefäß offenbar Sekretionspotenzen hat. Also das gesamte Gefäß liefert Harn zum Lumen, sowohl distal als auch proximal. Das hat auch BERKALOFF gezeigt — bis auf den distalen Abschnitt, den er noch nicht genau bearbeitet hat. Während der Durchschleusung des Harns wird bereits das Material intrazellular entweder in Vesikeln oder — wie die Bilder von Herrn STADHOUDERS beweisen — innerhalb des endoplasmatischen Retikulums auskristallisiert.

Thoenes: Der mitochondrienreiche Bürstensaum, den Sie zeigen, scheint mir eine bemerkenswerte Besonderheit der Epithelschicht der Malpighischen Gefäße zu sein. Wenn man die Bilder ansieht, hat man den Eindruck, daß sogar die Mehrzahl der Mitochondrien in den Mikrovilli sitzt und daß der Basisteil der Zelle ärmer an Mitochondrien ist. Das ist an sich eine umgekehrte Situation wie in dem proximalen Tubulusepithel der Säugerniere. Ich möchte Sie fragen, ob Sie sich das erklären können, vielleicht damit, daß im Lumen des Malpighi-Gefäßes eine sehr hohe Konzentration herrscht im Gegensatz zum proximalen Nierentubulus.

Wessing: Das Gefäß muß ja in der Region, wo reabsorbiert wird, sehr intensiv arbeiten, wenn es Stoffe aus dem Lumen zum Blut zurücktransportieren will, weil nämlich die Stoffe im Lumen teilweise schon in kristalliner Form vorliegen und hier gegen einen hohen osmotischen Druck gearbeitet werden muß. Dafür wird sehr viel Energie gebraucht und durch die Mitochondrien bereitgestellt.

Bücher: Mir ist die große Zahl der Ribosomen aufgefallen. Haben Sie das Gefühl, daß diese große Zahl der Ribosomen etwas mit der Funktion dieses Gefäßes zu tun hat?

Wessing: Was die Ribosomen innerhalb der Larvenzelle zu bedeuten haben, ist noch nicht bekannt. Hier sind sie vornehmlich an den Kanälen des endoplasmatischen Retikulums gelagert. Das Grundcytoplasma ist im Larvenzustand fast frei von Ribosomen. Dagegen wird fünf Stunden nach der Verpuppung das Cytoplasma geradezu überschwemmt mit Ribosomen. Während der ganzen Puppenzeit bleibt dieser Zustand bestehen. Wir haben zunächst einmal nach der Quelle der Ribosomen gesucht und die Untersuchung gerade abgeschlossen. Es zeigte sich, daß sich während dieser Zeit periphere Anteile des Nukleolus lösen, deren RNS-Grana durch die Poren der Kernmembran in das Cytoplasma ausströmen und das Cytoplasma sehr schnell mit Ribosomen versorgen. Wir können auf Grund dessen sagen, daß der Nukleolus eine Quelle der Ribosomen ist. Die Bedeutung dieser plötzlichen Überschwemmung des Cytoplasmas mit Ribosomen ist wahrscheinlich die, daß die Malpighischen Gefäße Anteil an der Synthese des Imaginalkörpers haben, weil sie nämlich bei Dipteren als eines der wenigen Organe die Puppenzeit überleben, fast alle anderen Organe fallen der Histolyse anheim. Es bleiben nur die Imaginalscheiben bestehen. Von da geht das Wachstum des imaginalen Körpers aus.

Physiology of salt glands

By

KNUT SCHMIDT-NIELSEN[1], Durham

With 2 Figures

Introduction

It aroused considerable interest when some six years ago it was discovered that the major route of electrolyte excretion in marine birds is not the kidney but glands in the head, known as the nasal glands. Similar glands are found in marine reptiles, but since the anatomical origin is orbital in some reptiles while it is nasal in birds, we have found it desirable to stress the functional similarity of these glands by the use of terms such as "salt secreting glands", or for simplicity, "salt glands". The subject was reviewed in 1960 [*56*] and the present review therefore deals mostly with developments since that time.

It will be observed that, in spite of several notable contributions, our understanding of the function of this gland is still incomplete and many major problems remain unsolved. The group at Duke University, where the discovery of extrarenal excretion was made, has since turned its attention to other fields and has made only minor and, in principle, insignificant contributions. The ultrastructure of the gland has been the subject of excellent studies in several laboratories; yet a clear understanding of the relation of function to structure is still lacking. Studies of cell metabolism in tissue slices are essentially uninformative, although challenging. Biochemical studies are most significant in the area of measurements of oxidation and phosphorylation rates of mitochondria in relation to the efficiency of sodium transport. Endocrine effects on the gland have been investigated, the results are interesting, but they have not yet definitely settled the relation between nervous control and endocrine influences on salt gland secretion.

The present review will deal with 1) general remarks on the function of the gland, 2) gland structure, 3) secretory function, and 4) tissue studies, cells, and cell components. Finally, a few remarks will be made on the osmotic regulation in each major vertebrate class, and special mention will be made of recent discoveries of secretory organs whose function was previously unknown.

---

[1] NIH Research Career Award 1-K6-GM-21,522. Work originating in the laboratory of the author was supported by NIH Grant HE-02228.

General remarks

The general characteristics of the nasal salt secreting gland of marine birds were reviewed in 1960 [56]. The conspicuously large paired gland is present in all marine birds so far investigated. In some birds the gland is located in the nasal region (e.g. pelicans), but in many it has been displaced to the supraorbital region of the skull (e.g. gulls) while the ducts

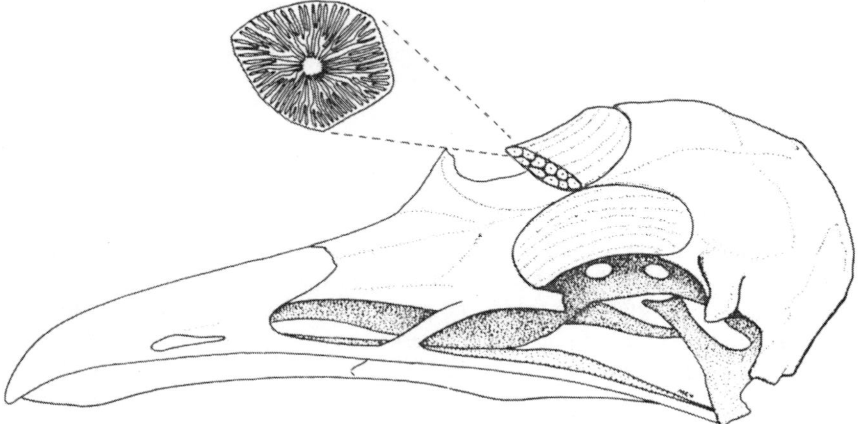

Fig. 1. The nasal salt-secreting gland of the gull (*Larus sp.*) *in situ*. In the gull and in many other (but not all) marine birds the paired nasal gland is located on top of the skull, above the orbit of the eye, in shallow depressions in the bone. Irrespective of the location of the gland, whether nasal, pre-orbital or supra-orbital, the secretion is discharged through ducts which open in the anterior part of the nasal cavity

still open in the nasal cavity. The general anatomy of the gland of the gull is visible from Fig. 1. In contrast to the kidney, the salt gland secretes only intermittently, the normal stimulus apparently being an excessive intake of sodium chloride. However, secretion seems to be induced by osmotic conditions rather than by an increase in sodium or chloride concentrations, for secretion can be induced by infusion of hypertonic sucrose. In the absence of osmotic stimulation no secretion takes place.

The secreted fluid always has a high concentration of sodium and chloride, a moderate concentration of potassium and bicarbonate, while other ions and organic material are practically absent (see Table 1). For any given species, the concentration is rather constant, ranging from about 500—600 mEq/l for the fish-eating cormorant to about 1000 to 1100 mEq/l for the oceanic plankton-eating petrel [56]. Further reports for a large

Table 1. *Typical composition of the fluid from the salt secreting gland of herring gulls (Larus argentatus). All concentrations in mEq/liter*

Cations		Anions	
Na^+	718	Cl^-	720
K^+	24	HCO_3^-	13
$Ca^{++} + Mg^{++}$	2	SO_4^{--}	0.68
Total	744	Total	734

pH = 7.03

number of species were given by McFARLAND [47], and several other investigators have reported on a smaller number of species. In each case the concentration of the secretion has been above that of sea water (about 500 mEq/l) and up to about twice this value.

Rate of secretion

The salt gland produces a fluid which not only has a surprisingly high salt concentration, comparable to mammalian urine in antidiuresis, but it also produces surprisingly large volumes of this fluid, making the total

Table 2. *Comparison of the rates of secretion from the salt gland of the herring gull and the human kidney in maximum water diuresis*

	Rate of secretion	
	per kg body weight	per gram gland
Kidney of man, water diuresis	0.24 ml/min	0.03 ml/min
Salt gland, herring gull	0.5 ml/min	0.6 ml/min

osmotic performance of the gland astonishingly high. If the rate of secretion is compared with the volumes of urine produced by the kidney of man, it appears that the relative rate of flow from the salt gland is some 20 times as high as that of the human kidney in maximum water diuresis (see Table 2).

The ability to secrete is present in the chick, and in at least some birds when the hatchling emerges from the egg. This has been demonstrated in the Adelie penguin, *Pygoscelis adeliae*, by DOUGLAS [12]. In the domestic duckling *(Anas platyrhynchos)* secretion from the salt glands began at three days' age after the ducklings were given sodium chloride to drink. For the following 10 days the sodium and potassium content of the secretion showed no consistent tendency either to increase or decrease in concentration, although it varied widely [24].

Although growing birds can tolerate sodium chloride loads and secrete from the salt gland, salt loads seem to retard the general growth (increase in body weight) of the bird. This has been found in glaucous-winged gulls *(Larus glaucescens)* [28] and on domestic ducklings [61]. The mechanism of growth retardation has not been clarified, but it seems to me that investigations involving the role of adrenal corticoid hormones may prove fruitful.

Repeated salt loads and secretory activity induce hypertrophy of the gland. This had been reported by HEINROTH [26] and by SCHILD-MACHER [55], although these authors were unaware of the role of the salt gland in excretion. At that time it was believed that the function of the gland was to produce a secretion which could protect the sensitive nasal membranes against the irritating effects of salt water. Hypertrophy of the gland has more recently been described and quantitative information has been given for the glaucous-winged gull [28] and for domestic ducklings [61, 18, 19].

Surgical removal of the salt glands has been carried out in Schmidt-Nielsen's laboratory by Maryanne Robinson Hughes. The interesting finding was that the gulls *(Larus argentatus)* did not become as sensitive to sodium chloride loads as would be expected in the absence of extrarenal salt excretion. However, in unpublished experiments it was found that some secretion still appeared from the region of the upper beak, its origin was at least in part from the supramaxillary gland, and the concentration was considerably higher than the plasma concentration, although not as high as that from the salt gland. Surgical removal has also been tried by Schwarz who found that mew gulls *(Larus canus)* survived extirpation well when released into their natural habitat [63, 64]. In laboratory studies he observed that after removal of the salt glands sodium chloride loads would still induce the secretion of a highly concentrated fluid which would appear at the external nares. These experiments were done with three species of gulls, *Larus argentatus, Larus canus,* and *Larus ridibundus.* Unfortunately, these experiments and those by Hughes are difficult to interpret because one lacks assurance that there was complete removal of all salt gland tissue, some regeneration might have taken place, and studies of total salt balance were not executed. Further studies on salt gland extirpation are much needed.

Secretory activity occurs in aquatic birds which inhabit alkaline waters of western and central North America, and Cooch has suggested a connection with the mass deaths of wild birds which are often observed in these areas [9, 10]. It is known that botulinus toxin (from *Clostridium botulinus* occurring in the often anaerobic waters) plays a major role in these deaths. Cooch suggested that a sublethal amount of the toxin, which blocks acetylcholine release and thus secretion from the salt gland, in conjunction with the salt load of the alkaline water, might still cause death of the birds. This would explain that except in extreme cases the treatment of paralyzed birds with fresh water is equally as effective as antitoxin in overcoming the "disease". Laboratory experiments with sublethal doses of toxin showed decreased secretion from the gland as well as increased lethality when combined with salt loads [10].

Gland structure

Anatomical and histological investigations of the salt gland were carried out by Marples [46], who also reviewed earlier work. After the excretory function of the gland had been discovered, further anatomical and histological studies were made by Fänge et al. [21].

A conspicuous feature in the anatomy of the gland is that the flow of blood in the capillaries surrounding the secretory tubules is in the opposite direction of the secreted fluid [57] (see Fig. 2). However, there is no possibility that this flow in opposite directions can function as a countercurrent multiplier system of the type that is so well known in connection with the concentrating process in the mammalian kidney. A countercurrent multiplier system depends on a single tube which is turned back on itself, forming a hairpin loop. Hairpin loops have not been found in the salt gland, and a countercurrent multiplier system therefore cannot be invoked in the explanation of the high salt concentrations found in the secreted fluid [57]. Further studies are needed of the microcirculation in the gland and its possible relation to the secretory process.

The ultrastructure of the salt gland has been studied by a number of investigators. DOYLE [13] demonstrated that in the black-backed gull

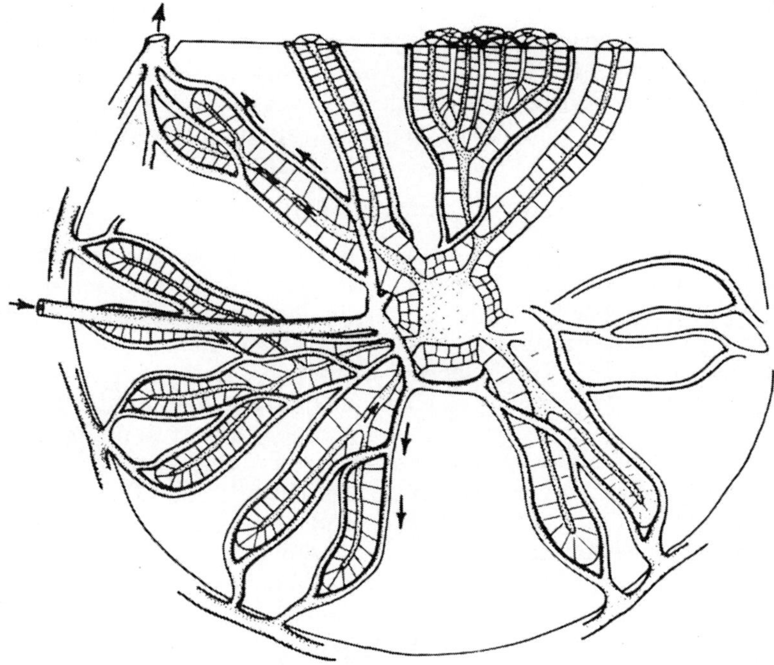

Fig. 2. Cross section (diagrammatic) of a single lobe from Fig. 1a. Note the central canal from which the branching tubular glands radiate towards the periphery. Blood capillaries are interspersed between the gland tubules, and the blood in these flows in the opposite direction of the secretion in the tubules (cf. text). From [57]

(Larus marinus) and petrel *(Oceanodroma leucorrhoa)* the secretory cells of the salt glands have extensive infoldings of the cell membranes which seem to be continuous from base to apex of the cell. Between these infoldings are densely packed mitochondria in large numbers.

Later investigations with increasingly sophisticated techniques have resulted in publications by ELLIS [16, 17], FAWCETT [23], and a series of papers by KOMNICK [40—45].

There is a striking similarity between the secretory cells of the salt gland and the tubular cell of the mammalian nephron, both in respect to the deep infoldings and the densely packed mitochondria (cf. [23, 54]). The many interesting aspects of the ultrastructure will be described later in this symposium by KOMNICK, who has made extensive studies of the salt gland.

An histochemical study was carried out by Ellis et al. [16], who compared growing ducklings which drank salt solutions and controls which drank only fresh water. In addition to hypertrophy of the gland in the birds which were drinking salt water, there was an increase in the size of the secretory cells, an apparent increase in the number of secretory cells, an elongation of the secretory tubules, and an increase in the size of the secretory lobules with concomitant dilution of the connective tissue. Histochemical differences were evident between the experimental and control birds, with the salt loaded glands having high succinic dehydrogenase and cytochrome oxidase activity. The secretory competence of the gland coincided with the appearance of abundant phospholipid and strong oxidative enzyme activity in the secretory cells. These observations are of interest in connection with biochemical studies by Hokin and by Chance to be mentioned later in this review.

Secretory function of the gland

One of the observations that several investigators have reported is that the concentration of the secretion is rather constant for each species of bird and independent of variations in the rate of flow. This independence of concentration and rate of flow was illustrated by Zaks and Sokolova [73]. The independence initially seems surprising to a person accustomed to studies of renal function, but the salt gland is not directly comparable to the far more complex kidney, and probably does what should be expected of a simpler secretory system.

It was proposed by Inoue that the concentration and the volume of the secreted fluid from the salt gland depend on two separate mechanisms [37]. In other words, salt and water transport should be independent processes. The supporting evidence was obtained from experiments on glands which were poisoned by retrograde injection of mercuric chloride into the lumen, as well as the injection of carbonic anhydrase inhibitor, which caused moderate decreases in salt concentration with maintenance of volume of secretion. In view of the current status of our knowledge of water transport it is difficult to evaluate whether the interpretation is correct.

The predominating cation of the salt gland secretion is always sodium while potassium is present in relatively small amounts. The Na : K ratio usually exceeds 20. However, there are much greater discrepancies between potassium values reported by various authors than there are in sodium values. Whether this is due to differences in analytical methods and procedures, or to real differences between the birds kept by various laboratories, remains uncertain. In my own laboratory Gideon Beer has attempted to influence the proportion of sodium and potassium in the secretion from the salt gland of herring gulls (unpublished data). Regardless of potassium loads the gulls excreted high sodium and low potassium, with the Na : K ratio always in excess of 12. If mecholyl stimulation was used instead of osmotic stimulation to induce secretion, the fluid had lower ion concentrations, but the Na : K ratio was unchanged.

This was true with potassium loads of up to 15 grams KCl per day in the food (a toxic dose), and also when the birds had been drinking a solution containing 350 mEq/l of KCl and 150 mEq/l of NaCl, instead of the usual 500 mEq/l of NaCl. These experiments therefore indicate that the relative amounts of sodium and potassium secreted by the salt gland of the herring gull are not affected by the relative need for elimination of these two ions.

Our knowledge of the secretion of anions is even more inadequate. In a study of the effects of various drugs NECHAY et al. [49] carefully examined the excretion of chloride and bicarbonate, and included a study of the effect of carbonic anhydrase inhibitors. It was found that sodium bicarbonate or moderate respiratory alkalosis had no effect on nasal fluid flow. It was again observed that carbonic anhydrase inhibitors reduce or abolish secretion from the nasal gland. However, it had earlier been shown by FÄNGE et al. [22] that in the presence of carbonic anhydrase inhibitors secretion will be resumed if the gland is directly stimulated by cholenergic chemicals. This was interpreted by FÄNGE et al. as meaning that the carbonic anhydrase is not directly involved in the secretory process, and it was suggested that the effect of carbonic anhydrase inhibition might be in the osmoreceptor reflex chain rather than in the gland itself (see also below, p. 277). I believe that there is still some differences of opinion in regard to the interpretation of these experiments.

The circulation to the gland which was briefly described by FÄNGE et al. [21] has been further studied by FÄNGE et al. [20] by means of polarographic oxygen electrodes. Vasodilation was observed during and after secretion had been induced by parasympathetic nerve stimulation, while sympathetic stimulation, adrenalin, nor-adrenalin, and serotonin had a vasoconstrictor effect.

Calculations of the expected blood flow to the gland were made by BORUT and SCHMIDT-NIELSEN [3] using two different approaches, one involving a calculation of the oxygen needed for performing the thermodynamically required work for sodium transport against an electrochemical gradient, the other involving the blood flow necessary to supply ions to the gland at the rate at which they occur in the secreted fluid. The amount of sodium transported by the gland when it secretes maximally in vivo requires energy which would necessitate a blood flow to the gland of at least 4.7 ml/gram gland/minute to supply the sodium pump. This figure is of the same magnitude as the blood flow to other organs with a high blood flow, such as the thyroid. This rate of blood flow, on the other hand, would necessitate an arteriovenous difference in sodium concentration as high as 90 mEq/l. Since this arteriovenous difference seems unreasonably high, it is likely that the blood flow exceeds by several fold that mentioned above. Obviously, a determination of the actual arteriovenous sodium differences in the secreting gland could be used to determine the blood flow to the gland. Such determinations are highly desirable, but they are technically very difficult to carry out because of the anatomical peculiarities of the gland. In my laboratory

18*

attempts at microcannulation of the blood vessels to the gland in herring gulls were unsuccessful, but it is possible that other species are better suited for this purpose.

The thought that iodide, which is secreted rapidly by many glands, could be used for a determination of blood flow of the avian salt gland did not prove feasible when this approach was explored [7]. The gland concentrated iodide to several times the plasma concentration, but chloride was slightly preferred above iodide. The hope that a particularly high iodide clearance could form the basis for a method of determining the blood flow to the gland was therefore not substantiated.

The effect of certain endocrines on salt gland secretion have been studied by Holmes and his collaborators. The total extrarenal secretion of sodium and potassium in response to a salt load was increased after treatment of ducks with adrenocortical steroids; the most potent hormone in this respect was aldosterone [29]. Although these experiments are readily interpreted as showing an endocrine effect on the salt gland, it is also possible that the effect is related to an increase in the plasma sodium concentration caused by the hormone. This could augment the total salt output of the gland without a direct endocrine effect on the gland.

A direct endocrine effect seems to be supported by further studies by Phillips et al. [52], who carried out adrenalectomy on domestic ducks. Total adrenalectomy resulted in complete obliteration of the extrarenal excretion of salt. Sub-total adrenalectomy resulted in diminished response to salt loads. The maintenance of adrenalectomized ducks on cortisone restored the extrarenal excretory pattern to near normal levels. These experiments suggest that not only osmotic stimulation and its effect via the secretory nerve is necessary for the onset of secretion, but that certain levels of adrenocortical hormones are also required. The complexity of these endocrine factors, in particular in relation to hypophyseal hormones such as ACTH and ADH, is so great that much more detailed studies are needed before clear conclusions can be drawn.

Studies of tissue slices

An extensive study of the respiration of salt gland slices using the Warburg technique was made by Borut and Schmidt-Nielsen [3]. The media used were varied to include metabolic substrates as well as wide variations in the ionic composition. The basic oxygen consumption of slices was approximately 10 μl oxygen/mg tissue dry wt/hour. The addition of methacholine, which stimulates the gland in vivo and mimics the normal parasympathetic control, caused an increase in respiration of about 50%. Increase in sodium chloride concentration in the medium inhibited respiration, while metabolic substrates (succinate) greatly augmented respiration.

A comparison of the observed oxygen consumption of slices to the energy thermodynamically required for secretion in the living bird, shows that the slices by no means consumed oxygen at a rate sufficient to transport sodium at the rates observed in the living animal [*3*]. The only possible conclusion is that tissue slices of this particular organ do not respire in the Warburg apparatus at a rate comparable to what they must do in vivo. This finding should caution against unqualified acceptance of Warburg respiration determinations, although good correspondence between such experiments and expected or calculated oxygen consumption has been found in many other cases.

Other attempts at studying the function of the secretory cells have likewise been discouraging. This is particularly true of attempts at determining the secretory potential and the transmembrane potential of the resting gland cell relative to the active cell. Use of microelectrodes was unsuccessful in establishing stable transmembrane potentials [*69*]. Presumably this is due to the peculiar structure of the cell with the extremely numerous and deep infoldings of the basal as well as the apical membranes of the cell. This means that a microelectrode inserted into a cell will cut across a number of infoldings which are in direct connection with the exterior of the cell. Thus, wherever the electrode penetrates it is, in a way, neither inside nor outside the cell.

Determinations of the potential between the duct and the blood were more successful and gave about 40–60 mv with the duct positive relative to the blood. This can, according to common practice, be interpreted as indicating a sodium pump as the primary secretory process. This is further supported by the abolishment of the potential by strophanthin (ouabain) [*69*], which also blocks secretion [*22*].

Another noteworthy observation was that the carbonic anhydrase inhibitor acetazolamide, which blocks secretion in the salt loaded gull, had no effect on the duct potential. Neither was the secretory response to nerve stimulation altered. This can again best be interpreted as an action of the drug not on the secretory cells but in the osmoregulatory reflex chain outside the gland (see above p. 275).

BONTING et al. have studied the Na-K activated adenosintriphosphatase (Na-K ATPase) in the salt gland of the herring gull [*2*] which was first described in this tissue by HOKIN [*36*].

It requires for activation both sodium and potassium, and is inhibited by the cardiac glycoside ouabain. The properties of this enzyme suggest that it is involved in the ouabain (strophanthine) sensitive salt (Na^+) secretion of the gland.

Herring gulls maintained for seven weeks on fresh water had only one-half the Na-K ATPase per gram gland of the freshly caught birds, while the gland weight had dropped to 65%. Calculations showed that the remaining activity was just sufficient for the average daily salt intake. The high molar activity of carbonic anhydrase, which was not significantly changed in the fresh water animals, indicates that this enzyme can play but a secondary non-limiting role in the salt secretion process [*2*]. The binding of sodium and potassium by microsomes from nasal salt gland and other glands of the sea gull *(Larus occidentalis)* has been examined [*48*].

Identification of sodium carrier

In 1959, Hokin and Hokin reported a high turnover rate for phosphatidic acid in salt gland slices when these were incubated with acetylcholine. The Hokins suggested that phosphatidic acid was an integral part of the sodium pump in the gland and that it might be the ionic transport agent itself [31]. Further evidence, again related to turnover rates, was presented in several following papers [32–35]. In the opinion of this reviewer the evidence presented by the Hokins is suggestive but circumstantial. Increased turnover rates in the presence of the agent which normally stimulates secretion in vivo (acetylcholine) may indicate a fundamental link to ion transport but does not provide proof. Further developments, hopefully with the use of new approaches, can be met with considerable anticipation.

Energy requirements of the pump

An elegant study has been made by Chance and his collaborators in which they examined the cytochrome content and respiratory activity of salt gland mitochondria [8]. The rate of respiration observed was in the magnitude of four times as high as that observed in tissue slices by Borut and Schmidt-Nielsen [3]. As mentioned above, Borut and Schmidt-Nielsen estimated that the highest respiratory rates they had observed were still insufficient to cover the thermodynamic requirement for work needed for secretion in vivo, and that their determinations therefore could not be used for estimations of transport efficiency. Chance et al. recalculated the thermodynamic feasibility of the transport using their new observations. On the basis of the oxygen utilization and the transport process they briefly discussed a number of current hypotheses for cation transport and concluded that their evidence is difficult to reconcile with the redox-pump hypothesis, the phosphatidic acid hypothesis, and the Na-K-activated ATPase hypothesis. The authors favor a consideration of the cation pump of the mitochondria, although they realize that even here the energetic considerations as well as the geometrical location of the mitochondria within the cell still leave difficulties unresolved.

Quite recently additional biochemical studies of the metabolic pathways in the salt gland have been published by van Rossum [70–72]. By examining the state of reduction of respiratory pigments and the simultaneous efflux of sodium from slices under the influence of methacholine, ouabain (strophanthin), and potassium, van Rossum evaluated the control of energy utilization in relation to cellular transport of sodium and potassium. Van Rossum has informed me that additional material is being prepared for publication (personal communication).

Non-avian salt glands

Elasmobranchs

The discovery by Burger that the rectal gland of sharks secretes a fluid which is essentially a sodium chloride solution with a concentration about twice that of the plasma and greater than that of sea water, clarified a major point in the osmotic regulation of elasmobranchs [6]. It was known from the work of HOMER W. SMITH that the elasmobranchs are practically in osmotic equilibrium with sea water due to a high concentration of urea in the body fluids [65]. However, it was not known how they handled an excess of salts, and BURGER's contribution is therefore notable. A later publication demonstrates that the secretion is produced in volumes sufficient to play a significant role in the salt economy [5]. Attempts to demonstrate a nervous control of the rectal gland were negative, acetylcholine (and several other substances) did not stimulate secretion, nor did atropine inhibit.

The rectal gland has been anatomically described by DOYLE [14] and by BULGER [4].

Teleost fish. Since the classical work of KEYS [38] it has been known that marine teleost fish, which habitually excrete a urine more dilute than the blood, manage to remain hypotonic to sea water through salt excretion by the gills. The particular cells which presumably are responsible for the salt transport were called "chloride cells" by KEYS and WILLMER [39]. Numerous studies have been made of these cells, most recently by employing the electron microscope (e.g. [11, 15, 53, 66]). Although the term "chloride cell" is still in use, there is no direct evidence that these particular cells are responsible for the salt transport. Neither is it certain that chloride rather than sodium is being transported. Therefore the term "chloride cell" should only be used with the explicit reservation that no decisive proof of their function in osmotic regulation has been obtained. I can indeed endorse the statement by DOYLE and GORECKI [15] that "The location in the epithelium and the interspecific distribution of these cells casts doubts on their supposed function as the principal site of chloride excretion (or absorption)".

Amphibians

A few amphibians are tolerant to brackish water, and at least one lives in full strength sea water, the crab-eating frog *(Rana cancrivora)* of the Gulf of Thailand [25]. As an adult this frog is essentially in osmotic equilibrium with the surrounding sea water, a major part of the osmotic concentration being due to urea. In this respect the crab-eating frog is similar to elasmobranchs. However, the retention of urea apparently is not due to an active tubular reabsorption of urea [62] as was previously found in the elasmobranch [65]. No evidence so far indicates any type of "salt gland" or similar organs or structures in the adult crab-eating frog. However GORDON and TUCKER have recently studied the tadpole of this frog and have shown that its osmotic adjustment to sea water is similar to that of the teleost fish (personal communication). In addition to the need for an ion transport mechanism analogous to the teleost gill, these observations raise important perspectives in regard to the relation between biochemical and morphological metamorphosis.

Mammals

There is no evidence that extrarenal salt excretion is of major importance in any mammal. The mammalian kidney is able to evolve into an organ which can produce extremely concentrated solutions. The highest observed concentrations seem to occur in the sand rat *(Psammomys obesus)* which has produced urine with 1920 mEq sodium/l, or about four times the concentration of sea water [58]. Although the information available about seals, whales, and sea cows is inadequate, there is no reason to believe that the kidney of these animals should be incapable of secreting any excess salt that may be taken in with food or water. I have tried to induce secretion from the nose or orbital region in the harbor seal *(Phoca vitulina)* by salt loads and methacholine injections, but the results were negative (unpublished experiments).

Reptiles

Salt secretion occurs from nasal glands in marine lizards and from glands in the orbit of the eye in marine turtles [60]. The turtle gland is apparently a modified lacrymal gland [51] and is definitely not of nasal origin.

Presumably, a brief histological description of "nasal" glands from several species of marine turtles by BENSON, PHILLIPS, and HOLMES [1] refer to the orbital gland. The authors also refer to a previous paper on the orbital salt secreting gland of the loggerhead turtle [60] as a study of the "nasal" gland. The same mistakes occur in an interesting review on the adrenocortical factors in adaptation to marine environments by HOLMES et al. [30], which among other subjects discusses electrolyte excretion in marine turtles. A more recent publication [27] presents the only study I know which adequately considers the quantitative role of the salt gland of any marine reptile. The Galapagos Iguana and other marine or semi-marine lizards have not, I believe, been studied in any detail beyond the original description that nasal salt excretion does take place [60].

However, a small number of papers describe secretion from the nasal region of terrestrial lizards. TEMPLETON [67] showed nasal salt secretion in "terrestrial iguanids" of primarily potassium in concentrations up to 750 mEq/l. Injection of sodium chloride caused secretion, but potassium levels remained high and sodium levels low. Later studies [68] on the terrestrial iguanas *Ctenosaura pectinata* and *Sauromalus obesus* showed concentrations up to 950 mEq/l potassium, or 190 times the plasma concentration. Again, injection of sodium chloride increased potassium but not sodium concentration of the nasal fluid. Sodium concentration in the fluid remained below that of the plasma.

Two additional species of *Sauromalus* (*S. hispidus* and *S. varius*) were examined by NORRIS and DAWSON [50]. Again, potassium far exceeded sodium in the nasal secretion and the principal anion was bicarbonate. The size of the nasal gland in *S. hispidus* was similar to that of the marine iguana (0.06% of body weight) and in *S. varius* about half this size. Structurally the gland is similar to other salt glands, branching secretory tubules are radially arranged around a central duct.

The lizards mentioned above are plant eaters and it is therefore reasonable that they excrete primarily potassium and that the principal anion is bicarbonate.

Similar results were previously reported by SCHMIDT-NIELSEN et al. [59] for secretion from two other desert lizards, *Dipsosaurus dorsalis* and *Uromastix aegyptius*, and in one tropical lizard, *Iguana iguana*. The latter is not particularly adapted to dry and arid conditions, and it is therefore interesting that this animal as well secretes salt from the nasal region, primarily potassium in the form of bicarbonate.

It was suggested that the extrarenal excretion of salts as observed in non-marine lizards and a few birds (desert partridge, *Ammoperdix heyi*, and ostrich, *Struthio camelus*) is related to the reabsorption of water in the cloaca [59]. In birds and lizards water reabsorption presumably occurs in the cloaca. It is generally considered that active transport of water is less feasible than movement of water by osmosis following a primary transport of salt. If the water reabsorption in the cloaca depends on a primary reabsorption of salts, the salts will of necessity have to be eliminated elsewhere, e.g. by the nasal gland. If this suggestion is correct, the evolutionary origin of the nasal gland may be in connection with the invasion of the terrestrial habitat by birds and reptiles, rather than specifically in association with the invasion of the marine habitat.

References

[1] BENSON, G. K., J. G. PHILLIPS, and W. N. HOLMES: J. Anat. (Lond.) **98**, 290 (1964).

[2] BONTING, S. L., L. L. CARAVAGGIO, M. R. CANADY, and N. M. HAWKINS: Arch. Biochem. (in press).

[3] BORUT, A., and K. SCHMIDT-NIELSEN: Amer. J. Physiol. **204**, 573 (1963).

[4] BULGER, R. E.: Anat. Rec. **147**, 95 (1963).

[5] BURGER, J. W.: Physiol. Zool. **35**, 205 (1962).

[6] —, and W. N. HESS: Science **131**, 670 (1960).

[7] CAREY, F. G., and K. SCHMIDT-NIELSEN: Science **137**, 866 (1962).

[8] CHANCE, B., C. P. LEE, R. OSHINO, and G. D. V. VAN ROSSUM: Amer. J. Physiol. **206**, 461 (1964).

[9] COOCH, F. G.: Canad. Wildlife Ser., Res. Prog. Rep., pp. 27 (1961).

[10] — Auk, **81**, 380 (1964).

[11] COPELAND, D. E., and A. J. DALTON: J. biophys. biochem. Cytol. **5**, 393 (1959).

[12] DOUGLAS, D. S.: Symposium de Biologie Antarctique, pp. 651. Paris:Hermann 1964.

[13] DOYLE, W. L.: Exp. Cell Res. **21**, 386 (1960).

[14] — Amer. J. Anat. **111**, 223 (1963).

[15] —, and D. GORECKI: Physiol. Zool. **34**, 81 (1961).

[16] ELLIS, R. A., and J. H. ABEL: Anat. Rec. **148**, 278 (1964).

[17] —, and J. H. ABEL: Science **144**, 1340 (1964).

[18] —, R. A. DELELLIE, and Y. H. KABLOTSKY: Amer. Zool. **2**, 406 (1962).

[19] —, C. C. GOERTEMILLER JR., R. A. DELILLIS, and Y. H. KABLOTSKY: Developmental Biol. **8**, 286 (1963).

[20] FÄNGE, R., J. KROG, and O. REITE: Acta physiol. scand. **58**, 40 (1963).

[21] —, K. SCHMIDT-NIELSEN, and H. OSAKI: Biol. Bull. **115**, 162 (1958).

[22] — —, and M. ROBINSON: Amer. J. Physiol. **195**, 321 (1958).

[23] FAWCETT, D. W.: Circulation **26**, 1105 (1962).

[24] GOERTEMILLER, C. C. JR., and R. A. ELLIS: Amer. Zool. 2, 525 (1962).
[25] GORDON, M. S., K. SCHMIDT-NIELSEN, and H. M. KELLY: J. exp. Biol. 38, 659 (1961).
[26] HEINROTH, O., and M. HEINROTH: Die Vögel Mitteleuropas in allen Lebens- und Entwicklungsstufen photographisch aufgenommen und in ihrem Seelenleben bei der Aufzucht vom Ei ab beobachtet. 3 Vols. Berlin: H. Bermühler 1926/28.
[27] HOLMES, W. N., and R. L. MCBEAN: J. exp. Biol. 41, 81 (1964).
[28] —, D. G. BUTLER, and J. G. PHILLIPS: J. Endocr. 23, 53 (1961).
[29] —, J. G. PHILLIPS, and D. G. BUTLER: Endocrinology 69, 483 (1961).
[30] — —, and I. C. JONES: In: Recent progress in Hormone Research 19, 619. New York, N. Y.: Academic Press 1963.
[31] HOKIN, L. E., and M. R. HOKIN: Nature (Lond.) 184, 1068 (1959).
[32] — — J. gen. Physiol. 44, 61 (1960).
[33] HOKIN, M. R., and L. E. HOKIN: Nature (Lond.) 190, 1016 (1961).
[34] HOKIN, L. E., and M. R. HOKIN: Fed. Proc. 22, 8 (1963).
[35] — — Biochim. biophys. Acta (Amst.) 71, 462 (1963).
[36] HOKIN, M. R.: Biochim. biophys. Acta (Amst.) 77, 108 (1963).
[37] INOUE, T.: Science 142, 1299 (1963).
[38] KEYS, A. B.: Z. vergl. Physiol. 15, 364 (1931).
[39] —, and E. N. WILLMER: J. Physiol. (Lond.) 76, 368 (1932).
[40] KOMNICK, H.: Protoplasma (Wien) 55, 414 (1962).
[41] — Protoplasma (Wien) 56, 274 (1963).
[42] — Protoplasma (Wien) 56, 385 (1963).
[43] — Protoplasma (Wien) 56, 605 (1963).
[44] — Protoplasma (Wien) 58, 96 (1964).
[45] —, u. U. KOMNICK: Z. Zellforsch. 60, 163 (1963).
[46] MARPLES, B. J.: Proc. Zool. Soc. Lond. 1932, 829 (1932).
[47] MCFARLAND, L. Z.: Nature (Lond.) 184, 2030 (1959).
[48] —, and H. SANUI: Proc. Soc. exp. Biol. (N. Y.) 113, 105 (1963).
[49] NECHAY, B. R., J. L. LARIMER, and T. H. MAREN: J. Pharmacol. exp. Ther. 130, 401 (1960).
[50] NORRIS, K. S., and W. R. DAWSON: Copeia (in press).
[51] PETERS, A.: Arch. mikr. Anat. 36, 192 (1890).
[52] PHILLIPS, J. G., W. N. HOLMES, and D. G. BUTLER: Endocrinology 69, 958 (1961).
[53] PHILPOT, C. W., and D. E. COPELAND: J. cell. Biol. 18, 389 (1963).
[54] RHODIN, J. A. G.: In: Diseases of the Kidney, p. 10 (M. B. STRAUSS and L. G. WELT, eds.). Boston: Little, Brown and Co. 1963.
[55] SCHILDMACHER, H.: J. f. Ornithol. 80, 293 (1932).
[56] SCHMIDT-NIELSEN, K.: Circulation 31, 955 (1960).
[57] — The Harvey Lectures, Series 58: 53 (1963).
[58] — Desert Animals. Physiological Problems of Heat and Water. Oxford: Clarendon Press 1964.
[59] —, A. BORUT, P. LEE, and E. C. CRAWFORD JR.: Science 142, 1300 (1963).
[60] —, and R. FÄNGE: Nature (Lond.) 182, 783 (1958).
[61] —, and Y. T. KIM: Auk 81, 160 (1964).
[62] —, and P. LEE: J. Exp. Biol. 39, 167 (1962).
[63] SCHWARZ, D.: J. Ornithol. 103, 180 (1962).
[64] —, and L. SPANNHOF: Naturwissenschaften 48, 414 (1961).
[65] SMITH, H. W.: Biol. Rev. 11, 49 (1936).
[66] STRAUS, L. P.: Physiol. Zool. 36, 183 (1963).
[67] TEMPLETON, J. R.: Amer. Zool. 3, 530 (1963).
[68] — Comp. Biochem. Physiol. 11, 223 (1964).
[69] THESLEFF, S., and K. SCHMIDT-NIELSEN: Amer. J. Physiol. 202, 597 (1962).
[70] VAN ROSSUM, G. D. V.: Biochem. biophys. Res. Commun. 15, 540 (1964).
[71] — Biochim. biophys. Acta (Amst.) 86, 198 (1964).
[72] — Biochim. biophys. Acta (Amst.) (in press).
[73] ZAKS, N. G., and M. M. SOKOLOVA: Sechenov physiol. J. USSR 47, 120 (1961).

Discussion

M. Hokin: The work which we would like briefly to discuss is concerned with the role of phosphatides in the secretory activity of the salt gland. In particular we would like to discuss some biochemical events which occur in the membranes of the salt gland cell when the Na$^+$ pumps in these membranes are turned on and off. Regulation of the activity of the Na$^+$ pump has been carried out primarily with the aid of three agents — acetylcholine, atropine and ouabain. Active extrusion of NaCl in this gland is initiated by acetylcholine and this effect of acetylcholine is abolished by atropine. The active transport of sodium in the stimulated salt gland is inhibited, as it is in other tissues, by cardiotonic steroids such as ouabain.

When the Na$^+$ pumps are turned on in salt gland slices by the addition of acetylcholine to the system, a compartment of phosphatidic acid is formed; the phosphate groups of this compartment of phosphatitic acid undergo continuous turnover at a rate sufficient to maintain isotopic equilibrium with the radioactive ATP of the tissue (Hokin, L. E., and M. R. Hokin, Federation Proc. **22**, 8, 1963). Phosphatidic acid can, of course, be a precursor for the formation of phosphatidylinositol and other phosphatides. However, the compartment of phosphatidic acid which we are discussing could not be a metabolic pool for the net synthesis of phosphatidylinositol or other phosphatides because the minimum turnover rate of the phosphate group is much greater than the rate of appearance of radioactivity in phosphatidylinositol and other phosphatides. There is no increased incorporation of glycerol-1-C^{14} during the formation of this compartment of phosphatidic acid, so that the synthetic reaction involved in the turnover of the phosphate group in this compartment of phosphatidic acid appears to be that in which diglyceride is phosphorylated using ATP:

$$\text{Diglyceride} + \text{ATP} \xrightarrow{\text{Diglyceride kinase}} \text{Phosphatidic Acid} + \text{ADP.}$$

Diglyceride kinase is present in the smooth membrane fraction of salt gland cells.

The compartment of phosphatidic acid which we are discussing is found in the smooth membrane fraction of the cell. It cannot be extracted without prior treatment with ethanol, so that it is presumably present in lipoprotein form. It amounts maximally to about 0.12 μmoles/g. of fresh tissue. This is about 8% of the total phosphatidic acid of the tissue. It is situated in the structure in such a way that it does not mix with the other phosphatidic acid of the tissue.

This compartment of phosphatidic acid is formed when NaCl secretion is initiated, it undergoes continuous turnover while stimulation of NaCl secretion is maintained, and it disappears when NaCl secretion is stopped, either by removal of acetylcholine or by addition of atropine. Under conditions in which a maximum amount of phosphatidic acid has been formed, the disappearance of this phosphatidic acid is associated with the appearance (measured either by P^{32} or by inositol-2-H^3 incorporation) of an equivalent amount of phosphatidylinositol (Hokin, M. R., and L. E. Hokin, in Metabolism and Physiological Significance of Lipids, London: John Wiley and Sons, 1964). These molecules of phosphatidylinositol do not undergo continuous turnover. However, they are lost again if the tissue is restimulated, and in their place phosphatidic acid appears. The data indicate that there is an acetylcholine-responsive lipoprotein in the salt gland membrane which has, as part of its lipid moiety, a molecule which exists as phosphatidylinositol when the tissue is not stimulated by acetylcholine. This phosphatidylinositol is converted to and exists as diglyceride or phosphatidic acid — these two forms are continually formed and reformed from each other — during the stimulated state when the tissue is secreting NaCl. The diglyceride part of the molecule is probably retained intact throughout. The overall picture is shown diagrammatically in Fig. 1.

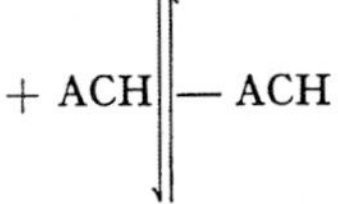

Fig. 1. The overall lipid interconversions associated with the turning on, the operation of, and the turning off of the Na$^+$ pumps in the salt gland.

On addition of atropine to salt gland slices which have been stimulated with any concentration of acetylcholine from 10^{-7}M upwards, approximately the same number of phosphatidylinositol molecules are formed after exposure to each concentration of acetylcholine. It appears therefore that all the possible sites are activated by even the lowest concentration of acetylcholine. However, if the stimulating concentration of acetylcholine is less than about 10^{-5}M., the amount of phosphatidic acid at any given time in the stimulated system is lower than the amount of phosphatidylinositol formed on reversion to the unstimulated state. At 10^{-5}M acetylcholine, the amount of phosphatidic acid found in the stimulated system and lost on addition of atropine is approximately equal to the amount of phosphatidylinositol

Table 1.

Effects of atropine ($10^{-5}M$) on phosphatidylinositol and phosphatidic acid after various concentrations of acetylcholine in goose salt gland slices

Initial acetylcholine concentration	Change after atropine (mμmoles[1]/g. of fresh tissue)		
	Gain in phosphatidylinositol (Total number of molecules which had been activated by acetylcholine before atropine)	Loss of phosphatidic acid (Steady state level of phosphatidic acid form in the activated state)	PI gain minus PA loss (Presumed to be the steady state level of the diglyceride form in the activated state)
10^{-7}M	60	33	27
10^{-6}M	72	40	32
10^{-5}M	73	67	6

[1] Calculated from the observed radioactivity. The calculation assumes isotopic equilibrium of the relevant lipid-P^{32} with the tissue ATP^{32}.

formed on addition of atropine. These observations can be accounted for by the assumption that all the possible sites are activated by even the lowest concentration of acetylcholine but that the steady state ratio of the diglyceride form to the phosphatidic acid form in the activated system is considerably influenced by the concentration of acetylcholine. This assumption would indicate that the relative rates of the reactions concerned with the interconversion of diglyceride and phosphatidic acid are such that with low concentrations of acetylcholine the steady state level favours the diglyceride form, but with higher concentrations of acetylcholine the steady state level favours the phosphatidic acid form (Table 1).

An observation which is of help in deciding the functional *location* of this responsive lipoprotein unit, is that ouabain also affects the steady state levels of the diglyceride form and the phosphatidic acid form in the stimulated system. In the presence of very low concentrations of acetylcholine, normally, as we have just seen, most of the responsive molecules would be in the diglyceride form at any given instant; however, if ouabain is added to the system under these conditions, virtually all of the responsive molecules are found in the phosphatidic acid form at any given instant (Table 2). Reversion to the phosphatidylinositol form on removal of the influence of acetylcholine is not affected by ouabain. It appears that in the stimulated tissue (i. e. in the presence of acetylcholine), when a molecule of ouabain sits on the site at which it acts, this causes, in the lipoprotein unit which is responsive to acetylcholine, either a slowing down of the reaction in which phosphatidic acid is converted to diglyceride, or an increase in the rate of conversion of the diglyceride form to the phosphatidic acid form, the net result being a shift in the ratio of the diglyceride form and the phosphatidic acid form to favor the phosphatidic acid form.

Since ouabain is a rather specific agent which acts at Na^+ pump sites, this provides some evidence that the lipoprotein unit with which we are dealing represents part of the Na^+ pump site in the membrane.

In summary it appears that there is a lipoprotein in the membranes of the salt gland cell which is responsive to acetylcholine and which is associated with Na^+ transport. When the tissue is resting, the lipid moiety of this lipoprotein exists as phosphatidylinositol; initiation of active transport of Na^+ involves the conversion of this molecule to diglyceride or phosphatidic acid. The continuous phosphorylation of the diglyceride form to give the phosphatidic acid form and dephosphorylation

Table 2.

Effect of ouabain on the steady state level of radioactive phosphatidic acid in the presence of different concentrations of acetylcholine in goose salt gland slices

Acetylcholine concentration	Increased phosphatidic acid radioactivity due to acetylcholine (P^{32} counts/minute/mg. of fresh tissue)	
	—	+ Ouabain (10^{-4}M)
10^{-7}M	125	460
10^{-6}M	267	496
10^{-5}M	513	552

of the phosphatidic acid form to give the diglyceride form during active Na^+ transport is compatible with the possibility that this is the transducer mechanism whereby the energy of ATP is used for active transport. This aspect however requires further investigation.

Tosteson: Did I understand correctly that ouabain produces increased labelling of phosphatidic acid and removes the label from phosphatidic acid and allows it to accumulate in phosphatidic acid again?

M. Hokin: Ouabain alone does not. I was talking about the effects of ouabain on an acetylcholine-activated system. In the presence of low amounts of acetylcholine where your steady-state level would normally favour diglyceride, in the presence of ouabain this is pushed over to give you the maximum amount of phosphatidic acid instead of diglyceride in the turning over system. The rates of turnover theoretically should be lower if this were part of the ion transport mechanism, but we cannot measure the rates of turnover because the time intervals with which we can deal are orders of magnitude higher than the turnover rates are likely to be. This is however a specific effect of ouabain on a sodium pump system which has been activated by acetylcholine. If you now take the acetylcholine away, the phosphatidic acid reverts to phosphatidylinositol, just as it would if ouabain were not there.

Tosteson: Even in the presence of the ouabain?

M. Hokin: Even in the presence of the ouabain. Ouabain does not affect the rate of conversion of the phosphatidic acid to the inactive phosphatidyl inositol form.

Tosteson: If you have ouabain present first, and then add acetylcholine —

M. Hokin: It does not affect the activation either, you still get the phosphatidyl inositol breaking down and giving rise to the diglyceride and phosphatidic acid forms.

Tosteson: So that ouabain does not inhibit the stimulation by acetylcholine?

M. Hokin: No, it does not have any effect.

Klingenberg: Is there another possibility of turning on the pump than by methacholine, for example, to turn it on by sodium? If this is an experimental possibility one should measure the initial rate of P-32-incorporation.

M. Hokin: Work from Dr. SCHMIDT-NIELSEN's laboratory has shown that sodium alone does not directly stimulate the pumps in the avian salt gland; they are exclusively under cholinergic control.

Klingenberg: If you pretreat with metacholine and then add sodium you have another condition to turn the pump on.

M. Hokin: The question is whether the pumps can be activated in the absence of sodium by cholinergic agents, and the answer is that in relation to the activation of this trigger mechanism for converting phosphatidyl inositol to diglyceride and phosphatidic acid, no, this does not occur in the absence of sodium in the medium. However, I do not think this has too much to do with the sodium transport in the salt gland because you also do not get acetylcholine effects on lipids in other systems in the absence of sodium. So that it may be that this is a red herring in that sodium is required for acetylcholine to trigger quite a lot of systems.

Klingenberg: I was actually driving to another point i. e. to measure the initial rate of the interconversion of the phosphatidyl inositol. It is very important for the overall turnover, to know the initial rates. Therefore, I think, it would be more appropriate to initiate them by the sodium than by the activator, metacholine.

M. Hokin: It is a good suggestion but we have the problem that actually the time intervals we measure are a function of the diffusion of the substances in the medium into all of the extracellular spaces which takes an appreciable time, so that the slope that you get is really a measure of the penetration of the agent that you are using, sodium or a particular drug, to all of the sites. Unfortunately, in this system where we have to use intact cells and slices we do have this very severe limitation, that we cannot make anything like quick changes to all the sites.

Klingenberg: I may give some comments on the metabolic aspects of the salt gland slices and also on the isolated mitochodria of the salt glands, particularly in view of the studies of CHANCE and VAN ROSSUM[1]. I myself have not worked with this material and thus I am only a referee and may add some comments of my own. Firstly, we may discuss the energy balance of the whole tissue. The most important figure is the cytochrome content of the salt gland as measured by CHANCE and VAN ROSSUM. The cytochrome content can be regarded as the oxidative capacity of the tissue since it will be proportional to the activity of the respiratory chain. The average content of cytochrome c of the salt gland as measured by CHANCE is $0.05 \cdot 10^{-6}$ mol/g fresh tissue.

It is of interest to compare the cytochrome content of the salt glands with that of other tissues. This is shown in the accompaning table.

Cytochrome content and tissue respiration[1]

Tissue	Cytochrome a-content	Oxygen uptake	Cytochrome a-turnover
	10^{-3} Mol/g fr	Atom 0/g.h	10^3/h
Thoracic M. (*Locusta*)	45	$5—10.10^3$	220—240
Heart M. (rat)	42	$1,6.10^3$	180
Liver (rat)	19	600	64
	Cytochrome c-content		Cytochrome c-turnover
Salt gland[2]	50	600	24

[1] cf. M. KLINGENBERG: In: „Funktionelle und Morphologische Organisation der Zelle", p. 69, Springer-Verlag 1963.

[2] Calculated from B. CHANCE et al.

Actively respiring muscles such as the insect flight muscle and the heart muscle have a high cytochrome content. The cytochrome content of the salt gland is roughly equal to these muscles. The cytochrome content is much lower in tissues as liver or brain.

On this basis, the comparison of maximum oxygen uptake is useful. By referring oxygen uptake to the cytochrome content the "cytochrome turnover" is obtained

[1] CHANCE, B., C. P. LEE, R. OSHINO, G. VAN ROSSUM, Am. J. Physiol. 206, 461 (1964).

(last column of table). The insect flight muscle has the highest cytochrome turnover and therefore can make use of its capacities more than the heart.

We will now apply this calculation to the salt gland. One of the highest values for the oxygen uptake of salt gland slices is 25 μl O_2/mg dry weight per hour.[1] As referred to the cytochrome content, this value gives the turnover of 24.10^3 per hour. This is a ten times smaller turnover than calculated for the insect flight muscle or heart muscle. The comparison shows that these salt gland slices run at a relatively low rate. They utilize only a small part of their oxidative capacity. It appears to be probable that the respiration of the slices does not reflect the in vivo respiration of the salt gland which has not yet been measured. Probably the slices run so low because they cannot perform so much ion transport in vitro as the tissue does in vivo when the highly organized membranes system is intact.

This also, of course, may explain some of the discrepancies in the Na^+/oxygen ratio which Dr. SCHMIDT-NIELSEN has pointed out. Based on this respiration of these slices there would be 24 Na^+ pumped per oxygen atom. The ratio appears too high and is probably due to the depressed oxygen uptake. It is of interest to further compare the cytochrome content and the content of the isolated mitochondria in salt gland with that of other tissues. As shown in the paper by Dr. KOMNICK the salt gland is loaded with mitochondria to 50% or more of its space. VAN ROSSUM and CHANCE have isolated mitochondria from these tissues and seen that they behave completely normally as mitochondria from other rat tissues kidney or heart. They also have a comparable cytochrome content. The high oxidative capacity of the tissues is only given by the high concentration of these mitochondria. As shown by CHANCE et al. these isolated mitochondria can easily be brought to fully active, maximum respiration by stimulating the oxidative phosphorylation. Thus they can fully use the oxidative capacity. From the respiration of isolated mitochondria one would arrive for the respiration of the salt gland tissues at a cytochrome turnover of $100 \cdot 10^3$ per hour or more. Thus there appears to be no deficiency in the mitochondria as a cause for the low oxidation rate of these slices. The mitochondria respond normally to the activation and inhibition of respiration correlated to the oxidative phosphorylation system.

Bücher: How much sodium is transported?

Schmidt-Nielsen: We have observed secretion rates in excess of 400 μEq Na per gram gland per minute. It can also be translated from the mitochondria; it will be the ratio of 8.

Klingenberg: May I make some further comments on the ratio Na^+/electron of 8, as calculated by CHANCE et al. This would be equivalent to 16 Na^+ per atom oxygen. This number has a bearing on the principal mechanism of the sodium pump. It eliminates any mechanism where the sodium is pumped against the electrons across the respiratory chain because there is no electrical compensation between 8 Na^+ and one electron. The ratio Na^+ to ATP is also too large to account for an ATP driven sodium pump when one ATP is used for one Na^+. Thus there appears to be no simple stiochiometric relation between the energy equivalents and the sodium pumped. It should be mentioned at this point that there are recent reports that the P/O ratio is up to 6. These reports are only from few laboratories and have to await further confirmation. But if we suppose, that this ratio would be correct, also then the use of one ATP per Na is not possible.

Mothes: For comparison, I think, it would be of some interest here, to draw attention to salt glands occurring in plants, particularly in salt plants or halophyta, which grow at the sea shore. All plants must take up much water, because evaporation is a pre-condition for assimilation of CO_2. This CO_2 can only enter the leaf cells, when the stomata are open, and at the same time water escapes into the air. Therefore plants cannot reduce the water uptake, and the problem the salt plants are faced with is that they have their roots in a salt-rich soil.

There are two types of salt plants. First, there are salt plants without salt glands. These are very succulent plants, which are able to take up water by their high concentration of organic substances — sugars, aminoacids, etc. They only have a small

---

[1] BORUT, A., and K. SCHMIDT-NIELSEN: Am. J. Physiol. **204**, 573 (1963).

amount of sodium chloride. And secondly there are salt plants with normal leaves, which have a high concentration of sodium chloride within the cell sap. They cannot diminish salt uptake, when taking up water. These plants are able to excrete salt by salt glands.

In the epidermis of the leaves there are salt glands consisting of a group of about 16 cells. You can observe these salt glands to be covered with crystallised sodium chloride. Initially plant physiologists believed that the salt glands were excreting a concentrated salt solution, but in 1915 RUHLAND was able to demonstrate that the salt concentration of the excreted solution is not higher than the concentration of the sodium chloride solution in the soil. He concluded that there is no concentration effect within the gland, but the concentration is made by the transpiration of water and by the air.

More recently this problem has been re-investigated by Dutch and English workers. It can be demonstrated that the concentration of sodium chloride within the cells of the salt glands is higher than the salt concentration of the solution in the medium. There must be a concentration effect in the salt gland cells, and the solution given up by these cells is diluted or not. Correctly speaking there must be a double effect, first the accumulation of sodium chloride within the cells, and secondly the excretion of a solution against a high concentration of salt by turgor pressure. We have no further investigations on this very interesting object, and I only wanted to sketch the possible simularity of problems, we are faced with in animal and plant salt glands.

Funktionelle Morphologie von Salzdrüsenzellen*

Von

Hans Komnick, Bonn

Mit 10 Abbildungen

Einleitung

Die erst vor wenigen Jahren entdeckte Funktion der Nasendrüsen
(Salzdrüsen) von marinen Sauropsiden [57−65, vgl. auch 45, 46, 67] und
der Rectaldrüsen von Elasmobranchiern [8−10] hat bereits den Anstoß
zu zahlreichen Untersuchungen[1] gegeben und auch ein erneutes Interesse
an der Morphologie dieser Exkretionsorgane erweckt (Salzdrüsen:
[15, 16, 20, 24, 26, 36, 38−42, 47, 69], Rectaldrüsen: [7, 17−19]). Diese
Organe sind, wie der vorhergehende Vortrag [59] gezeigt hat, auf die
Exkretion von Salzen in ziemlich hohen Konzentrationen spezialisiert.

Schon vor Kenntnis der Nasendrüsenfunktion war beobachtet
worden, daß die Drüsengröße bei den einzelnen Vogelarten mit der
Anpassung an die marine Lebensweise einhergeht [71] und bei der
gleichen Art stark variiert, je nachdem, ob die Tiere längere Zeit in Salz-
oder Süßwasserbiotopen gelebt haben [29, 56]. Diese von der Dauer der
funktionellen Beanspruchung abhängigen groben anatomischen Unterschiede, die durch neuere experimentelle Beobachtungen erhärtet werden
[vgl. Abb. 1a und b und 21, 22, 33, 38, 40], lassen auch funktionell
bedingte Unterschiede in der Mikromorphologie der exkretorischen
Zellen vermuten.

Ich habe daher den Versuch unternommen, den funktionellen
Strukturwandel der Epithelzellen der Salzdrüsentubuli von Silbermöwen
(Larus argentatus) zu verfolgen, um Hinweise für die zellphysiologische
Bedeutung einzelner Zellstrukturen vor allem im Hinblick auf den
Vorgang des transzellulären Ionentransportes zu gewinnen. In diesem
Zusammenhang ist es naturgemäß von Interesse, auch die Struktur
der Tubulusepithelzellen von Rectaldrüsen vergleichend zu beleuchten.

Zum besseren Verständnis der morphologischen Befunde sollen kurz
die Gemeinsamkeiten und die Unterschiede in der Physiologie und
Anatomie dieser beiden analogen Organe herausgestellt werden.

Vergleichende Physiologie

Die Rectaldrüse des Dornhaies scheidet eine NaCl-Lösung aus, deren
Konzentration (bezogen auf Cl⁻) etwa doppelt so groß ist wie die im

* Mit Unterstützung der Deutschen Forschungsgemeinschaft.
[1] Vgl. außer der später zitierten Literatur u. a. auch [2, 13, 23, 27, 35, 66, 68, 70].

Blutplasma und etwas höher liegt als im Seewasser. Die Tonizität des Exkretes entspricht jedoch der Osmolarität des Plasmas, welches gegenüber Seewasser leicht hypertonisch ist [8, 9]. Die NaCl-Konzentration des Salzdrüsen-Exkretes dagegen mit etwa 0,4—1,1 m je nach Tierart kann sogar 3—7mal größer als die im Blutplasma sein [45, 48, 49, 58]. Die Funktionsaktivität beider Organe kann durch osmotische Belastung stimuliert werden. Während jedoch die Salzdrüsen nach dem Alles-oder-Nichts-Gesetz arbeiten und nur nach einer osmotischen Belastung exkretorisch aktiv werden [45], reagiert die anscheinend dauernd tätige Rectaldrüse auf einen solchen Reiz lediglich mit einer höheren Exkretionsrate [8]. Während die Exkretionsaktivität der Salzdrüsen einer nervösen [25] und hormonellen [34, 53] Steuerung unterliegt und durch verschiedene Drogen beeinflußt werden kann [45], konnte bei der Rectaldrüse keine solche Steuerung und Beeinflußbarkeit nachgewiesen werden [8].

Aus diesen Gründen haben die Salzdrüsen zum Studium der Exkretionsmorphologie gegenüber der Rectaldrüse den Vorzug, da sich durch einfache Diätversuche Drüsen verschiedener Funktionszustände für elektronenmikroskopische Untersuchungen gewinnen lassen.

Material und Methoden

Zu diesem Zweck wurden die Salzdrüsen mehrerer 1—2jähriger Silbermöwen (*Larus argentatus* Brünn) 2—6 Monate lang durch perorale Verabreichung von Seewasser steigender Konzentration (bis 5% Seesalz) zur Exkretionsaktivität gereizt oder durch Verabreichung von Seewasser fallender Konzentration (bis 0,00005% Seesalz) zur Funktionsruhe gebracht [38, 40]. Das Drüsengewebe wurde nach Dekapitation der Vögel entnommen und unter vergleichbaren Bedingungen nach den heute gebräuchlichen Präparationsmethoden der elektronenoptischen Untersuchung zugeführt [38].

Zum histochemischen Nachweis der exkretpflichtigen Natrium- und Chloridionen wurde Salzdrüsengewebe eines exkretorisch aktiven Tieres in einem Gemisch von Osmiumtetroxyd und Kaliumhexahydroxoantimonat bzw. in einem solchen von Osmiumtetroxyd und Silberacetat oder -lactat fixiert [37, 42].

Zu vorläufigen morphologischen Vergleichszwecken diente die Rectaldrüse von Glatthaien (*Mustelus laevis* Risso). Dieses von einer Adria-Exkursion stammende Material wurde in 5% Glutaraldehyd [55] (0,04 in Cacodylatpuffer, pH 7,2) 7 Tage lang fixiert und transportiert und 1 Std lang in OsO₄/K₂Cr₂O₇ nach WOHLFARTH-BOTTERMANN [79] nachfixiert.

Vergleichende Anatomie

Auch in ihrer mikroskopischen Anatomie weisen die Salzdrüsen von Seevögeln und die Rectaldrüse von Haien Gemeinsamkeiten und Unterschiede auf. Die etwa bohnenförmigen Salzdrüsen setzen sich je nach Tierart aus 6—42 [47] mehr oder weniger parallel verlaufenden Drüsenlappen zusammen (Abb. 1a und b). Die Lobi (Abb. 1c) bestehen aus einem zentralen Sammelkanal, von dem verzweigte und am proximalen Ende blind geschlossene Drüsenschläuche ausstrahlen [24, 38, 41, 47]. Dieser Bauplan ist auch in der Rectaldrüse (Abb. 1d) von Haifischen verwirklicht [7], sodaß die gesamte etwa wurstförmige Drüse einem Lobus der Salzdrüsen vergleichbar ist.

Ein wesentlicher Unterschied besteht in der Blutversorgung der Salzdrüsenlobi und der Rectaldrüse. Im ersten Falle ziehen kleine Arterien
von der Peripherie der Lobi zum Sammelkanal, biegen dort um und

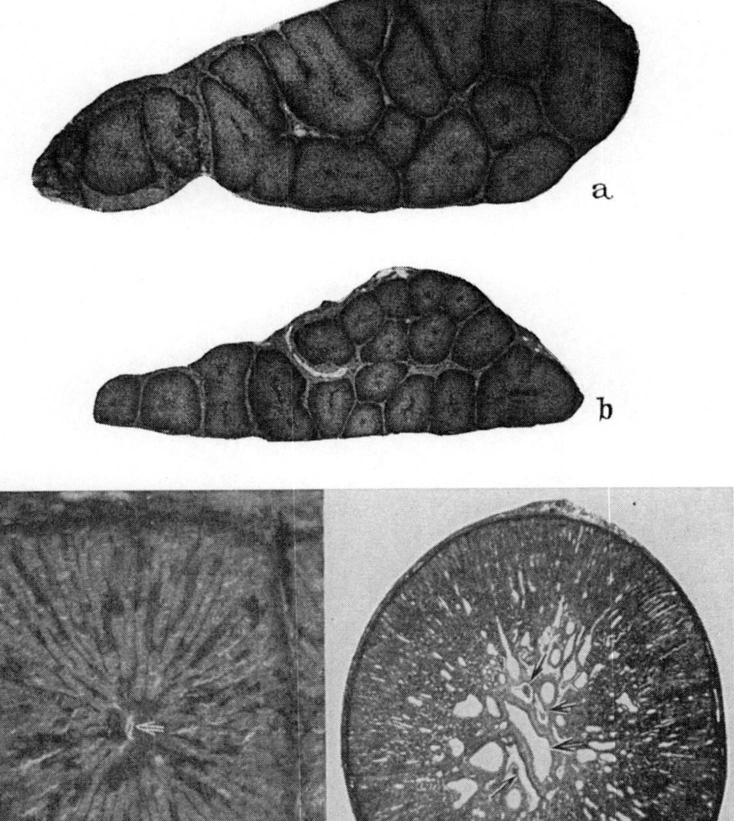

Abb. 1a u. b. Querschnitte durch die mittlere Region der Salzdrüse einer Silbermöwe, die
2 Monate lang bei salzreicher (a) und einer, die 2 Monate lang bei salzarmer (b) Diät gehalten wurde. Beachte den Größenunterschied der Drüsen und Lobi (etwa 60%)! Zenker;
Feulgen/Lichtgrün. Vergrößerung 9:1. c Querschnitt durch einen Lobus der Salzdrüse
(Silbermöwe). ⇐ Zentraler Sammelkanal. Zenker; Azan. Vergrößerung 46:1. d Querschnitt
durch die Rectaldrüse des Glatthaies. ⇐ zentraler Sammelkanal; ← venöse Blutgefäße.
Zenker; Azan. Vergrößerung 21:1 (Orig.)

verzweigen sich kapillar. Die Kapillaren laufen an den Drüsenschläuchen
entlang zur Peripherie zurück, wo sie in Sammelvenen einmünden [24].
Im zweiten Falle erfolgt die arterielle Blutzufuhr ebenfalls von der

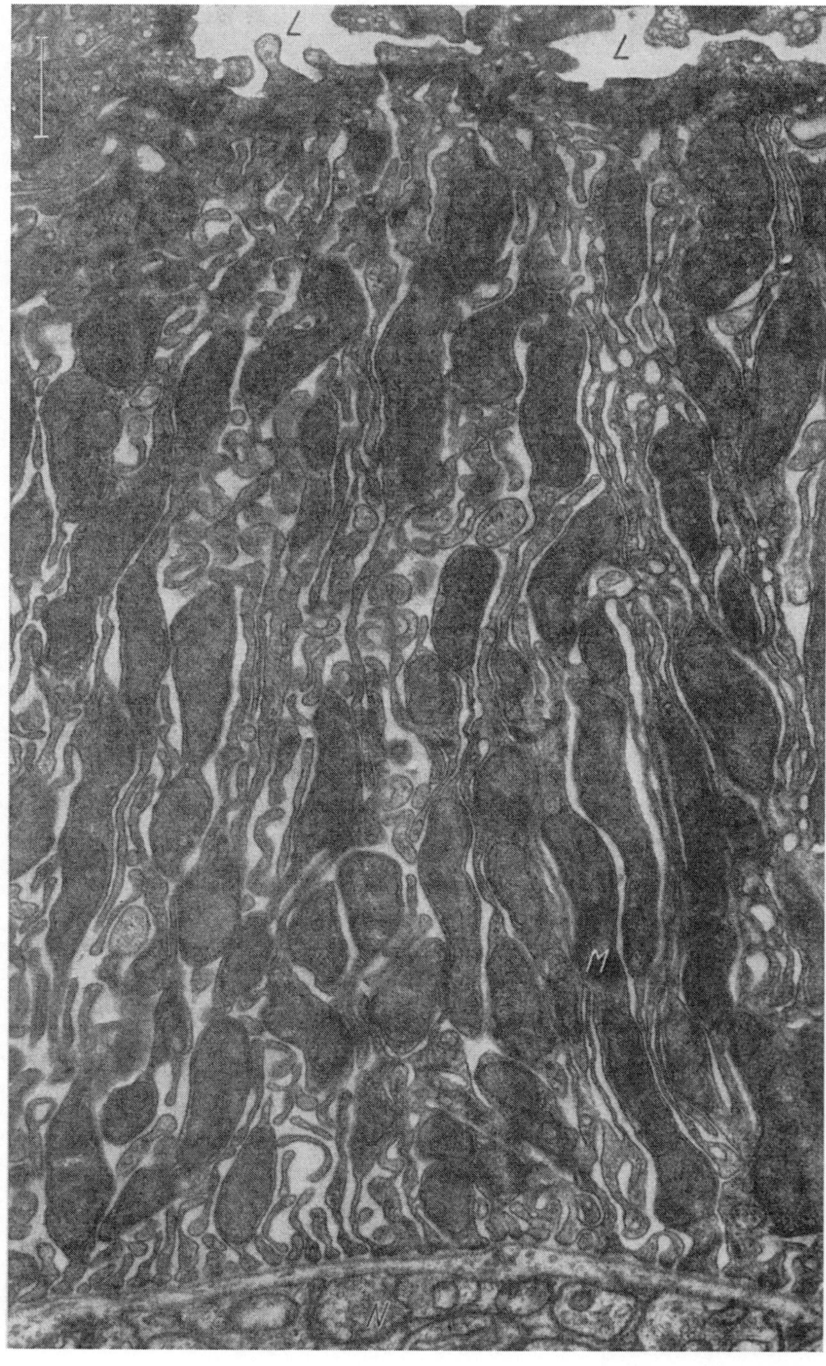

Abb. 2. Ausschnitt aus einem Tubuluslängsschnitt der aktiven Salzdrüse (Silbermöwe).
L = Tubuluslumen; M = Mitochondrien, N = markloser Nerv. Beachte die vielen von der
Epithelbasis bis zum Apex durchlaufenden extrazellulären Faltenkanäle! Vergrößerung
13 000:1 (Orig.)

Drüsenperipherie her, die venöse Ableitung der ebenfalls in den Zwickeln der Tubuluspackung verlaufenden Kapillaren jedoch im Zentrum der Drüse [7, 19]. Während bei den Salzdrüsen zwischen kapillarem Blut- und tubulärem Exkretfluß ein Gegenstrom besteht [24, 58], fließen Blut und Exkret in den entsprechenden Schlauchsystemen der Rectaldrüse in gleicher Richtung [7, 19].

Möglicherweise drückt sich in der Multiplikation der Salzdrüsenlobi und der Art ihrer Blutversorgung eine höhere morphologische Differenzierung dieses offenbar auch funktionell stärker differenzierten Organs aus, die, wie noch gezeigt wird, bei den Tubulusepithelzellen sicherlich zum Ausdruck kommt.

Der physiologische Befund der nervösen Steuerung der Salzdrüsenaktivität [25] spiegelt sich morphologisch in der reichen Versorgung mit marklosen Nervenfasern (Abb. 2, vgl. auch [22]) wieder. Die Nerven verlaufen ähnlich wie die Blutgefäße intertubulär, häufiger in engerer Nachbarschaft zu den Gefäßen als zu den Drüsenschläuchen. Synapsen konnten bisher nicht einwandfrei beobachtet werden.

Bei der Rectaldrüse, deren Tätigkeit allem Anschein nach keiner nervösen Steuerung unterliegt [8], konnte ein entsprechender Befund bisher nicht erhoben werden.

Vergleichende Mikromorphologie

Im elektronenoptischen Dünnschnittbild sind die Tubulusepithelzellen der Salzdrüsen durch periodische tiefe Einfaltungen vorwiegend der basalen Zellmembran in zahlreiche extrazelluläre Faltenkanäle, deren Gesamtheit als basales Labyrinth [54] bezeichnet wird, und in dünne Cytoplasmaleisten aufgefiedert (Abb. 2).

Das Tubulusepithel beherbergt zahlreiche langgestreckte Mitochondrien, die auf einen hohen oxydativen Energiestoffwechsel hindeuten und das morphologische Korrelat der osmotischen Arbeit der Zellen darstellen. Die Mitochondrien sind in den Cytoplasmaleisten aufgereiht und verleihen ihnen ein perlschnurartiges Aussehen. Die Zellmembranen schmiegen sich den Mitochondrien so eng an, daß diese innige Lagebeziehung für einen aktiven Stoffaustausch zwischen Zellinnerem und basalem Labyrinth nicht ohne Bedeutung sein dürfte (s. auch Abb. 9a).

Die Faltenstrukturen lassen eine strenge Orientierung von der Zellbasis zum Apex erkennen, worin sich besonders deutlich die polare Differenzierung der Zellen als morphologisches Korrelat des polaren, transepithelialen Stofftransportes manifestiert.

Eine derartige Differenzierung ist schon ziemlich früh von PEASE [52] als recht charakteristisch für Epithelzellen mit transzellulären Transportfunktionen erkannt worden. Während in anderen Exkretionsorganen diese Faltenstrukturen zumeist auf die basale Hälfte der Zelle beschränkt bleiben, sind sie in der Salzdrüse besonders stark ausgeprägt und durchlaufen den gesamten Zelleib von der Basis bis zum Apex, wo die extrazellulären Faltenkanäle über desmosomale Schlußleisten in das

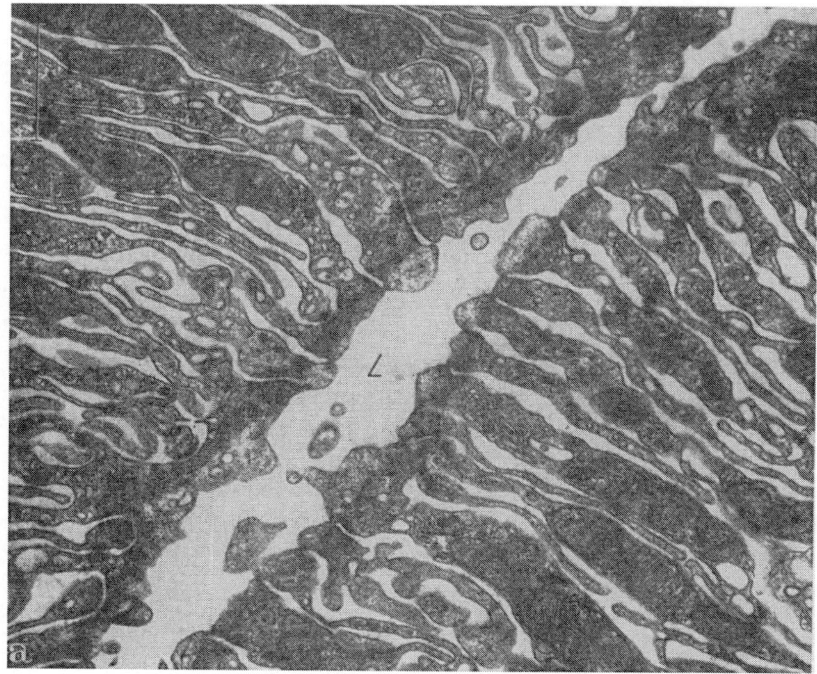

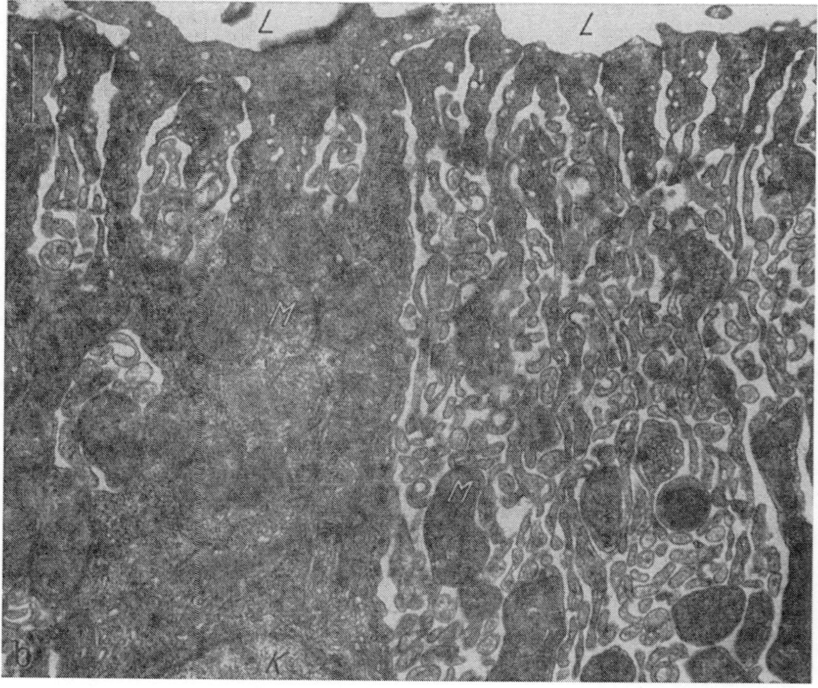

Abb. 3a u. b. Ausschnitte aus Tubuluslängsschnitten der aktiven Salzdrüse (Silbermöwe). K = Zellkern; L = Tubuluslumen; M = Mitochondrien. Beachte die zahlreichen in die Drüsenlichtungen einmündenden Faltenkanäle (in Abb. 3a) sowie die unterschiedlichen Strukturaspekte und Verzahnungen der Zellen (in Abb. 3b)! Vergrößerungen a 15100:1; b 11500:1 (Orig.)

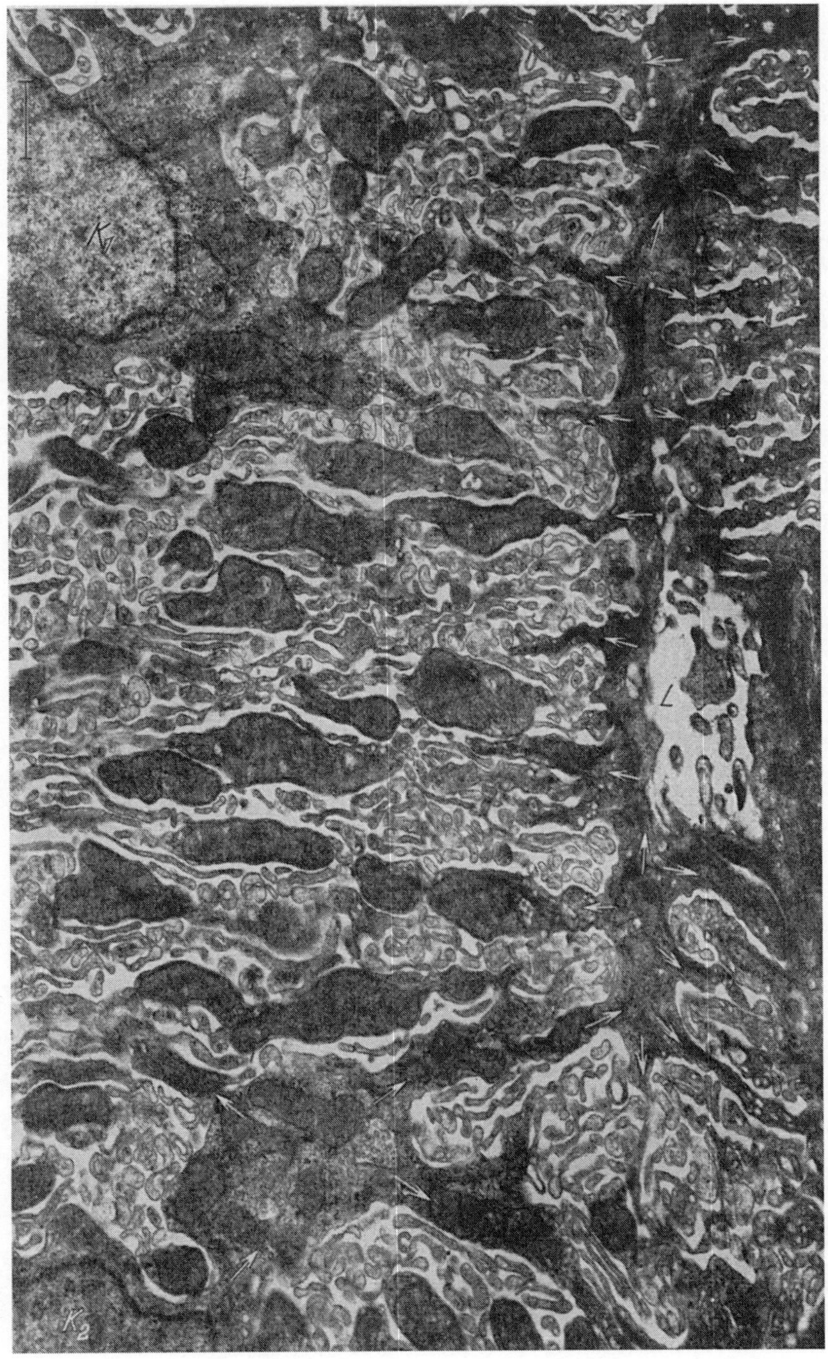

Abb. 4. Ausschnitt aus einem Tubuluslängsschnitt der aktiven Salzdrüse (Silbermöwe). K_1 und K_2 = Zellkerne; L = Tubuluslumen. Verfolge das dem Kern K_2 zugehörige Cytoplasma (*Pfeilmarkierungen*) und beachte die Polymorphie und Verzahnungen der Epithelzellen! Vergrößerung 10000:1 [aus *43*]

Tubuluslumen einmünden (Abb. 3a). Diese stärkere Differenzierung dürfte wiederum Ausdruck der hohen Transportleistung sein.

Die verwirrende Fülle der Faltenmembranen (Zellmembranen) läßt in der Regel keine Entscheidung über die Frage zu, wo eine Zelle aufhört und die Nachbarzelle beginnt, bzw. welche Cytoplasmaleiste zu welcher Zelle gehört (vgl. die Region zwischen den beiden Zellkernen K_1 und K_2 in Abb. 4). Nur in günstigen Schnitten wie in Abb. 4 kann das einem Zellkern (K_2) zugehörige Cytoplasma über größere Strecken verfolgt werden *(Pfeilmarkierungen!)*. Gleichzeitig läßt sich die Polymorphie dieser Epithelzellen ermessen, und es wird deutlich, daß ein großer Teil der Faltenstrukturen durch *Verzahnungen von dünnen Fortsätzen* (mit hellem und dunklem Cytoplasma, s. auch Abb. 3b) *verschiedener Zellen* zustande kommt [*38*].

Die Tubulusepithelzellen der Rectaldrüse (vgl. auch [*7*]) sind nicht so stark aufgegliedert. Die Zellen sind vielmehr nur durch breite scheibenartige Ausläufer ineinandergeschachtelt, die sich dann ihrerseits durch schmale und kurze Fortsätze seitlich miteinander verzahnen. Dadurch erhält man im Dünnschnittbild sich deutlich abhebende „Bänder" mit stark mäandrierenden Zellmembranen, die von der Basis bis zum Apex ziehen, sowie basale und apikale Faltenstrukturen, wenn die scheibenartigen Ausläufer anderer Zellen sich nur in die Basis oder den Apex einer Zelle schieben (Abb. 5b).

Die Mitochondrien liegen gruppenweise in den scheibenartigen Ausläufern. Die einzelnen Gruppen werden durch die „Bänder" der lateralen Verzahnungsleisten getrennt (Abb. 5b). In den Salzdrüsen beherbergen dagegen diese Verzahnungsleisten, die die ganze Epithelhöhe durchziehen können, selbst die Mitochondrien. Im Gegensatz zu den Salzdrüsen enthalten die Mitochondrien der Rectaldrüse relativ viele intramitochondriale Partikel (Abb. 5b, *P*), deren Anzahl bei Drüsen mit hoher Exkretionsrate nach Bulger [*7*] abnimmt. Diese Partikel werden heute entsprechend der Vorstellung von Weiss [*74*] als Kationen-Austauscherstrukturen angesehen und besitzen nach Peachy [*50*] ein besonderes Adsorptionsvermögen für divalente Kationen.

Im Vergleich zu den Salzdrüsentubuli sind die Lumina der Rectaldrüsentubuli von größerer und unterschiedlicher Weite (ihr Durchmesser kann die Epithelhöhe übersteigen), und die Apices besitzen Mikrovilli in wechselnder Zahl. Die apikale Zellmembran trägt auf der freien Oberfläche einen Belag von fadenförmigen Strukturen (Abb. 5c), die sog. *Antennulae microvillares* [*81*], die an den Mikrovilli einer Reihe von Epithelzellen [*5, 51*] sowie in ähnlicher Form an der Zellmembran von Amöben [*80*] gefunden wurden, dagegen nicht in den Salzdrüsentubuli nach den angewandten Präparationsmethoden beobachtet werden konnten. Dieser fädige Belag, der aus sauren Mucopolysacchariden bestehen soll [*7, 80*], wurde von verschiedenen Autoren im Zusammenhang mit den Transportfunktionen der jeweiligen Zellen diskutiert [*7, 11*]; sie sollen speziell die aus dem Lumen zu resorbierenden gelösten Substanzen binden (vgl. auch die experimentellen Befunde von Brandt und Pappas

[6] an Amöben und die vergleichende Darstellung von BRANDT [5]).
Im Lichte dieser Erkenntnis scheint der Belag der apikalen Zellmembran
auch auf eine resorptive Tätigkeit der Rectaldrüsenzellen hinzudeuten.

Weiterhin enthalten die Apices mancher Rectaldrüsenzellen im
Unterschied zu den Orbitaldrüsen tröpfchenartige Einschlüsse (Abb. 5a,
S), die an Schleimtröpfchen erinnern und die möglicherweise Ausdruck

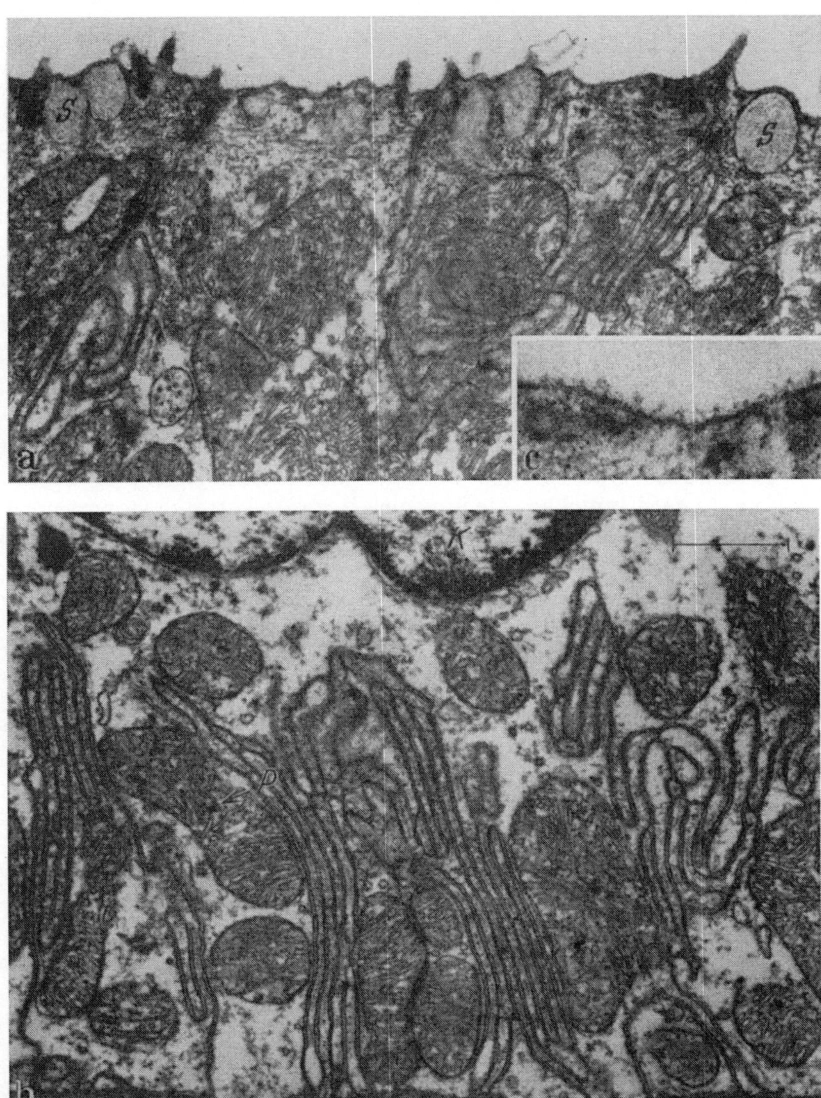

Abb. 5a u. b. Ausschnitt aus dem apikalen (a) und basalen (b) Tubulusepithel der Rectaldrüse (Glatthai). K = Zellkern; S = Sekrettröpfchen; P = intramitochondriale Partikel.
Beachte die sekundären Verzahnungsleisten der Zellmembranen und vergleiche die Epithelbasis in Abb. 5b mit der in Abb. 2! Vergrößerungen 15000:1. c fädiger Belag der apikalen
Zellmembran. Vergrößerung 55000:1 (Orig.)

einer weiteren Funktion dieser Zellen[1] sind oder lediglich das von Bulger [7] gesuchte Material zum Aufbau der *Antennulae microvillares* liefern [vgl. auch *51*].

In das relativ enge Tubuluslumen der Salzdrüsen ragen dagegen nur vereinzelte und häufig unregelmäßig gestaltete Zellzotten; ein ausgesprochener Bürstensaum, der in der Regel an der apikalen Oberfläche resorbierender Epithelien anzutreffen ist, fehlt den Salzdrüsentubuli völlig. Der spärliche Besatz mit Mikrozotten, denen heute zumeist eine resorptive Funktion zugemessen wird, deutet möglicherweise daraufhin, daß resorptive Prozesse bei der Exkretkonzentrierung in den Salzdrüsentubuli im Gegensatz zu den Verhältnissen in der Niere eine untergeordnete Rolle spielen. Da den Salzdrüsentubuli — im Gegensatz zu den Nephronen — den Glomeruli vergleichbare Druckfiltrationsapparate[2] zur Erzeugung eines Primärexkretes fehlen, würde eine alleinige Exkretkonzentrierung durch Reabsorption von Wasser aus dem Tubuluslumen einen Hin- und Rücktransport durch das Tubulusepithel bedeuten und damit die zelluläre Transportleistung vervielfältigen. Auch für die Verlegung der Exkretion und Reabsorption auf verschiedene Tubulusabschnitte haben wir keine sicheren Hinweise gefunden. Die morphologischen Verhältnisse legen daher den Verdacht nahe, daß die Exkretkonzentrierung im wesentlichen durch einen aktiven und selektiven Transport von NaCl durch das Tubulsepithel zur Drüsenlichtung erfolgt.

Als mögliche Transportwege springen dem Morphologen zwei Strukturen ins Auge: Einmal könnten die aus dem Kapillarraum stammenden exkretpflichtigen Stoffe durch die dünnen Cytoplasmaleisten, d. h. *intrazellulär*, zum anderen aber auch durch die Faltenkanäle, d. h. *extrazellulär*, zum Tubuluslumen wandern. Im letzteren Falle wäre zwischen dem Exkret im Tubuluslumen und dem Blut im Kapillarraum gar keine kontinuierliche zelluläre Trennwand vorhanden. Da das Kapillarendothel, wenn auch nur vereinzelt, Poren aufweist [*39*], wären als kontinuierliche Trennschichten lediglich die Basalmembranen des Endothels und Epithels und die apikalen Schlußleisten anzusehen.

Die für die Tubulusepithelzellen der Salzdrüsen typischen Differenzierungen sind nun keine starren, sondern fluktuierende Strukturen, die einen funktionell bedingten Strukturwandel erkennen lassen.

---

[1] *Anmerkung bei der Korrektur:* Eine nach Drucklegung dieses Referates an der Rectaldrüse des Grundhaies (*Galeorhinus galeus* L.) begonnene Untersuchung hat deutlich erwiesen, daß die Salzexkretion nicht die einzige Funktion dieses Organs darstellt. Vor allem die superfiziellen Epithelzellen des zentralen Sammelkanals besitzen offenbar eine *sekretorische* Funktion. Ihr Cytoplasma ist gekennzeichnet durch auffallend viele Golgi-Komplexe und beherbergt im apikalen Bereich zahlreiche Sekrettröpfchen. Manche Zellen werden fast völlig von Sekrettropfen angefüllt, so daß der Kern und Reste des Cytoplasmas an die basale Zellperipherie abgedrängt sind. Die chemische Natur des Sekretes ist noch unbekannt, doch dürfte es sich — nach der feinstrukturellen Organisation der Zellen zu urteilen — vermutlich um Schleimstoffe handeln.

[2] Die sieb- oder reusenartige Architektur der Tubuluswände in den Salzdrüsen würde zwar im Verein mit den Basalmembranen die morphologischen Voraussetzungen für eine Filtration erfüllen, die hohe Mitochondrienpopulation spricht allerdings gleichzeitig zu Ungunsten eines solchen passiven Vorganges.

Funktionelle Morphologie der Salzdrüsenzellen

Im Tubulusepithel der exkretorisch aktiven Salzdrüse (Abb. 2 und 6a) erscheinen die extrazellulären Faltenkanäle bzw. Interzellularen ziemlich weit und laufen in strenger Orientierung von basal nach apikal. Die länglichen Mitochondrien zeigen die gleiche Ausrichtung. Sie beherbergen dicht gepackte Cristae und besitzen im ganzen eine relativ große Elektronendichte.

Bei der ruhenden Drüse (Abb. 6b) erscheinen die Faltenkanäle zumeist kollabiert und ihre Orientierung ist mancherorts verlorengegangen. Die Mitochondrien sind in der Regel mehr oder minder geschwollen, zeigen eine Auflockerung ihrer Innenstrukturen und sind vergleichsweise im ganzen deutlich weniger elektronendicht.

Eine Erörterung über die Verläßlichkeit dieser im statistischen Sinne für den jeweiligen Funktionszustand typischen Strukturaspekte würde den Rahmen des Vortrages sprengen. Es erscheint jedoch angebracht zu vermerken, daß beispielsweise auch in der aktiven Drüse sich mitunter dicht benachbarte Bereiche des Tubulusepithels mit den typischen Merkmalen der Aktivität bzw. mit denjenigen einer kürzeren oder längeren Funktionsruhe finden lassen (Abb. 3b, vgl. auch [40]). Diese Strukturunterschiede können ihre Ursache nur in einem unterschiedlichen physiologischen Zustand der Zellen haben, was besagt, daß der Funktionsrhythmus nicht in allen Zellen synchron verläuft. Gleichzeitig werden durch solche Bilder aber auch die Befunde an Salzdrüsen verschiedener Funktionszustände (Abb. 6a und 6b) gestützt, bei denen der Funktionszustand — soweit überschaubar — die einzige Veränderliche ist. Bei standardisierten Präparationsbedingungen darf daher ein bestimmter Strukturaspekt zumindest als Ausdruck für den jeweiligen zellulären Funktionszustand gewertet werden. Speziell das unterschiedliche Aussehen der Mitochondrien in den beiden Funktionszuständen der Drüse kann als Indikator für ihre unterschiedliche Stoffwechselaktivität gelten [4].

Bei Salzdrüsen nach langer Funktionsruhe (6 Monate salzarme Diät) werden in manchen Zellen — vorwiegend der proximalen Tubulusabschnitte — die Faltenstrukturen völlig zurückgebildet, so daß jetzt die Grenzmembranen der tangierenden Zellen deutlich hervortreten (Abb. 6c). Außerdem wird die Anzahl der Mitochondrien reduziert.

Diese morphologischen Befunde finden eine Parallele in den histochemischen Beobachtungen von ELLIS et al. [22] an den Nasendrüsen junger Hausenten. Werden die Vögel nach dem Schlüpfen bei salzreicher Diät gezogen, dann zeigen die Tubulusepithelien vom 3. Tag an einen größeren Reichtum an Phospholipoiden und eine größere Aktivität an oxydativen Enzymen als diejenigen gleichaltriger Kontrolltiere, die bei salzarmer Diät gehalten wurden.

Bei beiden Versuchsgruppen wurden auch entlang der Drüsenschläuche histochemische Unterschiede gefunden. So war die Aktivität von Succinodehydrogenase und Cytochromoxydase in den acidophilen Zellen der distalen Tubulusabschnitte bedeutend größer als in den basophilen Zellen der proximalen Schlauchabschnitte. Die Autoren unterscheiden daher in den Drüsenschläuchen zwei verschiedene Zelltypen, die distal gelegenen *sekretorischen Zellen* und die proximal gelegenen, relativ undifferenzierten *Terminalzellen*, denen sie eine histogenetische Funktion zumessen.

Aus rein morphologischer Sicht ist in etwa eine solche Unterscheidung auch bei den Nasendrüsen der Silbermöwe möglich. Die Epithelzellen sind entlang der Drüsenschläuche zwar im Prinzip alle gleichartig differenziert, jedoch nimmt der Grad ihrer Differenzierung von proximal nach distal zu. Während die proximalen Zellen nur eine vergleichsweise geringe Mitochondrienpopulation besitzen und durch Faltenstrukturen weniger stark aufgegliedert sind, bleiben bei den distalen Zellen infolge der Auffiederung des Zelleibes und des dichten Besatzes der dünnen Cytoplasmafächer mit Mitochondrien die Ribosomen überwiegend auf mehr oder minder breite

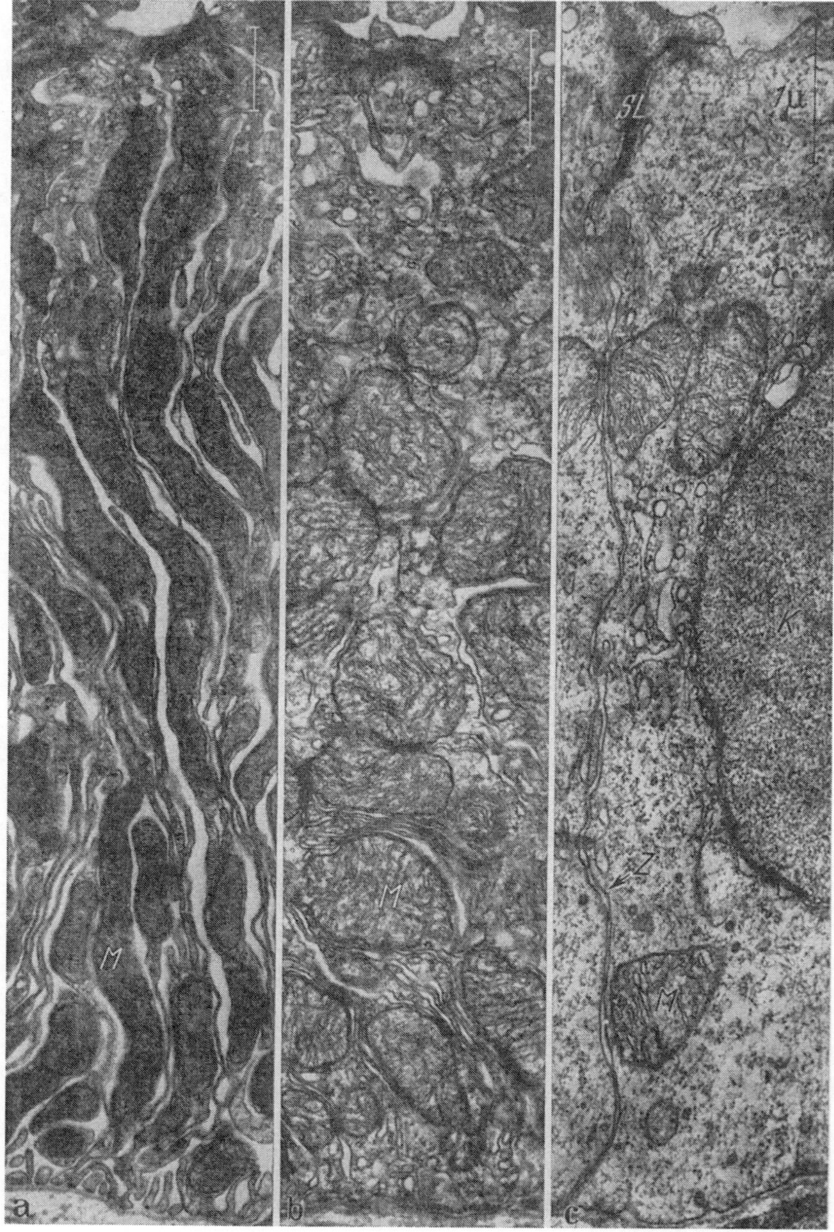

Abb. 6a—c. Ausschnitte aus dem Tubulusepithel der aktiven (a), der inaktiven (b), und einer Salzdrüse (Silbermöwe) nach 6 monatiger Funktionsruhe (c). K = Zellkern; M = Mitochondrien; Z = Zellmembranen zweier tangierender Epithelzellen. Beachte die Reduktion der Faltenstrukturen und die Veränderungen und Abnahme der Mitochondrien mit zunehmender Dauer der Funktionsruhe! Vergrößerungen a 11 000:1; b 16 000:1; c 18 500:1 (aus [43])

Zonen beschränkt, die vom Zellkern zum Apex ziehen und dem Tubulusquerschnitt das Bild eines basophilen Radspeichenmusters verleihen [38, 40]. Da die Verkürzung der Drüsenschläuche nach längerer salzarmer Diät (Abb. 1b) offenbar mit Rückdifferenzierungen in den proximalen Schlauchabschnitten (s. Abb. 6c) einhergeht, könnte die Verlängerung der Drüsenschläuche nach längerer salzreicher Diät (s. Abb. 1a) möglicherweise auf eine Proliferation in den proximalen Schlauchabschnitten beruhen.

Das Vorkommen des basalen Apparates (wie ich die Gesamtheit des basalen Labyrinthes, der Faltenmembranen und der Mitochondrien bezeichnen möchte) bei verschiedenen Tierklassen in einer Reihe phylogenetisch recht verschiedenartiger Organe mit transzellulären Transportfunktionen läßt heute keine Zweifel an seiner Bedeutung für die transzellulären Transportprozesse mehr zu. Als weitere Indizien für diese Ansicht dürfen seine extensive Ausbildung in den Salzdrüsen mit hoher Transportleistung gelten und vor allem seine unterschiedliche Ausprägung in Salzdrüsen verschiedener Funktionszustände in einer Deutlichkeit, wie sie beispielsweise auch in den Malpighischen Gefäßen von *Drosophila* im Larven-, Puppen- und Imaginalstadium beobachtet wurde [75, 76]. Über diese prinzipielle Erkenntnis hinaus, die auf PEASE [52] zurückgeht, sind bis heute allerdings außer der Druckwandlerhypothese von RUSKA, MOORE und WEINSTOCK [54] keine konkreten und gesicherten Vorstellungen über den Wirkungsmechanismus dieses Apparates entwickelt worden.

Nachweis der Ionentransportwege

Um nun weitere Hinweise für eine Korrelation von Struktur und Funktion zu gewinnen, haben wir versucht, die Transportwege durch das Tubulusepithel der Salzdrüse durch den topochemischen Nachweis der transportierten Natrium- und Chlorionen zu ermitteln [37, 42].

Bei der exkretorisch aktiven Salzdrüse läßt sich zumindest ein Lokalisationsort der Natrium- und Chloridniederschläge mit Sicherheit voraussagen, nämlich die Lichtungen der Sammelkanäle und Drüsenschläuche, in denen das Exkret abfließt. Das histochemische Ergebnis entspricht dieser Erwartung [42] und ist in Abb. 9b für das Natriumpräzipitat im Tubuluslumen dokumentiert.

Für das Epithel der Drüsenschläuche, durch das die Ionen transportiert werden, lassen sich keine Prognosen stellen. Eine Aussage über die transepithelialen Transportwege basiert allein auf der Lokalisationstreue der histochemischen Methoden, die nach verschiedenen hier nicht näher zu erörternden Kriterien [vgl. 42] aber gegeben ist.

Der Befund, daß die Natriumniederschläge in der Regel intrazellulär, d. h. in den dünnen Cytoplasmafächern und in den Mitochondrien (Abb. 7), die Chloridniederschläge hingegen extrazellulär in den Faltenkanälen (Abb. 8) liegen, spricht für eine getrennte Bewegung der beiden Ionenarten auf zwei verschiedenen Wegen. *Die Natriumionen werden offenbar intrazellulär, die Chloridionen dagegen extrazellulär durch das Tubulusepithel zur Drüsenlichtung transportiert,* wie es in Abb. 10 schematisch veranschaulicht ist.

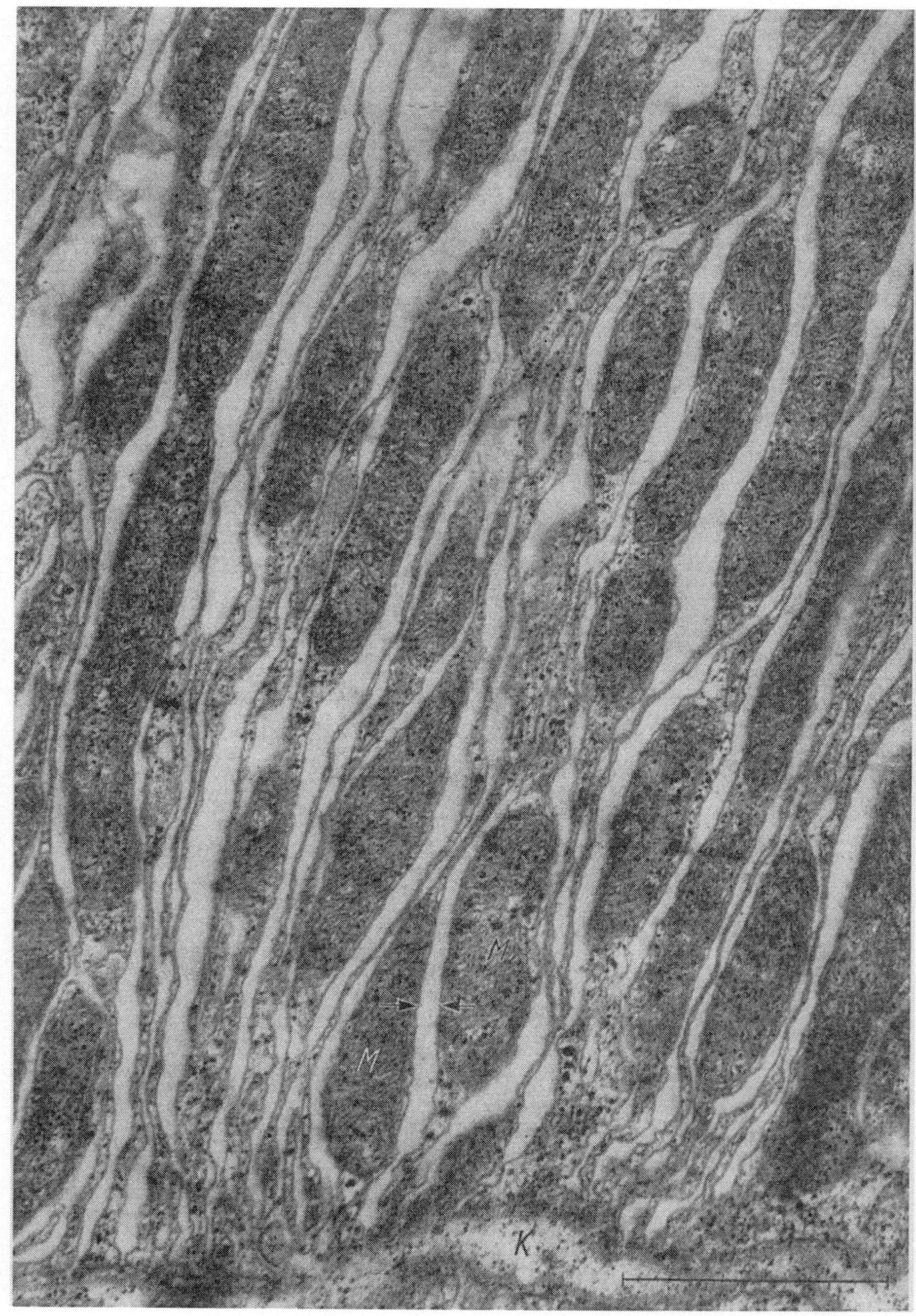

Abb. 7. Ausschnitt aus der Basis des Tubulusepithels einer aktiven Salzdrüse (Silbermöwe) mit intrazellulären Natriumniederschlägen. Die extrazellulären Spalträume des basalen Labyrinthes (→ ←) sind niederschlagsfrei. K = interstitielles Kollagen; M = Mitochondrien. Fix.: 1% OsO_4 + 2% K [Sb(OH)$_6$]. Vergrößerung 35 000:1 (aus [42])

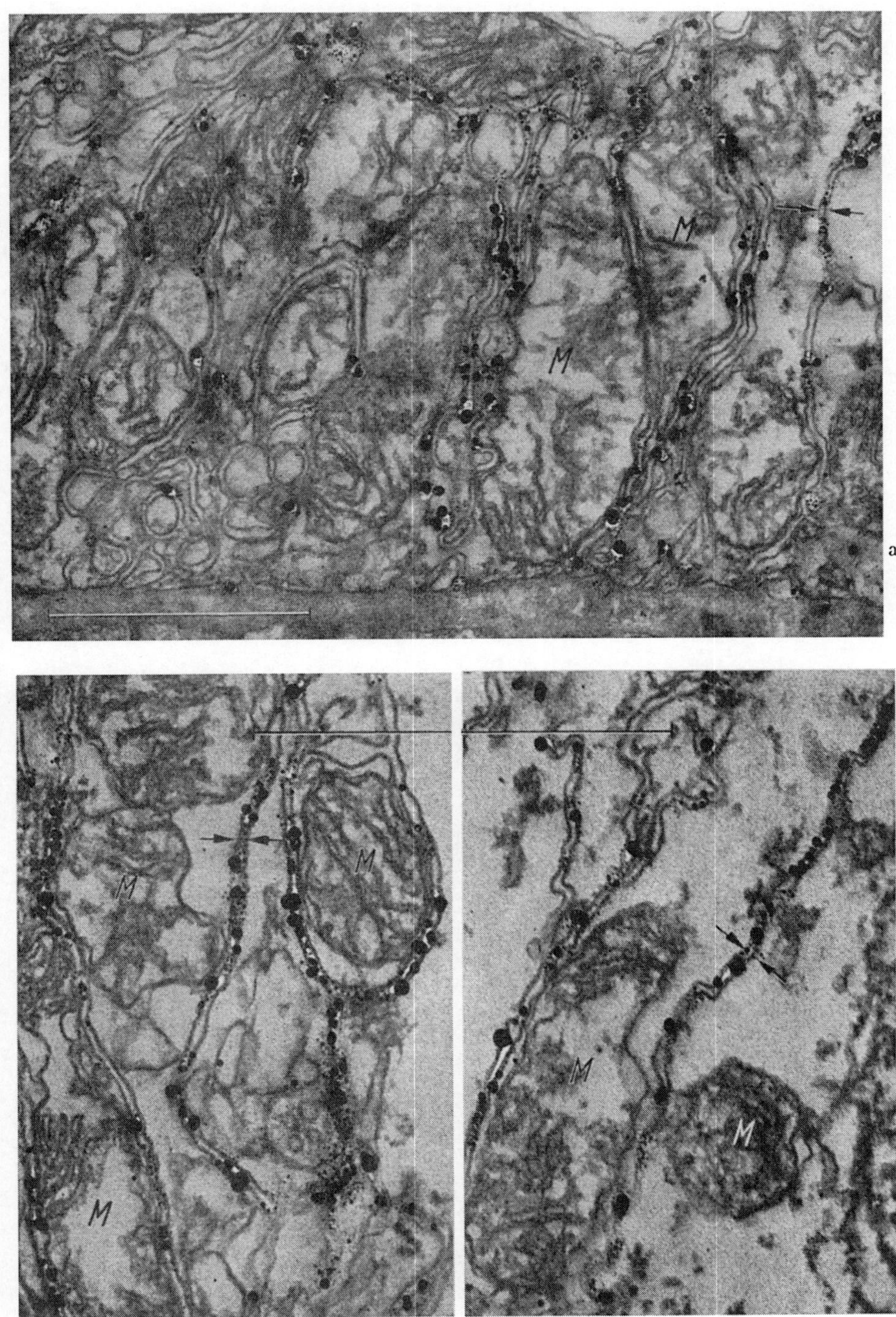

Abb. 8a—c. Ausschnitt aus basalen Bereichen des Tubulusepithels einer aktiven Salzdrüse (Silbermöwe) mit extrazellulären Chloridniederschlägen. Diese markieren deutlich den Verlauf der Faltenkanäle (→ ←). M = Mitochondrien. Fix.: 1% OsO_4 + 0,5% Ag-Lactat. Vergrößerungen a 36000:1; b und c 57000:1 (aus [42])

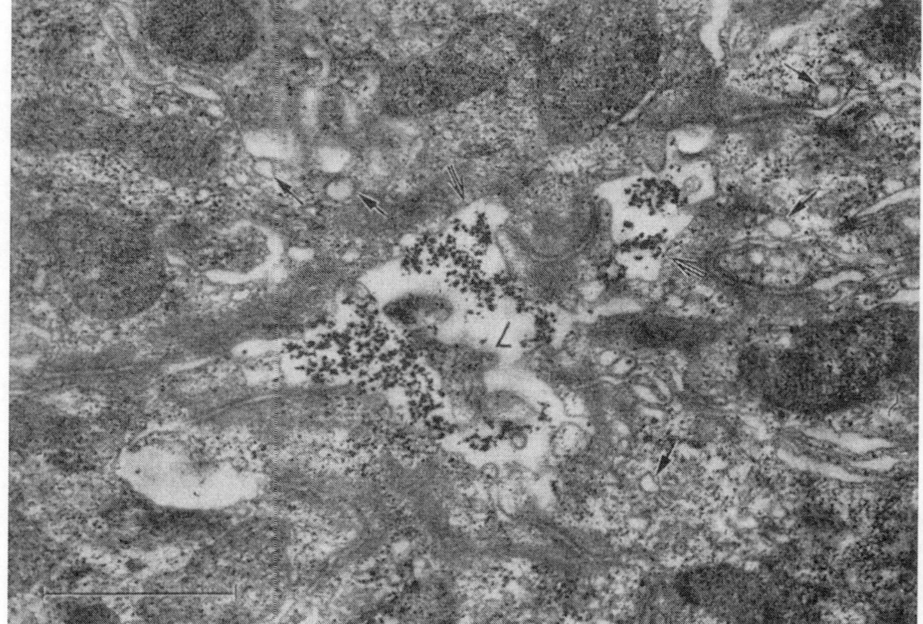

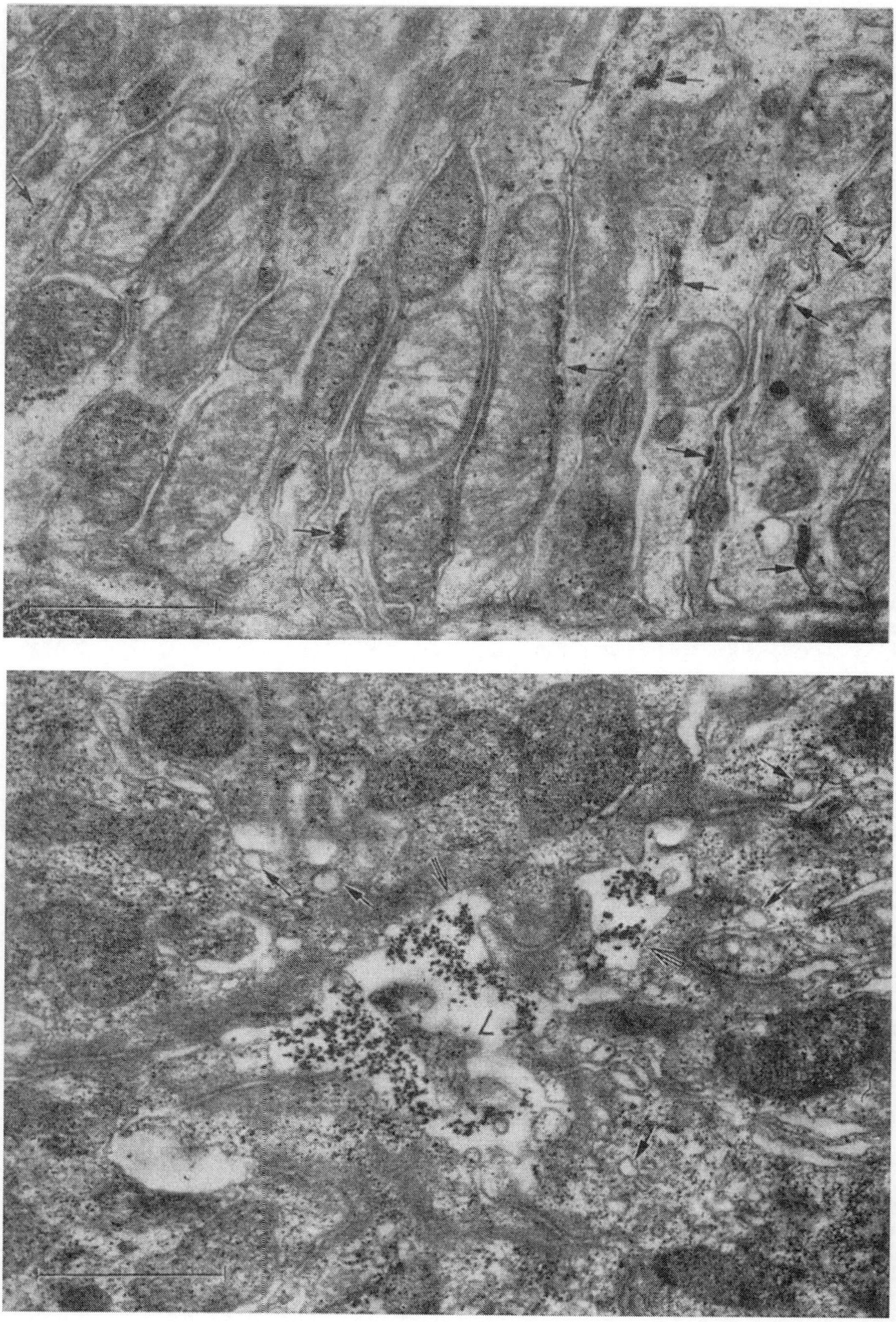

Abb. 9a. Ausschnitt aus der Basis des Tubulusepithels einer aktiven Salzdrüse (Silbermöwe)
mit extrazellulären Natriumniederschlägen (*Pfeilmarkierungen*) und einer unterschied-

Stellenweise findet man aber auch extrazelluläre Natriumpräzipitate in den Faltenkanälen und zwar immer dann, wenn die das basale Labyrinth begrenzenden Cytoplasmafächer und die darin liegenden Mitochondrien eine unterschiedliche Niederschlagsdichte aufweisen (Abb. 9a).

Auf den ersten Blick scheint die extrazelluläre Lokalisation der Natriumpräzipitate in Abb. 9a der obigen Feststellung über den intrazellulären Natriumtransport zu widersprechen. Bei Berücksichtigung der morphologischen Erkenntnisse über die Architektur des Tubulusepithels fügt sie sich jedoch sinnvoll in das entworfene Bild der Transportwege ein, ja ermöglicht darüber hinaus noch einen gewissen Einblick in den Transportmechanismus:

> Die Faltenkanäle können — wie gezeigt wurde (Abb. 3b, 4) — zumindest stellenweise durch innige Verzahnungen verschiedener Epithelzellen zustande kommen, d. h. sie werden in solchen Fällen von Fortsätzen verschiedener Zellen gesäumt, deren Funktionsrhythmus — wie ebenfalls gezeigt wurde (Abb. 3b) — nicht immer synchron zu verlaufen braucht.

Die Cytoplasmaleisten und Mitochondrien mit unterschiedlicher Niederschlagsdichte gehören wohl zu *verschiedenen* Zellen, und die aus der Regel fallende Präzipitatverteilung in Abb. 9a deutet auf eine unterschiedliche Transportaktivität der hier das basale Labyrinth begrenzenden Zellen hin. Dieser Befund wäre demnach so zu erklären, daß Natriumionen in einer zur Niederschlagsbildung ausreichenden Konzentration weiter in die Faltenkanäle hineindiffundieren können, wenn die angrenzenden Zellen nicht gleichzeitig optimal transportaktiv tätig sind, während sonst ihre Konzentration in diesen Räumen durch die rasche Aufnahme in die Zelle niedrig gehalten wird.

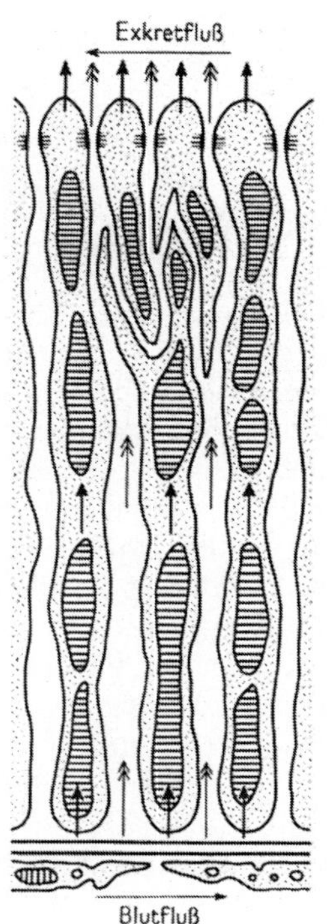

Abb. 10. Schematische Darstellung des Tubulusepithels der aktiven Salzdrüse (Silbermöwe) mit den Ionentransportwegen. ← Weg der Natriumionen; ⇇ Weg der Chloridionen (und vermutlich auch des Wassers) (aus [*43*])

Weiterhin läßt die unterschiedliche Niederschlagsdichte in den Mitochondrien deren unmittelbare Beteiligung am Natriumtransport vermuten. An der Fähigkeit dieser Zellorganellen zur osmotischen Arbeit kann heute nicht mehr gezweifelt werden [vgl. u. a. *1, 14, 28, 44*]. Der Sinn der dichten Lagebeziehung der Mitochondrien zu den Faltenmembranen

---

lichen Niederschlagsverteilung in den Mitochondrien. Fix.: 1% OsO_4 + 2% K [Sb(OH)₆]. Vergrößerung 25 300:1 (aus [*42*])

Abb. 9b. Ausschnitt aus einem Tubulusquerschnitt (apikaler bzw. zentraler Bereich) einer aktiven Salzdrüse (Silbermöwe) mit Natriumniederschlägen im Tubuluslumen (*L*). Die Doppelpfeile weisen auf kleine Kristalle, die in der apikalen Zellmembran zu liegen scheinen. Die präzipitatfreien Vakuolen (*einfache Pfeilmarkierungen*) werden als Strukturen der Wasserrückresorption gedeutet [vgl. *40, 42*]. Fix.: 1% OsO_4 + 2% K [Sb(OH)₆]. Vergrößerung 25 300:1 (aus [*42*])

könnte also seine Erklärung dadurch finden, daß die Aufnahme der Natriumionen aus dem basalen Labyrinth in die Zelle durch die mitochondriale Ionenpumpe [12] vermittelt wird.

Die histochemischen Präzipitatlokalisationen zwingen zu dem Schluß, daß die Passage der *Natriumionen* durch das Tubulusepithel *intrazellulär* verläuft, die der *Chloridionen* dagegen *interzellulär*.

Physiologische Einordnung

Der Nachweis der Ionentransportwege hat eine Reihe zellphysiologischer Konsequenzen, und es erhebt sich die Frage, inwieweit sich diese Befunde in physiologische Vorstellungen über den NaCl-Transport einordnen lassen. Der Befund des intrazellulären Natriumtransportes legt — wegen der Möglichkeit einer direkteren Beeinflußbarkeit durch die Zelle — die Vermutung nahe, daß dieser der aktive Teil des NaCl-Transportes (ähnlich wie in der Froschhaut [73]) ist. Zu der gleichen Feststellung kommen HOKIN und HOKIN [31] auf Grund ihrer biochemischen Untersuchungen an Salzdrüsen. Demzufolge werden die Natriumionen aktiv von den Epithelzellen in die Drüsenlichtung ausgeschieden, während die Chloridionen — durch die Faltenkanäle bzw. Interzellularen — dem elektrochemischen Gradienten des vortransportierten Natriums passiv nachfolgen. Diese Vorstellungen scheinen in den elektrochemischen Potentialmessungen von THESLEFF und SCHMIDT-NIELSEN [72] eine Stütze zu finden.

So gut der Begriff „aktiver Transport" von physiologischer Seite definiert ist [73], bleibt er im Grunde jedoch rein deskriptiv und verbirgt unser heutiges Unwissen über den biochemischen Mechanismus dieses Lebensphänomens. Dieser deskriptive Begriff erlaubt es, nach dem Transportergebnis den Transmembrantransport und den transzellulären Transport gemeinsam zu klassifizieren, obwohl der letztere sicherlich aus einer Reihe hintereinandergeschalteter Einzelprozesse bestehen dürfte, von denen einige sogar passiver Natur sein mögen.

Physiologische Befunde über Unterschiede in der Ionenkonzentration zwischen Zellinnerem und extrazellulärem Milieu sowie über ihre Abhängigkeit vom Energiestoffwechsel zwingen zu der Annahme sog. Ionenpumpen in den Zellmembranen [77, 78]. Die in dieser Bezeichnung zum Ausdruck kommende figurative Vorstellung stützt sich immer noch vorwiegend auf allgemeine Modellvorstellungen.

Ein Vorschlag zur Erklärung des aktiven Transmembran-Transportes unter Berücksichtigung der stofflichen Gegebenheiten und Umsetzungen wurde durch den HOKINschen Phosphatidsäurezyklus gemacht [30—32], der allerdings nicht ohne Kritik geblieben ist [78].

Der intrazelluläre Natriumtransport durch das Tubulusepithel setzt — wenn wir die Mitochondrien außer acht lassen — einen zweimaligen Transmembran-Transport voraus, nämlich den Transport durch die basale Zellmembran auf der Ingestionsseite und durch die apikale Zellmembran auf der Egestionsseite. Nach den HOKINschen Vorstellungen ist der Na-Influx passiv. Zur Erklärung des aktiven Na-Efflux wird eine bestimmte Lokalisation der den Phosphatidsäurezyklus katalysierenden Enzyme postuliert. Dagegen folgern BONTING et al. [3] aus ihren enzymatischen Studien an Salzdrüsen, daß der aktive Na-Efflux durch die in der apikalen Zellmembran vermutete Transport-ATPase vermittelt wird.

Nach den Niederschlagsverteilungen an der Epithelbasis (s. Abb. 7) muß jedoch auch ein aktiver Na-Influx angenommen werden. Dieser könnte rein formal, wenn die HOKINSCHEN Vorstellungen über die Natriumpumpe zutreffen, durch den wechselseitigen Austausch der beiden Enzyme erklärt werden. Ein solcher Austausch würde den Verlauf des Zyklus umkehren und damit einen Na-Influx bewirken [42]. Dieser Influx ließe sich aber auch möglicherweise durch die Tätigkeit der mitochondrialen Ionenpumpe erklären, so daß der eigentliche Transport durch die basale Zellmembran passiv wäre.

In diese Erörterung müssen auch die Untersuchungen von McFARLAND und SANUI [49] einbezogen werden, die eine Fähigkeit der „Mikrosomen" von Salzdrüsen zur Natrium- und Kaliumbindung nachweisen konnten. Völlig offen bleibt allerdings, was bei den Salzdrüsenzellen unter Mikrosomen morphologisch zu verstehen ist. Das endoplasmatische Retikulum allein dürfte in diesem Fall wohl kaum die Mikrosomenfraktion ergeben, sondern eher die Membranen des basalen Labyrinthes, d. h. die Zellmembranen. Eine Verbindung mit den HOKINSCHEN Vorstellungen könnte theoretisch durch die Annahme der Phosphatidsäure als Ursache für die Kationenbindung hergestellt werden. Da diese Bindungsfähigkeit nach den genannten Autoren anscheinend eine allgemeine Eigenschaft zellulärer Membranen ist und sich unabhängig von der Zellfunktion stets in der gleichen Größenordnung (bezogen auf g/N) bewegt, kann sie allein für die hohe Leistung der Salzdrüse nicht verantwortlich sein [49]. In Verbindung mit der durch die Faltenstrukturen bedingten Vergrößerung der Zellmembranfläche ließe sich jedoch ein Anhaltspunkt zur Erklärung der einseitigen Funktionen, der hohen Leistung und der einzigartigen Struktur dieser Zellen finden.

Die morphologischen und histochemischen Befunde erlauben keine Entscheidung über die Gültigkeit der einen oder anderen Hypothese über den Ionentransport in der Salzdrüse [3, 12, 16, 31, 49], sie schließen aber auch keine aus. Ihre Widersprüchlichkeiten könnten insofern scheinbar sein, als jeweils der Ton auf verschiedene Glieder einer längeren Kette liegt. Es scheint mir, daß in jeder Hypothese ein gewisser Wahrheitsgehalt liegt und lediglich der verbindende Gedanke fehlt. Die Kenntnis der Struktur des Tubulusepithels und der Ionentransportwege ist ein wertvoller Anhaltspunkt für weitere gezielte Studien zur Klärung dieses transzellulären Transportprozesses.

Literatur

[1] BARTLEY, W., and R. E. DAVIES: Biochem. J. **57**, 37 (1954).

[2] BERNARD, G. R., and J. F. HARTMANN: Anat. Rec. **137**, 340 (1960).

[3] BONTING, S. L., L. L. CARAVEGGIO, M. R. CANADY, and N. M. HAWKINS: Arch. Biochem. Biophys. **106**, 49 (1964).

[4] BORUT, A., and K. SCHMIDT-NIELSEN: Amer. J. Physiol. **204**, 573 (1963).

20*

[5] Brandt, P. W.: Circulation **26**, 1075 (1962).

[6] —, and G. D. Pappas: J. biophys. biochem. Cytol. **8**, 675 (1960).

[7] Bulger, R. E.: Anat. Rec. **147**, 95 (1963).

[8] Burger, J. W.: Physiol. Zool. **35**, 205 (1962).

[9] —, and W. N. Hess: Science **131**, 670 (1960).

[10] — — Anat. Rec. **136**, 173 (1960).

[11] Burgos, M.: Anat. Rec. **137**, 171 (1960).

[12] Chance, B., Ch.-P. Lee, R. Oshino, and G. D. V. van Rossum: Am. J. Physiol. **206**, 461 (1964).

[13] Carey, F. G., and K. Schmidt-Nielsen: Science **137**, 866 (1962).

[14] Davies, R. E.: Symp. Soc. exp. Biol. **8**, 453 (1954).

[15] Doyle, W. L.: Anat. Rec. **136**, 184 (1960).

[16] — Exp. Cell Res. **21**, 386 (1960).

[17] — Anat. Rec. **142**, 228 (1962).

[18] — In: Electron Microscopy **2**, YY-5. New York and London: Academic Press 1962.

[19] — Amer. J. Anat. **111**, 223 (1962).

[20] Ellis, R. A., and J. H. Abel: Anat. Rec. **148**, 278 (1964); Science **144**, 1340 (1964).

[21] —, R. A. de Lellis, and Y. H. Kablotzky: Amer. Zoologist **2**, 32 (1962).

[22] —, C. C. Goertemiller, R. A. de Lellis, and Y. H. Kablotzky: Development. Biol. **8**, 286 (1963).

[23] Fänge, R., J. Krog, and O. Reite: Acta physiol. scand. **58**, 40 (1963).

[24] —, K. Schmidt-Nielsen, and H. Osaki: Biol. Bull. **115**, 162 (1958).

[25] — —, and M. Robinson: Amer. J. Physiol. **195**, 321 (1958).

[26] Fawcett, D. W.: Circulation **26**, 1105 (1962).

[27] Goertemiller, C. C., and R. A. Ellis: Amer. Zoologist **2**, 525 (1962).

[28] Green, D. E.: In: Funktionelle und morphologische Organisation der Zelle. S. 86. Berlin-Göttingen-Heidelberg: Springer 1963.

[29] Heinroth, O., u. M. Heinroth: Die Vögel Mitteleuropas. Bd. III. Berlin: Vermühler 1927.

[30] Hokin, L. E., and M. R. Hokin: J. gen. Physiol. **44**, 61 (1960).

[31] — — Symposium on Membrane Transport and Metabolism, p. 204. Prague: Publishing House of the Czechoslovak Academy of Sciences 1961.

[32] — — Lab. Invest. **10**, 1151 (1961).

[33] Holmes, W. N., D. G. Butler, and J. G. Philips: J. Endocr. **23**, 53 (1961).

[34] —, J. G. Philips, and D. G. Butler: Endocrinology **69**, 483 (1961).

[35] Inoue, T.: Science **142**, 1299 (1963).

[36] Komnick, H.: Mikroskopie **17**, 46 (1962).

[37] — Protoplasma (Wien) **55**, 414 (1962).

[38] — Protoplasma (Wien) **56**, 274 (1963).

[39] — Protoplasma (Wien) **56**, 385 (1963).

[40] — Protoplasma (Wien) **56**, 605 (1963).

[41] — Protoplasma (Wien) **58**, 96 (1964).

[42] —, u. U. Komnick: Z. Zellforsch. **60**, 163 (1963).

[43] —, u. K. E. Wohlfarth-Bottermann: Fortschr. Zool. **17**, 1 (1964).

[44] Lehninger, A. L.: Biochem. biophys. Acta **48**, 324 (1961).

[45] McFarland, L. Z.: Nature (Lond.) **184**, 2030 (1959).

[46] — Anat. Rec. **138**, 366 (1960).

[47] — Anat. Rec. **137**, 376 (1960).

[48] — Amer. Zoologist **2**, 349 (1962).

[49] —, and H. Sanui: Proc. Soc. exp. Biol. (N. Y.) **113**, 105 (1963).

[50] Peachy, L. D.: J. Cell Biol. **20**, 95 (1964).

[51] —, and H. Rasmussen: J. biophys. biochem. Cytol. **10**, 529 (1961).

[52] Pease, D. C.: J. biophys. biochem. Cytol. Suppl. **2**, 203 (1956).

[53] Philips, J. G., W. N. Holmes, and D. G. Butler: Endocrinology **69**, 958 (1961).

[54] Ruska, H., D. H. Moore, and J. Weinstock: J. biophys. biochem. Cytol. **3**, 249 (1957).

[55] Sabatini, D. D., K. Bensch, and R. J. Barrnett: J. Cell Biol. **17**, 19 (1963).

[56] Schildmacher, H.: J. Ornithol. **80**, 293 (1932).

[57] Schmidt-Nielsen, K.: Sci. Amer. **200**, 109 (1959).
[58] — Circulation **21**, 955 (1960).
[59] — In: Sekretion und Exkretion. S.269. Berlin-Göttingen-Heidelberg-NewYork: Springer 1965.
[60] —, and R. Fänge: Fed. Proc. **17**, 142 (1958).
[61] — — Auk **75**, 282 (1958).
[62] — — Nature (Lond.) **182**, 783 (1958).
[63] —, C. B. Jörgensen, and H. Osaki: Fed. Proc. **16**, 113 (1957).
[64] — — — J. Physiol. (Lond.) **193**, 101 (1958).
[65] —, and W. J. L. Sladen: Nature (Lond.) **181**, 1217 (1958).
[66] Scothorne, R. J.: Nature (Lond.) **181**, 732 (1958).
[67] — Quart. J. exp. Physiol. **44**, 200 (1959).
[68] — Quart. J. exp. Physiol. **44**, 329 (1959).
[69] — J. Anat. **93**, 246 (1959).
[70] Schwarz, D., u. L. Spannhof: Naturwissenschaften **48**, 414 (1961).
[71] Technau, G. E.: J. Ornithol. **84**, 511 (1936).
[72] Thesleff, S., and K. Schmidt-Nielsen: Amer. J. Physiol. **202**, 597 (1962).
[73] Ussing, H. H.: In: Biochemie des aktiven Transportes. S. 1. Berlin-Göttingen-Heidelberg: Springer 1961.
[74] Weiss, J. M.: J. exp. Med. **102**, 783 (1955).
[75] Wessing, A.: Protoplasma (Wien) **55**, 264 (1962).
[76] — Protoplasma (Wien) **56**, 433 (1963).
[77] Wilbrandt, W.: Fortschr. Zool. **12**, 28 (1960).
[78] — Klin. Wschr. **41**, 138 (1963).
[79] Wohlfarth-Bottermann, K. E.: Naturwissenschaften **44**, 287 (1957).
[80] — Protoplasma (Wien) **52**, 58 (1960).
[81] Yamada, E.: J. biophys. biochem. Cytol. **1**, 445 (1955).

Summary

Despite their different ontogenetic origin, the salt glands of marine birds and the rectal glands of sharks are characterized by similarities in anatomy and in fine structure of their excretory cells. On the other hand, differences in physiology accompany morphological differences between these analogous organs.

In architecture the total rectal gland corresponds to one lobe of the salt gland. The blood supply shows a remarkable difference: between capillary blood and the tubular excretion there is counter-current flow in the salt gland and parallel flow in the rectal gland.

Electron micrographs of the tubular epithelium of active salt glands reveal an abundance of cytoplasmic leaflets with rows of mitochondria, alternating with extracellular channels running from the base to the tubular lumen. This feature is in reality a result of cell interdigitation and to a lesser extent, if at all, of simple infoldings of the basal cell membrane.

In principle, the tubular epithelium of the rectal gland exhibits the same construction, but varies in detail. Here the cells are interdigitated primarily by large spatulate processes containing groups of mitochondria, and secondarily by small microvillous extensions.

The rectal gland evidently has both an excretory and secretory function, the latter being indicated by secretory granules in some cells.

The tubule cells of the salt gland reveal structural fluctuations depending on alterations in the functional state. Glandular inactivity causes disorientation of the channels and swelling of the mitochondria. Finally, at least in the proximal tubules, a reduction of the basal labyrinth can be observed, indicating the retraction of cell processes.

Preliminary attempts to correlate cellular structure and function were made by histochemical precipitation of the transported ions. From the localization of precipitates it is concluded that the pathway of sodium is intracellular and that of chloride ions intercellular. The consequences of these results are discussed in connection with physiological hypotheses concerning sodium chloride transport through the tubular epithelium of the salt gland.

Diskussion

Niesel: Gibt es in der Gallenblase, bei der auch ein aktiver Natriumchloridtransport nachgewiesen ist, ähnliche Strukturen wie in der Salzdrüse, so daß ein ähnlicher Mechanismus vorliegen könnte?

Diamond: In der Gallenblase sieht es anders aus. Am luminalen Ende der Zelle hat die Gallenblase einen Bürstensaum; am basalen Ende ist es verschieden je nach Tierart. Bei Fischen findet man ein basales Labyrinth, allerdings nicht so stark entwickelt wie bei der Salzdrüse. Bei Kaninchen ist überhaupt kein Labyrinth vorhanden.

Kuyper: Sie haben gezeigt, daß das basale Labyrinth in den verschiedenen Stadien sehr verschieden sein kann. Muß man das deuten im Sinne einer Art von ,,Pseudopodienbewegung'' in den basalen Zellteilen?

Komnick: Ich glaube, man muß sich vorstellen, daß eine Entfaltung und eine Faltung dadurch zustande kommen, daß sich einmal diese Fortsätze retrahieren, zum anderen sich wieder vorschieben. So etwas gibt es auch in anderen Epithelien, vielleicht nicht in so starkem Maße. Denken Sie an das Dünndarmepithel der Maus nach den Untersuchungen von Ruska: im Durstzustand sind die lateralen Zellgrenzen extrem gefaltet, nach Wasseraufnahme entfaltet. Das ist im Prinzip das gleiche.

Wessing: Können Sie auf Grund Ihrer histochemischen Untersuchungen sagen, wie das Natriumchlorid aus dem Blutgefäß in die Epithelzelle hineinkommt? Werden dabei die Interzellularen benutzt, oder geht das Natriumchlorid durch die Endothelzelle?

Komnick: Das Endothel der Salzdrüsenkapillaren hat vereinzelt Poren, aber es ist überhaupt kein Vergleich mit den Glomerularkapillaren oder peritubulären Kapillaren in der Niere anzustellen. Ein Teil könnte möglicherweise durch Poren gehen. Wir finden aber regelmäßig in den Endothelzellen auch Niederschläge. Ich möchte annehmen, daß der überwiegende Teil durch die Zelle selbst geht. Man kann natürlich daran denken, daß durch die Poren ein erheblicher Flüssigkeitsstrom stattfinden könnte. Aber die Anzahl der Poren ist zu gering, um diese hohe Leistung zu erklären; denn es wird nicht nur konzentriert, sondern es wird auch eine erhebliche Menge Flüssigkeit abgeschieden. Es sind rund 5% Salz und 95% Wasser. Das Wasser wird immer außer Betracht gelassen. Es ist doch eine erhebliche Menge.

Staubesand: Haben die Poren Diaphragmen?

Komnick: Diese Frage kann ich nicht mit Sicherheit beantworten. Sie wissen, daß die Frage der Diaphragmen lange Zeit umstritten gewesen ist. Ich glaube, erst nach den klaren Aufnahmen, die Rhodin vorgelegt hat, tendiert man immer mehr zu Diaphragmen. Ich muß sagen, in der Salzdrüse sieht es manchmal aus, als seien welche da, manchmal nicht; ein Effekt, den Herr Thoenes früher einmal durch tangentiale Anschnitte erklärt hat. Diese Diaphragmen sind offensichtlich bei verschiedenen Fixierungsmethoden nicht immer gut darstellbar. Das haben wir auch im Referat von Herrn Kümmel gesehen. Es ist natürlich die Frage: Sind sie grundsätzlich vorhanden, oder sind sie nur in einigen Fällen vorhanden? Ich möchte aber doch annehmen, daß sich diese Diaphragmen in ihren Permeabilitätseigenschaften von der Zellmembran unterscheiden. Man kann natürlich nicht sagen, ob die Permeabilität erhöht oder erniedrigt ist.

Staubesand: Woher wissen Sie, daß sie sich deutlich unterscheiden?

Komnick: Weil es keine Zellmembran ist. Was es eigentlich ist, weiß man nicht. Es wird häufig so interpretiert, daß jeweils nur die äußeren Blätter der Zellmembran diese Diaphragmen bilden sollen. Wenn sich die Diaphragmen strukturell von der Zellmembran unterscheiden, liegt vermutlich auch ein funktioneller Unterschied vor. Dann findet man noch eine zentrale knopfartige Verdickung. Über die Funktion dieser Struktur ist überhaupt nichts bekannt.

Staubesand: Vielleicht kann man doch etwas sagen. Wenn man Kapillaren, die solche Poren mit Diaphragmen besitzen, mit Goldsol oder ähnlichen Markierungssubstanzen untersucht, dann findet man die Markierungssubstanzen so gut wie nie bevorzugt an der Stelle, wo sich die Pore befindet (Hammersen, 1964).

Komnick: Ich glaube, das ist kein Gegenbeweis. Die Passage von kolloidalen Partikeln erfordert doch eine größere ,,Porenweite'' als diejenige von Ionen oder

Wassermolekülen. Das ging doch auch aus der Diskussion im Anschluß an den Vortrag von Herrn KÜMMEL und Ihrem Vortrag hervor.

Staubesand: Wenn ich Sie richtig verstanden habe, haben Sie ganz am Anfang Ihres Vortrages gesagt, daß im Bereich der desmosomalen Schlußleisten diese Spalträume mit dem Lumen kommunizieren. Man hat doch eigentlich die Vorstellung gehabt, daß dort, wo Desmosomen sind, sich eben keine Kommunikation befindet, sondern gerade eine Abdichtung ist.

Komnick: Die morphologische Kontinuität der Interzellularspalten ist zweifellos gegeben, aber dazwischen ist eine Kittsubstanz, die Desmosomen, so daß es fraglich ist, ob auch eine funktionelle Kontinuität da ist. Die Desmosomen erscheinen zwar häufig erweitert, aber da möchte ich mich nicht festlegen. Das könnte ein Artefakt sein.

Staubesand: — daß die Desmosomen sich getrennt haben?

Komnick: Ja, möglicherweise in diesem Fall. Die Erweiterungen könnten durch Spannungen während der Fixation oder Polymerisation verursacht worden sein. Andererseits ist die Fähigkeit der Desmosomen, sich zu lösen, hinreichend evident aus den Untersuchungen von PETRY, OVERBECK und VOGELL an der Vaginalschleimhaut. Da muß man eine sehr große Dynamik dieser Struktur zugrunde legen, was um so erstaunlicher ist, als ihre Funktion die Zellhaftung ist, der Zusammenhalt der Zellen im Epithelverband. Aber ich glaube doch, daß sie dynamische Strukturen sind und daß sie auch für Wasser und Ionen bis zu einem gewissen Grade durchlässig sind. Es erhebt sich jetzt allerdings die Frage nach der Zonula occludens, die den eigentlichen Verschluß der Interzellularspalten darstellt. Ich habe mit meinen Methoden diese Zonula occludens in der Salzdrüse bisher nicht beobachten können. Das schließt aber nicht aus, daß sie da ist. Ich finde häufig vor den Schlußleisten größere Chloridniederschläge. Das habe ich früher interpretiert in der Annahme, daß hier die Durchlässigkeit etwas geringer ist und daß ein gewisser Stau eintritt. Wenn wir uns das Bild der Zonula occludens vergegenwärtigen, sehen wir recht eindrucksvoll den morphologischen Verschluß der Interzellularfuge durch die dichte zentrale Schicht. Transponieren wir jedoch den molekularen Aufbau der Zellmembranen in dieses Bild, so haben wir in der Mitte zwei Proteinschichten; wir erhalten in etwa eine ähnliche Struktur wie sie von DANIELLI für die sog. Proteinporen der Zellmembran postuliert wurde. Das heißt, nach dem molekularen Aufbau der Zonula occludens zu urteilen, ist eine gewisse Durchlässigkeit für Ionen nicht ohne weiteres auszuschließen, wenn sie auch erheblich geringer sein mag als in den Interzellularspalten selbst.

Ich habe, wie gesagt, die Zonula occludens in der Salzdrüse bisher nicht beobachtet. Ihr Fehlen wäre an sich erstaunlich, da sonst keine Barriere zwischen dem konzentrierten Exkret im Tubuluslumen und der nicht konzentrierten Flüssigkeit an der Basis bestünde. Nehmen wir einmal an, sie sei in der Salzdrüse vorhanden; nehmen wir weiter an, sie sei völlig undurchlässig und besäße auch keine Dynamik, wie sie bei der strukturell in etwa vergleichbaren Myelinlamelle in Abhängigkeit von osmotischen Kräften gezeigt worden ist; dann müßte das Schema der Transportwege, das lediglich in grober Vereinfachung die große Linie zeigen soll, insofern modifiziert werden, als die Chloridionen auf dem letzten Ende ihres Weges doch noch intrazellular transportiert werden. Im Corneaendothel des Kaninchens hat KAYE 1962 einen ähnlichen Fall gezeigt. Dort werden kolloidale Partikel in basaler Richtung normalerweise durch die Interzellularen transportiert, und zur Umgehung der Desmosomen macht die Zelle — man könnte sagen — partielle Cytopempsis. Obwohl im apikalen Cytoplasma der Salzdrüsenzellen auch häufig Vesikel angetroffen werden, möchte ich hier aber weniger an einen Vesikeltransport denken als vielmehr an einen Transmembrantransport.

Klingenberg: You suppose that the sodium transport is primarily intracellular and the chloride transport is intercellular. Now, first of all, of course, one has always to have a cation and anion accompanying each other, so they cannot really be separated. There must be some cation which neutralizes the chloride if you have only chloride in the intercellular space. Which cation could it be?

Komnick: Das kann ich nicht sagen. Es könnte Kalium, Calcium und sogar teilweise Natrium sein. Ich habe nur Nachweismethoden für Chloridionen und Natrium, und auch die sind nicht quantitativ. Wenn wir in der Regel in den Interzellularspalten keine Natriumniederschläge sehen, so bedeutet das nicht, daß dort kein Natrium ist, sondern nur, daß dort die Konzentration geringer ist, so daß es durch die Methode nicht erfaßt wird. Diese Bilder des Chlorid- und Natriumnachweises können nicht gedeutet werden in dem Sinne, daß die Ionen quantitativ getrennt sind und daß im Falle der Interzellularen außer Chloridionen keine anderen Ionen vorhanden sind.

Klingenberg: Maybe in this respect one should have more details on the ion exchange mechanism which you indicated. One has the intracellular leaflet and the intercellular space running side by side from the basal to the apical end. One would possibly have an intercellular stream of chloride ions and then have a continuous increase of concentration in the apical direction. This would be maintained by a continuous active ion exchange of the sodium, which increases the intracellular sodium concentration.

The question is now, whether the mitochondria are directly involved in the ion transport. I think this is improbable also in view of the discussion following Prof. Schmidt-Nielsen's paper. They supply energy; in what way we do not understand; but I do not think that the role of the mitochondria is to directly pump the sodium. There is some unknown mechanism driving an ion exchange pump, which exists in the infolded membranes.

Komnick: Da stimme ich Ihnen zu; die Pumpe könnte auch in der Membran liegen. Aber kann man nach den morphologischen und histochemischen Befunden die Rolle der Mitochondrien nur auf die Energieversorgung beschränken? Es gibt zahlreiche Arbeiten in der Literatur, die die Mitochondrien als Ionenpumpen verantwortlich machen.

Klingenberg: Isolated mitochondria can do ion accumulation at an extremely high rate, but they always do it at a very low efficiency actually. However, the energy efficiency seen in these organs is always much higher than is seen in the model experiments with isolated mitochondria. I think the isolated mitochondria are „artificially" driven to accumulate ions by driving proton exchange, and this can be shown for the transport of potassium, calcium and magnesium, and more recently also for sodium. But I do not think this is a mechanism the mitochondria really perform in vivo, to drive intra/extracellular pumps. They may do this in the cell to maintain a certain high potassium concentration which the mitochondria always have.

Komnick: Daß sie es isoliert tun, beweist doch, daß sie prinzipiell die Fähigkeit haben. Chance interpretiert seine Befunde über die Cytochrombestimmungen auch dahingehend, daß vorwiegend die Mitochondrien als Kationenpumpen arbeiten. So habe ich ihn verstanden.

Klingenberg: Well, the efficiency of the transport mechanism in isolated mitochondria is always much lower than the efficiency of the unknown transport mechanism in the salt gland.

Solomon: I would like to speak, perhaps, to amplify what Dr. Klingenberg said in one sense, namely we have recently measured the uptake of potassium by mitochondria and though it is an extrordinarily interesting phenomenon it is very much slower indeed, by probably at least one order of magnitude and maybe two, than the processes which Prof. Schmidt-Nielsen has spoken of. The second thing that I want to say, is that I am profoundly disturbed unless evidence can be given to indicate that between the time that the tissue was killed and the section was taken, there was no opportunity whatsoever for the ions to move, and so far as I can see, the only way that this opportunity can be gainsaid is if the tissue is frozen at a temperature of the order of $-180°$ C because at room temperature ions can diffuse. The third thing I wanted to know is how you saw the sodium and why you did not see the potassium.

Komnick: Zur letzten Frage: weil ich keine Nachweismethode für Kalium habe. Die Frage der Diffusionsgefahr habe ich früher [42] einmal ausführlich diskutiert. Ich möchte jetzt nur grundsätzlich sagen: Wenn wir ein Gewebe fixieren oder anfixieren, dann hat es sich doch gezeigt, daß wir es nicht in allen Aktivitäten — sagen wir einmal grob — abtöten, sondern gewisse Enzymaktivitäten bleiben noch eine Zeitlang erhalten. Sonst könnten wir keine histochemischen Nachweise durchführen. Daß auch ein histochemischer Ionennachweis mit Hilfe der Elektronenmikroskopie möglich ist, haben nach meiner Ansicht prinzipiell die Phosphatasenachweise gezeigt. Der elektronenmikroskopische Nachweis z. B. der lysosomalen sauren Phosphatase oder der membrangebundenen ATPase im anfixierten Gewebe ist nach dem Reaktionsablauf nichts anderes als ein Ionennachweis und hat gezeigt, daß nicht alle Enzymaktivitäten durch die Fixation restlos zerstört werden. Wenn also noch gewisse Enzymaktivitäten, die letztlich für diese Konzentrationsunterschiede wohl verantwortlich sind, erhalten bleiben, scheint mir die Diffusionsgefahr — zumindest zwischen den verschiedenen Kompartimenten — nicht in diesem Maße bedenklich zu sein.

Diamond: Ich möchte mich zu dem Schluß äußern, daß der Durchfluß des Natriums innerhalb der Zelle erfolgt und der des Chlorids zwischen den Zellen. Dieser getrennte Fluß ist aus einfachen physikalischen Grundsätzen unmöglich. Wir gehen doch davon aus, daß überall Elektroneutralität aufrechterhalten werden muß. Nun ist es möglich, daß elektrische Ladungen getrennt werden können, wenn es sich um einen Plattenkondensator handelt. Aber die Ladungen, die man durch einen Kondensator trennen kann, sind gering, sie betragen nur 0,00001% von der Gesamtladung. Eine faßbare quantitative und noch weniger eine qualitative Trennung von Ladungen ist durch einen Kondensator nicht möglich. Es gäbe wohl die Möglichkeit, daß Elektroneutralität aufrechterhalten würde durch einen Fluß von anderen Ionen. Aber nach den Angaben SCHMIDT-NIELSENs wissen wir, daß das Exkret der Salzdrüse fast ausschließlich aus Natrium und Chlorid besteht. Nun ist es möglich, daß an der gleichen Stelle die Konzentration von Natrium und Chlorid verschieden ist, weil vielleicht noch andere Ionen vorhanden sind. Aber da wir schon wissen, daß nur Natrium und Chlorid einen richtigen Durchfluß haben, so muß der Durchfluß von Natrium und der Durchfluß von Chlorid über denselben Weg verlaufen. Das Elektronenmikroskop kann uns bei der Beantwortung der Frage helfen, wo dieser Weg liegt. Aber das Mikroskop kann uns nicht zu dem Schluß bringen, daß diese Wege verschieden sind, weil dieser Schluß von vornherein aus elektrischen Gründen ausgeschlossen ist.

Niesel: Diese Schlußfolgerung ist nicht ganz richtig. Ich glaube, die Befunde von Herrn KOMNICK können nur dadurch gedeutet werden, daß die Membran als Ionenaustauscher dient, und zwar an der Innenseite als Kationenaustauscher und an der Außenseite als Anionenaustauscher. Dann ist es durchaus möglich, daß sich Natrium und Chlorid trennen und die Elektroneutralität erhalten bleibt. Nur muß dann die Membran, wie Herr KOMNICK schon sagte, in apikaler Richtung fließen und ein Membranfluß mit den getrennten Ladungen in Richtung auf den Apex stattfinden. Dann müssen sich im Lumen des Ausführungsganges die beiden Ionen vereinigen. So bleibt alles elektrisch neutral. Ich möchte Herrn HOKIN fragen, ob das mit seinem Phosphatidexperiment übereinstimmt, ob es also sein kann, daß hier Membranfluß mit dem Phosphatid-Effekt gekoppelt ist.

L. Hokin: I am not one who adheres very strongly to membrane flow as a mechanism of mass movement of substances such as ions, but I do not think we can rule it out. I would personally feel that the various kinetic studies that we have done in the salt gland do not suggest such a mechanism.

With respect to the particular membranes which are involed in the active transport one can break the problem down into several possibilities. One can say that the transport occurs at the basal membrane and the movement across the apical membrane is passive. If one says this, then one has to admit that there would be very high concentrations of sodium chloride inside the cell. This worries me, since Prof. SCHMIDT-NIELSEN, I believe, has shown that high sodium chloride is inhibitory to the respiration of the salt gland cell. This suggests that any mechanism which involves an elevated intracellular sodium chloride concentration — such as a basal membrane transport mechanism — would be unlikely. The second possibility is that the transport is at the apical membrane. We have preferred this hypothesis because we found that the intracellular sodium very rapidly equilibrates with the extracellular sodium when salt gland slices are suspended in media containing various concentrations of sodium chloride, and along with other tissues such as the frog skin it would seem reasonable that the sodium chloride would diffuse across the basal membrane passively and then be actively transported at the apical membrane. However, I do not believe that there is any direct evidence for this mechanism. It is perhaps only a little more compatible with existing information.

I think there is a third possibility, which cannot be ruled out, and that is that the sodium chloride is pumped into vesicles in the basal region, and the vesicles move up and fuse with the apical membrane — a sort of cytopempsis mechanism. You do see many vasicles in the salt gland, and I wonder, perhaps, whether some comments might be made about them. One of the problems with this mechanism is whether the rates of vesiculation and vesicle flow would be rapid enough. We once did do some calculations based on the rates of water uptake in amoeba from Prof. Holter's work, and assuming the usual kind of Q_{10} of 2 we came up with values for the volume of sodium chloride secretion which were not unreasonable. This hypothesis would get around the problem of elevated intracellular sodium chloride, since the sodium chloride would be sequestered within vesicles. Again, I do not think there is very good experimental support for this hypothesis, but I would like to hear something from the morphologists about the possible significance of these vesicles in the salt gland and particularly whether you see more of them in the stimulated gland.

Niesel: I think, all the arguments you have given would conform with the model here. The membrane, if it has flown up to the apex, could go into pinocytosis, and the sodium and chloride could combine and be brought out. That could be possible; and if the active transport would be only at the apex it would not be reasonable that the mitochondria are placed throughout the cell.

Komnick: Es gibt im apikalen Cytoplasma häufig sehr viele Vesikel. Aber ihr Vorkommen ist zu unregelmäßig, als daß es bisher mit einem bestimmten Funktionszustand sicher korreliert werden konnte. Ich habe sie früher mit einer eventuellen Rückresorption in Verbindung gebracht.

Schmidt-Nielsen: I should like to add a few words to the various hypotheses which propose vesicles and mechanical movement of membranes. The gland secretes per minute approximately its own volume of fluid, and any models that are proposed should be reasonably compatible with this fact. This is particularly pertinent in regard to the vesicle hypothesis. I do not believe that we see enough vesicles in the pictures, whether from active or passive glands, to permit vesicle movement as a reasonable explanation.

Klingenberg: I think, the active transport must be there on the closest possible site to the site where the mitochondria are. Can we agree on this?

Various Speakers: No.

Ullrich: May I ask, whether, in the intracellular space together with the water also the ATP is swept to the apical part of the cell where the transport is going on.

Klingenberg: In that case the ADP has to diffuse back to the mitochondria to be rephosphorylated for a relative large distance against the water flow at the same rate as ATP is swept with the water flow. I think this is a very improper mechanism. It is much more efficient to have the change of ATP and ADP between the membranes of the folds and the closely neighbouring mitochondria.

Transportwege in den Harnkanälchen der Säugerniere

Von

WOLFGANG THOENES, Würzburg

Mit 16 Abbildungen[1]

Schon die rein *quantitative* Betrachtung des Harnbereitungsprozesses in der Säugerniere, in dessen Verlauf das Harnvolumen auf etwa 1% des Ausgangswertes reduziert wird, läßt die Bedeutung der transmuralen Austauschvorgänge erkennen, welche in dem an das Glomerulum anschließenden Kanälchensystem zwischen Tubulusharn und Kapillarblut ablaufen. Darüber hinaus hat uns die Nierenphysiologie gerade in den letzten Jahren ungeahnt detaillierte Informationen über *qualitative* Schritte der Harnbereitung vermittelt. Auch diese sind an transmurale Stoff- bzw. Flüssigkeitsbewegungen in den verschiedenen Kanälchenabschnitten gebunden.

Im folgenden möchte ich vom morphologischen Standpunkt aus die Wege betrachten, die beim Austausch zwischen Kanälchenlumen und Blutkapillare beschritten werden. Dabei sind in jedem Nephronabschnitt gleichermaßen drei Bauelemente zu unterscheiden: Epithel, (tubulokapilläre) Basalmembran, porenhaltiges Endothel. Wenn die tubuläre und kapilläre Basalmembran nicht miteinander verschmolzen sind, wie es im Nierenmark die Regel ist, kommt als viertes und zugleich komplizierendes Element der interstitielle Raum hinzu, der zwischen die beiden Basalmembranen eingeschaltet ist.

Während porenhaltiges Kapillarendothel und Basalmembran von Region zu Region keine grundsätzlichen Unterschiede aufweisen und in funktioneller Beziehung als relativ konstant einzuschätzen sind, variiert die Struktur des Epithels in weiten Grenzen. Funktionsunterschiede von einem Nephronabschnitt zum anderen sind danach in erster Linie auf das Epithel zu beziehen. Es wird daher auch im Mittelpunkt unserer Betrachtungen stehen.

Überblickt man das Kanälchensystem des Nephrons (einschließlich der Sammelrohre) unter cytomorphologischem Aspekt, so lassen sich grob zwei Gruppen von Harnkanälchen unterscheiden, solche mit *mitochondrienreichen* Epithelien (proximaler und distaler Tubulus,

---

[1] Die beigefügten elektronenmikroskopischen Aufnahmen stammen — bis auf die Abb. 5 und 7 (Maus) — von der Rattenniere. Fixierung 1% OsO$_4$, Einbettung Vestopal W, Nachkontrastierung (teilweise) mit Uranylacetat. Porter-Blum-Mikrotom. Elmiskop I. Zeichenerklärung für die Abbildungen s. S. 340.

jeweils mit Pars convoluta und Pars recta) und solche mit *mitochondrien-armen* Epithelien (dünner Teil der Henleschen Schleife, Sammelrohr). Es ist sicher kein Zufall, daß die ersteren ausschließlich in der Rinde, die letzteren ausschließlich im Mark (bzw. in den Markstrahlen) gelegen sind (vgl. die Abgrenzung einer corticalen und einer medullären *Funktions*einheit innerhalb der *anatomischen* Einheit des Nephron [*49, 50*]).

Neben anderen Strukturmerkmalen kommt der Größe des Chondrioms für die funktionsmorphologische Betrachtung von Zellen bzw. Zellpopulationen insofern eine entscheidende Bedeutung zu, als die Synthese energiereicher Phosphate zur Bestreitung des zellulären Energiehaushaltes im Säugerorganismus zu mehr als 95% über die Atmungskettenphosphorylierung erfolgt [*23*], welche nahezu ausschließlich an die Mitochondrien und hier wieder besonders an die Cristae mitochondriales gebunden ist (vgl. [*16*]). Von dem gesamten Energieaufkommen entfällt ein relativ kleiner Teil auf den zellulären Erhaltungs- bzw. Strukturumsatz [*27*], so daß Zahl, Größe und Cristareichtum der Mitochondrien in grober Näherung als Maß für den *Funktionsumsatz* einer Zelle zu werten sind. Unterschiede in der Chondriomgröße bekommen besonderes Gewicht, wenn es sich um ein in den Grundzügen einheitliches Parenchym handelt. Das trifft in besonderem Maße für das Nephronepithel zu, dessen Hauptaufgabe im Stoffaustausch zwischen Blut einerseits und Primär- bzw. Sekundärharn andererseits liegt. Unter den hier stattfindenden Austauschvorgängen spielt der Transport von Kristalloiden, insbesonderere Ionen, eine führende Rolle. Die Tatsache, daß es sich dabei vielfach um einen sog. Bergauf-Transport handelt, der stark energiezehrend ist, rechtfertigt die besondere Beachtung des Chondrioms als Zeichen der epithelialen Transportaktivität.

A. Harnkanälchen mit mitochondrienreichen Epithelien

Hierzu gehören proximaler und distaler Tubulus in ihren gewundenen und gestreckten Teilen. Die Gemeinsamkeiten in der Epithelstruktur, die diese Tubulusabschnitte (ausgenommen Pars recta I) miteinander verbinden, sind in Abb. 1 zum Ausdruck gebracht. Betrachten wir zunächst das

proximale Tubuluskonvolut,

in dessen Verlauf nach heutiger Kenntnis 60—80% der Primärharnmenge in das peritubuläre Kapillarsystem reabsorbiert werden. Nachdem lange Zeit der kolloidosmotische Druck des peritubulären Kapillarblutes für die Reabsorption allein verantwortlich gemacht worden war, kann nunmehr als gesichert gelten, daß die Reabsorption in erster Linie auf einem aktiven Transport von seiten des Tubulusepithels beruht ([*8, 44, 48, 54*], dort weitere Lit.). Es erhebt sich die Frage, wie die Epithelstruktur zu diesen zellulären Transportvorgängen in Beziehung gesetzt werden kann.

Zu ihrer Beantwortung ist es notwendig, einige in diesem Zusammenhang besonders wichtige Strukturmerkmale der Epithelien bzw. des Epithelverbandes im proximalen Konvolut (vgl. [*1, 31, 35, 43, 49*]) herauszustellen (Abb. 2). Das sind: 1. der *Bürstensaum* an der lumenwärtigen Zelloberfläche, der sich strukturell in nichts von dem Bürstensaum z. B. des Dünndarmepithels unterscheidet, 2. die tiefen *Einfaltungen der basalen Zellmembran*, die mit einer intensiven gegenseitigen *Verzahnung* der Tubulusepithelien verbunden sind (Abb. 3), 3. mit den unter 2. genannten Phänomenen unmittelbar zusammenhängend: die

Entwicklung eines extra- bzw. interzellulären „*basalen Labyrinthes*",
das als ein kompliziertes, aber einheitliches Spaltensystem zwischen
Epithelzelle und Kapillarwand eingeschaltet ist und 4. die Vielzahl
großer und cristareicher *Mitochondrien*, die zum überwiegenden Teil in
engstem Kontakt mit der basalen Zellmembran stehen. — Die Tubulus-

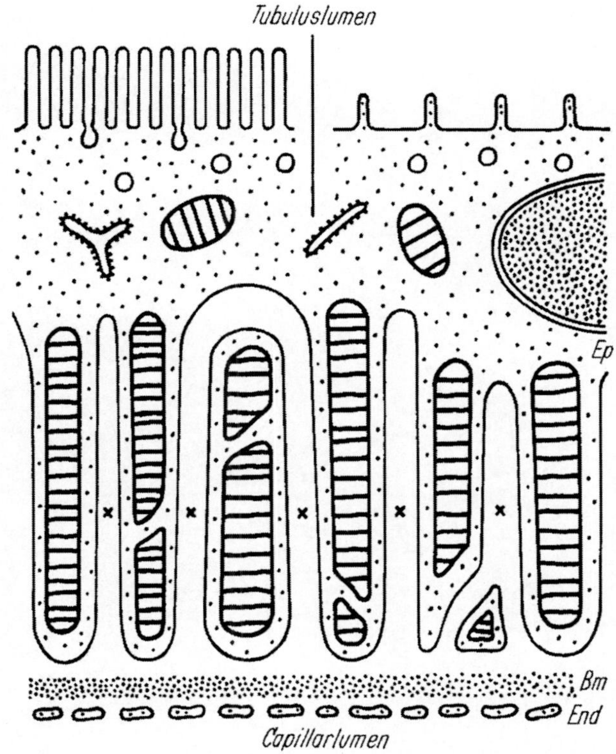

Abb. 1. Blut-Harn-Schranke in den Tubuluskonvoluten der Säugerniere. Charakteristika
der Epithelien (*EP*) des proximalen (links) und des distalen (rechts) Konvolutes in einem
Bild vereinigt. An der lumenwärtigen Zelloberfläche im proximalen Tubulus Bürstensaum,
im distalen einzelne Microvilli. Basalregion zeigt in beiden Fällen mitochondrienreiches
Fächer-Labyrinth-System. × = Basales Labyrinth, das vom Blutraum durch Basalmembran
(*BM*) und porenhaltiges Kapillarendothel (*END*) abgegrenzt ist [*49*]

epithelien sind regelmäßig — wie andere Epithelien auch — durch
Schlußleisten miteinander verbunden, die hier ausschließlich im lumennahen Zellbereich entwickelt sind.

Dem Verständnis dieses eigenartigen Konstruktionsprinzips könnte
folgende vergleichend-cytologische Betrachtung dienen: Als Musterbeispiel einer resorbierenden Zelle darf die vorhin kurz erwähnte *Dünndarmepithelzelle* gelten. Sie hat mindestens zwei Merkmale mit den
proximalen Konvolutepithelien der Niere gemein, nämlich den Bürstensaum und die — allerdings wesentlich geringere — Verzahnung mit den
Nachbarzellen [*28, 40, 56*]. In Ergänzung älterer lichtmikroskopischer
Befunde konnte gezeigt werden [*40*], daß bei der enteralen Wasserresorption der Interzellularraum erweitert wird, und zwar zunächst nur

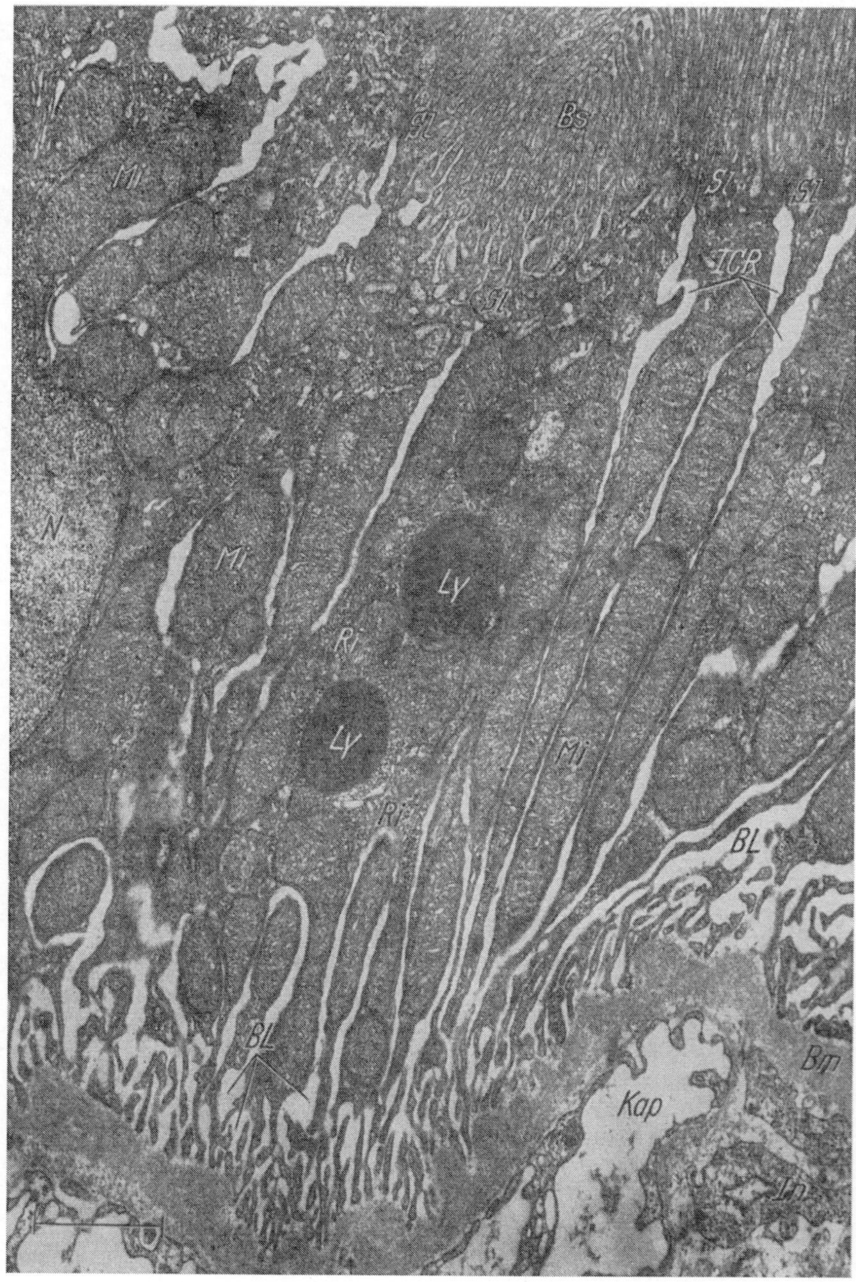

Abb. 2. Proximaler Tubulus — Pars contorta. Lumenwärts Bürstensaum, basal großes mitochondrienreiches Fächer-Labyrinth-System. Die Räume des „basalen Labyrinthes" stehen mit den lumennahen Interzellularräumen in einer anderen Schnittebene in Verbindung. 17500:1

im Basalbereich, im Extremfall aber — unter Lockerung oder vollständiger Aufhebung der Verzahnung — sogar bis nahe an den Bürstensaum bzw. an die apical gelegenen Schlußleisten (Abb. 4). Daraus muß geschlossen werden, daß die via Bürstensaum in das Zellinnere aufgenommene Flüssigkeit durch die laterale Zellmembran in den Interzellularraum abgegeben wird, ein Weg, der auch für höhermolekulare

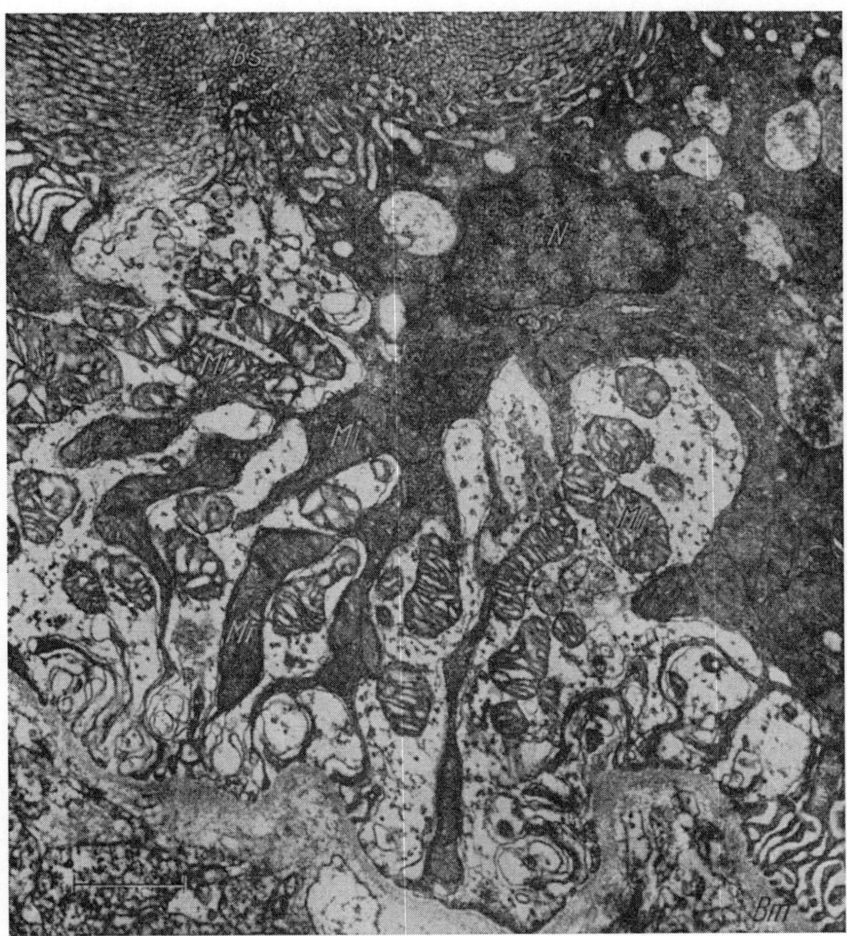

Abb. 3. Proximaler Tubulus — Pars contorta (nach 1stündiger Ischämie). Wegen des erhöhten Wassergehaltes der einen Zelle (Zellhydrops mit Mitochondrienschwellung vom Crista-Typ) wird die gegenseitige Verzahnung der Epithelien besonders deutlich. 15000:1

bzw. kolloiddisperse Stoffe nachgewiesen ist [29, 45]. Die Flüssigkeit gelangt also beim Verlassen der Zelle nicht sogleich in das Interstitium, sondern zuerst in einen der Basalmembran vorgeschalteten Raum, der apical von den Schlußleisten, lateral von den Zellmembranen und basal von der Basalmembran begrenzt ist. Erst in einem zweiten Schritt fließt

das Resorbat durch die Basalmembran in das kapillarenführende Interstitium ab. Es liegt auf der Hand, in diesem (interzellulären) Raum der
Dünndarmepithelschicht ein Äquivalent des „basalen Labyrinthes" im
Nierentubulus zu erblicken. Damit hätten beide Epithelarten im Prinzip
auch das oben unter 3. erwähnte Merkmal gemeinsam. Sie unterscheiden
sich in ihrer cytoplasmatischen Ausrüstung demnach — und zwar

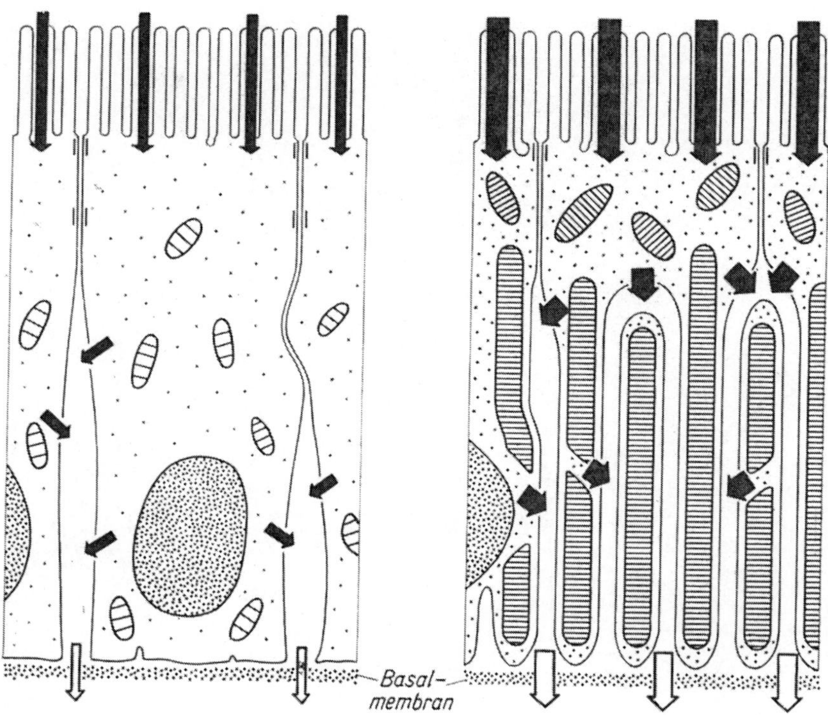

Abb. 4. Vergleich zwischen dem resorbierenden Epithel des Dünndarmes (links, in Anlehnung an [*56, 40*]) und des proximalen gewundenen Nierentubulus (rechts). Die ver-
einfachte schematische Darstellung berücksichtigt nur bestimmte zelluläre Bauelemente
(Zellmembran, Zellkern, im Cytoplasma Mitochondrien und freie Ribosomen) und nimmt
auf Größenunterschiede der Zellen keine Rücksicht. In beiden Fällen Bürstensaum an der
Lumenseite und erweiterungsfähige Interzellularräume an der Basis. Das unterschiedliche
Resorptionsvolumen (vgl. die Pfeildicke) korrespondiert mit Zahl, Größe und Cristareichtum der Mitochondrien, der Lagerungsdichte der Ribosomen und dem Grad der basalen
Oberflächenvergrößerung. Zwei Schritte im Resorptionsvorgang: transzellulärer Transport
(schwarze Pfeile) und „Endstreckentransport" (weiße Pfeile)

zugunsten der Tubulusepithelien — vor allem in zwei Punkten (Abb. 4):
(a) im *Ausmaß der Membranfaltung* und damit der Verzahnung unter
Ausbildung des umfangreichen „*Labyrinthes*" und (b) in der *Größe des
Chondrioms.*

Als plausibelste Erklärung für diese Differenz bietet sich die folgende an: Es
läßt sich überschlagsmäßig das Resorptionsvolumen pro Einheit resorbierender
Fläche (mm²) und Zeit (24 Std) im proximalen Nierentubulus (Werte vom Menschen) und im Dünndarm (Werte vom Hund) berechnen. Für den *proximalen
Tubulus* ergibt sich bei einer lichten Weite von 20 μ und einer Länge von 14 mm eine

resorbierende Innenfläche von rund 880000 $\mu^2 \sim 0,9$ mm². Von dem Filtratvolumen pro Nephron (90 μl/24 Std) werden 80% = 72 μl/24 Std im proximalen Tubulus reabsorbiert. Daraus ergibt sich, umgerechnet auf die Flächeneinheit, ein Resorptionsvolumen von *80 μl/mm²/24 Std*. Dabei handelt es sich um die normale antidiuretische Niere, also nicht um einen oberen Grenzwert für die Resorptionskapazität des Tubulus. — Für den *Dünndarm* ist ein Resorptions-Höchstwert mit 8 μl/cm² Darmschleimhaut/min (= 11,5 ml/cm² Darmschleimhaut/24 Std) gemessen worden [13]. Da die Dünndarmschleimhaut pro mm² 18—40 Zotten, im Mittel also etwa 30 Zotten aufweist (vgl. [30]), entfallen auf den cm² Schleimhaut etwa 3000 Zotten. Die Oberfläche einer Zotte (Durchmesser an der Basis 0,2 mm, Länge 0,5—1 mm) kann im Mittel mit 0,4 mm², die *resorbierende* Oberfläche eines 1 cm² großen Schleimhautstückes (mit 3000 Zotten) daher mit etwa 1100 mm² angenommen werden. Auf dieser Grundlage errechnet sich ein maximales Resorptionsvolumen im Dünndarm von etwa *10 μl/mm²/24 Std*. Damit ergäbe sich für die Resorptionsvolumina pro resorbierender Flächeneinheit in Darm und Tubulus ein Verhältnis von etwa 1:8. Dieses vergrößert sich zugunsten des Tubulus um ein mehrfaches, wenn man es auf die einzelne *Zelle* bezieht; denn der Epithelbesatz ist im Darm nach eigenen Messungen etwa 10mal dichter als im Tubulus. Aus dieser überschlagsmäßigen Berechnung leitet sich eine wesentlich höhere Resorptionsleistung der einzelnen Tubulusepithelzelle gegenüber der Darmepithelzelle ab.

Es scheint mir naheliegend, die vergleichsweise hohe Transportleistung der Tubulusepithelien für die vorhin (unter a und b) genannten Strukturunterschiede zwischen Darm- und Nierenzellen verantwortlich zu machen: a) Da der transzelluläre Transport energiebedürftig ist, muß ein großes Chondriom als Energiespender vorhanden sein. b) Die außerordentlich starke Faltung der seitlichen und basalen Zellmembran kann verstanden werden als Ausdruck einer für rasche Flüssigkeitsabgabe notwendigen Oberflächenvergrößerung — unter der Voraussetzung, daß die Durchtrittsrate für Solute und Wasser pro Flächeneinheit Membran eine naturbedingte obere Grenze hat.

Als weiterer Effekt dieser Membranfaltung ergibt sich eine stärkere Aufgliederung jenes Raumes, der zwischen Epithelzellen und Kapillare eingeschaltet ist und hier als „basales Labyrinth" [43] bezeichnet wird. Nach ersten Befunden von CAULFIELD und TRUMP [3] erscheint es möglich, daß die Labyrinthräume des Nierentubulus bei kräftiger Reabsorption erweitert werden, worin eine Parallele zur Ausweitung der Interzellularspalten im resorbierenden Darmepithel zu erblicken wäre. Im letzteren Falle wird besonders deutlich, daß dazu ein hydromechanischer Druck erforderlich ist; denn das Volumen der Darmepithelzellen nimmt während der Resorption nicht ab, sondern zu. Es liegt nahe, einen solchen Druckanstieg auch im basalen Lybarinth des Nierentubulus anzunehmen. Damit führen die Überlegungen zu der von RUSKA [41, 43] vertretenen „Druckwandler"-These, welche dem basalen Labyrinth eine entscheidende Bedeutung im tubulären Reabsorptionsprozess zuweist: Den Soluten (insbesondere Na$^+$), die im Verlauf der Reabsorption aus dem Zellinneren in das Labyrinth transportiert werden, strömt aufgrund ihrer osmotischen Kraft Wasser passiv nach. Dadurch wird im basalen Labyrinth ein hydromechanischer Druck erzeugt, der einen Fluß des Resorbates via Basalmembran in Richtung Kapillare bewirkt. Unter dieser Voraussetzung würden beide Teilvorgänge der Reabsorption — *transzellulärer Transport* (vom Tubuluslumen in das basale Labyrinth)

und „*Endstreckentransport*" (vom basalen Labyrinth in die Kapillare)
[*49*] — allein von der Aktion des Tubulusepithels abhängig sein.

Zur Bedeutung des *kolloidosmotischen Druckes* im Reabsorptionsprozeß wäre vom morphologischen Standpunkt aus zweierlei zu bemerken: 1. Die im basalen Labyrinth befindliche Flüssigkeit ist, nach
dem elektronenmikroskopischen Bild zu urteilen, im Regelfall eiweißfrei

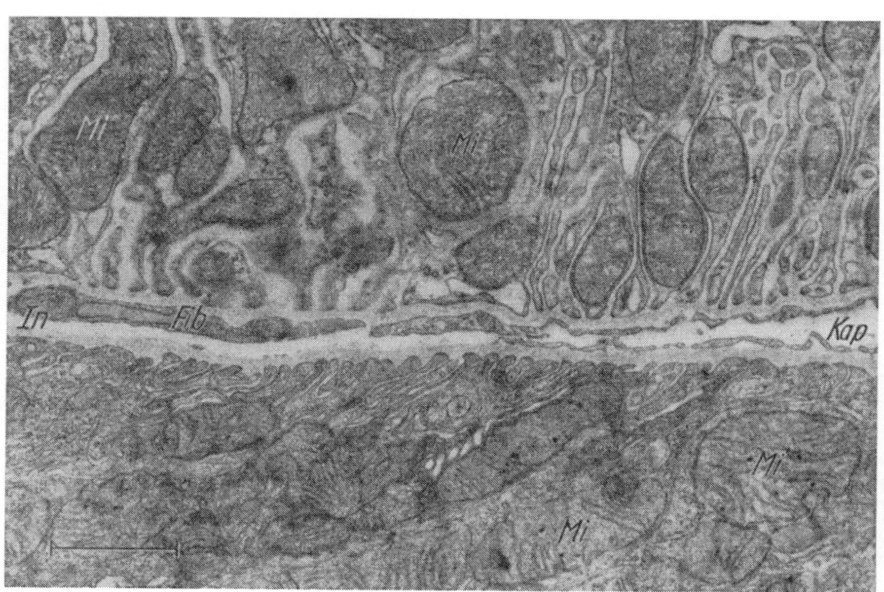

Abb. 5. Basis zweier benachbarter Tubuli contorti, rechts an eine Kapillare, links an den
interstitiellen Raum (mit Fibrocyten) grenzend. 17800:1

oder doch extrem eiweißarm. Für das Blutplasma in den peritubulären
Kapillaren muß dagegen ein Eiweißgehalt von mindestens 8 g-% angenommen werden. Zwischen beiden Flüssigkeiten befinden sich nur die
tubuläre bzw. tubulokapilläre Basalmembran und das porenhaltige
Kapillarendothel, beides für niedermolekulare Substanzen äußerst
permeable Schranken, durch welche hindurch sich die kolloidosmotische
Druckdifferenz (von etwa 30 mm Hg) auswirken dürfte. 2. Die Circumferenz des Tubulus grenzt nicht überall direkt an die Kapillare, sondern
streckenweise auch an einen echten interstitiellen Raum, in den das
Resorbat aus diesem Sektor des Tubulus zunächst abfließt (Abb. 5).
Es muß also von hier aus noch in das Kapillarlumen gelangen (sofern
es nicht in das Lymphsystem abdrainiert wird, vgl. [*46*]), wobei ebenfalls
die kolloidosmotische Druckdifferenz (zwischen Kapillarlumen und
Interstitium) in Rechnung zu stellen ist. Unter diesen Umständen muß
angenommen werden, daß der kolloidosmotische Druck des Blutplasmas
in den peritubulären Kapillaren für den Abfluß des Resorbates aus dem
basalen Labyrinth und dem Interstitium von Bedeutung ist.

Danach sind als Kräfte für den „Endstreckentransport" [49] der Reabsorption einmal der (durch die Aktion des Tubulusepithels erzeugte) hydrostatische Druck und zum anderen der kolloidosmotische Druck des Kapillarblutes in Betracht zu ziehen. Es ist zu hoffen, daß durch kombinierte physiologische und mikromorphologische Untersuchungen weiterer Aufschluß über den Mechanismus des Endstreckentransportes erhalten werden kann.

Aus den Abb. 1 u. 6 geht hervor, daß der auffälligste Unterschied zwischen den Epithelien des

distalen Tubulus (Pars contorta et recta)

und denen des proximalen Konvolutes die Ausgestaltung der lumenwärtigen Zelloberfläche betrifft: statt des Bürstensaumes finden sich hier nur einige Microvilli. Ganz unvoreingenommen würde man danach im distalen Tubulus, verglichen mit dem proximalen, eine geringere Resorptionsleistung erwarten. Dem entspricht, daß in diesem Tubulusabschnitt bei etwa gleicher Länge (13,6 cm, zusammen für Pars recta und convoluta [26], die sich in der Epithelstruktur prinzipiell gleichen, s. Abb. 6) höchstens 20% der Primärharnmenge (gegenüber 60–80% im proximalen Tubulus) reabsorbiert werden. Übereinstimmend damit hat GERTZ [8] gefunden, daß der Auswärtsfluß im distalen Tubulus nur $^1/_3$ von dem im proximalen Tubulus beträgt.

Trotzdem ist der enorme Unterschied in der luminalen Oberfläche zwischen diesen beiden Tubulusabschnitten nicht befriedigend erklärt, solange man ihn nur auf das Resorptions*volumen* (proximal:distal ~ 4:1) bezieht; denn das Oberflächenverhältnis zwischen proximalem und distalem Tubulus ist mit Sicherheit höher zu veranschlagen als das Verhältnis der Resorptionsvolumina. (Für die durch den Bürstensaum bewirkte Flächenzunahme ist der Faktor 50 berechnet worden [38]). Es liegt nahe, diese Diskrepanz mit *qualitativen* Unterschieden im Reabsorptionsgeschehen zu erklären, und zwar mit der Tatsache, daß die Reabsorption z. B. von Glucose und Aminosäuren, die an besondere Ferment- und Carriersysteme gebunden ist (vgl. [14, 53]), nur im proximalen Tubulus erfolgt. Hier klingt abermals die vorhin gezogene Parallele zwischen proximalen Nierenepithelien und Darmepithelien an: Beide Zellarten resorbieren – außer Salzen und Wasser – Glucose und Aminosäuren, beide Zellarten sind trotz unterschiedlichen Resorptionsvolumens mit einem Bürstensaum von gleicher Gestalt und praktisch gleichem Fermentmuster ausgestattet. Daraus kann geschlossen werden, daß das Ausmaß der Oberflächenvergrößerung, das der Bürstensaum beinhaltet, nicht so sehr durch das Volumen als vielmehr durch die *stoffliche* Zusammensetzung des Resorbates bestimmt wird.

Das bisher Gesagte bezieht sich auf die Reabsorption vieler *niedermolekularer* Substanzen, die nacheinander die apicale und die laterobasale Zellmembran der Tubulsuepithelien in derzeit unsichtbarer Form durchsetzen. Höher-, insbesondere *makromolekulare* Stoffe können die Zellmembran offenbar in der gleichen Form nicht durchdringen. Sie werden auf pinocytotischem Wege in das Zellinnere aufgenommen [24, 45, 51] und hier mit Hilfe zelleigener Fermente in kleinere Moleküle zerschlagen, die dann wie die anderen niedermolekularen Substanzen durch die basale Zellmembran in das basale Labyrinth und schließlich in die Kapillare gelangen. Das bekannteste Beispiel sind die sog. hyalinen

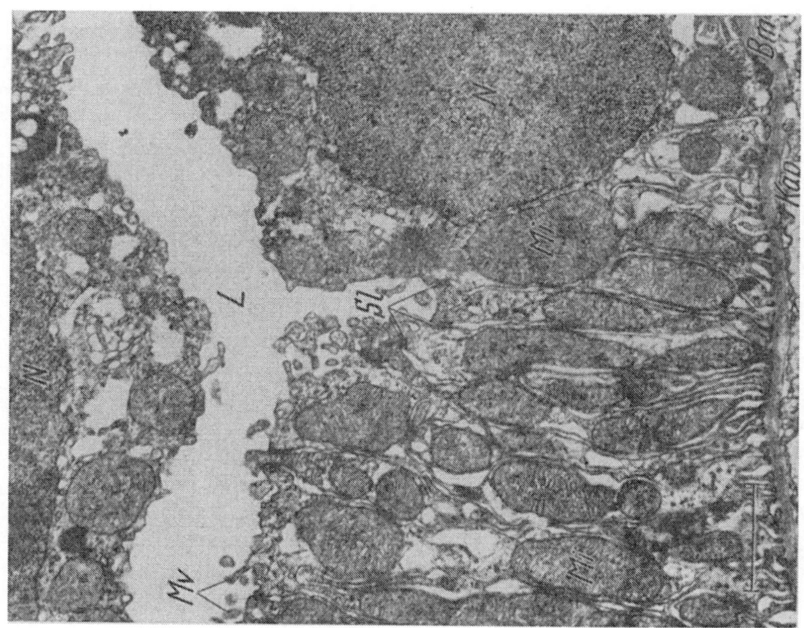

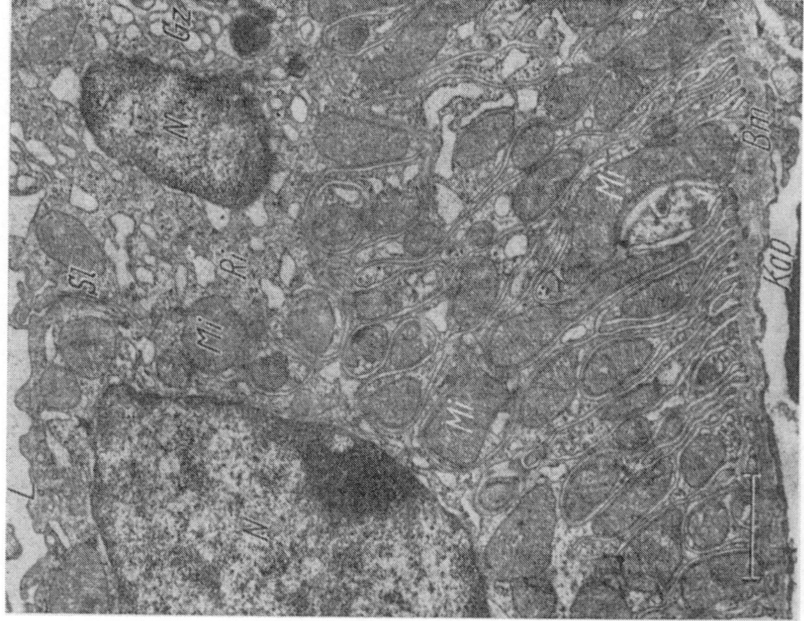

Abb. 6. Distaler Tubulus — Pars contorta (*a*) und Pars recta (*b*). In der Pars contorta Epithel durchschnittlich höher, Kern im apicalen Zellbereich gelegen, in der Pars recta durchschnittlich geringere Zellhöhe, Kern mehr in Zellmitte gelegen. 15000:1

Eiweißtropfen, deren Entstehung und weiteres Schicksal in ihrer Beziehung zu den Lysosomen am Beispiel der Hämoglobinniere eingehend untersucht worden sind [24, 25].

Schließlich soll noch ein theoretisch denkbarer dritter Transportweg durch den Tubuluswall in seiner Problematik wenigstens angedeutet werden. Es wurde vorhin erwähnt, daß das basale Labyrinth bei geeigneter Schnittführung bis an die lumennahen Schlußleisten zu verfolgen

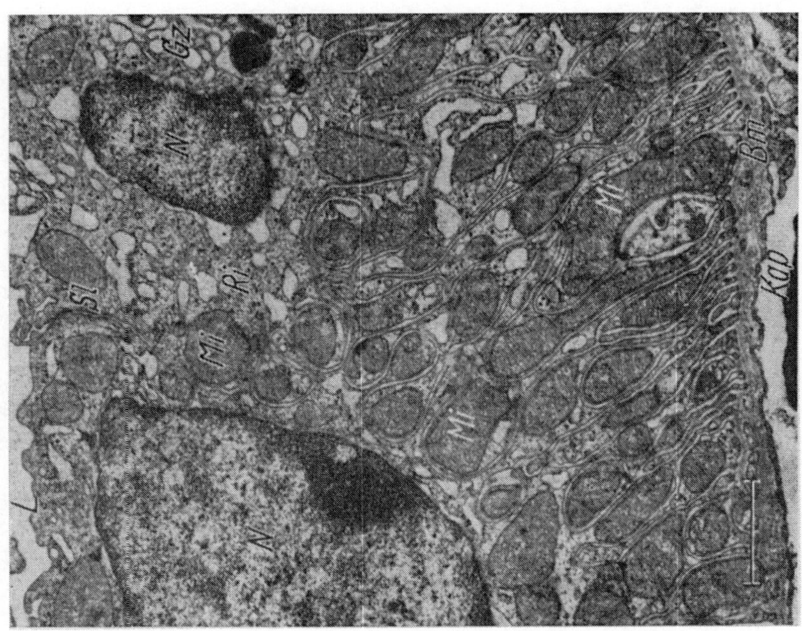

Abb. 7. Proximaler Tubulus — Pars contorta. Die leicht erweiterten Intercellularspalten sind bis an die Schlußleisten unterhalb des Bürstensaumes zu verfolgen. Die große Zahl von Schlußleisten auf engem Raum ist Ausdruck der intensiven „groben" (siehe Fußnote S. 332) Verzahnung der Konvolutepithelien. 35000:1

ist (Abb. 7). Den Schlußleisten kommt daher eine zentrale Bedeutung für die Frage nach der Existenz eines rein *extra- bzw. interzellulären transmuralen Austauschweges* zwischen Tubuluslumen und Interstitium zu; denn Basalmembran und porenhaltiges Kapillarendothel, welche das Labyrinth unten gegenüber dem Kapillarraum begrenzen, sind für Diffusionsbewegungen praktisch frei durchgängig. Daß die Schlußleisten für *höher*molekulare Stoffe, wie Eiweiß, eine feste Barriere darstellen, kann nach den Untersuchungen von MILLER [24] als gesichert gelten. Ihre Durchlässigkeit für *nieder*molekulare, diffusible Substanzen ist dagegen umstritten. KOMNICK [18] hat sich im Falle der Salzdrüse für eine [„wenn auch relativ schlecht(e)"] Ionenpermeabilität der Schlußleisten ausgesprochen. Auf der anderen Seite haben FARQUHAR und PALADE [7] lumenwärts von dem eigentlichen Desmosom ("Macula adhaerens") bzw. der vorgeschalteten „Zonula adhaerens", in deren

Bereich die Plasmamembranen der benachbarten Zellen durch eine Kittsubstanz verbunden sind, eine Verschmelzungszone der Plasmamembranen (ohne Kittsubstanz!) nachgewiesen (Abb. 8). In dieser „Zonula occludens" — als integrierendem Bestandteil des gesamten Schlußleistenkomplexes — fehlt also ein interzellulärer Raum herkömmlichen Sinnes. Aus diesem Grunde vermuten die Autoren, daß eine Diffusion entlang dem Interzellularspalt nicht möglich ist. Das Wesentliche der Befunde von FARQUHAR und PALADE liegt wohl darin, daß durch die Membranfusion eine *rein zelluläre* Barriere zwischen Tubuluslumen einerseits und Labyrinth bzw. Interstitium andererseits hergestellt wird, so als sei das gesamte Tubuluslumen von einer geschlossenen Zellmembran begrenzt. Diese Vorstellung erscheint wesentlich im Hinblick auf den von physiologischer Seite geführten Nachweis transtubulärer Potentiale als Ausdruck aktiver, vom Tubulusepithel getragener Transportvorgänge (Lit. [*6, 9*]).

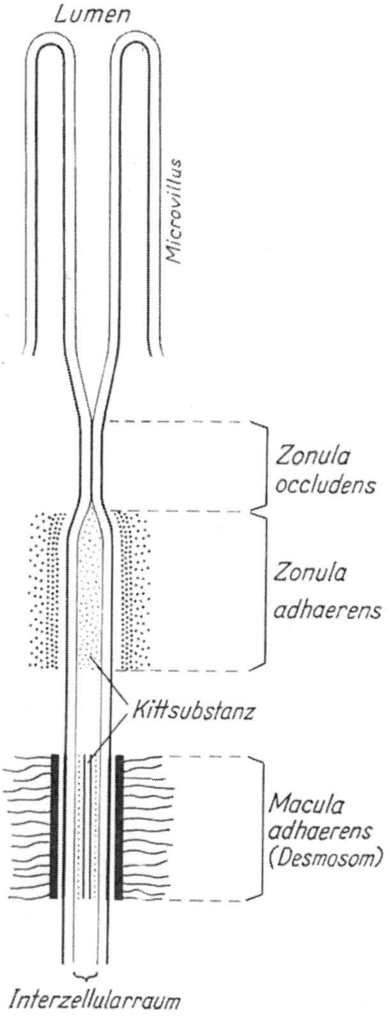

Unter den mitochondrienreichen Harnkanälchen nimmt die

Pars recta des proximalen Tubulus

morphologisch eine gewisse Sonderstellung ein. Nach ersten lichtmikroskopischen Hinweisen [*47, 57*] konnte elektronenmikroskopisch gezeigt werden [*34, 49*], daß der Epithelwall dieses Tubulusabschnittes strukturell in einigen Punkten von dem des proximalen Konvolutes abweicht. Wir haben dieses Ergebnis, das sich mit früheren Befunden von ROLLHÄUSER und VOGELL [*38*] nicht ganz deckt, noch einmal an mehreren Rattennieren überprüft, und zwar unter Kontrollbedingungen und während einer (nach Angaben von ROLLHÄUSER [*36*]) erzeugten kräftigen Phenolrotexkretion, kenntlich an einer starken Rotfärbung besonders des Außenstreifens der äußeren Markzone, weniger der Markstrahlen (15 und 20 min nach subcutaner Injektion von 2 ml einer 0,6%igen Phenolrotlösung). In beiden Fällen ist die Epithelstruktur des distalen gestreckten

Abb. 8. Schematische Darstellung des „Schlußleisten-Komplexes" — gezeichnet nach Befunden von FARQUHAR und PALADE [*7*] (Original)

Hauptstückabschnittes übereinstimmend folgendermaßen zu charakterisieren (Abb. 9, 10):

Während an der Lumenseite ein regelrechter Bürstensaum entwickelt ist, der sich von dem der Konvolutepithelien nicht unterscheidet, zeigt die Basalregion ein abweichendes Verhalten: Die basale Zellmembran ist streckenweise glatt, streckenweise auch eingefaltet, wobei jedoch die Faltenhöhe häufig nicht mehr als den Querdurchmesser eines Mitochondrions beträgt. Die Folge ist, daß die Epithelien untereinander nur mit zarten Fortsätzen und ganz basisnah verzahnt sind, wenn man von einigen kompakteren Zellfortsätzen absieht, die sich in die Nachbarzelle einbuchten oder zwischen zwei andere Zellen einschieben. Das basale Fächer-Labyrinthsystem ist also gegenüber dem Konvolut stark reduziert, besonders in den terminalen Abschnitten im Bereich der äußeren Markzone, nicht ganz so stark im Markstrahlbereich. Übereinstimmend damit verlaufen die seitlichen Zellmembranen weitgehend gestreckt. Außer der lumennahen Schlußleiste finden sich hier meist noch weitere desmosomale Kontaktstellen zwischen den Nachbarzellen. Im Cytoplasma liegen cristareiche Mitochondrien in erheblicher Zahl und unregelmäßiger Verteilung sowie zahlreiche freie Ribosomen und vor allem reichlich Profile des Endoplasmatischen Retikulums (ER), die teils glatt, teils unregelmäßig mit Ribosomen besetzt sind. Eine oder mehrere, relativ große Golgi-Zonen liegen in Kernnähe.

Auf der Suche nach einem morphologischen Substrat des transzellulären exkretorischen Transportes ist uns in den Phenolrotnieren lediglich eine mäßige, aber diffuse Weitstellung des ER aufgefallen, wobei der Binnenraum des öfteren ein lockeres, schwach osmiophiles Material enthält. Daß es sich um ER und nicht um cytopemptische Bläschen handelt, geht aus der unregelmäßigen Größe und Form der membranbegrenzten Hohlräume und dem (unregelmäßigen) Besatz mit Ribosomen hervor (Abb. 10, Ausschnitt). Die Profile des ER grenzen des öfteren hart an die basale oder laterale Zellmembran und finden sich häufig auch unterhalb des Bürstensaumes; eine Kommunikation mit dem Extrazellulärraum war jedoch nirgends sicher zu erkennen. Allerdings ist ein ähnliches Verhalten des ER nicht selten auch in Kontrollnieren zu beobachten. Das gilt gleichfalls für den Nachweis runder membranbegrenzter Bläschen, die im basalen Cytoplasma einzelner Zellen vorkommen.

Bei aller Verschiedenheit der Epithelstruktur im gewundenen und gestreckten Teil des proximalen Tubulus, die durch diese Befunde erneut belegt wird, dürfen doch die Gemeinsamkeiten nicht übersehen werden: In beiden Fällen ist sowohl ein Bürstensaum — als sicheres morphologisches Zeichen resorptiver Tätigkeit — als auch ein basales Fächer-Labyrinthsystem vorhanden. Das letztere ist im gewundenen Teil stark ausgeprägt, im gestreckten Teil dagegen nur schwach entwickelt. Der Unterschied zwischen Pars convoluta- und Pars recta-Epithelien ist also bezüglich dieser Strukturmerkmale ein gradueller.

Damit stellt sich abschließend die Frage, welche verallgemeinernden Schlüsse aus dem Dargelegten für die funktionelle Interpretation der Zellstrukturen in den mitochondrienreichen Harnkanälchen der Säugerniere zu ziehen sind. Die Koinzidenz von hoher reabsorptiver Leistung einerseits und starker Ausprägung des basalen mitochondrienreichen Fächer-Labyrinthsystems andererseits im proximalen und distalen Konvolut kann nur dahingehend verstanden werden, daß die charakteristische Basalstruktur der Epithelien hier in erster Linie im Dienste der Reabsorption steht. Das bedeutet vice versa: Überall dort — nämlich außer in den Konvoluten auch im gestreckten Teil des distalen Tubulus —, wo die basale Zelloberfläche unter Bildung eines mitochondrienreichen

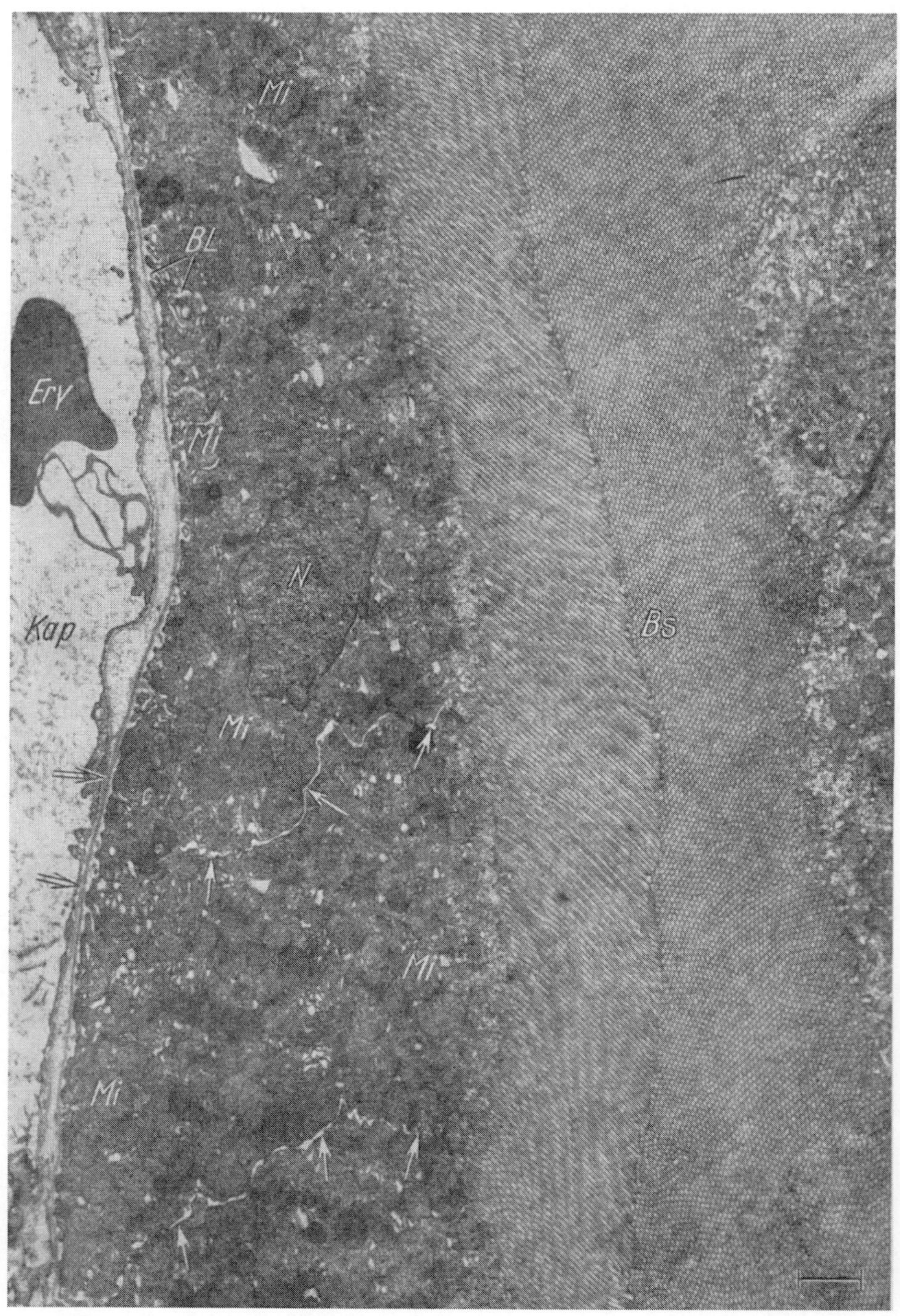

Abb. 9. Proximaler Tubulus — Pars recta (präterminaler Abschnitt). Regelmäßiger Bürstensaum an der Lumenseite. Flaches basales Fächer-Labyrinth-System an der Zell-basis. Wegen der vorwiegend basisnahen Verzahnung relativ gestreckter Verlauf der seitlichen Zellgrenzen (→) und weite Distanz der Schlußleisten. Tubuläre und kapilläre Basalmembran nur streckenweise verschmolzen (⇒). 8800:1

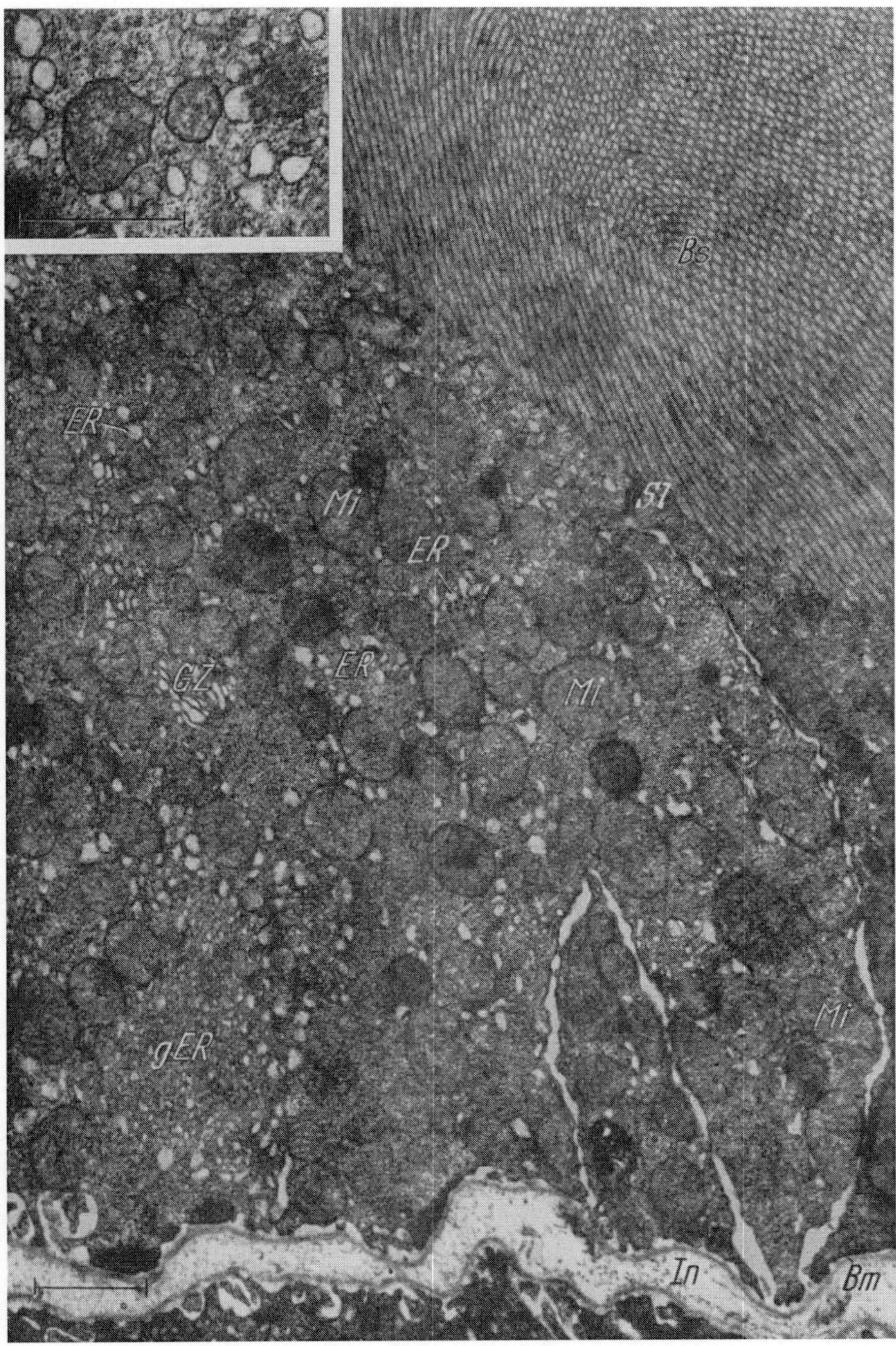

Abb. 10. Proximaler Tubulus — Pars recta (präterminaler Abschnitt) 15 min nach subcutaner Applikation von 12 mg Phenolrot. Im Cytoplasma viele cristareiche Mitochondrien
und freie Ribosomen. Diffuse Weitstellung des Endoplasmatischen Retikulums, das unregelmäßigen Besatz mit Ribosomen aufweist (s. Ausschnitt). Außerdem Zonen fein-
maschigen glatten Endoplasmatischen Retikulums (links unten). 15000:1, Ausschnitt
22500:1

Fächer-Labyrinth-Systems stark vergrößert ist, kann mit einer hohen resorptiven Leistung gerechnet werden. Innerhalb des *proximalen* Tubulus würde man daher aufgrund der Epithelstruktur für den gewundenen Teil eine hohe und für den gestreckten Teil eine geringere Resorptionsrate anzunehmen haben.

Es erscheint jedoch fraglich, ob die Epithelstruktur im gestreckten Teil des proximalen Tubulus allein durch eine — gegenüber dem Konvolut reduzierte — Reabsorptionsleistung hinreichend erklärt ist; denn zwischen dem Umfang und der „Qualität" (Cristareichtum) des Chondrioms einerseits und dem geringen Grad der basalen Membranfaltung andererseits besteht eine Diskrepanz, die ohne die Annahme zusätzlicher energiebedürftiger Funktionen meines Erachtens schwer verständlich ist. Außerdem sind Unterschiede im Fermentgehalt zwischen Pars contorta und Pars recta des proximalen Tubulus festgestellt worden [*15*].

Aussagen über den *exkretorischen* (von der Kapillare zum Tubuluslumen gerichteten) Transport in seiner Beziehung zur Epithelstruktur sind wesentlich problematischer. Wir haben davon auszugehen, daß exkretorische Prozesse grundsätzlich im gesamten proximalen Tubulus (z. B. Phenolrot [*22*, *37*, *39*], PAH [*4*]) sowie im distalen Tubulus (Kalium$^+$ [*32*]) möglich sind, also auch in den Konvoluten, die mit dem großen Fächer-Labyrinth-System ausgestattet sind. Das bedeutet, daß das Vorhandensein eines tiefgreifenden Fächer-Labyrinth-Systems einen basal-apical gerichteten Transport in der Säugerniere nicht ausschließt — eine Feststellung, die im Hinblick auf die Befunde an der Salzdrüse [*17*] wichtig erscheint. Unklarheit besteht nur über den Transportmechanismus. Die früher von ROLLHÄUSER [*36*] geäußerte Vermutung, daß das Phenolrot den Epithelwall innerhalb der Faltenkanäle passiert, würde nach heutiger Kenntnis (s. oben) einen rein *inter*zellulären Transport bzw. einen Transport entlang der Zellmembran beinhalten. Es spricht jedoch vieles dafür (vgl. [*22*]), daß der Farbstoff zur Exkretion in das Zell*innere* aufgenommen wird. Ob er bei seinem Weg durch das Cytoplasma im Raum der cytoplasmatischen Matrix (Grundplasma) wandert oder bestimmte Strukturen (wie z. B. das Endoplasmatische Retikulum, s. Abb. 10) benutzt, muß offenbleiben. Für einen cytopemptischen Transport gibt es bisher keine sicheren Anhaltspunkte.

B. Harnkanälchen mit mitochondrienarmen Epithelien

Die hier zu besprechenden Harnkanälchen — dünner Teil der Henleschen Schleife und Sammelrohr — liegen ausschließlich im Nierenmark, wenn man von den initialen Sammelrohren in den Markstrahlen absieht. Der zwischen die Partes rectae des proximalen und distalen Tubulus eingeschaltete

dünne Teil der Henleschen Schleife,

der in der Theorie des Haarnadelgegenstrom-Mechanismus eine zentrale Stellung einnimmt, wird von außerordentlich flachen Epithelien gebildet. Im kernfernen Teil beträgt die Epithelhöhe etwa 1μ. Hervorzuheben ist die relativ einfache cytoplasmatische Organisation (Abb. 11): die

Mitochondrien sind klein, cristaarm und gering an Zahl; Profile des Endoplasmatischen Retikulums und freie Ribosomen sind nicht sehr zahlreich. Die lumenwärtige Zelloberfläche trägt kurze Microvilli. Das bei den Rindentubuli erwähnte Prinzip der Verzahnung ist auch hier beibehalten. Dabei muß unterschieden werden zwischen einer — wenn

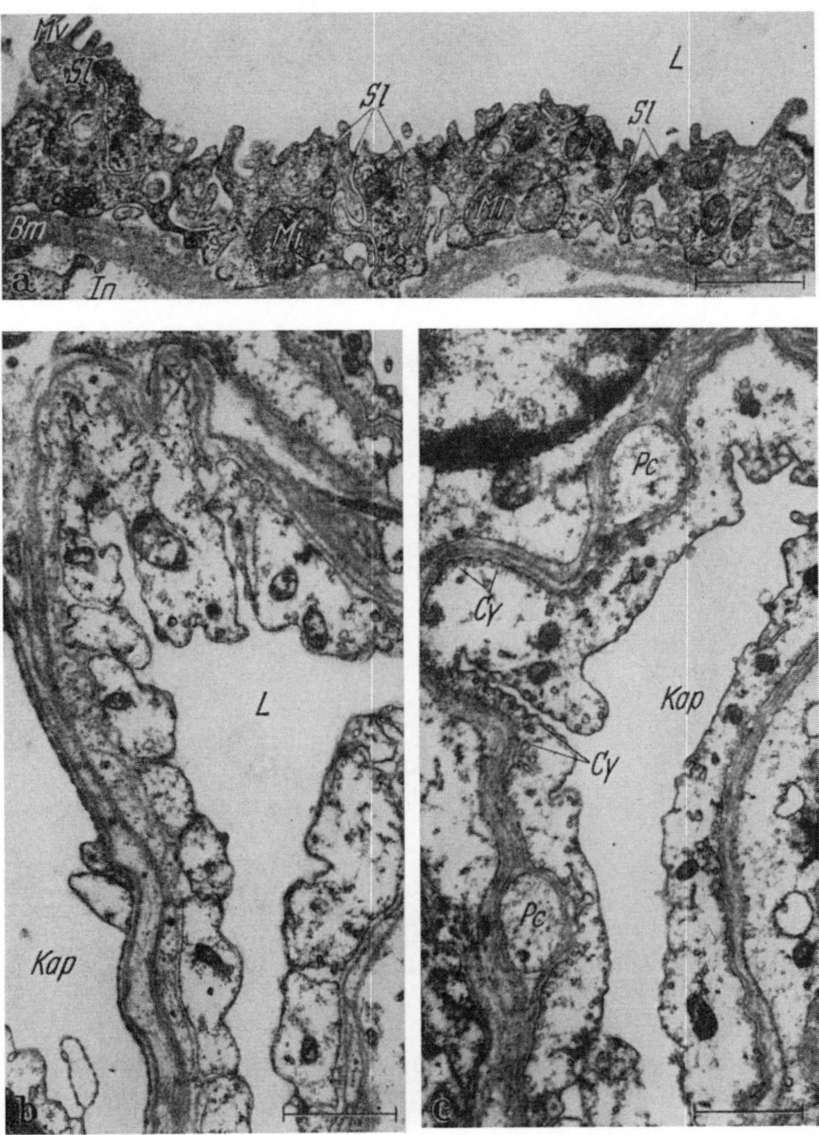

Abb. 11. Dünner Teil der Henleschen Schleife. *a*: Absteigender Schenkel im Innenstreifen der äußeren Markzone. *b*: Schleifenquerschnitt ohne Microvilli und praktisch fehlender basaler Verzahnung aus der inneren Markzone. *c*: Kapillare mit geschlossenem Endothel und Pericyten (*Pc*) aus der inneren Markzone. 16000:1

ich so sagen darf — „groben" und einer „feinen" (basalen) Verzahnung[1]. Die erstere beruht auf einer die gesamte Zellhöhe betreffenden Aufgliederung der seitlichen Zellabschnitte in vergleichsweise grobe Fortsätze, mit denen die benachbarten Schleifenepithelien ineinandergreifen [35]. Aus diesem Grunde sind im Querschnittsbild in der Regel mehrere Epithelien getroffen, die durch typische Desmosomen in Lumennähe miteinander verbunden sind (Abb. 13a). Es ist offensichtlich, daß durch die Verzahnung die Strecke der an das Kanälchenlumen

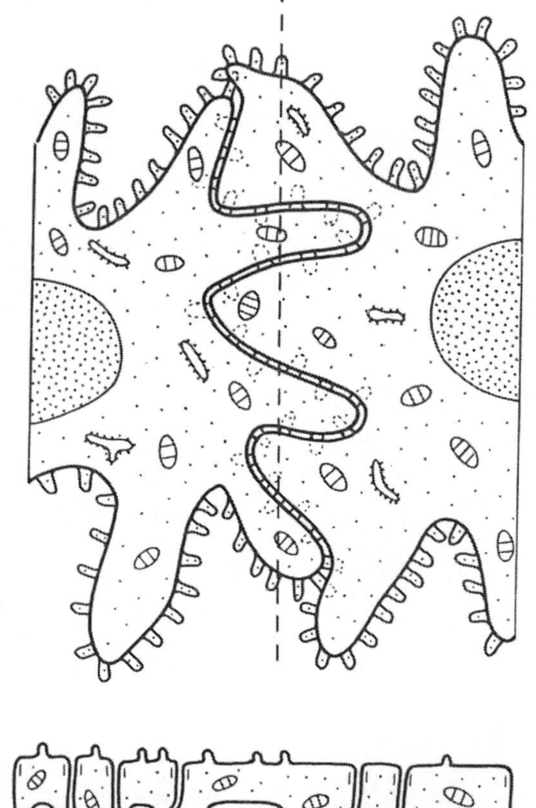

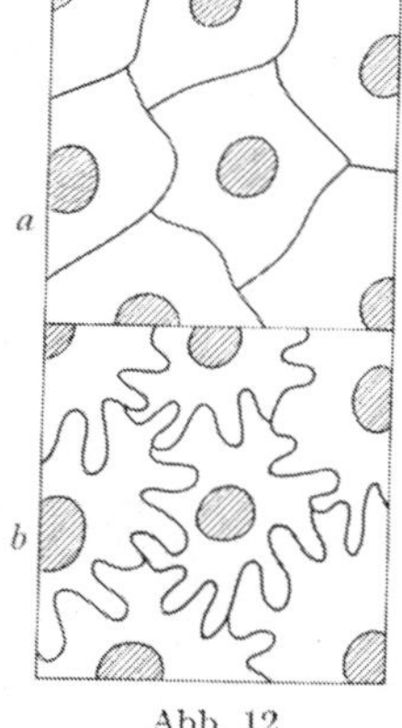

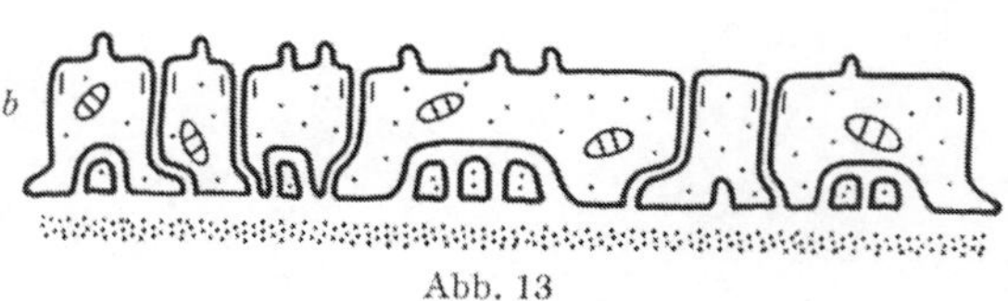

Abb. 12

Abb. 13

Abb. 12. Strecke der an das Lumen angrenzenden Interzellularspalten bei glatten seitlichen Zellkonturen (a) und bei „grober" Epithelverzahnung (b)

Abb. 13. Schematische Darstellung der beiden Verzahnungstypen des Epithels im dünnen Teil der Henleschen Schleife. a: Ansicht von der Lumenseite. Von den „groben" Ausläufern, mit denen die Epithelien ineinandergreifen, gehen kleinere, vorwiegend basale Fortsätze aus, die wechselseitig unter die Nachbarzelle geschoben sind. b: Ansicht im Schnitt (entsprechend der gestrichelten Linie in a)

angrenzenden Interzellularspalten wesentlich größer ist, als wenn die Zellen seitlich glatt konturiert wären (Abb. 12). Die „feine" (basale) Verzahnung beruht dagegen auf der Ausbildung zahlreicher, sehr zarter und schmaler, vorwiegend basaler Zellfortsätze, die nur Grundplasma und

---

[1] Es sei erwähnt, daß das doppelte Verzahnungsprinzip auch in den beiden Konvoluten und in der Pars recta des distalen Tubulus verwirklicht ist. Bei der Besprechung wurde dort nur auf die basale Verzahnung eingegangen.

einige Ribosomen enthalten und in größerer Zahl wechselseitig unter die
jeweilige Nachbarzelle geschoben sind (Abb. 13). Auf diese Weise entsteht − wie in dem später zu besprechenden Sammelrohr − ein „basales
Labyrinth en miniature" [50].

Die Intensität der Epithel-Verzahnung stellt sich im elektronenmikroskopischen Bild je nach der Schnittebene (kernfern−kernnah)
und je nach der Höhe, in der die Schleife getroffen ist, unterschiedlich
dar. LAPP und NOLTE [21] haben darauf hingewiesen − wir können
diesen Befund bestätigen −, daß die in der äußeren Markzone gelegenen
dünnen (initialen absteigenden) Schleifenabschnitte relativ viele Microvilli an der Lumenseite sowie Membranfaltungen an der Zellbasis aufweisen (Abb. 11a). Im oberen Teil der inneren Markzone, wo nebeneinander *ab*- und *auf*steigende Schenkel liegen, finden sich außerdem
Schleifenquerschnitte, deren Epithelien keine Microvilli und praktisch
keine basale Membranfaltung erkennen lassen (Abb. 11b). In den unteren Papillenregionen stellt sich das Schleifenepithel dagegen ziemlich
einheitlich mit wenig Microvilli und durchschnittlich geringer Verzahnung dar. Die Autoren schließen aus diesem Etagenunterschied auf
eine unterschiedliche Beschaffenheit der beiden Schleifenschenkel.

In dem Gedanken, daß sich ein Strukturunterschied zwischen ab-
und aufsteigendem Schenkel am deutlichsten in dem Anfangs- und
Endabschnitt der dünnen Schleife abzeichnen müßte, haben wir jeweils
die Übergangszone zwischen dem dicken und dünnen Teil der beiden
Schleifenschenkel untersucht (Abb. 14, 15). Dabei fanden sich in beiden
Regionen Zellen mit nur wenigen und solche mit relativ dicht stehenden
Microvilli; auch die gegenseitige Verzahnung der Epithelien kann in der
betreffenden Schnittebene weitgehend fehlen, sie kann aber auch recht
intensiv sein, und zwar in der vorhin erwähnten „groben" und „feinen"
(basalen) Form (Abb. 15). Danach zeigen die Epithelien im Initial- und
Terminalabschnitt der dünnen Schleife ein ziemlich übereinstimmendes
Bild. Es muß vorerst offenbleiben, ob diese Gemeinsamkeit Ausdruck
der in beiden Fällen gegebenen Übergangssituation des Epithels ist.
Immerhin wird durch diese Befunde eine endgültige Stellungnahme zur
Frage der Strukturdifferenz zwischen ab- und aufsteigendem Teil der
dünnen Schleife erschwert. Auch das Verhalten der Basalmembran
(homogene Beschaffenheit oder Schichtung) ist nach unseren Befunden
wechselhaft (s. Abb. 14, 15). Soviel kann jedoch sicher gesagt werden,
daß ein Unterschied zwischen den Schenkeln nicht auf einem abrupten
Formationswechsel an der Schleifenspitze beruht; dafür stellt sich das
Kollektiv der Schleifenquerschnitte in den tieferen Papillenregionen zu
einheitlich dar. Als gesichert darf weiterhin gelten, daß im *ab*steigenden
dünnen Schenkel papillenwärts die Zahl der Microvilli und die Intensität
der groben und feinen Verzahnung abnimmt.

Hinsichtlich der funktionellen Bewertung der geschilderten Strukturen sind drei Phänomene besonders hervorzuheben: die auffallend
geringe Höhe bzw. Dicke des Epithels, die Verzahnung mit Ausbildung
eines flachen basalen Labyrinthes und die Armut an Mitochondrien und
freien Ribosomen. Das Beispiel der corticalen Tubuli zeigt, daß die

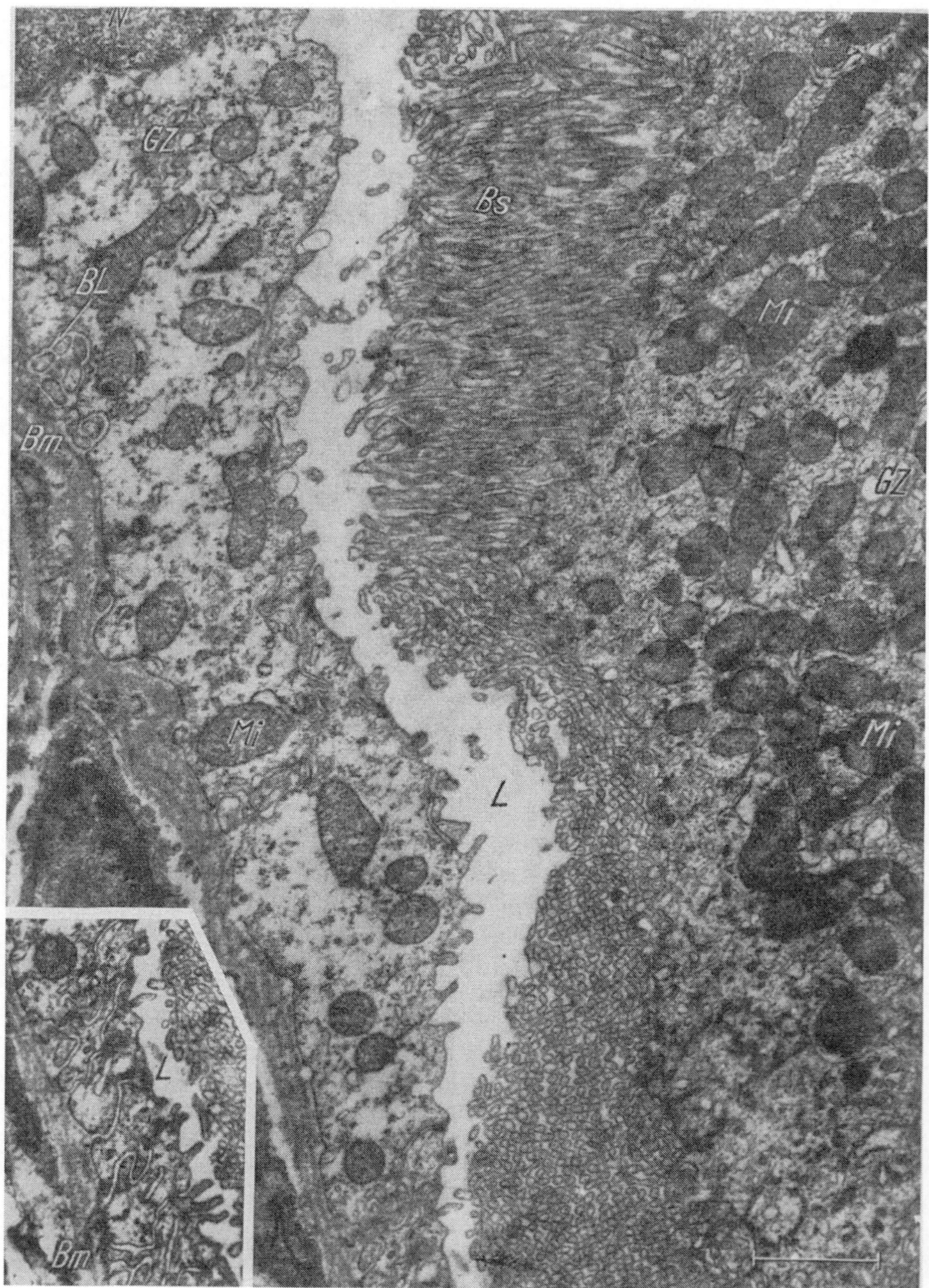

Abb. 14. Übergangszone: Pars recta des proximalen Tubulus (rechte Bildhälfte) — absteigender Schenkel der dünnen Henleschen Schleife (linke Bildhälfte). Ausschnitt: Fortsetzung des dünnen Schleifenepithels über die untere Bildgrenze hinaus. Hier intensive („grobe" und „feine") Verzahnung. Deutliche Schichtung der Basalmembran. 16000:1

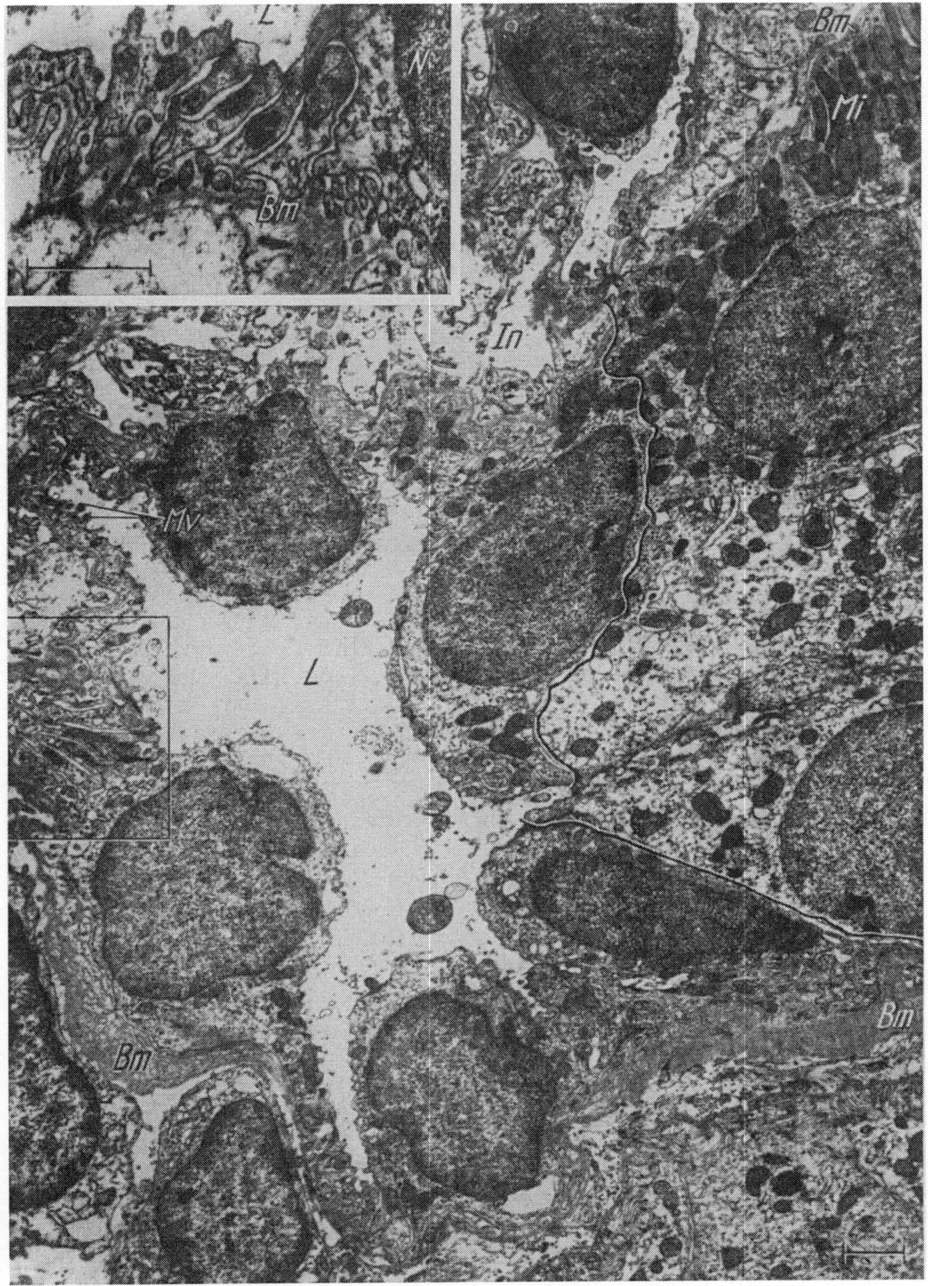

Abb. 15. Übergangszone: Aufsteigender Schenkel der dünnen Henleschen Schleife (linke Bildhälfte) — Pars recta des distalen Tubulus (rechte Bildhälfte). Grenze durch Linie markiert. Rechts oben mitochondrienreiches Fächer-Labyrinth-System in der Pars recta des distalen Tubulus. 7000:1. Ausschnitt: Teilvergrößerung aus dem dünnen Schleifenteil (linker Bildrand Mitte, Parallelschnitt). Intensive („grobe" und „feine") Verzahnung. Stellenweise deutliche Schichtung der Basalmembran. 16000:1

Epithelien mit starker Ionen-Transportaktivität über ein großes
Chondriom — als Zeichen hohen oxydativen Stoffwechsels und damit
Energieumsatzes — verfügen. Daran gemessen, wird man den Epithelien
der dünnen Schleife nur eine geringe Transportaktivität zutrauen,
selbst unter der Voraussetzung, daß die in der inneren Markzone gelegenen Harnkanälchen Energie möglicherweise auch durch anaerobe
Glykolyse gewinnen [5, 11]; denn nach Berechnungen von KRAMER,
THURAU und DEETJEN [19] liegt der Energieumsatz im Nierenmark bei
Berücksichtigung beider Energiequellen (oxydativ und glykolytisch)
immer noch unterhalb 10% desjenigen der Rinde. Andererseits gibt der
morphologische Befund mit dem Nachweis der Microvilli und des — wenn
auch vergleichsweise kümmerlichen — basalen Labyrinthes nicht die
Handhabe, jegliche Transportaktivität der dünnen Schleifenepithelien
zu leugnen.

Damit wird die Frage nach den Vorgängen im Haarnadelgegenstrom-
system des Nierenmarkes berührt, die Gegenstand zahlreicher Hypothesen geworden sind. Die ursprünglich entwickelten Vorstellungen
arbeiten mit einem unterschiedlichen funktionellen Verhalten der
beiden Schleifenschenkel, wobei entweder ein (passiver) Wasserausstrom
aus dem absteigenden [12] oder ein (aktiver) Salzausstrom aus dem aufsteigenden Schenkel [20, 55] diskutiert werden. Nach dem vorhin Gesag-
ten kann aus dem morphologischen Befund eine Vorrangstellung gerade
des aufsteigenden Schenkels im Sinne einer besonderen Transportaktivität für Salze (Ionen) nicht begründet werden; denn im Differenzierungsgrad (äußere Zellform, Mitochondriengehalt) sind die Epithelien
des aufsteigenden Schenkels denen des absteigenden sicher nicht über-
legen. (Beim dicken Teil des aufsteigenden Schenkels im Innenstreifen
der äußeren Markzone sind sowohl morphologisch [50] als auch funktionell [52] andere Verhältnisse gegeben.) Geht man davon aus, daß ein
aktiver Wassertransport in der Biologie als unwahrscheinlich gilt
(vgl. [2]), so könnte ein (Netto-)Wasserausstrom im absteigenden
Schenkel nur durch andere Kräfte, z. B. durch hydrostatischen Druck
erklärt werden. Da über die Permeabilität der Kittleisten in der dünnen
Schleife noch nichts Sicheres bekannt ist — es wäre z. B. zu prüfen, ob
auch hier eine Zonula occludens vorhanden ist —, muß offenbleiben, ob
ein solcher hydromechanisch bedingter Flüssigkeitsausstrom etwa über
die reichlich vorhandenen Interzellularen erfolgen kann. Bei der geringen
Dicke der Epithelschicht müßte man schließlich auch einen *transzellulären* Flüssigkeitsdurchtritt in Erwägung ziehen, wobei die Epithelien passiv durchströmt würden.

Mindestens vom cytomorphologischen Standpunkt aus erscheinen die
letztgenannten Überlegungen deshalb nicht ganz abwegig, weil bei den
Blutkapillaren mit geschlossenem Endothel (z. B. im quergestreiften
Muskel, aber auch im Nierenmark, s. Abb. 11c), welche mit den dünnen
Schleifen strukturell manche Gemeinsamkeit haben, eine ganz ähnliche
Problematik besteht [33, 42]. Ist doch auch für die Kapillarschranke
noch nicht endgültig geklärt, ob die unter dem Einfluß des Blutdruckes
austretende Flüssigkeit den Weg durch die Endothelzellen oder auch

durch die verkitteten Interzellularspalten nimmt. Aber auch dann, wenn man — unter Hinweis auf das Beispiel der Kapillare — einen passiven Flüssigkeitsaustritt aus der dünnen Schleife für möglich und — als zweite Voraussetzung — einen ausreichenden Druckgradienten für gegeben hält, wäre noch problematisch, wie man sich einen überwiegenden *Wasser*ausstrom, zudem beschränkt auf den absteigenden Schenkel, erklären sollte.

Aus alledem geht hervor, daß die beiden erörterten Mechanismen zur Erzeugung der Konzentriereinzeleffekte im Bereich der dünnen Schleife mit den morphologischen Gegebenheiten bislang nicht ohne weiteres zur Deckung zu bringen sind. Überblickt man das gesamte Nephron, so imponiert von allen strukturellen Eigenschaften der dünnen Schleife am meisten die starke Reduzierung der Zellhöhe. Es liegt nahe anzunehmen, daß dadurch der Austausch zwischen Schleifenlumen und Interstitium erleichtert wird, gleichgültig, wie diese Austauschprozesse im einzelnen geartet sind.

Daß neben den übrigen Strukturmerkmalen der Harnkanälchenepithelien auch die *Zellhöhe* ein funktionell zu bewertender Faktor sein könnte, drängt sich auf, wenn man den Epithelwall im

Sammelrohr

zum Vergleich heranzieht (Abb. 16). Es wurde früher darauf hingewiesen [50], daß die dünnen Schleifenepithelien und die typischen (hellen) Sammelrohrepithelien viele und wesentliche Eigenschaften gemein haben: kleine kurze Microvilli an der Lumenseite; basisnahe („feine") Verzahnung mit Ausbildung eines sehr flachen basalen Labyrinthes, das in der Diureseniere erweitert sein kann [21]; Verbindung der Epithelien durch typische Schlußleisten in Lumennähe; wenige kleine, cristaarme Mitochondrien sowie spärliche freie Ribosomen und Profile des Endoplasmatischen Retikulums im transparenten Cytoplasma. Es fehlt die „grobe" Verzahnung, weshalb die seitlichen Zellgrenzen gestreckt verlaufen. Ein wesentlicher Unterschied liegt außerdem in der 5—8fach größeren Zellhöhe der Sammelrohrepithelien.

Ähnlich wie beim dünnen Schleifenepithel läßt die cytoplasmatische Organisation der Sammelrohrzellen, insbesondere die Kleinheit ihres Chondrioms, auf eine vergleichsweise geringe Transportaktivität schließen. Das gilt für das gesamte Sammelrohr, und zwar für die papillenspitzennahen Abschnitte mehr als für die initialen; denn die cytoplasmatische Ausrüstung (Mitochondrien, Endoplasmatisches Retikulum, Ribosomen) der Sammelrohrepithelien wird von proximal nach distal spärlicher (Abb. 16). Daraus erhellt, daß im Sammelrohr die Osmolarität des Harnes (bei Antidiurese) und der zelluläre Differenzierungsgrad der Epithelschicht papillenwärts sich gegensinnig verhalten, d. h. gerade dort, wo der Harn seine hohe Endkonzentration erreicht, werden im cytomorphologischen Bild Zeichen für eine *stärkere* Transportaktivität vermißt.

Für die vorliegenden Betrachtungen über die Transportwege in den Harnkanälchen der Säugerniere wurde als *tertium comparationis* der

Mitochondriengehalt der Epithelien gewählt. Es ist uns keine anatomische Einheit des Parenchyms in den Säugerorganen bekannt, in der der Mitochondriengehalt der Epithelien von Abschnitt zu Abschnitt so stark

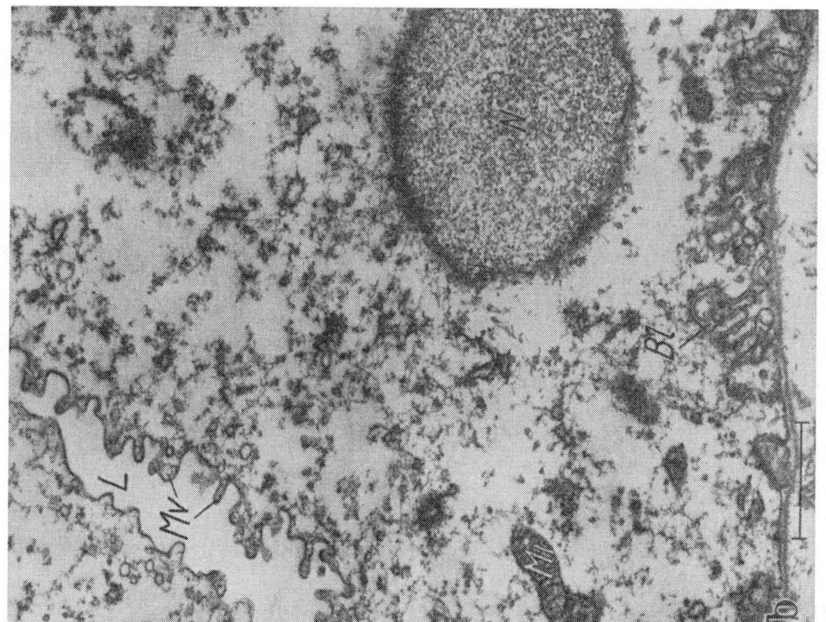

Abb. 16. Sammelrohr. *a*: im Bereich der äußeren Markzone, *b*: im Bereich der inneren Markzone (etwa mittlere Höhe der Papille) 16000:1.

wechselt wie im Nephron. Hinzu kommt als Besonderheit die charakteristische Reihenfolge und Anordnung der mitochondrienreichen und mitochondrienarmen Kanälchenabschnitte, von denen die ersteren in der Rinde, die letzteren im Mark liegen. In Übereinstimmung damit ist der O_2-Verbrauch in der Rinde um ein Vielfaches höher als im Mark [10, 19]. Danach wendet die Niere zur Bewältigung der resorptiven und exkretorischen Transportvorgänge in den corticalen Tubuli erhebliche Energiemengen auf, obwohl diese Transporte weitgehend in blutisotonischer Form erfolgen. Es ist ein bemerkenswertes Faktum, daß die Endkonzentrierung des Harnes gerade im Bereich solcher Harnkanälchen vor sich geht, deren Epithelien nur über einen vergleichsweise geringen Funktionsumsatz verfügen.

Literatur

[1] BARGMANN, W., H. KNOOP u. TH. H. SCHIEBLER: Z. Zellforsch. **42**, 386 (1955).
[2] BURCK, H.: Klin. Wschr. **40**, 761 (1962).
[3] CAULFIELD, J. B., and B. F. TRUMP: Amer. J. Pathol. **40**, 199 (1962).
[4] DEETJEN, P., u. H. SONNENBERG: Persönliche Mitteilung.
[5] DICKENS, F., and H. WEIL-MALHERBE: Biochem. J. **30**, 659 (1936).
[6] EIGLER, F. W.: In: Nierensymposion Göttingen 1959. Stuttgart: Thieme 1960.
[7] FARQUHAR, M. G., and G. E. PALADE: J. Cell Biol. **17**, 375 (1963).
[8] GERTZ, K. H.: Pflügers Arch. ges. Physiol. **276**, 336 (1963).
[9] GIEBISCH, G.: In: Nierensymposion Göttingen 1959, S. 49. Stuttgart: Thieme 1960.
[10] GRUPP, G., u. K. HIERHOLZER: Z. Biol. **109**, 197 (1957).
[11] GYÖRGY, P., W. KELLER u. TH. BREHME: Biochem. Z. **200**, 356 (1928).
[12] HARGITAY, B., u. W. KUHN: Z. Elektrochem. **55**, 539 (1951).
[13] HEIDENHAIN, R.: Pflügers Arch. ges. Physiol. **43** (Suppl.), 1 (1888).
[14] HEINZ, E.: In: Biochemie des aktiven Transportes. 12. Coll. Ges. Physiol. Chemie, S. 167. Berlin-Göttingen-Heidelberg: Springer 1961.
[15] KISSANE, J. M.: J. Histochem. Cytochem. **9**, 578 (1961).
[16] KLINGENBERG, M.: In: Funktionelle und morphologische Organisation der Zelle, S. 69. Berlin-Göttingen-Heidelberg: Springer 1963.
[17] KOMNICK, H.: Protoplasma (Wien) **56**, 605 (1963).
[18] —, u. U. KOMNICK: Z. Zellforsch. **60**, 163 (1963).
[19] KRAMER, K., K. THURAU u. P. DEETJEN: Pflügers Arch. ges. Physiol. **270**, 251 (1960).
[20] KUHN, W., u. A. RAMEL: Helv. chim. Acta **42**, 293 (1959).
[21] LAPP, H., u. A. NOLTE: Frankfurt. Z. Path. **71**, 617 (1962).
[22] LIETZ, H.: Z. Zellforsch. **62**, 374 (1964).
[23] MARTIUS, C.: Klin. Wschr. **35**, 223 (1957).
[24] MILLER, F.: J. biophys. biochem. Cytol. **8**, 689 (1960).
[25] —, u. G. E. PALADE: Verh. dtsch. Ges. Path. **47**, 346 (1963).
[26] MÖLLENDORFF, W. VON: In: Handb. mikr. Anat. d. Menschen. Bd. VII, 1. S. 1. Berlin: Springer 1930.
[27] OPITZ, E.: Verh. dtsch. Ges. Path. **33**, 18 (1949).
[28] PALAY, S. L., and L. J. KARLIN: J. biophys. biochem. Cytol. **5**, 363 (1959).
[29] — — J. biophys. biochem. Cytol. **5**, 373 (1959).
[30] PATZELT, V.: Der Darm. In: Handb. mikr. Anat. d. Menschen. Bd. V, 3. S. 1. Berlin: Springer 1930.
[31] PEASE, D. C.: Anat. Rec. **121**, 723 (1955).
[32] PITTS, R. F.: In: Diurese und Diuretica, S. 143. Berlin-Göttingen-Heidelberg: Springer 1959.
[33] RENKIN, E. M.: Verh. Ges. Dtsch. Naturf. u. Ärzte **102**, 57 (1962/63).
[34] RHODIN, J.: Proc. Stockholm Conf. Electron Microscopy 1956, S. 180. Stockholm: Almqvist u. Wiksell 1957.
[35] — Int. Rev. Cytol. **7**, 485 (1958).

[36] ROLLHÄUSER, H.: Z. Zellforsch. **46**, 52 (1957).
[37] — Z. Zellforsch. **51**, 348 (1960).
[38] —, u. W. VOGELL: Z. Zellforsch. **47**, 53 (1957).
[39] — — Z. Zellforsch. **52**, 549 (1960).
[40] RUSKA, C.: Z. Zellforsch. **52**, 748 (1960).
[41] RUSKA, H.: Marburger Sitz. Berichte **82**, 3 (1960).
[42] — In: Bad Oeynhausener Gespräche VI, 27. Berlin-Göttingen-Heidelberg: Springer 1964.
[43] —, D. H. MOORE, and J. WEINSTOCK: J. biophys. biochem. Cytol. **3**, 249 (1957).
[44] SCHATZMANN, H. J., and E. E. WINDHAGER: Amer. J. Physiol. **195**, 570 (1958).
[45] SCHMIDT, W.: Z. Zellforsch. **58**, 573 (1962).
[46] SIADAT-POUR, A.: Beitr. path. Anat. **120**, 382 (1959).
[47] SJÖSTRAND, F. S.: Acta anat. (Basel) **1**, Suppl. 1, 1 (1944).
[48] SOLOMON, A. K.: In: The methods of isotopic tracers applied to study of active ion transport, S. 106. Oxford: Pergamon Press 1959.
[49] THOENES, W.: Klin. Wschr. **39**, 504 (1961).
[50] — Klin. Wschr. **39**, 827 (1961).
[51] TRUMP, B. F.: J. Ultrastruct. Res. **5**, 291 (1961).
[52] ULLRICH, K. J., G. PEHLING u. M. ESPINAR-LAFUENTE: Pflügers Arch. ges. Physiol. **273**, 562 (1961).
[53] WILBRANDT, W.: In: Biochemie des aktiven Transportes. 12. Coll. Ges. Physiol. Chemie, S. 112. Berlin-Göttingen-Heidelberg: Springer 1961.
[54] WINDHAGER, E. E., and G. GIEBISCH: Nature (Lond.) **191**, 1205 (1961).
[55] WIRZ, H.: Helv. physiol. pharmacol. Acta **11**, 20 (1953).
[56] ZETTERQVIST, H.: The ultrastructural organization of the columnar absorbing cells of the mouse jejunum. Stockholm: Aktiebolaget Godvil 1956.
[57] ZIMMERMANN, K. W.: Arch. mikr. Anat. **78**, 199 (1911).

Zeichenerklärung für die Abbildungen:

BL = Basales Labyrinth; *Bm* = Basalmembran; *Bs* = Bürstensaum; *End* = Endothel; *Ery* = Erythrocyt; *ER* = Endoplasmatisches Retikulum; *Fib* = Fibrocyt; *gER* = Glattes Endoplasmatisches Retikulum; *GZ* = Golgi-Zone; *ICR* = Intercellularraum; *In* = Interstitium; *Kap* = Kapillarlichtung; *L* = Kanälchenlumen; *Ly* = Lysosomen; *Mi* = Mitochondrien; *Mv* = Microvilli; *N* = Zellkern; *Ri* = Ribosomen; *Sl* = Schlußleisten.

Summary

The transport route for the exchange of substances between the lumen of the kidney tubules and the peritubular blood capillaries consists of at least three structural elements in all tubular segments: tubular epithelium, basement membrane, and porous capillary endothelium. The tubular epithelium plays the most important role in the transmural exchange of substances. The mitochondria content is taken as a criterion for the transporting activity of the epithelium:

A. *Mitochondria-rich tubules* (proximal and distal tubule with straight and convoluted segment). The epithelium of the proximal convoluted tubule has four important structural features: 1. brush border, 2. deep infoldings of the laterobasal cell membrane with mutually interdigitating epithelial cells, 3. extensive "basal labyrinth", 4. many large mitochondria rich in cristae. A cyto-morphological comparison with the absorption epithelium of the intestine (Fig. 4) suggests that the epithelial cells of the proximal convoluted tubule and of the whole distal tubule have a very high absorptive capacity. The brush border of the proximal tubule is related not to the amount but to the nature of the absorbate. The absorptive process involves two steps: 1. transcellular transport (from the tubular lumen into the basal labyrinth), and 2. "termal transport" (from the basal labyrinth into the capillary). In the straight segment of the proximal tubule (Fig. 9, 10) the basal labyrinth is markedly reduced. It may be assumed that the absorptive rate in this segment of the tubule is lower than in the convoluted segment. However, the relatively high mitochondrial content and the large number of free ribosomes imply additional, still unknown functions in this part of the tubule. Epithelia with a brush border and basal labyrinth also carry out excretory transport (from capillary to tubule lumen). The morphological picture does not permit a definite statement concerning the transport mechanism. Based on FARQUHAR and PALADE's description of the "Zonula occludens" in the junctional complex (Fig. 8), the problem of *intercellular* transport through the epithelial wall is discussed.

B. *Mitochondria-poor tubules* (thin segment of the loop of HENLE, collecting duct). The very squamous cells of the thin segment of the loop of HENLE are poor in cytoplasmic organelles and especially poor in mitochondria. The cells are doubly interdigitated (Fig. 13), as in some other tubular segments. The structure of the epithelium is identical in the initial and terminal segments of the thin loop (Fig. 14, 15). The transport activity of the epithelium in the thin loop may be estimated as very low. Current hypotheses about the origin of the primary concentrating effect in the countercurrent system of the thin loop agree poorly with the morphological findings. The cells of the collecting duct also have scanty cytoplasmic equipment. The chondriom is small. In addition, the number and size of the mitochondria decrease in a proximal-to-distal direction. This suggests that the osmolarity of the urine (in antidiuresis) and the transport activity of the cells in the collecting ducts vary inversely.

Diskussion

Schmidt: Haben Sie die Meinung, daß die Rückresorption zum größeren Teil über Pinocytose oder zum größeren Teil durch Mikrovilli erfolgt?

Thoenes: Ich würde glauben, daß sich das nach der Molekülgröße richtet. Je größer das Molekül ist, um so mehr kommt die Pinocytose in Frage. Wir wissen das für Trypanblau — wie Sie selbst gezeigt haben — und für Eiweißstoffe. Aber ich glaube, daß die niedermolekularen Substanzen, wie eben vor allem auch Ionen, die uns hier besonders interessieren, durch die Zellmembran gehen, entweder durch den Bürstensaum im proximalen Tubulus oder durch die Mikrovilli bzw. die nicht ausgestaltete Luminaloberfläche im distalen Tubulus.

Zschiesche: Wie hoch würden Sie den Prozentsatz der Flüssigkeit einschätzen, der nicht direkt in die Kapillaren geht, sondern in die Interstitien und über die Lymphgefäße, die ohne Zweifel auch eine Rolle spielen.

Thoenes: Darüber kann man im Moment noch keine sichere Aussage machen, und zwar deshalb, weil diese Bilder alle noch von nicht intravital fixierten Nieren stammen. Die intertubulären Kapillaren laufen nach Abtrennung der Niere vom Gefäßsystem leer. Alle Messungen, die wir jetzt an einer solchen *postmortal* fixierten Niere machen, können keine Auskunft über den intravitalen Zustand geben. Aber wenn man das intravitale Bild der Niere vor Augen hat, und zwar unter dem Stereomikroskop, würde man diesen Anteil sicher sehr gering einschätzen.

David: Wir haben den Ureter unterbunden und dann im Elektronenmikroskop häufig gefunden, daß hochgradig erweiterte Lymphgänge im Interstitium liegen. Das würde wahrscheinlich doch unter anderen Bedingungen durchaus möglich sein.

Ullrich: Nach direkten Messungen von LE BRIE (Proceedings 11th Annual Conference on the Nephrotic Syndrome 1959, Ed. J. METCOFF, p. 52) ist der Lymphfluß kleiner als $1^0/_{00}$ des Plasmadurchflusses durch die Niere.

Zschiesche: Dann würde man annehmen müssen, daß bei der Unterbindung der Lymphgefäße ein Teil der Interstitialflüssigkeit direkt aus den Blutkapillaren stammt und nicht aus dem basalen Labyrinth.

Ullrich: Ich möchte in diesem Zusammenhang etwas zur Interpretation histologischer Bilder sagen. Bei Lebendbeobachtung der Niere kann man sehen, daß die Nierentubuli nach Unterbindung der Blutversorgung innerhalb weniger Sekunden kollabieren, da die Tubulusflüssigkeit supravital aus dem Lumen rückresorbiert wird. Die Histologen haben bis vor kurzem kollabierte Tubuli auf Grund ihrer Bilder für normal gehalten. Veränderungen nach Unterbindung der Durchblutung zeigt auch der Bürstensaum. Diese supravitalen Veränderungen kann man durch intravitale Vorgabe von nicht resorbierbaren Stoffen z. B. von Mannitol in das Tubuluslumen voraussagbar steuern.

Außerdem zeigt das Beispiel der Sammelrohre, wie lange ein Fehlurteil auf Grund morphologischer Kriterien nachwirken kann. HEIDENHAIN hat 1874 den Sammelrohren wegen ihres hellen Epithels nur die Rolle ableitender Wege zugeschrieben. Erst 80 Jahre später hat man gefunden, daß den Sammelrohren eine wichtige Funktion bei der Harnkonzentrierung zukommt und die Endanpassung der Natriumrückresorption und der Wasserstoffionensekretion dort geschieht. Das histologische Aussehen, vor allem das Vorhandensein nur weniger Mitochondrien, verleitete zunächst zu einer Fehleinschätzung der biologischen Bedeutung der Sammelrohrzellen. Bei näherem Zusehen ist aber doch eine Beziehung zwischen Struktur und Funktion gegeben: Der Nettotransport von NaCl ist nämlich in den mitochondrienreichen Zellen des proximalen und distalen Konvolutes viel größer als in den Sammelrohrzellen, wo durch relativ kleinen NaCl-Nettotransport entgegen hohen Konzentrationsunterschieden der NaCl-Stoffwechsel fein einreguliert wird. Es scheint somit eine Beziehung zwischen dem Nettotransport von NaCl, dem O_2-Verbrauch und dem Mitochondriengehalt einer Nierenzelle zu bestehen.

Thoenes: Das Kollabieren der Nephronlichtung hat auf die Epithelstruktur, die Grundstruktur des Epithels, über die ich hier gesprochen habe, keinen Einfluß.

Zu dem letzten Punkt darf ich daran erinnern, daß ich nur gesagt habe, daß die Sammelrohrepithelien eine vergleichsweise geringe Transportaktivität haben müßten, eine erheblich geringere als die Konvolutepithelien. Sonst müßten andere Energiequellen zur Verfügung stehen, und ich glaube, daß die Glykolyseaktivität, die da immer als Ausweg angeführt wird, nicht ausreicht.

Diamond: Wir haben während dieser Tagung öfter gehört, daß dort, wo die Energiequellen zu sehen sind, auch der Mechanismus des aktiven Transports vorhanden sein muß. Ich möchte nur ein Beispiel geben, das zeigt, daß das nicht immer der Fall sein muß. In der Gallenblase des Fisches sind die Mitochondrien an der luminalen Grenze der Zelle aufgehäuft. Bei diesem Organ aber wissen wir, daß der Mechanismus für den aktiven Transport des Natriumchlorids am entgegengesetzten basalen Ende der Zelle liegt, wo nur wenige oder keine Mitochondrien vorkommen. Wenn man also nur nach der Morphologie schließen würde, wo der Transportmechanismus liegt, würde man einen falschen Schluß ziehen.

Solomon: I would like to confirm what Dr. DIAMOND has said because if, in necturus kidney, for example, you would say that where the mitochondrion was, there was the work done, you would be making a mistake because the mitochondrion in the species contains no cytochrome oxidase and is unable to carry out the ordinary oxydation. Unless you know that the mitochondria are really the source of energy, then it is hard to make this correlation.

Beziehungen zwischen Zellstruktur und Stofftransport in der Niere[1]

Von

HELLMUTH SITTE, Homburg-Saar

Mit 11 Abbildungen

A. Vorbemerkungen

Eine Korrelation zwischen den Zellstrukturen der Niere und dem Stofftransport in diesem Organ stößt nach wie vor auf große Schwierigkeiten. Jede einzelne Niere weist im Rahmen des tubulären Apparates eine Vielzahl verschiedenartig strukturierter Epithelien auf, welche miteinander in Wechselwirkung stehen. Die einzelnen Elemente dieses tubulären Apparates sind von Tier zu Tier mengenmäßig unterschiedlich verteilt und topographisch unterschiedlich angeordnet [42, 80, 97, 108, 114]. Morphologische und physiologische Untersuchungsergebnisse sollten daher nur dann korreliert werden, wenn die Befunde am gleichen Objekt unter identischen Bedingungen erhoben wurden — ein Tatbestand, der heute noch kaum realisiert ist. Bei diesem Vergleich ergibt sich ein weiteres Problem: Es ist im Moment schwierig — wenn nicht unmöglich, die zentralen Bereiche des Organes in einem Zustand zu erfassen, der dem Zustand in vivo mit Sicherheit entspricht. Diese zentralen Bereiche umschließen die gesamte Außenzone des Markes, welche einen sehr regen Stoffwechsel aufweist (Mitochondrienzahl! Vgl. [102, 119, 122]), eine von der Rinde abweichende Struktur zeigt und zudem mit etwa 20 % des Gewichtes [124] einen nicht zu vernachlässigenden Teil der Niere darstellt. Schließlich können auch die oberflächennahen Strukturen der Rinde und Papille nicht ohne besondere Kunstgriffe lebensecht fixiert werden.

Bei allen *morphologischen Arbeiten* an der Niere ist zu berücksichtigen, daß jedes Abweichen vom normalen Fließgleichgewicht innerhalb weniger Sekunden zu einem signifikanten Strukturwandel führen kann; dies gilt insbesondere für jede Änderung des Blutdurchflusses und betrifft vor allem die besonders stoffwechselaktiven Kanälchenabschnitte in der Rinde und der Außenzone des Markes; vgl. [36, 44, 57, 75, 98, 116, 117]. Man ist dadurch gezwungen, möglichst rasch, bzw. bei intaktem Blutkreislauf zu fixieren; dies ist jedoch bei den zentralen Zonen des Organes relativ schwer zu realisieren. Die meisten elektronenoptischen Befunde an der normalen Niere beziehen sich daher leider auf ein Material, das nicht restlos lebensgetreu fixiert ist [16, 18, 26, 64, 83, 84, 89, 95, 96, 107, 119, 120, 122]. Relativ selten findet man Fixationsmethoden verwendet, welche speziell den Gegebenheiten bei der Niere angepaßt sind [27, 62, 77, 78]; die in diesen Arbeiten vorgelegten qualitativen Befunde beschränken sich jedoch auf die Ratte und erfassen damit nur eines

---

[1] Mit Unterstützung der Deutschen Forschungsgemeinschaft.

von vielen gängigen Versuchstieren der Physiologie. Das Problem wird dadurch schließlich noch weiter kompliziert, daß supravitale und postmortale Veränderungen offensichtlich bei verschiedenen Ausgangsbedingungen vollkommen verschiedenartig ablaufen [9, 15, 36, 89, 98, 138]; morphometrische Analysen [5, 153] sind daher in diesem Rahmen von größter Bedeutung und können wertvolle Hinweise für die Funktion einzelner Strukturelemente liefern.

Als Bindeglied zwischen Morphologie und Physiologie spielt die *Vitalmikroskopie* [115, 116, 118, 138] eine nicht zu unterschätzende Rolle; ihr Einsatzbereich ist jedoch auf eine sehr geringe optische Auflösung sowie die Organoberflächen (Rinde und Papille) beschränkt. Letzteres gilt leider auch für die wichtigsten *direkten physiologischen Verfahren:* Hier haben eine äußerst verfeinerte Technik der Mikropunktion [34, 55, 81, 133, 136, 137, 146, 147, 148], -perfusion [101, 112, 128, 142] und -katheterisierung [41, 45] sowie das Arbeiten mit Mikroelektroden [31, 143], Farbstoffen [115] und radioaktiven Substanzen [54, 55, 58, 110] in der jüngsten Zeit wertvolle Aufschlüsse erbracht. Andere Verfahren ermöglichen zwar Rückschlüsse auf zentrale Objektpartien, liefern aber nur indirekte, summarische Aussagen, wie z. B. die Clearance-Verfahren [55, 82, 100, 109, 111, 128], die Stop-Flow-Methode [59, 60, 145], die Gewebsschnitt-Kryoskopie [6, 150] oder andersartige Arbeiten an Gewebsschnitten [47, 56, 129, 130, 131]. Ein kompletter, gesicherter Überblick scheint daher auch in diesem Sektor im Augenblick schwer zu realisieren.

Ein sehr elegantes und einleuchtendes Beispiel für die Wechselbeziehung zwischen Struktur und Funktion könnte das *Kuhnsche Haarnadelgegenstrom-Modell* für die Harnkonzentrierung nach dem Multiplikationsprinzip darstellen (vgl. z. B. [37, 38, 49—53, 146—150]): Die Modellstruktur entspricht weitgehend dem Feinbau der Henleschen Schleifen — der Konzentrationsanstieg im Nierenmark deutet auf eine Gegenstrommultiplikation in diesem Bereich des Organes. Als treibende Kraft wurde zunächst der hydrostatische Druck in den Tubuli in Erwägung gezogen; später nahm man statt dessen einen aktiven Elektrolyttransport an (,,Salztransportniere'' [52]). Im Anschluß daran wurde die Möglichkeit diskutiert, daß die aktiven Transportprozesse vorwiegend oder ausschließlich in der Außenzone des Markes lokalisiert sind [124, 125, 132] — eine Arbeitshypothese, welche sich mit den morphologischen Fakten gut deckt. Nach Messungen beim Wüsten-Nagetier *Psammomys* findet jedoch auch in den Überleitungsstücken der Innenzone ein Konzentrationsanstieg zur Papillenspitze hin statt [35, 97, 134]. Es ergibt sich damit die Frage, welche Kräfte einen Multiplikator in der Innenzone des Markes in Gang halten können. Ein aktiver Elektrolyttransport konnte in diesem Abschnitt bislang ebensowenig nachgewiesen werden [127] wie gröbere Strukturunterschiede zwischen ab- und aufsteigenden Überleitungsstücken [122]. Nach diesen negativen Befunden bieten im Moment lediglich Druckdifferenzen zwischen ab- und aufsteigenden Schenkeln [126] oder neuartige Modellvorstellungen und -versuche auf der Basis der Diffusion [69, 88] eine Möglichkeit einer Deutung. Submikroskopische Strukturanalysen von *Psammomys*-Nieren liegen noch nicht vor; auch sie könnten unter Umständen neue Fakten zutage fördern und damit zu einer Klärung der Situation beitragen. Leider wird die Diskussion der Gegenstromhypothese zusätzlich dadurch erschwert, daß verbindliche Aussagen über die Osmolarität in den verschiedenen, durch lipoidische Membranen getrennten Mikrobereichen der Außenzone fehlen. So ist bis heute noch unklar, auf welche Art und Weise ein derartiger Multiplikator im einzelnen arbeitet; eine endgültige Korrelation von Struktur und Funktion erscheint daher auch bei diesem Modellfall derzeit noch nicht möglich.

Unsere gängigen Anschauungen über die *Physikochemie und Biochemie des Flüssigkeitstransportes durch Membranen* [4, 17, 33, 70, 144] beruhen im wesentlichen noch auf Arbeitshypothesen. Eine Erklärung von Wasserverschiebungen auf der Basis der Wasserüberführung (= Kodiffusion [19]) scheint nicht zulässig [20]. Versuche mit bimolekularen lipoidischen Modellmembranen konnten in jüngster Zeit entscheidende Fortschritte verzeichnen [123], haben aber scheinbar noch nicht dasjenige Stadium erreicht, welches Vergleiche zwischen den Modellen und den enzymatisch aktiven Membranen lebender Zellen erlauben könnten.

Schließlich erscheint noch nicht endgültig abgeklärt, in welchem Umfang *osmotische Gradienten* zwischen den Zellen und den angrenzenden extrazellulären Räumen existieren und aufrecht erhalten werden [21, 46]. Man denkt in diesem

Zusammenhang unwillkürlich an die sicherlich bestehenden Gradienten zwischen Blut-Epithel-Intermediärharn im Bereich der Macula densa; das gleiche gilt wohl für das System Galle-Gallenblase-Blut, sowie für Süßwasserprotozoen [*10, 46*]. Die zu diesem Fragenkomplex geäußerten Meinungen widersprechen sich vielfach [*3, 8, 13, 14, 24, 25, 70, 86, 113*]; eine endgültig verbindliche Stellungnahme scheitert auch in diesem Sektor offensichtlich an unseren begrenzten Möglichkeiten und Kenntnissen.

So kann man zusammenfassend feststellen, daß unsere morphologischen, wie physiologischen Befunde trotz erstaunlicher Fortschritte in jüngster Zeit [*5, 119, 124, 125, 134, 135*] noch immer sehr bruchstückhaft sind und klare, verbindliche Korrelationen zwischen Zellstruktur und Organfunktion auf einer hinreichend breiten Basis noch nicht ermöglichen.

Man wird daher wiederum die Frage aufwerfen müssen, auf welchem Weg neue konkrete Resultate erhoben werden können, die geeignet sein könnten, neue Aspekte zur alten Frage „Struktur und Funktion" [*2, 7, 61, 63, 71, 80, 89, 119*] zu liefern. Von morphologischer Seite aus scheinen hierfür vorzugsweise Studien geeignet, welche sich neuer Fixationsmethoden bedienen, quantitative Aussagen ermöglichen sowie vergleichend morphologischen Charakter aufweisen. Die *Methodik* sollte zunächst so weit verbessert werden, daß alle Bezirke der Niere in einwandfreiem Erhaltungszustand mit hoher und höchster Auflösung untersucht werden können. Vergleiche identischer Bereiche nach unterschiedlicher Fixation können hierbei Aufschlüsse über den supravitalen Strukturwandel und in weiterer Folge über die damit verbundenen Vorgänge erbringen. Dadurch ergeben sich weitere Indizien zur Funktion verschiedener Abschnitte und konkrete Hinweise für indirekte physiologische Methoden.

Aufbauend auf eine derart verbesserte, hinreichend verläßliche Fixationsmethodik können die entscheidenden Strukturelemente (Zelloberfläche, Endomembransystem, Mitochondrien) im el. opt. Ultradünnschnittbild morphometrisch erfaßt werden; über entsprechende „*stereologische Methoden*" verfügt man heute; vgl. [*22, 23, 39, 40, 48, 140*]. Sie können in einfacher Weise für die vorliegenden Zwecke modifiziert werden [*102, 103*]. Es ist anzunehmen, daß sich auf diese Weise im Verein mit vergleichend morphologischen Studien im Laufe der Zeit ein tieferer Einblick in die Beziehungen zwischen dem Feinbau des Organes und dem daran gebundenen Stoffwechsel erreichen läßt.

Im folgenden wird über den Stand derzeit laufender Fixationsstudien sowie morphometrischer Untersuchungen berichtet; im Anschluß daran wird versucht, einige Fragen zu diskutieren, welche sich aus den Beziehungen zwischen Feinbau und Funktion ergeben.

B. Fixation der Niere

I. Standardfixation im Blöckchen
nach Abtrennen vom Blutkreislauf

Im allgemeinen werden die Objekte für die Licht- und Elektronenmikroskopie unmittelbar nach der Entnahme aus dem Organismus in eine geeignete Fixierungsflüssigkeit (Aldehyd- oder OsO_4-Lösung [*73, 94, 104*]) eingebracht. Die begrenzte Diffusionsgeschwindigkeit der

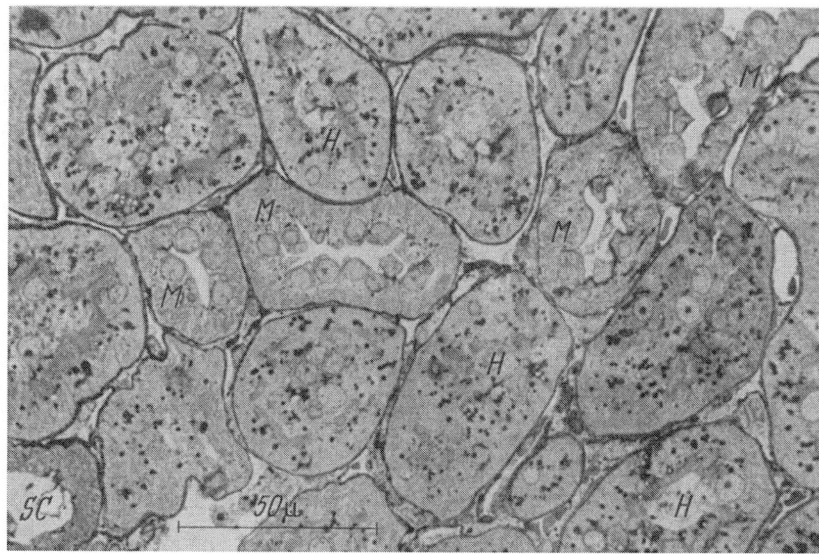

Abb. 1. Rattenniere, Dünnschnitt, Silberimprägnation. Standardfixation nach Unterbrechen des Blutkreislaufes in OsO₄-Lösung. Sämtliche Erythrozyten ausgewaschen, Blutkapillaren und Hauptstücke (H) kollabiert. Kollabierte Lumina der Mittelstücke (M) erst nach der Fixation wieder geöffnet. Lediglich ein Schaltstück (SC) offen. Plan-Achromat 40:1

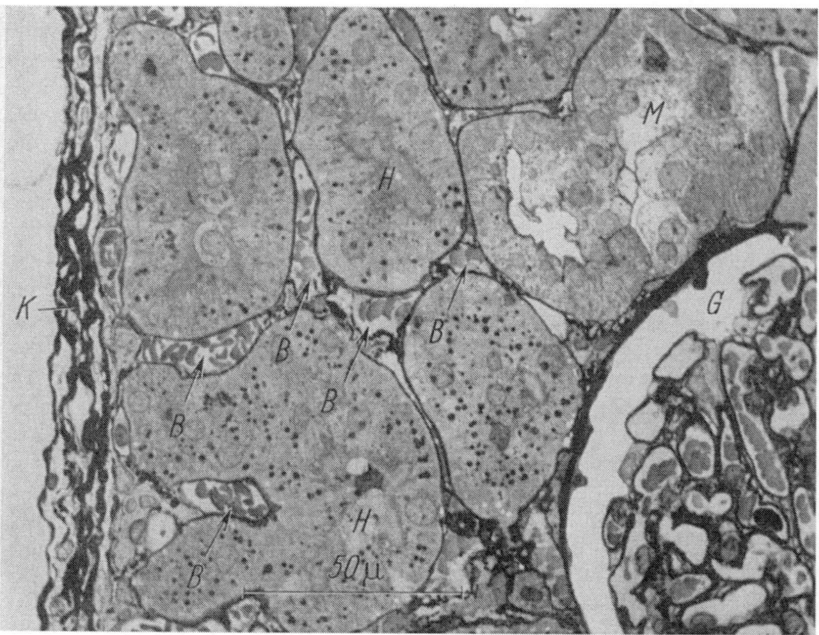

Abb. 2. Rattenniere, Dünnschnitt, Silberimprägnation. Fixation der Niere in toto nach Abklemmen der Gefäße am Hilus. Kapillaren (B) mit Blut gefüllt. Links Nierenkapsel (K), rechts ein Mittelstück (M). Hauptstücke (H) wie in Abb. 1 kollabiert; (G) Glomerulum. Plan-Achromat 40:1

Fixierungsmittel [87, 105, 151] bringt es mit sich, daß zentrale Bereiche des Objektes erst nach Einsetzen der Autolyse vom Fixans erreicht werden und ihre Feinstruktur dementsprechend schlecht erhalten wird; demgegenüber sind die Randzonen bei den meisten tierischen und menschlichen Objekten gut erhalten, soweit sie nicht bei der Objektentnahme mechanisch beschädigt werden [73]. Für die Niere trifft dies jedoch nicht zu [36, 116, 117, 139]: Makroskopisch zeigt ein augenblickliches Erblassen der Rinde nach Durchtrennen der Gefäße, daß die Erythrozyten aus der Rindenschicht abfließen. Dies dürfte einerseits auf die elastische Wirkung der Kapsel zurückzuführen sein, welche nach Wegfall des Blutdruckes nicht mehr kompensiert wird; andererseits spült wohl das noch weiter aus den Harnkanälchen entstehende Resorbat die Erythrozyten aus. Der nach Sistieren des Blutkreislaufes fortgesetzte transtubuläre Flüssigkeitstransport führt binnen kürzester Frist zu einem Kollaps aller Hauptstücke [116, 138], sowie eines begrenzten Bereiches der Mittelstücke beiderseits der *Macula densa*. Man erhält daher durch die Standardfixation ein räumlich verzerrtes Bild der Rindenstruktur (Abb. 1). Bereits lichtoptisch ist der Kollaps aller Blutkapillaren deutlich zu erkennen – Erythrozyten fehlen praktisch vollständig. Hierbei muß man sich vor Augen halten, daß die kortikalen Tubuli *in vivo* von prall gefüllten Blutkapillaren umgeben sind [75, 117]; verschließt man die Gefäße am Hilus vor dem Abtrennen des Organes mit einer Klemme und fixiert die Niere ohne weiteren Eingriff *in toto*, so erhält man hiervon einen vagen Begriff (Abb. 2). Die Mittelstücke zeigen im Bereich der Kontaktstelle am Glomerulum eine kantige, im Querschnitt sternförmige Begrenzung des Lumen (vgl. hierzu Abb. 3a und b), welche den sicheren Rückschluß erlaubt, daß die Abschnitte beim Eintreffen des Fixans ebenfalls vollkommen leer gesaugt sind und sich erst im Verlaufe der weiteren Präparation vor der Polymerisation des Einschlußmediums wieder öffnen. Hierbei behalten die fixierten Ränder ihre eingepreßten Formen, welche *in vivo* an Zelloberflächen sicherlich nicht existieren. Die Zellen der Haupt- und Mittelstücke sind nach dieser Fixation relativ hoch, das Chondriom liegt vorzugsweise basal (Abb. 4a); die Zellstrukturen (Kerne, Mitochondrien, Grana usw.) der kortikalen Tubuli bleiben hierbei jedoch im allgemeinen gut erhalten – es ist lediglich ihre topographische Anordnung verschoben.

Demgegenüber treten in der Außenzone des Markes – vornehmlich an den *partes rectae* der Haupt- und Mittelstücke – sehr häufig starke Artefakte auf. Diese Zellen sind hierbei defekt und unregelmäßig gequollen – ihre Membranen aufgerissen. In den geöffneten Lumina der Mittelstücke finden sich stark gequollene, ausgestoßene Zellbestandteile (Abb. 5a). Die Blutkapillaren sind auch in diesem Bereich kollabiert, enthalten aber reichlich eingeklemmte, durch Zusammenpressen deformierte Erythrozyten (Abb. 5b); soweit die Kapillaren ein offenes Lumen aufweisen, ist dies meist sekundär nach der Fixation entstanden, wie dies oben bei der Besprechung des Mittelstückes diskutiert wurde. Demgegenüber zeigen die Überleitungsstücke (dünne Abschnitte der Henleschen Schleifen) im Bereich der Außenzone ebenso wie in der

Innenzone des Markes keine derartigen Schäden. Die Innenzone des Markes zeigt als einziger Bereich der Säugerniere auch nach der Standardfixation normal mit Erythrozyten gefüllte Blutkapillaren im Verein mit normal dimensionierten Harnkanälchen.

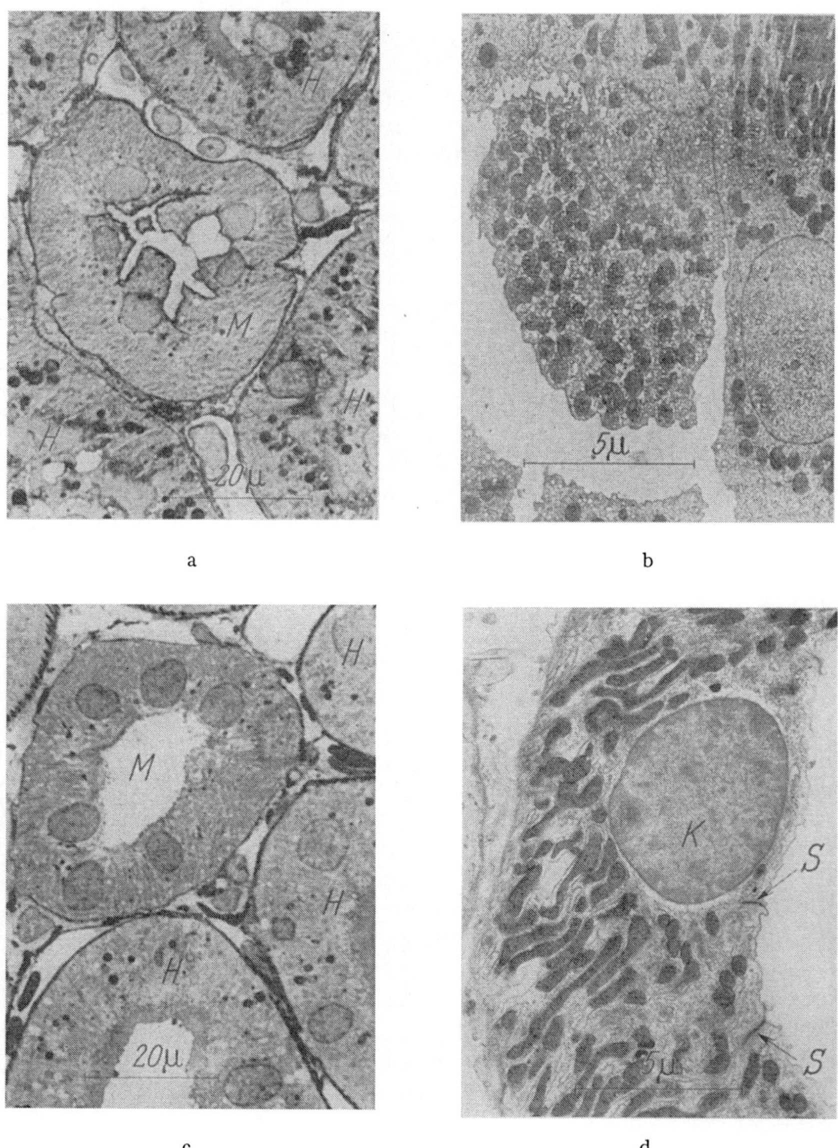

Abb. 3a—d. Rattenniere. Mittelstücke nach verschiedenartiger Fixation im licht- und el. opt. Bild. a Standardfixation eines Gewebeblöckchens nach B.I.; Lumen zum Zeitpunkt der Fixation geschlossen — später wieder geöffnet. Hierdurch kantige artefizielle Lumengrenze. b desgl. im Elektronenbild; el. opt. Vergr. 2600:1. c Mittelstück (*M*) und Hauptstücke (*H*) nach Anfixieren der durchbluteten Nierenrinde. Präparat und Optik wie a. d desgl. im Elektronenbild. Epithel relativ flach, Kern (*K*) kaum vorgewölbt; Schlußleisten (*S*) gut sichtbar. El. opt. 2600:1

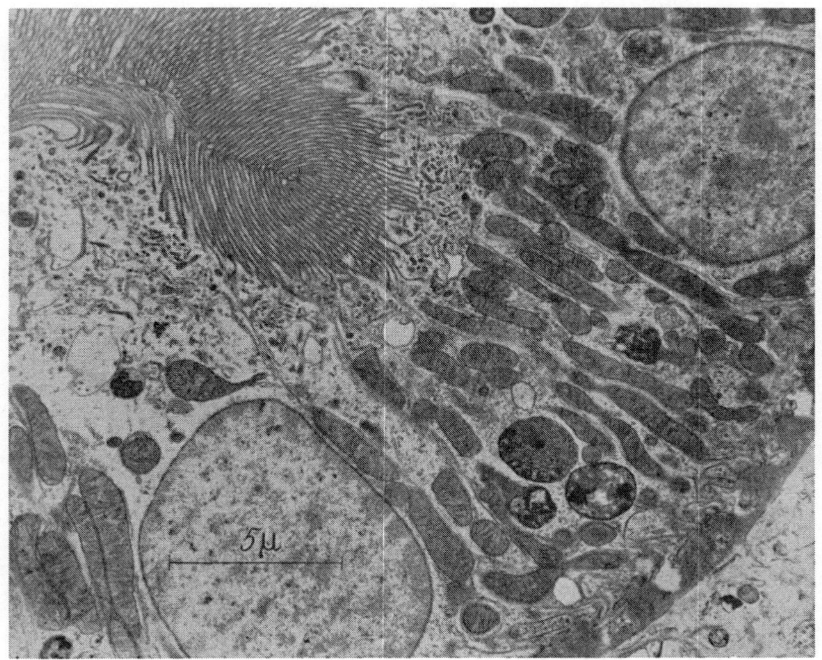

a

b

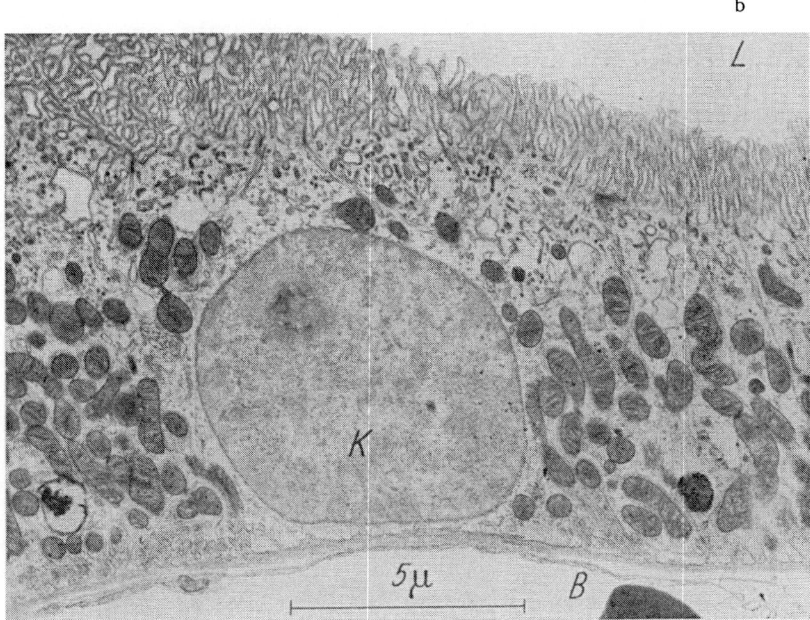

Abb. 4a u. b. Rattenniere. Hauptstücke nach unterschiedlicher Fixation im Elektronenbild, el. opt. 2600:1. a Standardfixation nach Unterbrechen des Blutkreislaufes. Kanälchenlumen geschlossen. b Anfixieren des durchbluteten Organes. Kanälchenlumen offen; Epithel kaum höher als Zellkern (K); Blutkapillaren (B) geöffnet

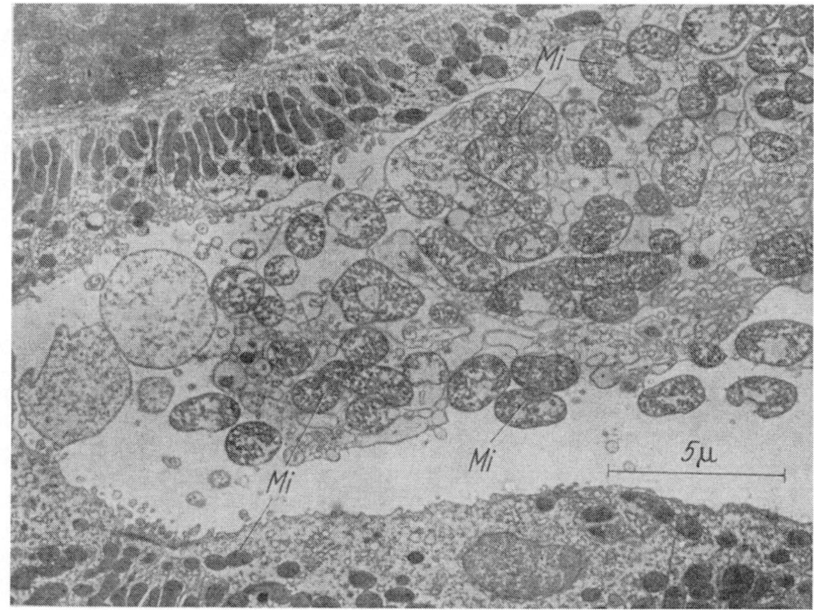

a

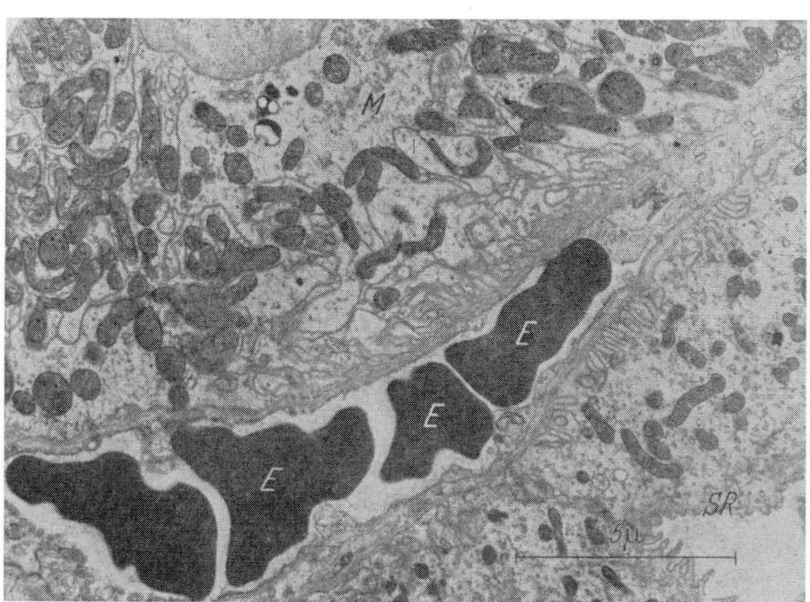

b

Abb. 5a u. b. Rattenniere. Artefakte in der Markzone nach Standardfixation. El. opt. 2600:1. a Gequollene Mitochondrien (*Mi*) und Zellbestandteile, welche bereits vor der Fixation in ein Mittelstücklumen abgestoßen worden sind. b „trocken gelegte" Blutkapillare, welche sich nach der Fixation wieder etwas geöffnet hat; vgl. Profil der Erythrozyten (*E*) und Form der Kapillarwand. Im rechten unteren Eck ein Sammelrohr (*SR*) mit typischer Differenzierung der basalen Zellmembran. *M* = Mittelstück

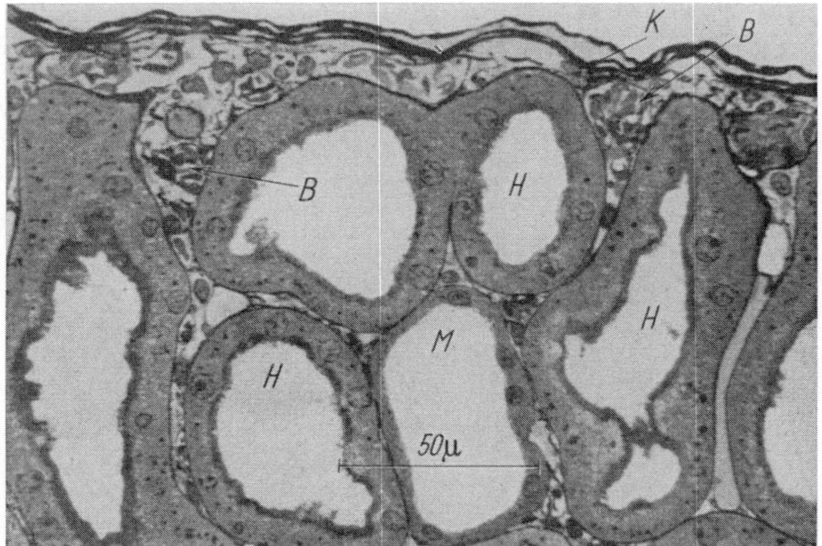

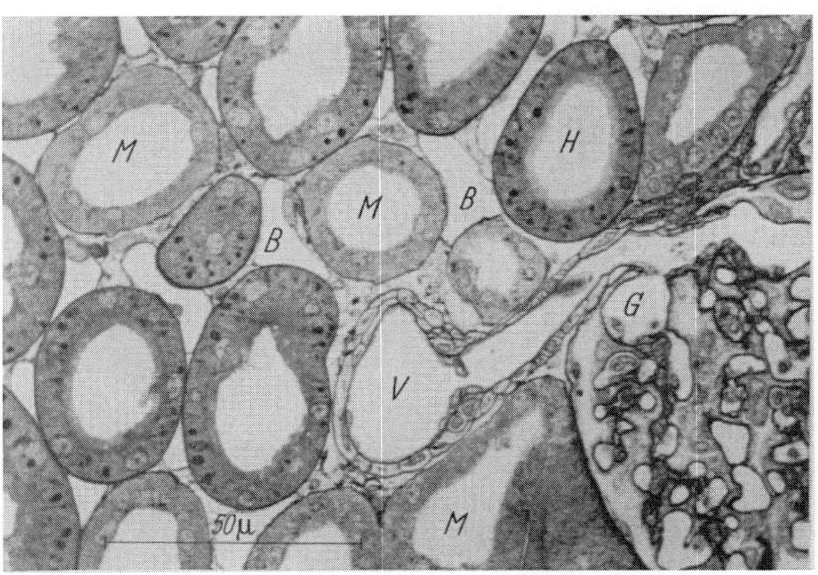

Abb. 6a u. b. Rattenniere. Einwandfreie Fixation. Dünnschnitt, Silberimprägnation, Plan-Achromat 40:1. a Anfixieren der durchbluteten Rinde. Am oberen Bildrand die Nierenkapsel (K). Kapillaren (B) mit Blut gefüllt. Zahlreiche Hauptstücke (H) — ein Mittelstück (M). b Fixation durch Perfusion mit OsO₄-Lösung. Im Vas afferens (V) des Glomerulum (G) und in den Blutkapillaren (B) keine Blutkörperchen, Haupt- (H) und Mittelstücke (M) gut erhalten

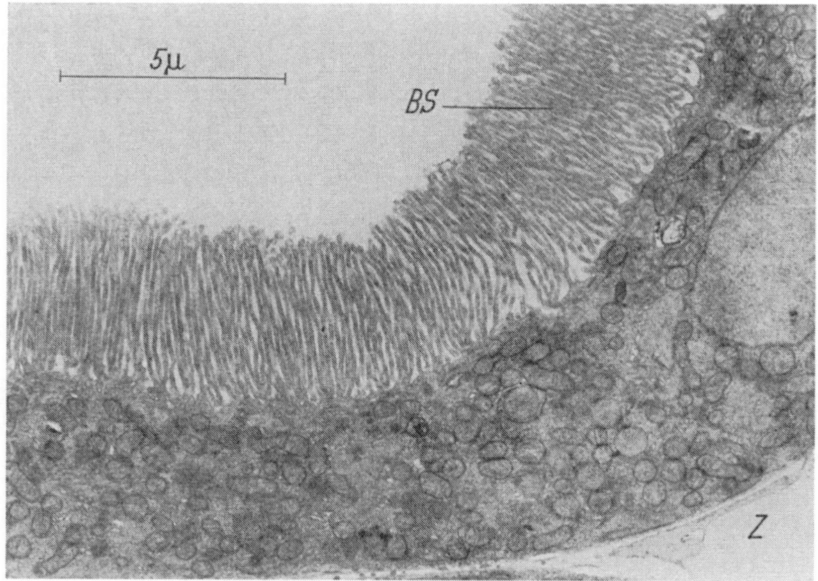

a

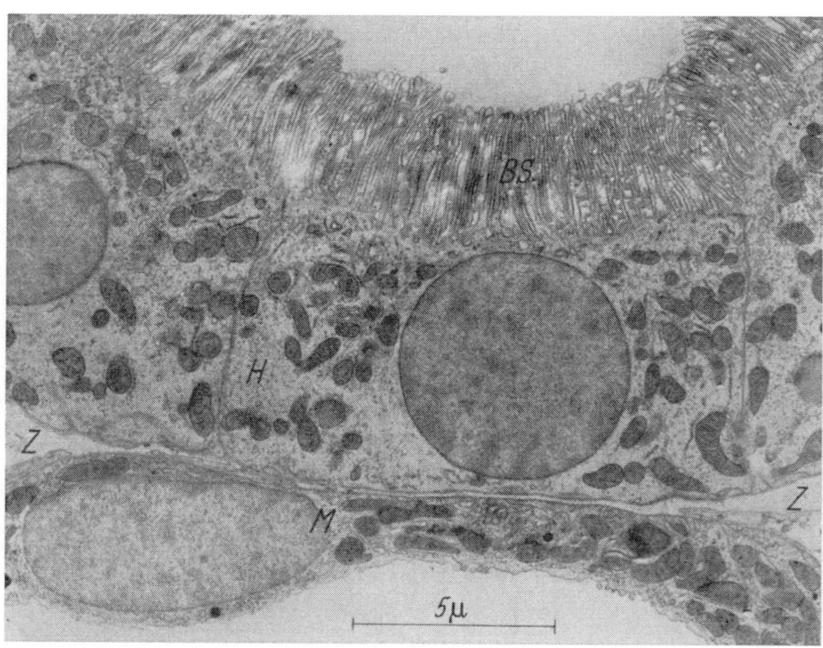

b

Abb. 7a u. b. Rattenniere, Hauptstücke im Bereich des Außenstreifens im Nierenmark, Fixation durch Perfusion mit OsO$_4$-Lösung. El. opt. 2600:1. a Abschnitt mit besonders hohem Bürstenbesatz (*BS*). b Abschnitt an der Grenze zwischen Außen- und Innenstreifen. Zwischen Haupt- (*H*) und Mittelstück (*M*) befindet sich ein lockeres Zwischengewebe (*Z*)

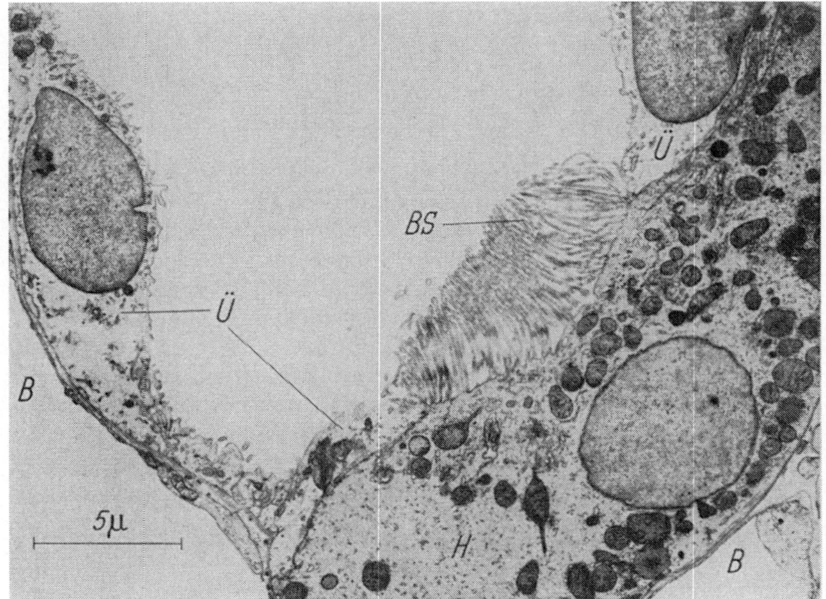

a

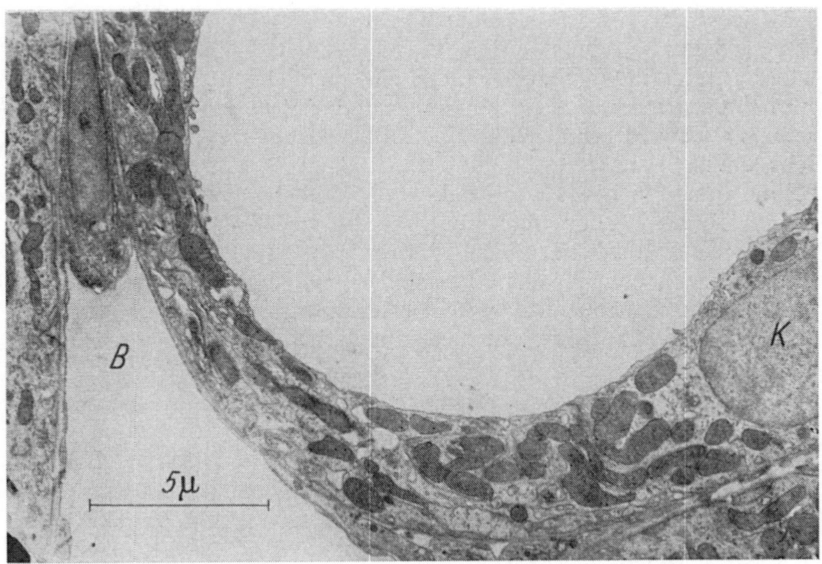

b

Abb. 8a u. b. Harnkanälchen der Rattenniere an der Grenze zwischen Innen- und Außenstreifen der Außenzone des Markes. Fixation durch Perfusion mit OsO₄-Lösung. El. opt. 2600:1.—a. Übergang vom Hauptstückepithel (*H*) mit letzten Bürstensaumstäbchen zum niederen Epithel des Überleitungsstückes (dünner Teil der Henleschen Schleife; *Ü*). Im Bild zwei Blutkapillaren (*B*). b Mittelstück (Aufsteigender dicker Schenkel der Henleschen Schleife) mit sehr flachem Epithel und Kern (*K*). Zwischen den Kanälchen eine Blutkapillare (*B*).

II. Fixation der durchbluteten Nierenrinde

Bereits 1955 legte Pease [77,78] Bilder mit offenen Hauptstücken nach äußerlichem Anfixieren der durchbluteten Rinde vor und zeigte, daß der Erhaltungszustand bei diesem Vorgehen optimal ist; die Befunde wurden jedoch kaum beachtet und die Methode erst um vieles später [62] wieder aufgegriffen. Nach dem oberflächlichen Anfixieren zeigen die Hauptstücke, Mittelstücke und Rindensammelrohre ein wesentlich flacheres Epithel, in dem die Kerne kaum vorgewölbt sind (Abb. 6a). Der Mitochondrienreichtum dieser Epithelien springt jetzt besonders deutlich ins Auge. Die Mitochondrien selbst sind praktisch gleichmäßig über das Zytoplasma verstreut (Abb. 4b). Die Blutkapillaren sind geöffnet und mit Erythrozyten gefüllt, wenn auch ihre Abmessungen nicht genau dem Zustand *intra vitam* entsprechen. Der große Nachteil dieser Methode liegt analog zur Mikropunktion darin, daß nur die äußerste Rindenschicht hiermit erfaßt werden kann; die durch Artefakte am schwersten betroffene Außenzone des Markes kann hiermit nicht erreicht werden.

III. Perfusion der Niere mit Fixans

Die tieferen Bereiche des Organes können u. a. durch eine Perfusion [141] der gesamten Niere mit Fixierungsflüssigkeit im Anschluß an eine kurze Spülung mit einem physiologischen Medium fixiert werden. Ein derartiges Vorgehen hat sich bereits bei anderen fixationslabilen Strukturen [74] gut bewährt und führt auch hier zu einer wesentlichen Reduktion der Artefakte (Abb. 6b, 7 und 8). Wie nach der Fixation nach B. II. erscheint auch hier die Mitochondrienkonzentration im Gewebe besonders hoch. Durchmesser und Höhe der Bürstensaumstäbchen sind starken Schwankungen unterworfen und nicht so gleichmäßig, wie nach der Standardfixation B. I. Mittelstücke und Sammelrohrepithelien sind flach — ihre Kerne kaum vorgewölbt (Abb. 3d und 8b). Die sonst für die Außenzone des Markes charakteristischen Artefakte fehlen in gut fixierten Bereichen vollkommen; allerdings sind neben diesen in Mark und Rinde fleckförmige Zonen mit einem schlechten Erhaltungszustand zu beobachten, welche jedoch an den kollabierten Hauptstücken leicht zu erkennen sind.

IV. Andere Fixierungsverfahren

Neben den oben angeführten Verfahren B. II und B. III bestehen weitere, welche die Möglichkeit zu einer lebensechten Fixation bieten: So kann die freigelegte, durchblutete Niere beispielsweise mit einer großlumigen Menghini-Nadel punktiert werden [5, 43, 44, 68, 76]; es ist dabei möglich, die Punktionszylinder binnen weniger Sekunden in das Fixans zu aspirieren oder auszustoßen. — An Stelle der gesamten Niere kann auf dem Wege der Mikroperfusion [112] ein einzelnes Nephron mit Fixans durchspült werden [121]; die gesamte Umgebung dieses Nephrons bleibt während der Fixation optimal durchblutet — ein Umstand, der bei den Methoden nach B. II und B. III nicht immer gegeben ist. Zudem erlaubt die Mikroperfusionsanordnung ein unmittelbar vorangehendes Durchspülen dieses Nephrons mit verschiedenen Medien. — Eine weitere Möglichkeit besteht im Injizieren von Fixantien in die freigelegte, durchblutete Niere [27, 77]. — Bei sämtlichen angeführten Verfahren können an die Stelle der OsO_4-Lösungen rascher diffundierende Aldehyd-Lösungen [93, 94, 104, 105] treten, welche unter Umständen noch größere Bereiche einwandfrei erhalten. — Schließlich sind alle Verfahren von Interesse, welche ein rasches Einfrieren garantieren [36, 75, 117]: Hierbei stehen die He-II-Methoden [30] im Vordergrund des Interesses, erfordern aber einen relativ großen Aufwand. Die weitere Verarbeitung des eingefrorenen Objektes könnte entweder nach der Gefrierätztechnik [66, 67] oder nach einer Einbettung mit dem Ultramikrotom erfolgen [12, 29, 30, 99, 106]. — So bestehen offensichtlich zahlreiche weitere Möglichkeiten, welche im Rahmen der vorliegenden Studien noch nicht vergleichend geprüft werden konnten.

C. Morphometrie

I. Grundlagen

Der Stoffaustausch in den verschiedenen Nierenbereichen hängt weitgehend von den Mitochondrien als Energietransformatoren, den lipoidischen Zellmembranen als Diffusionsbarrieren sowie Orten aktiver Transportprozesse und schließlich von der räumlichen Anordnung dieser Elemente zueinander ab. Unabhängig von den verschiedenen Arbeitshypothesen für den aktiven Transport von Partikeln durch die Zellmembranen scheint gesichert, daß diese Prozesse im wesentlichen durch den Umsatz energiereicher Phosphorsäure-Ester finanziert werden; ebenso scheint gesichert, daß diese Ester zum überwiegenden Anteil aus den Mitochondrien stammen. Im Rahmen des gestellten Themas ist daher primär der Umfang der zellulären Membranflächen und des Chondrioms von Interesse. *Membranflächen* können im Schnittpräparat nach der „Nadelmethode" vermessen werden [11, 40]. Im Prinzip werden hierfür die Schnittpunkte zwischen den Membranen und einer über sie gezogenen oder projizierten Schar paralleler Linien gezählt. Absolutwerte erhält man nach der Multiplikation mit geeigneten Faktoren [40, 103]; Relativwerte können wesentlich einfacher ermittelt werden, indem man die Zahl der Schnittpunkte der Linienschar mit verschiedenen Membranabschnitten oder Membranen direkt zueinander in Beziehung setzt und dadurch das Flächenverhältnis „Membran$_1$: Membran$_2$: Membran$_3$ ... usw." erhält. *Volumina* werden auf ähnliche Weise mit der „Treffermethode" [32] ermittelt, indem man über das Meßfeld einen gleichmäßigen Raster zeichnet oder projiziert [22, 23, 39, 40, 48, 102, 103, 140]. Auch hier vereinfacht sich das Auswerten wesentlich, wenn nur Relativwerte („Volumen$_1$: Volumen$_2$: Volumen$_3$ ... usw.") benötigt werden. Schließlich können Flächen- und Volumenbestimmungen kombiniert in einem Arbeitsgang ausgeführt werden [103]. Hierdurch ist es beispielsweise möglich, die Oberfläche der Zelle, sowie verschiedene Areale ihres Endomembransystemes mit dem Gesamtvolumen oder bestimmten Teilvolumina (z. B. Mitochondrienvolumen) in Bezug zu setzen.

Das Prinzip dieser Methoden ergibt sich am besten an Hand eines Modells. Abb. 9 zeigt eine apolare Zelle, von der für diesen Zweck nur Zell-, Kern- und Mitochondrienmembranen dargestellt sind. Es können an diesem Modell fünf funktionell

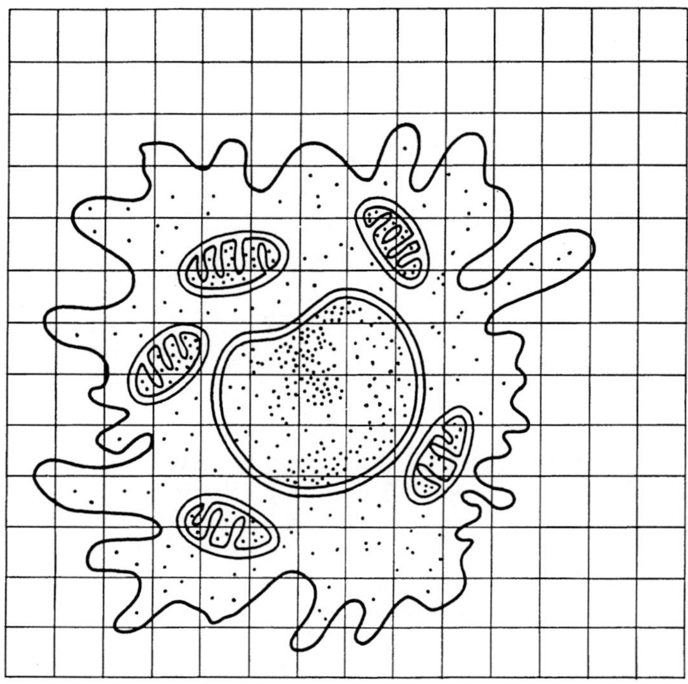

Abb. 9. Schema einer Zelle, deren Flächen- und Volumen-Relationen nach der „Nadel-" bzw. „Treffermethode" bestimmt werden; vgl. hierzu Text

unterschiedliche Membransysteme (Zellmembran, Kernhülle sowie äußere, innere und Cristae-Membran der Mitochondrien) neben drei Phasen (Hyaloplasma, Karyoplasma und Mitochondrienplasma) vermessen werden, welche eine unterschiedliche Zusammensetzung aufweisen; das Volumen der Kern- und Mitochondrienhülle kann im allgemeinen vernachlässigt werden. Das gesamte Bildfeld wird durch ein Koordinatennetz bedeckt. Die Linien dieses Netzes bilden Schnittpunkte (S) mit den verschiedenen Membranen, deren Zahl sowohl für die vertikalen, wie für die horizontalen Linienzüge ermittelt wird (vgl. Tabelle 1.). Die Schnittpunkte, welche die sich kreuzenden Geraden des Rasters untereinander bilden, fallen wiederum auf verschiedene Bereiche der Zellen in verschiedener Häufigkeit; die Zahl der Treffer (T) auf jede der drei angeführten Phasen wird nun ebenfalls ermittelt (Tab. 1). Da Schnittpunkte (S) und Treffer (T) mit ein- und demselben Koordinatenraster bestimmt werden, können sie zueinander in Beziehung gesetzt werden. So erscheint z. B. für passive Austauschvorgänge Zelle-Interstitium die Relation S_z/T_{ges} (entspr. Zelloberfläche zu Zellvolumen) von Bedeutung; für den aktiven, durch Mitochondrienmetabolite finanzierten Ionentransport ist das Verhältnis T_m/S_z (entspr. Mitochondrienvolumen : Membranfläche; u. U. detailliert für bestimmte Membranabschnitte; vgl. unten), für die Mitochondrienleistung u. U. die Beziehung S_{cm}/T_m (Cristae mitochondriales pro Volumeneinheit Mitochondrien)

Tabelle 1

Strukturelement	Abkürzung[1]	Resultat nach Abb. 9	
Äußere Zellmembran	S_z	81	40%
Kernhülle.	S_k	32	16%
Außenmembran der Mitochondrien . .	S_{am}	30	15%
Innenmembran ohne Cristae	S_{im}	28	13,5%
Cristae mitochondriales	S_{cm}	32	15,5%
Gesamtfläche	S_{ges}	203	100%
Hyaloplasma	T_h	46	68%
Karyoplasma	T_k	13	19%
Mitochondrieninhalt	T_m	9	13%
Gesamt-Trefferzahl.	T_{ges}	68	100%

[1] S für Schnittpunkte („Nadelmethode"); T für Treffer („Punktzählung").

von Bedeutung. Beim Vermessen polarer Strukturen (z. B. einzelne Epithelzellen) müssen mehrere Messungen unter verschiedenen Winkellagen durchgeführt werden, da die Vorzugsrichtung sonst zu Fehlern führt. Die an dieser Stelle lediglich grob skizzierten Verfahren der Morphometrie lassen sich praktisch beliebig modifizieren und dadurch auch sehr speziellen Fragestellungen anpassen.

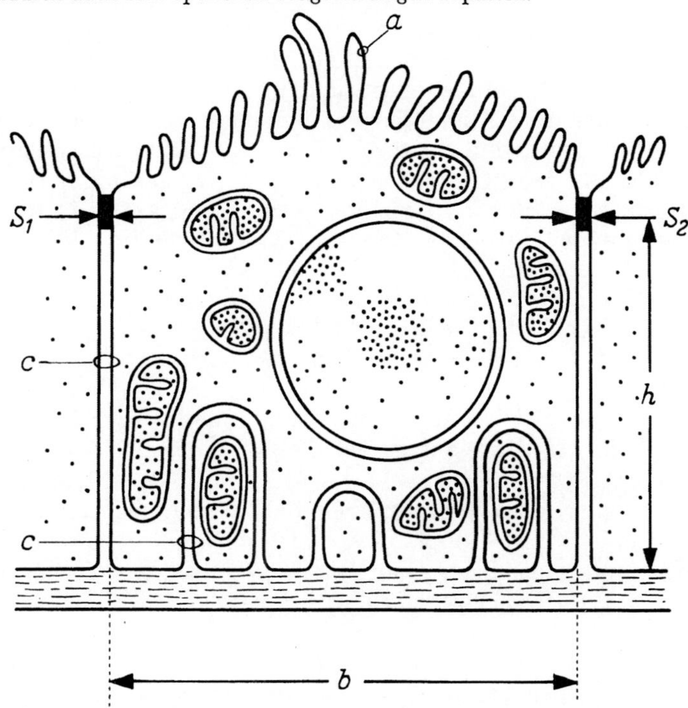

Abb. 10. Schema einer Tubulus-Epithelzelle mit Schlußleisten (S_1) und (S_2); vgl. hierzu Text

Es ist evident, daß den verschiedenen Membranbereichen der polar gebauten Kanälchenepithelien auf Grund ihrer unterschiedlichen Anordnung, sowie der unterschiedlichen Zusammensetzung der angrenzenden extrazellulären Medien eine unterschiedliche Funktion zukommt. Man muß bei einer morphometrischen Analyse daher zumindest drei Abschnitte der äußeren Zellmembran getrennt erfassen: Einerseits die *apikale Zellgrenze (a)* gegen das Kanälchenlumen, welche sich von Schlußleiste (S_1) zu Schlußleiste (S_2) erstreckt; vgl. Abb. 10. Andererseits den *Abschnitt (b), welcher der Basalmembran anliegt* und grob vereinfacht im ebenen Schema die Länge „*b*" aufweist; schließlich diejenigen *Bereiche (c), in denen die Membranen verzahnter Zellen aneinander* liegen. Der prozentuale Anteil der einzelnen Bereiche kann an Hand der Schnittpunkte $(S_a, S_b$ und $S_c)$ mit dem Koordinatenraster nach Abb. 9 ermittelt und in der oben dargelegten Weise mit der Trefferzahl T_{ges} sowie T_m in Bezug gesetzt werden.

Die damit gegebenen Möglichkeiten für quantitative morphologische Studien am Nephron sind im Moment erst in relativ bescheidenem Umfang genützt, da zunächst noch die morphometrische Kontrolle vergleichender Fixationsstudien im Vordergrund des Interesses steht. Immerhin können im folgenden bereits einige Richtwerte angegeben werden.

II. Angewandte Methodik und erste Ergebnisse

Die in Tab. 2 zusammengestellten Befunde wurden durchgehend an der Rattenniere erhoben, welche nach B. II. (Rinde) bzw. B. III. (alle Zonen) in gepufferter OsO_4-Lösung fixiert und in Araldit eingebettet wurde. Für das quantitative Auswerten dienten meist Aufnahmen von Pb-kontrastierten Ultradünnschnitten bei el. opt. Direktvergrößerungen von 2600 : 1; die meisten Schnitte zeigen hierbei alle interessierenden Membranen. Nach der Treffermethode wurde zunächst die „*Mitochondrienkonzentration*" *(MI)* der Zellen nach der Definition $MI = 100 \times T_m/T_{ges}$ (%) ermittelt. In einem Arbeitsgang hiermit wurde die Größe S_z/T_{ges} (entspr. Zelloberfläche zu Zellvolumen) ermittelt, wobei S_z alle Membranbereiche einschließt $(S_z = S_a + S_b + S_c)$; die Einzelbereiche wurden hierbei prozentual in der oben angegebenen Weise erfaßt.

Erwartungsgemäß weisen die kortikalen Haupt- und Mittelstücke besonders hohe *MI*-Werte auf, zuweilen bis 35% des Zellvolumens unter Einschluß des Bürstensaumes und des Zellkernes. Bei weiteren Schlußfolgerungen aus diesen Prozentsätzen darf jedoch nicht außer Acht gelassen werden, daß Haupt- und Mittelstückmitochondrien — abgesehen von wahrscheinlich vorhandenen Unterschieden im Chemismus — eine unterschiedliche Feinstruktur aufweisen, welche sehr wohl bei identischen *MI*-Werten unterschiedliche Leistungen möglich erscheinen ließe.

Besondere Bedeutung wird der *Abnahme des F_c-Wertes* vom proximalen zum distalen Bereich des *Hauptstückes* zugemessen. An der Grenze zwischen Außen- und Innenstreifen der Außenzone des Markes (vgl. Abb. 8a sowie eng benachbarte Stelle in Abb. 7b) findet man eine glatte basale Zellmembran ohne Invaginationen bzw. Verzahnungen mit benachbarten Zellen; hierdurch steht F_a mit einem Anteil bis zu 97%

der äußeren Membranfläche besonders stark im Vordergrund (Bürstensaum!). Gänzlich andersartig verhalten sich die einzelnen *Membranabschnitte im Mittelstück*, in dem F_c über den gesamten Bereich mit Werten bis 95% den weitaus größten Teil an der Zelloberfläche einnimmt. Erstaunlich sind weiterhin die hohen *MI*-Werte der Mittelstücke im Mark. Bei den *Sammelrohren* stehen der Flächenbestimmung – offenbar

Tabelle 2

Abschnitt	Meßwert	Rinde	Mark		Innenzone des Markes
			Außenzone des Markes	Innenstreifen	
			Außenstreifen		
		in %	in %	in %	in %
Hauptstück	MI	20—35	10—20	—	—
	S_{ges}/T_{ges}	5—8	3—5	—	—
	F_a	60—80	80—97	—	—
	F_b	3—5	1—4	—	—
	F_c	10—40	2—10	—	—
Mittelstück	MI	30—35 (20)[1]	25—30		
	S_{ges}/T_{ges}	5—6 (2,5)	2—5		
	F_a	2,5—5	10—15		
	F_b	2,5—5	5—10		
	F_c	86—95	75—80		
Sammelrohr	MI	10—20	Typ I: 10% Typ II: 4%		5% 2%
Überleitungsstück (Dünner Teil der Henleschen Schleife)	MI	—	unter 2%		

[1] Eingeklammerte Werte für Übergang zu Sammelrohr (Schaltstück)

durch die wechselnde Vakuolisierung der dunklen Zellen – derzeit noch gewisse Schwierigkeiten entgegen; ebenso ist der *MI*-Wert in der Rinde starken Variationen unterworfen, wogegen sich im Mark klar zwei Zelltypen mit verschiedenem Mitochondriengehalt trennen lassen; vgl. Tab. 2. Die *Überleitungsstücke* (dünne Abschnitte der Henleschen Schleifen) wurden im Hinblick auf ihren augenscheinlich minimalen Stoffwechsel (*MI* unter 2%) bislang noch nicht näher untersucht, da an ihnen aktive Prozesse in größerem Umfang nicht erwartet werden können.

D. Diskussion

Im Gegensatz zur Standardfixation B. I. scheinen die Mitochondrien in den stoffwechselaktiven Abschnitten des Nephrones nunmehr nach Fixation B. II. und B. III. (vgl. oben) gleichmäßig in der Zelle verteilt; die polare Anordnung dieser Organellen kann also kaum mehr als Grund für die Transportrichtung angesprochen werden. Trotzdem weisen sich die Zellen klar als polare Bauelemente aus: Dies durch die polare Anordnung der Schlußleisten und durch die unterschiedliche Differenzierung der apikalen und basalen Abschnitte der Zellmembran. Auf die *Bedeutung der Schlußleisten* wurde bereits verschiedentlich hingewiesen; vgl. [28]. Sie scheinen nicht nur einen äußerst festen Zusammenhalt der Zellen zu garantieren [72, 90], sondern auch einen interzellulären Stofftransfer zu unterbinden [64]. Die „zonula occludens" im Sinne von Farquhar und Palade [28] entspricht strukturell einer "external compound membrane" nach Robertson's Konzept [85]. Es ist wahrscheinlich, daß auch für Wasser und kleine Molekel der Weg über die Zelle einschließlich zweier Membranpassagen geringere Energien benötigt, als der Weg zwischen den kollabierten („fusionierten" [28]) Membranteilen der benachbarten Epithelzellen im Bereich der Schlußleiste. In diesem Zusammenhang soll kurz darauf hingewiesen werden, daß die *Clearance-Verfahren* bei jedem umfangreicheren extrazellulären Stofftransfer aus dem Harn- in den Blutraum unter Umgehen der Epithelmembranen ihren Sinn verlören. Es darf nach dem heutigen Stand unserer Kenntnisse angenommen werden, daß die Clearance-Stoffe durch die Endothelporen und Epithelschlitze unter Umgehen jeder Zellmembran in die Bowmansche Kapsel gelangen können, da das glomeruläre Deckepithel abweichend von den Kanälchenepithelien keine Schlußleisten aufweist. Vom Harnpol des Glomerulum bis zur Papillenspitze wird demgegenüber eine Rückdiffusion in den Blutkreislauf durch die immer vorhandenen Kittleisten zwischen allen Epithelzellen ausgeschlossen. Bei apikaler Lage der Kittleisten (vgl. nochmals Abb. 10) rechnen demnach auch die Seitenwände der Zelle in der Höhe „h" zum basalen Membranabschnitt. Jedes Molekel, welches die Zellmembran in diesem „basalen Abschnitt" verläßt, kann zwar auf extrazellulärem Weg in den Blutkreislauf, nicht aber zurück in das Kanälchenlumen gelangen. Sind die Seitenwände benachbarter Zellen verzahnt oder die basalen Zellmembranen eingefaltet („Basales Labyrinth" im Sinne von H. Ruska; vgl. [86, 91, 92, 119]), so entsteht eine sehr große Oberfläche. Man kann sich leicht vorstellen, daß eine derart beschaffene Fläche vorwiegend für eine Sekretion in das Blut in Frage kommt, da die engen Öffnungen der Schlitze an der Zellbasis im Gegensatz zu den Räumen zwischen den einzelnen Bürstensaumstäbchen das Eindiffundieren größerer Stoffmengen aus dem Blutkreislauf nicht erlauben; für eine Ausscheidung ist daher die Fläche $F_c + F_b$, für eine Stoffaufnahme an der Basis jedoch nur die Fläche F_b verfügbar — eine Umkehr des Stoffaustausches unter identischen Bedingungen erscheint an dieser gleichrichterartigen Struktur so lange nicht möglich, als die Interzellularen geschlossen sind.

Die Lage kompliziert sich weiter, wenn man die aktiven Prozesse in den Rahmen dieser Diskussion einbezieht. Hierfür ist zunächst die hypothetische Annahme notwendig, daß der Innenseite der Zellmembran ubiquitär in gleichmäßiger Verteilung diejenigen Enzymsysteme anhaften, welche im Verein mit den energiereichen Mitochondrienmetaboliten den aktiven Ionentransport in Gang halten; es ist dabei zunächst gleichgültig, auf welche Weise die Ionen transportiert werden. Mölbert et al. [65] konnten zeigen, daß Phosphatasen gleichmäßig die gesamte

innere Zellwand der Hauptstücke auskleiden. Nimmt man mit [1]
Analoges für die hier in Frage stehenden Enzymsysteme an, so könnte
man z. B. für eine Hauptstückzelle folgende Betrachtung anstellen:
Man vereinfacht das Problem zunächst dadurch, daß man die Zellen
als kleine Würfel ausbildet. In einer derartigen Zelle erhielte bei zentraler
Lage des Kernes und gleichmäßiger Verteilung der Mitochondrien jede

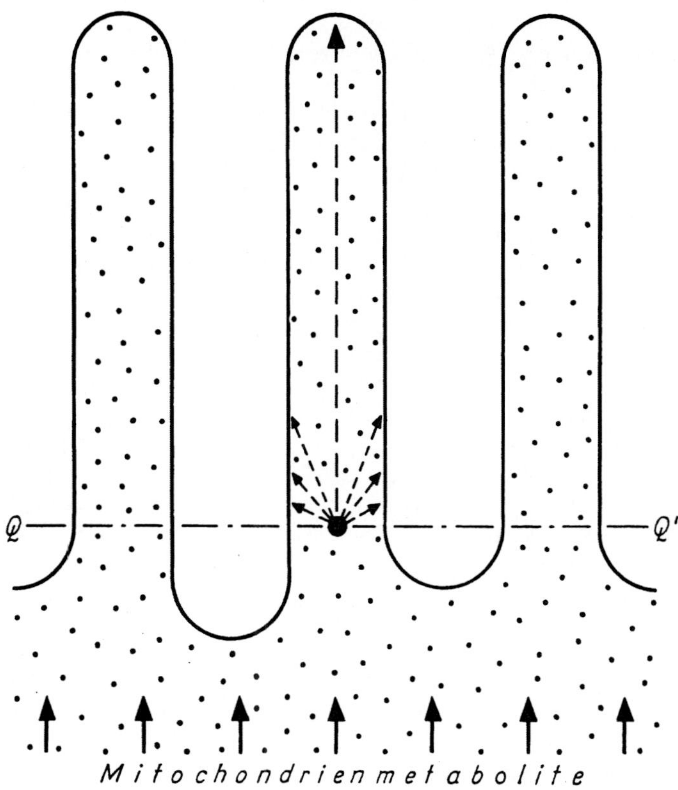

Abb. 11. Diffusion energiereicher Phosphorsäure-Ester in den Bürstensaum; vgl. hierzu
Text

Würfelwand die gleiche Menge energiereicher Phosphate. Die Grenzfläche gegen das Lumen (F_a) erhielte somit $^1/_6$ der Gesamtmenge, die
Grenze gegen die Basalmembran (F_b) ebenfalls $^1/_6$ – den Rest von $^2/_3$
erhielten die Seitenwände. Es stünde demnach bei der vorgegebenen Lage
der Schlußleisten in erster grober Näherung bereits fest, daß etwa $^5/_6$
der Energie für aktive Transportprozesse der Zellbasis zur Verfügung
gestellt wird – eine Größenordnung, welche sich kaum verschiebt, wenn
wir von unserem Würfelmodell auf die Kanälchenzelle zurückgehen.
Setzt man die Betrachtung fort, so ist man nunmehr gezwungen, die
strukturelle Differenzierung der einzelnen Membranabschnitte zu berücksichtigen – vorzugsweise den *Bürstensaum*. Auf Grund eines Quer-
schnittes $(QQ'$; Abb. 11) wird mit der Treffermethode zunächst fest-

gestellt, welcher Prozentsatz energiereicher Metabolite überhaupt in das Innere der Bürstensaumstäbchen diffundiert; es ist dies nur ein relativ kleiner Bruchteil des Sechstels, welches der gesamten apikalen Grenzfläche zur Verfügung steht. Die weitere Verteilung der eindiffundierenden Molekel hängt von der Wahrscheinlichkeit ab, mit der bei Berühren der Wand eine Reaktion eintritt; ist die Zerfallsrate hoch, so darf man erwarten, daß der größte Teil der Bürstensaumfläche einen *praktisch leblosen Membranabschnitt* darstellt, an dem trotz der vorhandenen Enzyme aus Brennstoffmangel kaum aktive Prozesse stattfinden können. Diesem Umstand kann u. U. wesentlich größere Bedeutung zukommen, als der numerischen Vergrößerung der Oberfläche[1]. Der umgekehrte Fall tritt beim *basalen Labyrinth* ein: Hier stehen Energieträger in relativ großen Mengen zur Verfügung — ein Umstand, auf den bereits von mehreren Seiten hingewiesen wurde [*18, 89, 91, 92, 119*]; zweifellos ist an dieser Stelle die „Natriumpumpe" lokalisiert.

Wahrscheinlich wird es in Zukunft notwendig werden, bei der näherungsweisen Kalkulation des Energiehaushaltes der Membranen drei mögliche Fälle zu trennen, in denen unterschiedliche Funktionstypen zu erwarten wären: (1) Bereiche ohne nennenswerte Energieversorgung (z. B. Bürstensaum); (2) Bereiche, in denen ein gut entwickeltes Chondriom relativ große Flächen mit Energie versorgt (z. B. basales Labyrinth), sowie (3) Bereiche, in denen kleine Flächen von einer Vielzahl von Mitochondrien versorgt werden (z. B. F_a des Mittelstückes; F_b der *pars recta* des Hauptstückes). Man ist im allgemeinen nicht geneigt, den unter (3) aufgeführten, nicht besonders auffallenden Flächen eine besondere Aktivität zuzuschreiben. Wenn man die oben skizzierten Verfahren dazu nützt, die umgesetzten oder zur Verfügung stehenden Energiebeträge abzuschätzen, kann man sich vom Gegenteil überzeugen[2]. Ein Beispiel soll dies erläutern: Die Zellen im distalen Bereich des Hauptstückes enthalten knapp vor Beginn des Überleitungsstückes nur mehr rd. 10% Mitochondrien — gegenüber rd. 30% im Konvolut; dafür haben die basalen Zelloberflächen nur eine Ausdehnung von rd. 5% der Gesamtoberfläche — gegenüber 30% im Konvolut. Eine präzise Aussage erhält man schließlich durch den Vergleich der Größe $(S_b + S_c)/T_m$, welche ergibt, daß trotz reduzierter Mitochondrienkonzentration in der *pars recta* verglichen mit dem Konvolut pro Zeit- und Flächeneinheit etwa die zehnfache Menge energiereicher Verbindungen verfügbar sein müßte. Ähnliche Verhältnisse — jedoch in umgekehrter Richtung — ergeben sich beim Vergleich der apikalen und basalen Flächen der Mittelstücke: Die verfügbaren Energiemengen pro Zeit und Flächeneinheit sind lumenseitig wesentlich höher

---

[1] Sollten diese Annahmen zu Recht bestehen, so würden die Phosphatasen gerade an den Orten, wo sie der Lichtmikroskopiker infolge der mehrfachen Überschichtung von Zellmembranen leicht nachweisen kann und oft mit der Resorption in Verbindung setzt, kaum eine funktionelle Bedeutung erlangen.

[2] Hier, wie bei den vorangehenden Ableitungen, ist natürlich stets zu berücksichtigen, daß man viele Parameter nicht kennt und manches daher schwer abschätzen kann. Umso weniger scheinen jedoch im Augenblick exakte Berechnungen notwendig oder berechtigt, da sie bei hohem Aufwand kaum mehr erbringen können.

als an der Zellbasis — vorausgesetzt, daß alle vorweg gemachten Annahmen zu Recht bestehen; wäre dies der Fall, so könnte man auf diese Weise Mikrobereiche erfassen, welche weder einem Experiment, noch einer Messung zugänglich sind.

Es mag genügen, an diesem Ort das Prinzip einer derartigen Betrachtungsweise zur Diskussion zu stellen. Es versteht sich dabei von selbst, daß auch in dem Fall, daß derartige Überlegungen zu konkreten Resultaten führen, noch eine beträchtliche Arbeit investiert werden muß, um die verschiedenen Aussagen auf eine breitere Basis zu stellen. Einstweilen kann auf Grund der durchgeführten Messungen bestätigt werden, daß zwischen den gewundenen und gestreckten Abschnitten der Haupt- und Mittelstücke erhebliche Unterschiede bestehen; dies betrifft sowohl den Feinbau der Epithelzellen, die Topographie der verschiedenen Harnkanälchen und Kapillaren sowie die Häufigkeit der verschiedenen Strukturen (Volum-%).

Die *Unterschiede zwischen Konvolut und pars recta* betreffen neben dem Mitochondriengehalt „*MI*" vor allem den prozentualen Anteil der apikalen (F_a) und basalen Zelloberflächen (F_b, F_c). Bei annähernd konstanten Absolutwerten von F_b sinkt der Anteil F_c in der pars recta des Hauptstückes praktisch auf Null ab (vgl. Abb. 7b u. 8a, sowie Tab. 2); auf diesen Tatbestand machte bereits THOENES aufmerksam [*119, 122*]. Die Unterschiede zwischen pars contorta und recta des distalen Tubulus sind ebenfalls signifikant, wenn auch bedeutend weniger ausgeprägt. Neben der Zellstruktur muß die Lagebeziehung der Tubuli sowie die topographische Anordnung von Kapillaren und Zwischengewebe die Funktion des Systemes beeinflussen. Im Gegensatz zum Konvolut verlaufen im Bereich der partes rectae offensichtlich proximale und distale Tubuli über lange Strecken in enger Nachbarschaft und sind hierbei lediglich durch ein lockeres Zwischengewebe (*Z* in Abb. 7b) getrennt, welches die Diffusion niedermolekularer Flüssigkeitsanteile nicht behindert. Blutkapillaren sind im Vergleich zur Rinde weniger häufig anzutreffen; dies entspricht der relativ geringeren Durchblutung. Die gestreckten Anteile von Haupt- und Mittelstücken liegen im Gegensatz zum Konvolut im Außenstreifen im Verhältnis 1:1 vor. Rückschlüsse von denjenigen Abschnitten des Konvolutes, welche der Mikropunktion zugänglich sind, auf die gestreckten Haupt- und Mittelstücke im Mark sind aus diesen Gründen — wie schon von anderer Seite häufig geäußert — sicherlich unzulässig. Man darf vielmehr annehmen, daß mit der unterschiedlichen Zellstruktur und Topographie der einzelnen Elemente auch eine unterschiedliche Funktion der Epithellagen verbunden ist. Nimmt man ferner auf Grund der Mikropunktions-Befunde am distalen Konvolut [*34, 55, 137, 148*] an, daß der aufsteigende dicke Schenkel der Schleife vorwiegend Na$^+$ aktiv in das Interstitium befördert, so müßten davon direkt angrenzende Haupt- und Überleitungsstücke zwangsläufig betroffen werden. Bei einer Gegenstrom-Multiplikation dieser Art stehen morphologisch vollkommen verschiedene, z. T. sehr stoffwechselaktive Elemente über ein lockeres Zwischengewebe in Wechselwirkung. Man darf hierbei nicht ohne weiteres annehmen, daß alle beteiligten Räume (gestreckte Anteile von Haupt- und Mittelstück; Überleitungsstücke; Blut- und Lymphkapillaren; Interstitium) in gleichen Höhenlagen stets annähernd gleiche osmotische Drucke aufweisen[1]. Besonderes Interesse verdienen schließlich in diesem Zusammenhang jene Nieren, denen die Innenzone des Markes vollständig fehlt [97, 114, 124]: Sie liefern den Beweis, daß das System der Außenzone imstande ist, Harn zu konzentrieren.

Außer dem oben diskutierten System in der Außenzone ist aber offenbar auch die Innenzone in der Lage, Konzentrationseffekte zu erzielen [35, 97]. Es ist zunächst noch nicht gesichert, daß Überleitungsstücke, Kapillaren, Sammelrohre und Interstitium der Innenzone bei *Psammomys* ebenso strukturiert sind, wie bei der genauer untersuchten Rattenniere; wäre dies der Fall, so wären die Wände der beiden im Austausch stehenden Schleifenschenkel (ab- und aufsteigendes Überleitungsstück) identisch aufgebaut, ontogenetisch gleichwertig und man dürfte folglich Permeabilitätsunterschiede zwischen ab- und aufsteigendem Schenkel nicht erwarten. Enthielten die Überleitungsstücke von Psammomys ebenso wenig Mitochondrien, wie diejenigen der Ratte, so könnte man zudem kaum an aktive Elektrolyttransporte in einem nennenswerten Umfang glauben; dies stünde mit der Tatsache in Einklang, daß bislang ein aktiver Na^+-Transport an diesem Epithel nicht nachgewiesen werden konnte [127, 135]. Es ist daher vorläufig schwer, den gemessenen Konzentrationsanstieg in der Innenzone zu erklären. Auf der einen Seite bliebe die Möglichkeit, daß relativ geringe Druckdifferenzen zwischen ab- und aufsteigendem Schenkel eine Gegenstrom-Multiplikation in Gang halten [126] — womit man auf das erste Konzept dieses Systemes zurückkommen würde. Auf der anderen Seite gibt es Modelle [69, 88], welche offensichtlich Konzentriereffekte im Gegenstrom ohne aktive Ionentransporte ermöglichen; indessen bleibt nach wie vor zu berücksichtigen, daß diese Modelle mit einer lebenden Niere wohl nicht ohne weiteres in Bezug gesetzt werden können.

Ohne Rücksicht auf unsere unzureichenden Kenntnisse — insbesondere auf dem Sektor der vergleichenden Nierenmorphologie — läßt sich scheinbar eine Tatsache heute bereits abgrenzen: *Nieren ohne Innenzone* (100 % kurze Nephrone [114, 124]) sind imstande, Harn zu konzentrieren. Gleiches gilt für Nieren, welche *ausschließlich lange Nephrone* und damit eine verschwindend kleine Außenzone neben einer *relativ sehr großen Innenzone* aufweisen (*Psammomys:* Überleitungsstück entspricht 65 % der Gesamtlänge des Nephrons [97, 114]). Nach Messungen an *Psammomys* [35] besteht kaum ein Zweifel, daß auch die Innenzone konzentrieren kann. Man muß also folgern, daß *zwei vollkommen verschieden strukturierte Systeme* (Innenzone, Außenzone) Konzentrationsarbeit leisten können; die Annahme liegt nahe, daß dermaßen verschiedenartige Strukturen diese Arbeit *nicht auf gleiche Weise* vollbringen. Man wird daher aus dem kontinuierlichen Anstieg der Na^+-Konzentration über den

---

[1] Ein Nachweis der Osmolarität in diesen Mikrobezirken ist mit Hilfe der Gewebsschnitt-Kryoskopie [150] bekanntlich nicht möglich, da dieses Verfahren nur Summenwerte über größere Bezirke liefert [149].

Gesamtumfang der Außen- und Innenzone einer Niere nicht folgern dürfen, daß die Natriumpumpe im Überleitungsstück in der gleichen Weise (aktiv) arbeitet, wie im aufsteigenden dicken Schleifenschenkel [97].

In jedem Fall wird es in Zukunft notwendig sein, in parallel geführten Studien verläßliche vergleichend-morphologische Ergebnisse mit gleichzeitig erhobenen physiologischen Befunden zu korrelieren; erst dadurch dürfte es möglich sein, verschiedene Strukturbilder der Niere mit Hilfe morphometrischer Analysen bestimmten Funktionszuständen zuzuordnen. Morphometrische Arbeiten dieser Art sind sehr zeitraubend; dies gilt insbesondere dann, wenn methodische Artefakte mit hinreichender Sicherheit ausgeschlossen, die Werte statistisch abgesichert und durch vergleichend morphologische Arbeiten verschieden strukturierte Nieren unter verschiedenen funktionellen Bedingungen erfaßt werden sollen. Darüber hinaus wären ähnlich gekoppelte Untersuchungen an anderen Strukturen, welche ebenfalls größere Flüssigkeitsmengen transportieren (Speicheldrüsen, Gallenblase, Darm, Plexus chorioideus) sinnvoll und nützlich. Bis zu einer Korrelation derartiger Befunde mit besser fundierten physiko- und biochemischen Vorstellungen über die Funktion biologischer Membranen dürfte noch geraume Zeit verstreichen. Einstweilen wird man daher bis auf weiteres mit den gängigen Arbeitshypothesen vorlieb nehmen müssen, deren Wahrheitsgehalt noch zu beweisen bleibt.

Meinen Kollegen, Herrn Dr. U. Fritz, Dr. M. Steinhausen und Dr. G. Werner bin ich für ihre Mitarbeit bei den Fixierungsstudien wie am Elektronenmikroskop zu wärmstem Dank verpflichtet.

Literatur

[1] Ashworth, C. T., F. J. Luibel, and S. C. Stewart: J. Cell Biol. 17, 1 (1963).

[2] Bargmann, W.: Zbl. inn. Med. 60, 881 (1939).

[3] Beament, J. W. C.: Symp. Soc. exp. Biol. 8, 94 (1954).

[4] Bell, D. J., and J. K. Grant (Edit.): Fine structure and function of the membranes and surfaces of cells. Biochem. Soc. Symposia. Cambridge University Press 1963.

[5] Bohle, A., u. J. Jahnecke: Klin. Wschr. 42, 1 (1964).

[6] Bray, G. A.: Amer. J. Physiol. 199, 915 (1960).

[7] Brody, T. G., and J. J. Mackenzie: Proc. roy. Soc. B 87, 593 (1914); zit. nach Jahnecke u. Mitarb., l.c.

[8] Brodsky, W. A., W. S. Rehm, W. H. Dennis, and D. G. Miller: Science 121, 302 (1955).

[9] Brunner, H., G. Kuschinsky, H. Peters u. H. Vorherr: Naunyn-Schmiedebergs Arch. exp. Path. Pharmak. 233, 57 (1958).

[10] Buddenbrock, W. v.: Vergleichende Physiologie. Bd. III. Ernährung, Wasserhaushalt und Mineralhaushalt der Tiere. Basel-Stuttgart: Birkhäuser-Verlag 1956. Vgl. insbes. p. 443ff.

[11] Buffon, G. L. L.: Essai d'arithmétique morale. Suppl. à l'Histoire Nature IV. Paris (1877); zit. nach Hennig, l.c.

[12] Bullivant, St.: J. biophys. biochem. Cytol. 8, 639 (1960).

[13] Burck, H. C. in: Kleinzeller, A., u. A. Kotyk: Membrane transport and metabolism. Proc. Symp. Prag 1960, p. 578. London: Academic Press 1961.

[14] Burck, H. C., u. H. Netter: Klin. Wschr. 38, 359 (1960).

[15] Caulfield, J. B., and B. F. Trump: Amer. J. Path. 40, 199 (1962).

[16] Clark, S. L. jr.: J. biophys. biochem. Cytol. 3, 349 (1957).

[17] DANIELLI, J. F.: Symp. Soc. exp. Biol. **8**, 502 (1954).
[18] DAVID, H.: Dtsch. Gesundh.-Wes. **18**, 257, 311 (1963).
[19] DIAMOND, J. M.: J. Physiol. (Lond.) **161**, 503 (1962).
[20] — Diskussionsbemerkung: 2. Symp. Ges. Dtsch. Naturforscher u. Ärzte „Sekretion und Exkretion", Reinhardsbrunn 1964. Berlin-Heidelberg-New York: Springer-Verlag 1965.
[21] DICK, D. A. T.: Int. Rev. Cytol. **8**, 387 (1959).
[22] ELIAS, H. (Herausgeber): Verh. 1. Int. Kongr. Stereologie, Wien 1963. Wien VI, Kaunitzg. 33: Congressprint 1963.
[23] — A. HENNIG, and P. M. ELIAS: Z. wiss. Mikr. **65**, 70 (1961).
[24] ERNST, E., u. P. MITCHELL: Diskussionsbemerkungen zu H. C. BURCK in: KLEINZELLER, A., and A. KOTYK (Edit.): "Membrane transport and metabolism", Proc. Symp. Prag 1960, pp. 594—596. London: Academic Press 1961.
[25] —, u. L. HOMOLA: Acta physiol. Acad. Sci. hung. **3**, 487 (1952); zit. nach NETTER, l.c.
[26] FARQUHAR, M. G., R. L. VERNIER, and R. A. GOOD: Schweiz. med. Wschr. **87**, 501 (1957).
[27] — S. L. WISSIG, and G. E. PALADE: J. exp. Med. **113**, 47 (1961).
[28] —, and G. E. PALADE: J. Cell Biol. **17**, 375 (1963).
[29] FEDER, N., and R. L. SIDMAN: J. biophys. biochem. Cytol. **4**, 593 (1958).
[30] FERNÁNDEZ-MORÁN, H.: Science **129**, 1284 (1959); Ann. N. Y. Acad. Sci. **85**, 689 (1960); Circulation **26**, 1039 (1962).
[31] GIEBISCH, G.: J. cell. comp. Physiol. **51**, 221 (1958).
[32] GLAGOLEFF, A. A.: On the geometrical methods of quantitative mineralogic analysis of rocks. Trans. Inst. Econ. Min., Moskau, **59** (1933); zit. nach HENNIG, l.c.
[33] GLYNN, I. M.: Int. Rev. Cytol. **8**, 449 (1959).
[34] GOTTSCHALK, C. W., and M. MYLLE: Amer. J. Physiol. **185**, 430 (1956); **189**, 323 (1957); **196**, 927 (1959).
[35] — W. E. LASSITER, M. MYLLE, K. J. ULLRICH, B. SCHMIDT-NIELSEN, G. PEHLING, and R. O'DELL: Excerpta med. (Amst.) **29**, 43 (1960).
[36] HANSSEN, O. E.: Acta path. microbiol. scand. **49**, 280, 297 (1960).
[37] HARGITAY, B., u. W. KUHN: Z. Elektrochem. **55**, 539 (1951).
[38] — — u. H. WIRZ: Experientia (Basel) **7**, 276 (1951).
[39] HENNIG, A.: Mikroskopie **11**, 1 (1956).
[40] — Zeiss Werk-Z. Heft 30 (1958).
[41] HILGER, H. H., J. D. KLÜMPER u. K. J. ULLRICH: Pflügers Arch. ges. Physiol. **267**, 218 (1958).
[42] HOLLATZ, W.: Z. Anat. Entwickl.-Gesch. **65**, 482 (1922).
[43] IVERSEN, P., and C. BRUN: Amer. J. Med. **11**, 324 (1951).
[44] JAHNECKE, J., B. KOMMERELL u. A. BOHLE: Klin. Wschr. **40**, 227 (1962).
[45] JARAUSCH, K. H., u. K. J. ULLRICH: Pflügers Arch. ges. Physiol. **264**, 88 (1957).
[46] KITCHING, J. A.: Symp. Soc. exp. Biol. **8**, 63 (1954).
[47] KLEINZELLER, A. in: KLEINZELLER, A., and A. KOTYK (Edit.): "Membrane transport and metabolism", Proc. Symp. Prag 1960, p. 527—542. London: Acad. Press 1961.
[48] KÜGELGEN, A. v., u. B. BRAUNGER: Z. Zellforsch. **57**, 766 (1962).
[49] KUHN, W.: Klin. Wschr. **37**, 997 (1959); Ref. in: „Glomeruläre und tubuläre Nierenerkrankungen", Int. Nierensymp. Würzburg 1960, p. 3. Herausgeber E. WOLLHEIM. Stuttgart: G. Thieme-Verlag 1962.
[50] —, u. H. MARTIN: Z. physikal. Chem. (A) **189**, 317 (1941).
[51] —, u. K. RYFFEL: Hoppe-Seylers Z. physiol. Chem. **276**, 145 (1942).
[52] —, u. A. RAHMEL: Helv. chim. Acta **42**, 628 (1959).
[53] —, u. H. J. KUHN: Z. Elektrochem. **65**, 426 (1961).
[54] LASSEN, N. A., J. B. LONGLEY, and L. S. LILIENFIELD: Science **128**, 720 (1958).
[55] LASSITER, W. E., C. W. GOTTSCHALK, and M. MYLLE: J. clin. Invest. **39**, 1004. (1960); Amer. J. Physiol. **200**, 1139 (1961).
[56] LJUNGBERG, E.: Acta med. scand., Suppl. **186** (1947).

[57] LONGLEY, J. B., and M. S. BURSTONE: Amer. J. Path. 42, 643 (1963).
[58] — N. A. LASSEN, and L. S. LILIENFIELD: Fed. Proc. 17, 99 (1958).
[59] MALVIN, R. L.: Progr. cardiovasc. Dis. 3, 432 (1961).
[60] — W. S. WILDE, and L. P. SULLIVAN: Amer. J. Physiol. 194, 135 (1958).
[61] MARSHALL JR., E. K.: Physiol. Rev. 14, 133 (1934).
[62] MAUNSBACH, A. B., S. C. MADDEN, and H. LATTA: J. Ultrastruct. Res. 6, 511 (1962).
[63] METCOFF, J.: Pediat. Clin. N. Amer. 6, 43 (1959); zit. nach CAULFIELD u. TRUMP, l.c.
[64] MILLER, F.: J. biophys. biochem. Cytol. 8, 689 (1960).
[65] MÖLBERT, E., F. DUSPIVA u. O. V. DEIMLING: Histochemie 2, 5 (1960).
[66] MOOR, H.: Z. Zellforsch. 62, 546 (1964).
[67] — C. RUSKA u. H. RUSKA: Z. Zellforsch. 62, 581 (1964).
[68] MUEHRCKE, R. C., R. M. KARK, and C. L. PIRANI: New Engl. J. Med. 253, 537 (1955).
[69] NIESEL, W., u. H. RÖSKENBLECK: Pflügers Arch. ges. Physiol. 276, 555 (1963); 277, 302 (1963).
[70] NETTER, H.: Theoretische Biochemie — Physikalisch-chemische Grundlagen der Lebensvorgänge. Berlin-Göttingen-Heidelberg: Springer-Verlag 1959; vgl. insbes. pp. 92, 111.
[71] OLIVER, J.: J. Mt. Sinai Hosp. 15, 173 (1948); Bull. N. Y. Acad. Med. 37, 81 (1961).
[72] OVERTON, J., and J. SHOUP: J. Cell Biol. 21, 75 (1964).
[73] PALADE, G. E.: J. exp. Med. 95, 285 (1952).
[74] PALAY, S. L., S. M. McGEE-RUSSELL, S. GORDON, and M. A. GRILLO: J. Cell Biol. 12, 385 (1962).
[75] PARKER, M. V., H. G. SWANN, and J. C. SINCLAIR: Tex. Rep. Biol. Med. 20, 425 (1962).
[76] PARRISH, A. E., and J. S. HOWE: J. Lab. clin. Med. 42, 152 (1953).
[77] PEASE, D. C.: Anat. Rec. 121, 723 (1955).
[78] — J. Histochem. Cytochem. 3, 295 (1955).
[79] — J. biophys. biochem. Cytol. 2/4, Suppl., 203 (1956).
[80] PETER, K.: Bau und Entwicklung der Niere (1909, 1927); zit. nach ULLRICH (1959).
[81] RICHARDS, A. N.: Methods and results of direct investigations of the function of the kidney. The Beaumont Found. Lect., Ser. B. Baltimore: Williams and Wilkins 1929; zit. nach ULLRICH (1960).
[82] — B. B. WESTFALL, and P. A. BOTT: Proc. Soc. exp. Biol. (N. Y.) 32, 73 (1934).
[83] RHODIN, J.: Correlation of ultrastructural organization and function in normal and experimentally changed proximal convoluted tubule cells of the mouse kidney. Thesis, Anat. Dept., Karolinska Inst., Stockholm 1954.
[84] — Int. Rev. Cytol. 7, 485 (1958); Amer. J. Med. 24, 661 (1958); Verh. dtsch. Ges. Path. 43, 274 (1958); Proc. 10th Annual Conf. on Nephrotic Syndrom, p. 30 (1959); Ref. in „Glomeruläre und tubuläre Nierenerkrankungen", Herausgeber E. WOLLHEIM, Int. Nierensymposion, Würzburg 1960, p. 92. Stuttgart: Thieme-Verlag 1962.
[85] ROBERTSON, J. D.: Biochem. Soc. Symp. 16, 3 (1959); Verh. 4. Int. Kongr. Elektronenmikroskopie, Berlin 1958, Bd. II. p. 159. Berlin-Göttingen-Heidelberg: Springer-Verlag 1960.
[86] ROBINSON, J. R.: Symp. Soc. exp. Biol. 8, 42 (1954).
[87] ROMEIS, B.: Mikroskopische Technik. 15. Aufl. München: Leibniz-Verlag 1948; vgl. insbes. §§ 193—197.
[88] RÖSKENBLECK, H., u. W. NIESEL: Pflügers Arch. ges. Physiol. 277, 316 (1963).
[89] ROTTER, W., H. LAPP u. H. ZIMMERMANN: Dtsch. med. Wschr. 87, 669 (1962).
[90] RUSKA, C.: Z. Zellforsch. 52, 748 (1960).
[91] RUSKA, H.: Marburger Sitzungsber. 82/2, 3 (1960); zur Druckwandlerhypothese, vgl. insbesondere Abb. 18 sowie Text hierzu auf p. 19.
[92] — D. H. MOORE, and J. WEINSTOCK: J. biophys. biochem. Cytol. 3, 249 (1957).

[93] SABATINI, D. D., K. G. BENSCH, and R. J. BARRNETT: Anat. Rec. 142, 274 (1962); J. Cell Biol. 17, 19 (1963).

[94] —, F. MILLER, and R. J. BARRNETT: J. Histochem. Cytochem. 12, 57 (1964).

[95] SAKAGUCHI, H., u. Y. SUZUKI: Kejo J. Med. 7, 17 (1958).

[96] SAMPAIO, M. M., A. BRUNNER JR., and B. OLIVEIRA-FILHO: J. biophys. biochem. Cytol. 4, 335 (1958).

[97] SCHMIDT-NIELSEN, B., and R. O'DELL: Amer. J. Physiol. 200, 1119 (1961).

[98] SCHUBERT, G. E., E. BUERGER u. A. BOHLE: Z. Kreisl.-Forsch. 51, 1014 (1962).

[99] SENO, S., and K. YOSHIZAWA: J. biophys. biochem. Cytol. 8, 617 (1960).

[100] SHANNON, J. A., and H. W. SMITH: J. clin. Invest. 14, 393 (1935).

[101] SHIPP, J. C., I. B. HANENSON, E. E. WINDHAGER, H. J. SCHATZMANN, G. WHITTEMBURY, H. YOSHIMURA, and A. K. SOLOMON: Amer. J. Physiol. 195, 563 (1958).

[102] SITTE, H.: In ELIAS, H. (Herausgeber): Verh. 1. Int. Kongr. Stereologie, Wien 1963, Ref. 24. Wien VI. Kaunitzgasse 33: Congressprint 1963.

[103] — Volumen- und Flächenbestimmung nach Schnitt- und Abdruckbildern. Ref. in: Seminar 2470 „Grundlagen und Methoden der Elektronenmikroskopie". Techn. Akad. Esslingen (1964).

[104] SITTE, P.: Protoplasma (Wien) 49, 447 (1958).

[105] — Histochemie 2, 76 (1960).

[106] SJÖSTRAND, F. S., and R. F. BAKER: J. Ultrastruct. Res. 1, 239 (1958).

[107] —, and J. RHODIN: Exp. Cell Res. 4, 426 (1953).

[108] SMITH, H. W.: From fish to philosopher. The story of our internal environment. New York: Oxford University Press 1951; CIBA Edition (1959).

[109] — The kidney — Structure and function in health and disease. New York: Oxford University Press 1951.

[110] SOLOMON, A. K.: Ion and water transport in single proximal tubules of the necturus kidney. The method of isotopic tracers applied to the study of active transport. In: 1er Colloque de Biol. de Saclay. Oxford: Pergamon Press 1959; zit. nach ULLRICH (1960).

[111] — D. L. MAUDE, I. H. SHEHADEH, and W. N. SCOTT: 2. Symp. Ges. Dtsch. Naturforsch. u. Ärzte „Sekretion und Exkretion", Reinhardsbrunn 1964, p. 378. Berlin-Heidelberg-New York: Springer-Verlag 1965.

[112] SONNENBERG, H., u. P. DEETJEN: Pflügers Arch. ges. Physiol. 278, 669 (1964).

[113] SPANNER, D. C.: Symp. Soc. exp. Biol. 8, 76 (1954).

[114] SPERBER, J.: Zool. Bidr. Uppsala 22, 249 (1944).

[115] STEINHAUSEN, M.: Pflügers Arch. ges. Physiol. 277, 23 (1963); 279, 185 (1964).

[116] — I. IRAVANI, G. E. SCHUBERT, R. TAUGNER u. techn. Mitarb.: Virchows Arch. path. Anat. 336, 503 (1963).

[117] SWANN, H. G.: Tex. Rep. Biol. Med. 18, 566 (1960).

[118] THURAU, K., u. P. DEETJEN: Nachr. Akad. Wiss., Göttingen, II. Math.-Physik. Kl. H. 2, 27 (1961).

[119] THOENES, W.: Klin. Wschr. 39, 504, 827 (1961).

[120] — Z. Zellforsch. 55, 486 (1961).

[121] — Ref. 3. Symp. Ges. Nephrologie, Berlin Sept. (1964).

[122] — Ref. 2. Symp. Dtsch. Ges. Naturforsch. Ärzte „Sekretion und Exkretion", Reinhardsbrunn 1964, p. 315. Berlin-Heidelberg-New Nork: Springer-Verlag 1965.

[123] TOSTESON, D. C.: Pers. Mitteilung (1964).

[124] ULLRICH, K. J.: Ergebn. Physiol. 50, 434 (1959).

[125] — Ref. in BUCHBORN, E., u. K. D. BOCK (Herausg.): „Diurese und Diuretica", Int. Symp., p. 21—38. Berlin-Göttingen-Heidelberg: Springer-Verlag 1959; Dtsch. med. Wschr. 84, 1197 (1959); Ref. in KRAMER, K., u. K. J. ULLRICH (Herausg.): Nierensymposium, Göttingen 1959, p. 76—81. Stuttgart: Georg Thieme-Verlag 1960.

[126] — Sammelref. Physiol. Kongr. Leiden 1962; hektographierter deutscher Sprechtext.

[127] — Pflügers Arch. ges. Physiol. 281, 9 (1964); vgl. insbes. p. 11.

[128] ULLRICH, K. J.: Ref. 2. Symp. Ges. Dtsch. Naturforsch. u. Ärzte „Sekretion und Exkretion", Reinhardsbrunn 1964, p. 392. Berlin-Heidelberg-New York: Springer-Verlag 1965.
[129] — F. O. DRENCKHAHN u. K. H. JARAUSCH: Pflügers Arch. ges. Physiol. 261, 62 (1955).
[130] —, u. K. H. JARAUSCH: Pflügers Arch. ges. Physiol. 262, 537 (1956).
[131] —, u. G. PEHLING: Pflügers Arch. ges. Physiol. 267, 207 (1958).
[132] — — Pflügers Arch. ges. Physiol. 267, 207 (1958).
[133] — — u. M. ESPINAR-LAFUENTE: Pflügers Arch. ges. Physiol. 273, 562 (1961).
[134] — K. KRAMER, and J. W. BOYLAN: Progr. cardiovasc. Dis. 3, 395 (1961).
[135] —, and D. J. MARSH: Ann. Rev. Physiol. 25, 91 (1963).
[136] WALKER, A. M., and J. OLIVER: Amer. J. Physiol. 134, 562 (1941).
[137] — P. A. BOTT, J. OLIVER, and M. C. MacDOWELL: Amer. J. Physiol. 134, 580 (1941).
[138] WALTHER, D.: Münch. med. Wschr. 105, 1370 (1963).
[139] WEAVER, A. N., C. Y. McCARVER, and H. G. SWANN: J. exp. Med. 104, 41 (1958).
[140] WEIBEL, E.: Lab. Invest. 12, 131 (1963).
[141] WEISS, CH., H. PASSOW, and A. ROTHSTEIN: Amer. J. Physiol. 196, 1115 (1959).
[142] WHITTEMBURY, G.: J. gen. Physiol., Symposion Caracas 1959, 43 Suppl., 43 (1960).
[143] WILBRANDT, W.: J. cell. comp. Physiol. 134, 562 (1941).
[144] — Int. Rev. Cytol. 13, 203 (1962).
[145] WILDE, W. S., and R. L. MALVIN: Amer. J. Physiol. 195, 153 (1958).
[146] WIRZ, H.: Helv. physiol. pharmacol. Acta 11, 20 (1953).
[147] — Helv. physiol. pharmacol. Acta 13, 42 (1955).
[148] — Helv. physiol. pharmacol. Acta 14, 353 (1956).
[149] — Mod. Probl. Pädiat. 6, 86 (1960); vgl. insbes. p. 90.
[150] — B. HARGITAY u. W. KUHN: Helv. physiol. pharmacol. Acta 9, 196 (1951).
[151] WÜSTENFELD, E.: Z. wiss. Mikr. 62, 241 (1955); 63, 7, 86 (1956).
[152] ZIMMERMANN, H.: Verh. dtsch. Ges. Path. 45, 299 (1960).
[153] — A. GERSON u. G. EHLERS: Frankfurt. Z. Path. 73, 338 (1964).

Summary

Correct comparisons between cell structure and transport processes under different functional conditions seem only to be possible if differences in volumes and surface areas are measured by means of quantitative morphometric methods. Methods of this kind ("stereological" ones) have been elaborated in the past [22, 23, 39, 40, 48, 102, 103, 140]. They allow absolute or relative measurements of volumes and surface areas on twodimensional pictures — for instance, electron micrographs. Also, one is able to correlate volumes with surface areas without great difficulties. It is e.g. possible to measure the volume of the mitochondria (V_{mi}) and to correlate this (V_{mi}) with the amount of membrane surfaces (S) limiting these cells against the interstices. This possibility of calculating (V_{mi}/S) seems important in connection with active transport, which is influenced by the oxidative capacity of the mitochondria and the area of the cell surface membranes. In other cases it is simple to compare different sectors of the membrane system of the cell — e.g., basal and apical surfaces of the tubular epithelia or cell surface and endoplasmic reticulum. The stereological methods, briefly discussed in the present paper, are very variable and can easily be adapted for a lot of problems concerning comparative studies.

Stereological methods as mentioned above seem to be unprofitable if the object is not fixed and studied under conditions which represent absolutely or approximately the situation in vivo. It is very difficult to fulfil this condition in kidney morphology. For instance, alterations in structure occur immediately if the normal circulation of blood is reduced or stopped [5, 9, 15, 36, 44, 57, 75, 89, 98, 116, 117, 138, 153]. Hence normal fixation of kidney tissue [73, 94] using little cubes seems not to be very useful. But there exist many specially adapted methods [5, 27, 30, 66, 67, 77, 121] for kidney fixation. Two techniques have been used with good results as the basis for the morphometric work presented here. In the first case, the kidney surface is fixed by dipping into fixative or dropping fixative on [62, 77, 78]. In the second, the whole kidney is perfused with the fixative after a short period of washing blood plasma and cells out of the vessels, using a slightly modified physiological method [141].

Preliminary morphometric results indicate (see Table 2) that the concentration of mitochondria in the proximal tubule drops from 35% in the neighbourhood of the glomerulum down to 20% at the end of the pars recta. In the same direction the basal cell surface is diminished from approximately 40 to 2% of the total cell surface as a result of the practically total disappearance of the "basal labyrinth" [91, 92, 119, 122] at the end of the tubuli recti I, which is doubtless of functional importance. The differences between the straight and convoluted portion of the distal tubule are not so impressive but statistical significant; in this element the concentration of mitochondria is always $\geq 25\%$. There is a great difference between the epithelia of the proximal and distal tubules (including the cortical collecting ducts) and the thin portions of the loop of HENLE, which only contain very small amounts of mitochondria ($\leq 2\%$). Differences between the descending and ascending limb of the thin portion could not be detected; compare also [122]. The medullar part of the collecting duct contains two different types of cells clearly separated by morphometry (Type I and II, Table 2). With the aid of these measurements a rough — and somewhat speculative — calculation of the diffusion of metabolites seems to be possible. Provided that ATP diffuses from the mitochondria to the different surface membranes and that these surfaces incorporate enzymes involved in active transport in uniform distribution [1, 65], one could perhaps conclude, e.g. for the proximal tubule, that the brush-border extensions have a very poor and the "basal labyrinth" a very efficient energy supply ("sodium pump"). Similar speculations concern the other portions of the nephron and are presented for discussion.

Kidneys without the inner zone, which contain 100% short-looped nephrons, are able to produce a concentrated urine [97, 114, 108, 124], and it is therefore reasonable to conclude that the outer zone of the medulla is able to concentrate. On the other hand, in kidneys containing 100% long-looped nephrons [97, 114, 124], measurements have demonstrated that a considerable increase in urine osmolarity occurs in the inner zone [35, 97]. One is therefore forced to conclude that the inner zone is also able to perform the work of concentration. But the fine structure is totally different in the inner and outer zone of the medulla [119, 122], provided that the inner zone of *Psammomys* has essentially the same ultrastructure as the inner zone of the relative well known mouse and rat kidneys. The outer zone includes the straight portions of the proximal and distal tubule with a very high oxidative metabolism and (especially in the distal tubule) a very effective sodium pump [124]. Therefore, in the outer zone a good morphological correlate to the countercurrent function is demonstrable. In contrast to the above-mentioned situation in the outer zone, the inner zone seems not to be able to perform significant active transport (mitochondria concentrations $\leq 2\%$), which is in agreement with present physiological knowledge [126, 127]. The actual processes in the inner zone are quite obscure, and there exist at present only hypotheses, fictions, and models [69, 88, 126, 127]. Only one point seems in the author's opinion to be quite clear: *different structures are normally working in a different manner*. In other words: If the outer and the inner zone of the kidney medulla are both able to enhance the osmolarity of the urine, then it is very probable that the outer zone does this work in a way different from the inner zone.

For a better and detailed correlation of kidney structure and function, further studies are urgently needed, especially combined physiological and morphometric work under the same functional conditions and naturally on the same object.

Diskussion

Schmidt: Die apikale Zelloberfläche, die mit Mikrovilli besetzt ist, welche nach der vorhin genannten Hypothese praktisch nur bis zu ihrer Wurzel mit Energie versorgt werden, zeigt am Grunde zwischen den Mikrovilli Pinocytosevorgänge, die sicherlich Energie erfordern. Bei der Amphibienniere, z. B. bei Urodelen, erfolgt die Pinocytose nicht nur zwischen, sondern auch am seitlichen Abschnitt der Mikrovilli. Bei der Amphibienniere sind auch nach meinen Untersuchungen die Mitochondrien immer gleichmäßig über das Cytoplasma verteilt gewesen. Hat sich eine Amphibienniere mit Farbstoffen so belastet, daß der Stoffwechsel wesentlich herabgesetzt wird, ist die Pinocytose nur noch im unteren Bereich zwischen den Mikrovilli erkennbar.

H. Sitte: Ich wollte sagen, daß die Pinocytoseblasen in wesentlich geringerem Umfange auftreten, wenn man intravital fixiert. Mir scheint das unter Umständen ein Indiz dafür, daß beim Zusammenklappen der Struktur der Rest des Lumeninhaltes, der im Moment nicht resorbiert werden kann, in die Bürstensaumbasis verschwindet. Ich weiß nicht, ob Sie die Bilder von Miller kennen, der mit Hämoglobin gearbeitet hat. Da findet sich ein Labyrinth unter dem Bürstensaum. Wir haben Hämoglobin gespritzt und dann versucht, intravital zu fixieren. Wir haben bei derselben Konzentration natürlich überhaupt keinen nennenswerten Kontrast bekommen, weil nicht der gesamte Tubulusinhalt auf ein relativ geringes Volumen zusammengepreßt wurde, und wir haben auch nicht dieses Labyrinth gefunden.

Bücher: Wenn wir die morphologischen Befunde vergleichen, dann fällt auf, daß bei *Drosophila* die Mitochondrien in den Mikrovilli sitzen und beim Warmblüternephron nicht. Der Stoffwechsel der Dipteren ist — allerdings kann ich das nur für die Muskeln sagen und nicht für die Malpighischen Gefäße — extrem kohlenhydratabhängig. Orthopteren haben dagegen, zumindest was ihre Flugmuskulatur anbetrifft, weitgehend fettsäureabhängige Zellatmung. Das heißt, sie haben in den Mitochondrien einen hohen Spiegel der Enzymaktivität des Fettsäureabbaus, und sie extrahieren — wie das menschliche Herz das auch vermag — den Sauerstoff weitgehend auf der Basis einer Fettsäureverbrennung. Ich glaube nicht fehlzugehen in der Annahme, daß auch die Niere bei den Warmblütern zu den Organen gehört, die eine relativ hohe Potenz der Fettsäureverbrennung haben. Fettsäuren sind aber schwerer zu transportieren als Zucker. Es könnte also sein, daß diese Lage der Mitochondrien bei *Drosophila* in den Mikrovilli möglich ist, weil dort Energie gebraucht wird, gleichzeitig aber auch die Brennstoffe zur Verfügung stehen, daß sie aber bei der Niere des Warmblüters dort nicht stecken können, weil die Versorgung mit dem Brennstoff Schwierigkeiten bereitet. Das ist hypothetisch. Direkte Beziehung zu Ihren Überlegungen hat der Gedanke, daß Sie nicht nur an die Energieversorgung denken dürfen, sondern daß wir hinsichtlich der Verteilung der Mitochondrien auch fragen müssen: Wie werden sie mit Sauerstoff versorgt, und wie werden sie mit dem notwendigen Brennmaterial versorgt? Herr Kuyper, liegen die Mitochondrien bei den Malpighischen Gefäßen von *Periplaneta*, also bei Orthopteren, auch in den Mikrovilli?

Kuyper: Es gibt bei *Periplaneta* auch im Bürstensaum sehr viele Mitochondrien, aber die Malpighischen Gefäße sind sehr reich an Glykogen. Das würde also stimmen.

Niesel: Ich darf noch einmal auf das Generalproblem der Vorträge von Herrn Thoenes und Herrn Sitte zurückkommen, auf die Frage nämlich, ob es in den Harnkanälchen möglich ist, ohne aktive Transportprozesse einen konzentrierten Harn zu schaffen. In diesem Zusammenhang möchte ich über Modellversuche berichten, die das Haarnadelgegenstromprinzip betreffen.

Gegenstromkonzentrierungsprozesse ohne Mitwirkung aktiver Transportvorgänge

W. Niesel und H. Röskenbleck

Es kann heute als sicher gelten, daß die Henleschen Schleifen im Nierenmark als Gegenstrommultiplikatoren arbeiten, die kleine Konzentriereffekte multiplizieren und dadurch hohe osmotische Konzentrationen im Nierenmark aufbauen. In Abb. 1 ist das Schema einer konzentrierenden Haarnadelgegenstromvorrichtung nach Kuhn [1] wiedergegeben. Zwischen den Schenkeln F_1 und F_2 wird eine kleine Konzentrationsdifferenz erzeugt und aufrechterhalten. Diese Konzentrationszunahme im zufließenden Schenkel des Gegenstromsystems wird durch den Flüssigkeitsstrom kontinuierlich in den Scheitel transportiert und erhöht dort die Konzentration, bis sich ein hohes Konzentrationsgefälle von der Basis bis zum Scheitel des Gegenstromsystems aufgebaut hat.

Nach KUHN und RAMEL [2] soll die Konzentrationsdifferenz zwischen den Schenkeln der Henleschen Schleifen — der sog. Einzelkonzentriereffekt — durch

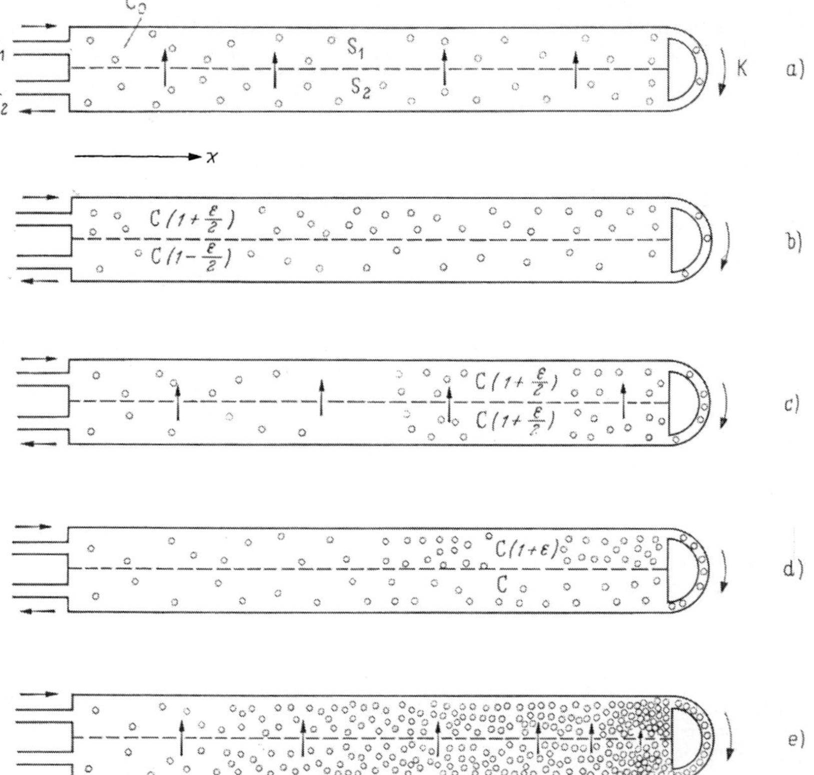

Abb. 1. Schema einer konzentrierenden Haarnadelgegenstromvorrichtung nach KUHN (1959). Die Abbildung soll die Vervielfältigung eines kleinen Einzelkonzentriereffektes veranschaulichen. a—d Langsame Zunahme der Konzentration in der Schleife des Gegenstromsystems durch Multiplikation eines Einzeleffektes. e Konzentrationsgefälle nach erfolgter Multiplikation

aktive Transportvorgänge zustandekommen. Bei einem solchen Konzentrierungsmechanismus muß während eines Harndurchlaufes durch die Schleife jeweils die gesamte konzentrierte Stoffmenge vom herauslaufenden in den hineinlaufenden Schenkel aktiv transportiert werden.

Neben Mikropunktionsbefunden [3] sprechen vor allem nach THOENES und SITTE [4, 5] morphologische Befunde gegen diesen „aktiven" Konzentrierungsmechanismus: Denn im Gegensatz zu anderen ionentransportierenden Zellen sind die Epithelien der Henleschen Schleifen im inneren Mark, in dem die hauptsächliche Konzentrierung erfolgt, mitochondrienarm.

Vom energetischen Standpunkt aus günstiger arbeiten Multiplikationssysteme, die durch Diffusion oder — wie das in Abb. 2 dargestellte Modell — durch Osmose konzentrieren. Die Trennwände zwischen den Gegenstromschenkeln sind in diesem Modell wasserpermeabel. Gelangt in den abführenden Schenkel eine osmotisch wirksame Substanz, so wird dieser hyperton und entzieht dem zuführenden Schenkel durch Osmose Wasser. Hierdurch wird im zuführenden Schenkel ein Einzelkonzentriereffekt für alle Stoffe, die nicht durch die Trennwand permeieren können, erzeugt. Durch Gegenstrommultiplikation bauen sich wiederum wie im Modell der Abb. 1 hohe Konzentrationen auf.

Theoretische Überlegungen [6] aufgrund von Untersuchungen von MOREL, B. SCHMIDT-NIELSEN und O'DELL u. a., [7, 8, 9] sprechen dafür, daß die NaCl-

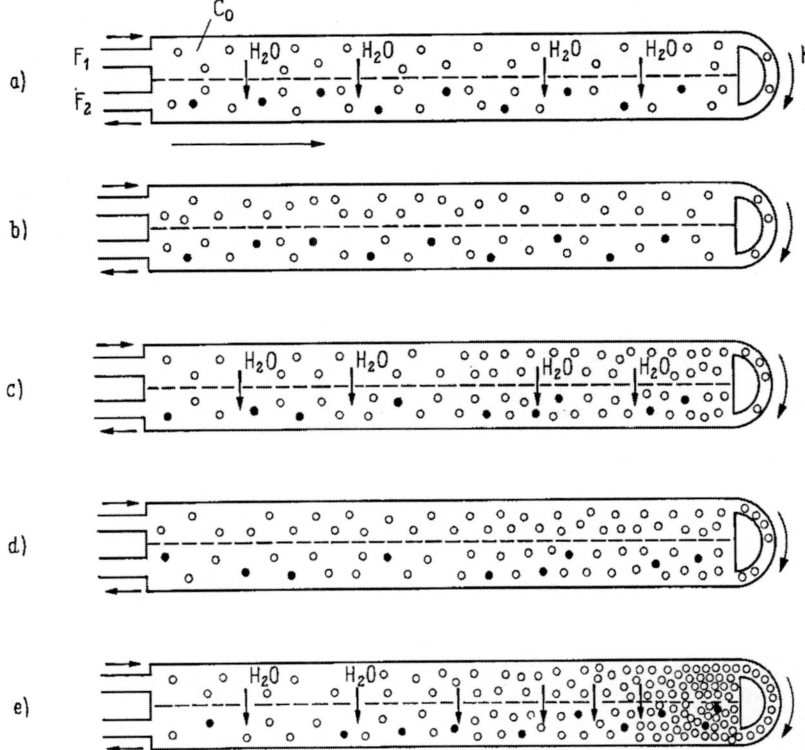

Abb. 2. Schema einer konzentrierenden Haarnadelgegenstromvorrichtung mit einem osmotischen Gradienten als Einzeleffekt, in der gleichen Darstellungsweise wie in Abb. 1. Im Gegensatz zum Kuhnschen Multiplikator wird hier die Konzentrierung nicht durch Transport von gelösten Stoffen von F_2 nach F_1, sondern durch osmotische Wasserverschiebung in entgegengesetzter Richtung erzeugt. Die Kreise stellen die Substanz dar, die im Gegenstrom konzentriert wird, die schwarzen Punkte die zugeführte osmotisch wirksame Substanz, die den osmotischen Einzelkonzentriereffekt bewirkt

Konzentrierung in der Niere in der Tat durch eine solche osmotische Multiplikation erfolgt und daß der Harnstoff die hierbei wirksame osmotische Substanz sein könnte.

Der Harnstoff wird im distalen Konvolut der Tubuli gegenüber NaCl, das aktiv resorbiert wird, relativ angereichert und beschreibt nach ULLRICH, KRAMER und BOYLAN [10, 11, 12] in der Niere einen Kreislauf (Abb. 3), dadurch, daß ein Teil des Harnstoffes aus den Sammelrohren in die Henleschen Schleifen diffundiert und über das distale Konvolut wiederum den Sammelrohren zugeführt wird. Durch diesen Kreislauf wird der Harnstoff sowohl im Nierenmark als auch in den Sammelrohren angereichert.

Dem Nierenmark wird also über die Sammelrohre eine überwiegend harnstoffhaltige Lösung und über die Tubuli eine überwiegend NaCl-haltige Lösung zugeführt. Da KUHN und RYFFEL [13] bereits 1942 zeigten, daß aus isotonen Lösungen unterschiedlicher Zusammensetzung hypertone Lösungen erzeugt werden können, liegt der Gedanke nahe, daß die Konzentrierung in der Niere aufgrund der unterschiedlichen Zusammensetzung von Sammelrohrstrom und Tubulusstrom zustandekommen könnte.

So vermögen z. B. die in Abb. 4 dargestellten Gegenstrommodelle in Zusammenwirken zwischen dem NaCl-haltigen Gegenstromsystem der Tubuli und dem harnstoffhaltigen Sammelrohrstrom hypertone Lösungen zu erzeugen, wenn die Trennwände zwischen den Schenkeln der Stromgebiete geeignete Permeabilitäten besitzen [14].

Im ersten Modell sind die Wände der Sammelrohre impermeabel für alle Stoffe, die der Tubuli permeabel für Harnstoff; im Scheitel wird der gesamte Sammelrohrstrom resorbiert (das Modell arbeitet also — wie die übrigen dargestellten Modelle — ohne Harnentnahme).

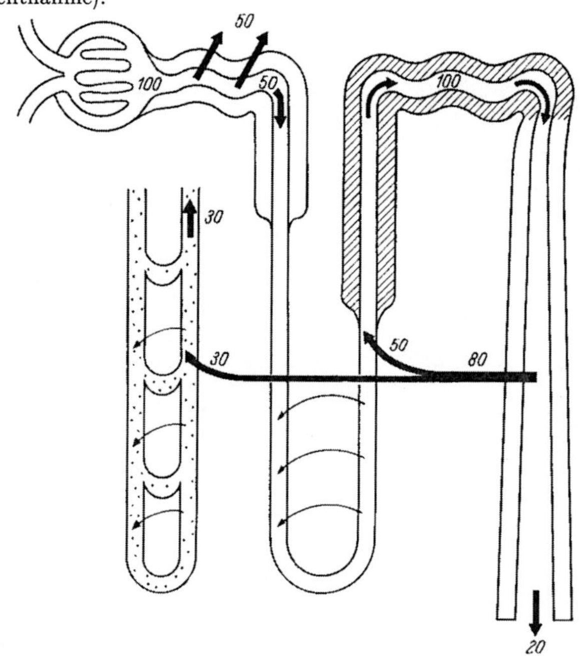

Abb. 3. Harnstoffkreislauf in der konzentrierenden Niere nach ULLRICH, KRAMER, BOYLAN [12]. Es bedeuten: Dünne Pfeile = Gegenstromdiffusion von Harnstoff; Dicke Pfeile = Nettofluß von Harnstoff. Der von den Sammelrohren in die Henleschen Schleifen diffundierende Harnstoff fließt über die ascendierenden Schenkel in die Nierenrinde zurück, um erneut über die Sammelrohre dem Nierenmark zugeführt zu werden. Das Abfließen des Harnstoffes über das Blutgefäßsystem wird im wesentlichen durch Gegenstromdiffusion verhindert. Vom abfiltrierten Harnstoff wird etwa 50% proximal resorbiert, der Rest dem Nierenmark zugeführt

Da der Harnstoff aus dem abfließenden in den zufließenden Schenkel des Tubulussystems diffundiert, reichert er sich im Stromgebiet der Tubuli an, so daß seine Konzentration bis auf die Konzentration des Sammelrohrstromes ansteigen kann. Eine eigentliche Gegenstrommultiplikation im Sinne von KUHN findet jedoch nicht statt.

Durch die Gegenstromdiffusion im Modell der Abb. 4a wird erreicht, daß die beiden Tubulusströme gegenüber dem Sammelrohrstrom hyperton werden. Wird, wie es im Modell der Abb. 4b dargestellt ist, die Trennwand zwischen Sammelrohrstrom und einem der Tubulusströme wasserpermeabel gewählt, so tritt ein osmotischer Wasserfluß aus dem Sammelrohrstrom in den Tubulusstrom auf, der den Sammelrohrstrom konzentriert. Dieser Konzentriereffekt multipliziert sich im Gegenstrom.

Im Modell der Abb. 4b tritt ein Wasserkurzschluß zwischen Sammelrohrstrom und abführendem Tubulusstrom, im Modell der Abb. 4c, bei dem die Trennwände

zwischen zuführendem Tubulusstrom und Sammelrohrstrom wasserpermeabel sind,
ein Wasserkurzschluß zwischen Sammelrohr- und zuführendem Tubulusstrom auf.

Um NaCl konzentrierende Systeme zu erhalten, brauchen die Permeabilitäten
der Ströme lediglich umgekehrt zu werden. So wird im Modell der Abb. 4d NaCl
durch Gegenstromdiffusion zwischen den Sammelrohren und den Tubuli zurück-

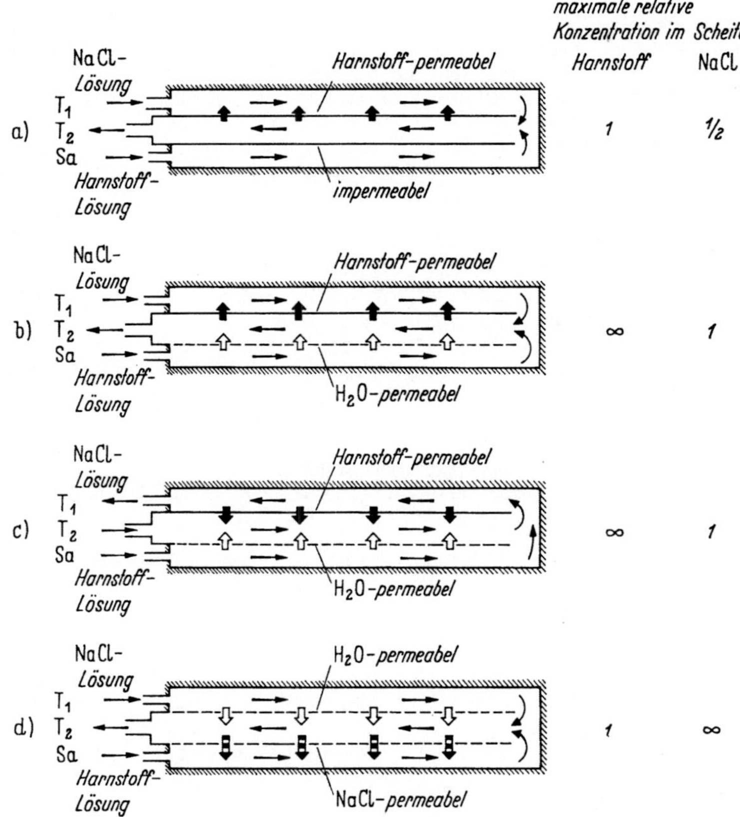

Abb. 4. Konzentriersysteme für Harnstoff und NaCl ohne Harnentnahme. Es bedeuten: T_1 und T_2 zuführende
und abführende Schenkel der Henleschen Schleifen; Sa Sammelrohre; *schwarzer Pfeil* Diffusionsstrom des
Harnstoffes; *gestreifter Pfeil* Diffusionsstrom des NaCl; *weißer Pfeil* osmotischer Wasserfluß. zu a) Das
Modell konzentriert Harnstoff durch Gegenstromdiffusion und kann keine hohen Konzentrationen erzeugen. zu b) u. c) Harnstoff wird durch Gegenstromdiffusion zurückgehalten und erhöht die osmotische
Konzentration in T_2 bzw. T_1, so daß durch osmotischen Wasserfluß ein Konzentriereffekt für Harnstoff in
den Sammelrohren erzeugt wird. zu c) NaCl wird durch Gegenstromdiffusion zurückgehalten und erhöht die
osmotische Konzentration in T_2, so daß durch osmotischen Wasserfluß ein Konzentriereffekt für NaCl im
zuführenden Tubulusstrom T_1 erzeugt wird

gehalten und dadurch eine Konzentrationserhöhung im aufsteigenden Tubulus
geschaffen, die durch osmotischen Wasserentzug aus dem absteigenden Tubulus den
Konzentriereffekt für NaCl setzt.

Die Konzentrierungssysteme der Abb. 4 können in der dargestellten Form in der
Niere nicht verwirklicht sein, da sie im Gegensatz zur Niere ohne Harnentnahme
arbeiten und nur entweder Harnstoff oder NaCl konzentrieren. Dennoch könnte der
Konzentrierungsmechanismus der Niere auf demselben Prinzip beruhen; nämlich
darauf, daß NaCl oder Harnstoff durch Gegenstromdiffusion einen hypertonen
Strom schaffen, der durch osmotischen Wasserfluß in einem zuführenden Strom-
schenkel einen Konzentriereffekt setzt, der sich im Gegenstrom multipliziert.

Die Frage nach dem Vorliegen des hier dargestellten Konzentrierprinzips in der Niere wird noch dadurch erschwert, daß das Nierenmark von zwei Gegenstromsystemen – dem der Tubuli und dem der Blutgefäße – durchzogen wird, die beide am Konzentriervorgang mitwirken könnten.

Da jedoch der Harnstoff im Nierenmark durch Gegenstromdiffusion zurückgehalten wird und die Sammelrohre sowohl wasserpermeabel als auch harnstoffpermeabel sind [15, 16], ist es sehr wahrscheinlich, daß der Harnstoff im Nierenmark nach der Modellvorstellung der Abb. 4b konzentriert wird. Der Weg, der bei der osmotischen Konzentrierung des NaCl in den Tubuli eingeschlagen wird, bleibt jedoch offen. Es wäre denkbar, daß die Konzentrierung von Harnstoff *und* NaCl durch eine Kombination verschiedener Konzentrierungsmechanismen, von denen einer Harnstoff, der andere NaCl konzentriert, zustandekommt.

Alle geschilderten Konzentrierungsmechanismen benötigen keine energieverbrauchenden Prozesse. Aktive Prozesse sind lediglich in der Nierenrinde erforderlich, um im distalen Konvolut den Harnstoff gegenüber NaCl durch selektive NaCl-Resorption relativ anzureichern.

Literaturverzeichnis

[1] KUHN, W.: Klin. Wschr. 37, 996 (1959).
[2] —, u. A. RAMEL: Helv. chim. Acta 42, 628 (1959).
[3] MARSH, D., u. K. J. ULLRICH: In Vorbereitung.
[4] THOENES, W.: 2. Int. Symposium der „Gesellschaft Deutscher Naturforscher und Ärzte" 11.—14. Oktober 1964. Dieser Band S. 315.
[5] SITTE, H.: 2. Int. Symposium der „Gesellschaft Deutscher Naturforscher und Ärzte" 11.—14. Oktober 1964. Dieser Band S. 343.
[6] NIESEL, W., u. H. RÖSKENBLECK: Pflügers Arch. ges. Physiol. 276, 555 (1963).
[7] MOREL, F. F., F. F. GUINNEBAULT et C. AMIEL: Helv. physiol. pharmacol. Acta 18, 183 (1960).
[8] SCHMIDT-NIELSEN, B., R. O'DELL, and H. OSAKI: Amer. J. Physiol. 200, 1125 (1961).
[9] O'DELL, R. J. SCHLEGEL, and J. CUELLAR: Proc. of the Intern. Union of Physiol. Scien. XXII Intern. Congr. Leiden Vo. II, No. 270 (1962).
[10] LEVINSKY, N. G., and W. R. BERLINER: J. clin. Invest. 38, 741 (1959).
[11] SCHMIDT-NIELSEN, B.: Proc. of the Intern. Union of Physiolog. Scien. XXII. Intern. Congr. Leiden, Vol. 1 Part 1 P, 377 (1961), Lectures and Symposia 1—X.
[12] ULLRICH, K. J., K. KRAMER, and J. W. BOYLAN: Progr. cardiovasc. Dis. 111, No. 5 (1961).
[13] KUHN, W., u. R. RYFFEL: Hoppe-Seylers Z. physiol. Chem. 276, 145 (1942).
[14] NIESEL, W., u. H. RÖSKENBLECK: In Vorbereitung.
[15] KLÜMPER, J. D., K. J. ULLRICH u. H. H. HILGER: Pflügers Arch. ges. Physiol. 267, 238 (1958).
[16] HILGER, H. H., J. D. KLÜMPER u. K. J. ULLRICH: Pflügers Arch. ges. Physiol. 267, 218 (1958).

Summary

Countercurrent concentrating mechanisms are described, which produce hypertonic solutions simply by diffusion and osmosis from isotonic NaCl and isotonic urea solutions. Assuming appropriate permeabilities of the collecting ducts and the loops of HENLE, the urea diffusing out of the collecting ducts would produce by countercurrent diffusion and osmosis the primary concentration effect that is multiplied by countercurrent flow.

Bücher: In Ihrem neuen Modell gehen Sie davon aus, daß bestimmte Teile für Harnstoff impermeabel sein müssen?

Niesel: Bevor wir überhaupt an diese Modelle herangegangen sind, haben wir geprüft, wie stark Harnstoff relativ zu Natriumchlorid osmotisch wirksam sein kann.

Wir haben festgestellt, daß der Harnstoff dieselbe Schrumpfung der Nierenzelle hervorruft wie eine isoosmotische Natriumchloridlösung. Der Harnstoff kann also nicht frei in die Nierenzellen hinein bzw. durch die Zellen hindurch diffundieren.

Ullrich: Die Permeabilität des proximalen und distalen Konvolutes für Harnstoff ist von uns direkt gemessen worden. — Ich werde in meinem Vortrag eine Tabelle darüber zeigen. — Es kommt jedoch bei Ihren Berechnungen nicht auf die absolute, sondern auf die relative Permeabilität von Harnstoff, Kochsalz und Wasser an. Wir wissen heutzutage trotz vieler Bemühungen noch nicht, was das "primum movens" für die Harnkonzentrierung in der inneren Markzone ist. Alles deutet aber darauf hin, daß es nicht ein NaCl-Transport ist. Möglicherweise spielen unterschiedliche Permeabilitäten an verschiedenen Stellen des Gegenstromsystems die ausschlaggebende Rolle. Ich glaube auch, daß uns zum gegenwärtigen Zeitpunkt nur Modelle weiterbringen. Man muß nur sorgfältig prüfen, ob sie den Verhältnissen in der Niere wirklich entsprechen.

Thoenes: Man kann wohl sagen, daß das, was Sie jetzt als Modell gezeigt haben, von den Modellen, die bisher erörtert worden sind, am besten zu den morphologischen Befunden paßt, und das ist — glaube ich — ein großer Fortschritt.

Inulin Absorption by *Necturus* Proximal Tubule and some Consequences thereof

By

A. K. SOLOMON, D. L. MAUDE, I. H. SHEHADEH, and W. N. SCOTT, Boston

With 4 Figures

We have been engaged for many years in a systematic study of the mechanism of ion and water transport by the proximal tubule of *Necturus* kidney [6,7,12,15,17,18,19]. All of our results have been based on the premise that inulin was an acceptable reference substance for the measurement of water movement, being neither absorbed nor secreted by the epithelial cells in the tubule. Recently we have obtained compelling evidence that this premise does not obtain because inulin is absorbed by the proximal tubule of *Necturus*. It is the purpose of the present communication to recount this evidence and to reexamine our older conclusions in the light of this finding.

The use of inulin as a measure of the glomerular filtration rate has been discussed carefully and extensively by SMITH [16] who points out that the evidence that this substance is not absorbed by the kidney is inferential. It is worth quoting his views directly on this point; "it is in regard to tubular reabsorption and tubular excretion that the greatest difficulty is encountered, for there is no way to examine these processes directly either in single nephrons or in the intact animal, and all conclusions must be based upon the relative behavior of several substances under a variety of conditions". The most important arguments in favor of the inference that inulin is not absorbed appear to be: i) the identity of clearances measured by inulin and by creatinine in a number of species including frog, *Necturus* [1], dog and rabbit; ii) the direct dependence of inulin excretion on its plasma concentration in man and dog; iii) the observation that the rate of inulin clearance provides a limiting rate for the clearance of a wide variety of substances; and iv) the finding of RICHARDS, BOTT and WESTFALL [11] that inulin perfused through the renal-portal system of the frog kidney does not enter the undamaged tubular lumen either by secretion or diffusion. Although the weight of this evidence is impressive, it is important to realize that none of the arguments exclude a one way process of absorption from the lumen into the blood by a mechanism which does not discriminate between any of the molecules present in the tubule and absorbs them all at the concentration in which they are present in the lumen. A process possessing these characteristics would have escaped detection by any of the criteria described by SMITH.

The most important evidence that inulin is a satisfactory reference substance for water movement in *Necturus* comes from the demonstration by BOTT [1] that the tubular fluid/plasma concentration ratios of creatinine and inulin are exactly the same all the way along the proximal and distal tubules as well as in the urine. We made use of her finding in the first paper of our series in which we reported that creatinine clearance was exactly the same as C^{14} inulin clearance. In view of SMITH's discussion and the general acceptance of inulin as a measure of glomerular clearance in many species, we concluded that C^{14} inulin was a satisfactory

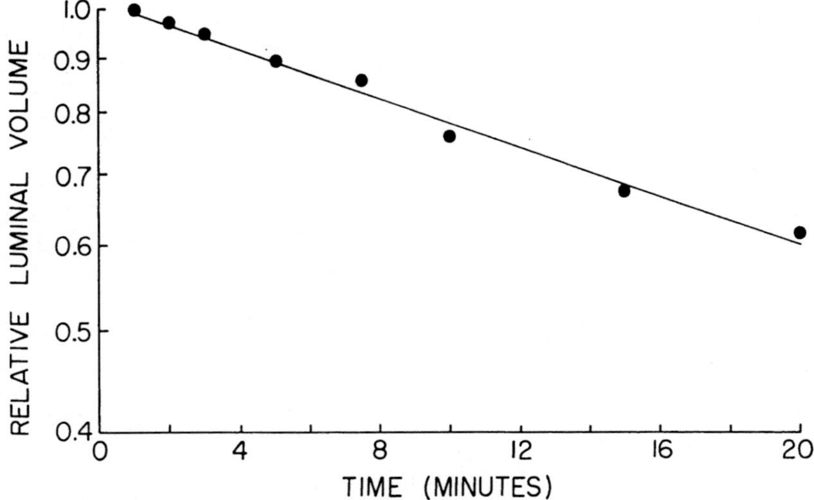

Fig. 1. Time course of fluid absorption by *Necturus* proximal tubule, as measured by sequence photography

reference molecule for the measurement of water movement in stopped flow microperfusion. However our most recent evidence indicates that this conclusion is not a necessary consequence of either BOTT's or our own observations.

This unexpected finding arose from a study of the time course of fluid absorption from *Necturus* proximal tubule [13] making use of sequential photography following the technique described by GERTZ [3]. Photographs are taken of columns of fluid isolated between oil droplets in those segments of the proximal tubule which are visible on the surface of the kidney. When either the most proximal or the most distal segment of the proximal tubule is filled with 100 mM NaCl (isosmotic with *Necturus* plasma) water absorption follows a logarithmic time course, as shown in Fig. 1. On the average 50% of the fluid is absorbed in a 20 minute perfusion period, almost twice the 27% figure obtained in our previous measurements [12, 18, 19] using C^{14} inulin as a reference molecule. Such a discrepancy is outside of the experimental errors of the two methods.

We first considered the possibility that the difference might be ascribed to biological variation and that the present group of *Necturi* were characterized by an ability to transport fluid at a rate greater than previously observed. This possibility was ruled out by a new series of stopped flow microperfusion experiments. Water absorption measured with C^{14} inulin was 20% at the end of a 20 minute period, smaller than our characteristic figure of 27%. In this set of experiments the technique was modified slightly in order to provide additional information about the quantity of fluid absorbed from the kidney. The volume of perfusion fluid injected into the tubule at the beginning of the experiment was measured exactly; at the close of the experiment all of the fluid remaining in the tubule was collected and its volume and C^{14} inulin content were measured. The results of this part of the experiment agreed in general with the photographic measurements, in that 67% of the initial fluid content of the tubule and 69% of the initial C^{14} inulin could not be accounted for in the collected perfusate. In an experiment of this sort it is necessary to rule out the possibility that the material which had disappeared from the tubule might have leaked out into the kidney as a result of gross damage at the sites of tubular puncture or might have been left behind as a result of incomplete collection. Accordingly, the portion of the kidney containing the perfused tubule was excised, macerated and extracted by a technique which had been shown by Page and Solomon [8] to recover 98.6% of the C^{14} inulin in cat papillary muscle. Only 13% of the initial C^{14} was recovered from the excised kidney by this procedure which showed that the preponderant fraction of the C^{14} inulin lost from the tubule had been transported away from the site of injection.

Though it seemed unlikely that the sample of C^{14} inulin used in this set of experiments contained fragments small enough to diffuse out of the tubule, it was important to exclude this possibility by direct experiment. For this purpose, we used the doubly perfused kidney initially described by Cullis [2] and used by Richards, Bott and Westfall [11] in their demonstration that inulin did not enter the lumen of frog tubules either by diffusion or secretion. When the portal system of the *Necturus* kidney was perfused with C^{14} inulin in our experiments, the concentration of C^{14} in the urine was only 0.3% of that in the perfusing fluid. This result was in good agreement with the findings of Richards, Bott and Westfall and demonstrated that our sample of C^{14} inulin did not contain significant amounts of radioactive fragments small enough to diffuse from the portal system into the tubular lumen and thus make their way into the urine. Since inulin is excreted in considerable amounts in clearance experiments, it seems most unlikely that a leakage of fragments into the tubule would have been obscured by subsequent complete absorption in more distal portions of the system.

The experiments with the doubly perfused preparation were continued for a further period in order to obtain information about the movement of inulin in the other direction, from tubular lumen to blood. The preparation was washed until no radioactivity appeared in the wash solution. Several proximal tubules were blocked with oil and filled with a 100 mM NaCl solution containing C^{14} inulin of high specific activity.

When all the fluid that had perfused the vascular system over a 40 minute period was collected, it contained 48% of the radioactivity initially injected into the tubules. This portion of the experiment confirms our finding that C[14] inulin is absorbed from the tubules, and indicates that C[14] originally placed in the tubule can make its way into the blood stream. Furthermore, the two parts of the experiment, taken together, indicate

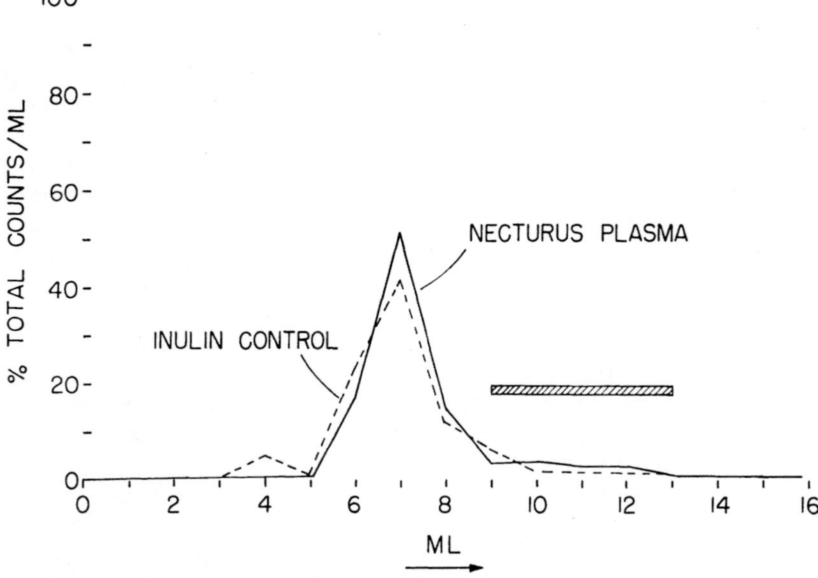

Fig. 2. Sephadex fractionation of radioactivity in *Necturus* plasma following 60 minute free flow perfusion of a single kidney with C[14] inulin. For comparison, the C[14] inulin itself has been subjected to fractionation on the same Sephadex column. The horizontal bar to the right of the peak indicates the fractions in which added glucose molecules are found

that there is a very high degree of rectification in the transport process, since many times more radioactivity is absorbed from the tubule, than is returned either by diffusion or secretion.

It remains to be shown whether the C[14] which appears in the blood remains in the chemical form of inulin, or whether it has suffered some gross metabolic alteration in its passage. We first confirmed that appreciable amounts of C[14] passed into the blood in conventional stopped flow microperfusion *in situ* experiments using high specific activity inulin. At the end of the experiments 5—15% of the total C[14] injected into the tubule could be found in one ml of plasma. In subsequent experiments inulin was perfused through a freely flowing tubule for 60 minutes to provide enough plasma radioactivity for fractionation on a Sephadex column. Fig. 2 shows that the peak of the C[14] activity in the plasma corresponds with the peak obtained when C[14] inulin is itself fractionated on a Sephadex column. This indicates that the molecular weight of the material that has entered the blood stream is similar to that of the inulin initially injected into the tubule.

In sum, these several kinds of experiments show that inulin can be transported from the tubule to the blood stream by what is essentially a one way process. Furthermore, by one method of fractionation at least, there appears to be no essential change in the molecular weight of the inulin in its passage from the lumen to the blood. As we have discussed earlier the characteristics of this absorption process are just those which would have escaped detection by the classical criteria imposed by Smith, particularly if the process can be shown to be nonspecific and apply to other molecules chemically dissimilar from inulin.

The discrepancy between the amount of water absorbed from the tubule as measured with C^{14} inulin and that measured either volumetrically or photographically may therefore be ascribed to the absorption of inulin itself. Since the difference between the two estimations is almost a factor of two, it is clear that water fluxes measured previously may have been grossly underestimated and that inulin can no longer be accepted unquestionably as a satisfactory molecule for clearance measurements in *Necturus*.

Smith points out that inulin is physiologically inert and that there are no enzymes in the blood that are capable of hydrolyzing it; our evidence that it is transported into the blood with no significant change of molecular weight is consonant with this. Consequently, it seemed most unlikely that a mechanism would exist that was specialized for the carriage of inulin, and we expected that other molecules would probably be absorbed by the same route. Analogous experiments to those performed with inulin were therefore carried out with I^{131} human serum albumin. 53% of the injected I^{131} albumin was recovered from the tubule, appreciably more than the 31% for C^{14} inulin. Only 18% of the I^{131} could be accounted for in the kidney tissue which means that 29% should be found in the rest of the animal. Since the average content of I^{131} in each ml of plasma was 5% of the total radioactivity injected into the tubule much of the missing 29% had reached the plasma, assuming the fractional blood volume of *Necturus* to be similar to that measured in the frog by Prosser and Weinstein [10].

Free flow perfusion of single tubules provided plasma samples with radioactivity high enough to permit analysis by gel electrophoresis. Fig. 3 shows that the radioactivity in the plasma migrated at a rate typical of human serum albumin and appeared in a zone separate from the region in which *Necturus* serum proteins are found. Thus, like inulin, albumin has been transported into the blood by a process which causes no essential alteration in its physical characteristics, measured in this case by electrophoresis.

Although more inulin is removed from the tubule than albumin in 20 minutes, it is possible that the presence of albumin in the tubule slows down absorption, since we have previously found [18] that when albumin is present at a concentration of 35 g/liter in conventional stopped flow microperfusion experiments, water absorption is reduced from 27% to 15%. Since the present albumin concentration was 20 g/liter we cannot exclude such an effect. The question as to whether albumin and inulin are absorbed at the same or different rates can best be settled by experiments in which both substances are measured simultaneously, as was done by BOTT [1] when she showed that *Necturus* tubules did not discriminate between inulin and creatinine. We may conclude that the transport system absorbs the protein, albumin, at a rate similar to, but not necessarily identical with, the rate at which it absorbs the polysaccharide, inulin, and the imide, creatinine.

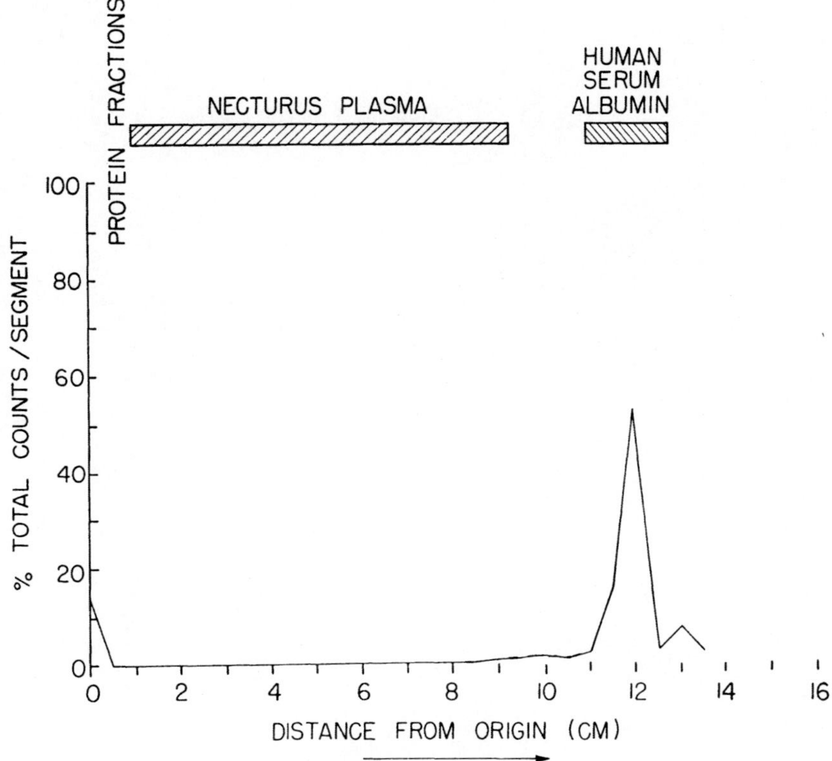

Fig. 3. Gel electrophoresis of plasma radioactivity following 60 minute free flow perfusion of a single kidney tubule with I[131] human serum albumin. For comparison samples of *Necturus* plasma and I[131] human serum albumin were measured chemically following electrophoresis on a similar gel. It is seen that the albumin peak coincides with the peak in radioactivity and is separated by a gap from the band containing the *Necturus* plasma proteins

The question next arises as to the possible nature of the non-specific process responsible for the transport at similar rates of three such dissimilar molecules. Though present information is far from complete, the results suggest that a process such as pinocytosis could be involved. In an electron microscopic study of ferritin absorption by *Necturus* proximal tubule MAUNSBACH [5] has presented evidence for the formation of pinocytotic vacuoles at the luminal membrane. One of his photomicrographs is shown in Fig. 4. We consider the first step in pinocytosis to be the engulfment of a quantum of solution by a vacuole which subsequently moves into the cell carrying each solute at exactly the same concentration as was present in the lumen of the tubule. We do not know whether the vacuole moves across the cell with its contents intact, or whether some or all of its contents enter the cytoplasm. For the present

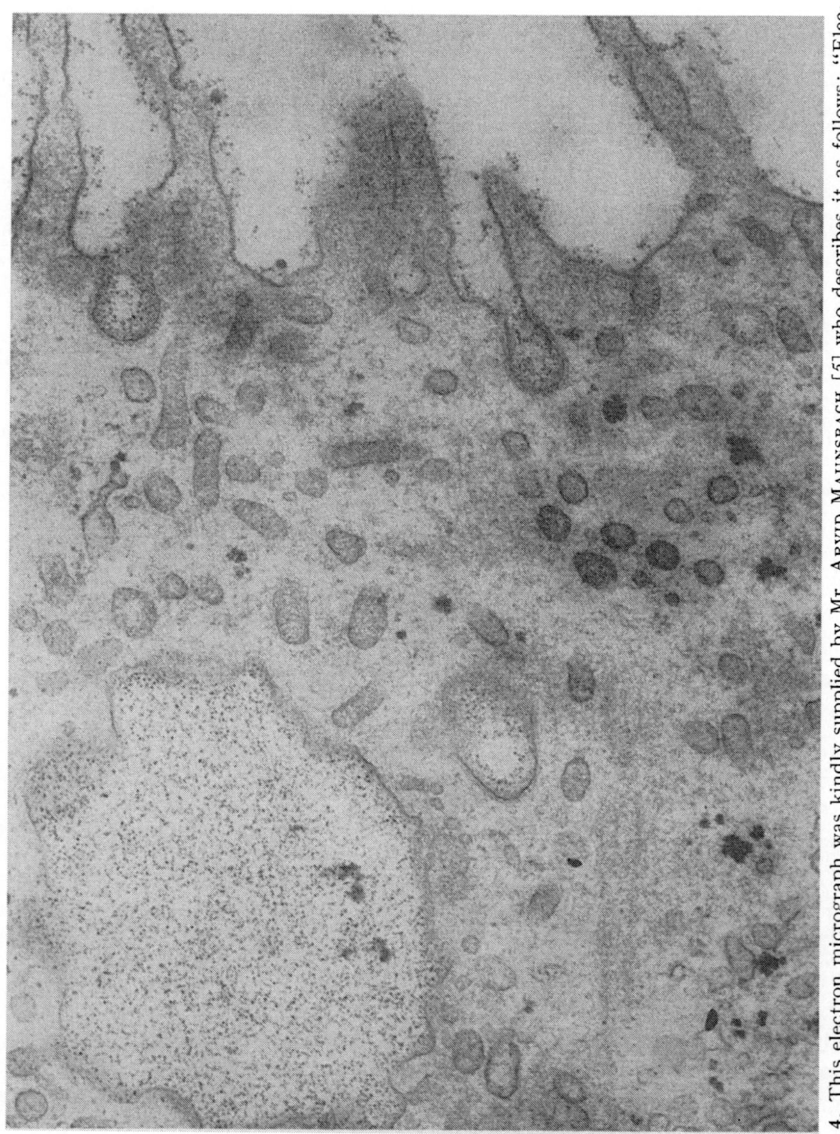

Fig. 4. This electron micrograph was kindly supplied by Mr. Arvid Maunsbach [5] who describes it as follows: "Electron micrograph of luminal part of a cell from a *Necturus* proximal tubule which was microperfused with a solution containing ferritin. In the tubule lumen, around obliquely sectioned brush border projections, are seen numerous small dark dots, each representing one ferritin molecule. The cell membrane in two places forms invaginations which enclose many ferritin molecules. In the cytoplasm are seen ferritin-containing vacuoles which presumably originate from pinched-off cell membrane invaginations. Magnification 47,000 ×."

purpose, we assume that the rate limiting step is at the cell membrane and that the contents of the vacuole are not returned to the lumen.

A process satisfying this description would account qualitatively for all the evidence presented concerning the *Necturus* tubule including the one way nature of the transport. We have already pointed out that inulin, creatinine, and human serum albumin are transported at similar rates. To these molecules, Na may be added. The evidence for the inclusion of Na comes from the observation that the Na concentration in the free flowing tubules remains unchanged and equal to the plasma concentration. Since the volumetric and photographic methods demonstrate an extra absorption of water, this must have contained Na and the Na/inulin ratio in this extra fluid must be the same as in the perfusion fluid[1]. When an ion is added to the list of substances transported by the new mechanism, it is even more difficult to imagine a system other than pinocytotic transport to fit with the observations. Such a transport system would agree with the observation in this and other species that inulin provides a limiting rate for the clearance of a wide variety of substances.

The question next arises as to the effect of the present finding on our previous conclusions concerning the nature of salt and water transport out of *Necturus* proximal tubule. The major conclusions drawn in our earlier papers are:

i) Net H_2O flux out of the proximal tubule is dependent upon the Na concentration within the lumen.

ii) Net H_2O flux is linearly dependent upon net solute flux, going to zero when net solute flux goes to zero. The relationship between these variables corresponds to isosmotic absorption, and the fluid within the tubule remains isosmotic with the plasma.

iii) The osmotic pressure of the serum proteins is not the driving force for water absorption form the proximal tubule.

iv) Na is actively transported out of the tubule into the blood. On the basis of measurements of electrical potential differences and cell Na concentration, the active step for Na transport is located at the peritubular face of the epithelial cell.

v) There is evidence for an active mechanism transporting K from the cell to the tubular lumen, presumably located at the luminal face of the cell.

Two of these conclusions are not affected by the presence of a pinocytotic mechanism. The active transport of Na and the localization of

---

[1] *Note added in proof:* This particular argument is based on the implicit assumption that the extra absorption of water is characteristic of all populations of *Necturus*. Recent observations indicate that the extra absorption is much less in "winter" *Necturi* than in the adult "summer" *Necturi* on which the experiments described in the text were carried out. In a series of 7 animals, obtained in November 1964, only 0.6 % of the I^{131} albumin was found in each ml of plasma, far less than the 4.9 % in the experiments on summer *Necturi* described in the text. Two summer *Necturi* remained and gave results of 2.6 and 5.7 %, comparable to those obtained previously on the same series of animals. This suggests that the quantitative importance of this process is subject to biological variation. Therefore, to establish rigorously that Na is also absorbed by this process, it would be necessary to demonstrate that the Na concentration in tubular fluid from summer *Necturi* is isotonic with plasma, as it is in winter ones. It seems likely that this is indeed the case since some albumin is absorbed in winter animals in which the proximal tubular Na is isotonic with plasma [7].

this step at the peritubular face of the cell are independent of the amount of fluid transported since they rest upon the demonstration that Na is transported up an electrochemical potential gradient. The finding that water absorption takes place against an adverse protein osmotic pressure gradient is also independent of the amount of fluid absorbed, and it is valid to conclude that protein osmotic pressure is not the driving force for water absorption from the proximal tubule.

In order to examine whether our other conclusions need modification it is necessary to consider the fate of the vacuole as it moves through the cell. Two limiting cases may be considered. In one the vacuole moves across the cell with its contents wholly undisturbed and discharges them all at the peritubular face. This would be a separate and parallel process which would result merely in the addition of a second mechanism of salt and water transport to the one which has already been studied and described in detail in our previous papers. At the other extreme we might imagine that some or all of the contents of the vacuole would be discharged into the cytoplasm, to be pumped out of the cell by the Na pump at the peritubular face. The quantitative importance of the pinocytotic mechanism in fluid transport would depend upon the fraction of water moved by the pump. The observation of albumin in the plasma suggests that at least part of the original vacuole travels all the way across the cell. The distinction between these two limiting cases will be difficult to make experimentally, since the amount of water transported is the same in both instances. Some help will surely be provided by electron microscopic study of the progression of vacuoles across the cells. Alternatively, if one process were characterized by streaming potentials, and the other not, as are the two mechanisms of water movement in gall bladder described by Pidot and Diamond [9], it might be possible to separate them by direct experiment.

What lies ahead is fascinating and exciting. We are faced with a transport process which is far less amenable to description in exact biophysical terms than those that have previously been studied. To learn the nature of the physical and chemical forces involved is a challenging task that should lead to greater understanding of the molecular basis of cell function.

References

[1] Bott, P. A.: Amer. J. Physiol. **168**, 107 (1952).

[2] Cullis, W. C.: J. Physiol. (Lond.) **34**, 250 (1906).

[3] Gertz, K. H.: Pflügers Arch. ges. Physiol. **276**, 336 (1963).

[4] Giebisch, G., R. M. Klose, G. Malnic, W. J. Sullivan, and E. E. Windhager: J. gen. Physiol. **47**, 1175 (1964).

[5] Maunsbach, A.: J. Cell Biol. **19**, 48 A (1963).

[6] Oken, D. E., and A. K. Solomon: Amer. J. Physiol. **204**, 377 (1963).

[7] — G. Whittembury, E. E. Windhager, and A. K. Solomon: Amer. J. Physiol. **204**, 372 (1963).

[8] Page, E., and A. K. Solomon: J. gen. Physiol. **44**, 327 (1960).

[9] Pidot, A. L., and J. M. Diamond: Nature (Lond.) **201**, 701 (1964).

[10] Prosser, C. L., and S. J. F. Weinstein: Physiol. Zoöl. **23**, 113 (1950).

[11] Richards, A. N., P. A. Bott, and B. B. Westfall: Amer. J. Physiol. **123**, 281 (1938).

[*12*] Schatzmann, H. J., E. E. Windhager, and A. K. Solomon: Amer. J. Physiol. **195**, 570 (1958).

[*13*] Scott, W. N., D. L. Maude, I. H. Shehadeh, and A. K. Solomon: Physiologist **7**, 249 (1964).

[*14*] Shehadeh, I. H., D. L. Maude, W. N. Scott, and A. K. Solomon: Physiologist **7**, 254 (1964).

[*15*] Shipp, J. C., I. B. Hanenson, E. E. Windhager, H. J. Schatzmann, G. Whittembury, H. Yoshimura, and A. K. Solomon: Amer. J. Physiol. **195**, 563 (1958).

[*16*] Smith, H. W.: The Kidney. Chapter 3, pp. 47—52, Chapter 9. New York: Oxford University Press 1951.

[*17*] Solomon, A. K.: Amer. J. Physiol. **204**, 381 (1963).

[*18*] Whittembury, G., D. E. Oken, E. E. Windhager, and A. K. Solomon: Amer. J. Physiol. **197**, 1121 (1959).

[*19*] Windhager, E. E., G. Whittembury, D. E. Oken, H. J. Schatzmann, and A. K. Solomon: Amer. J. Physiol. **197**, 313 (1959).

Discussion

Ullrich: Bei vielen Nierenphysiologen haben Ihre Untersuchungen große Aufregung hervorgerufen. Es war wohl seit jeher eine offene Frage, ob Inulin eine zuverlässige glomeruläre Testsubstanz ist. Meine Mitarbeiter und ich hatten uns deshalb öfter darüber unterhalten, ob man diese Frage in Perfusionsexperimenten überprüfen soll. Aber wir haben es unterlassen, weil Dr. Gertz mit seiner Tröpfchenresorptionsmethode und der Farbstoffpassagezeit zu gleichen Werten für den proximal rückresorbierten Filtratanteil kam wie Giebisch und Gottschalk, die den Anstieg der Inulinkonzentration in proximal entnommenen Mikroproben ihrer Berechnung zugrunde legten.

Als wir Ihre beiden Abstracts [Physiologist **7**, 249 und 254 (1964)] bekamen, haben wir doch Untersuchungen angestellt, um zu sehen, ob auch aus den Tubuli der Ratte Inulin herausdiffundiert.

Sonnenberg und Deetjen [Pflügers Arch. ges. Physiol. **278**, 669 (1964)] haben eine hochpräzise Mikroperfusionspumpe konstruiert, mit deren Hilfe wir ein einzelnes Nephron perfundiert haben. Die Infusionsrate wurde so gewählt, daß sie der Größe des normalen Glomerulumfiltrates entsprach. Als Lösung wurde eine isotone NaCl-Lösung verwendet, der mit C_{14} markiertes Inulin zugesetzt war. Die Infusion lief über 10 min. Während und nach der Infusion wurde in 10-min-Perioden von beiden Nieren getrennt der Ureterurin gesammelt. Zunächst wurde gemessen, wieviel von dem infundierten Inulin überhaupt im Urin wiedergefunden wurde. Die Kalibrierung der infundierten Inulinmenge wurde dadurch erreicht, daß die Radioaktivität gemessen wurde, die in 10 min von der Pumpe in einen auf einem Parafilmblättchen befindlichen Urintropfen infundiert wurde. Diese Menge gleich der im Experiment in einen proximalen Tubulus infundierten Inulinmenge gesetzt, ergab, daß im Mittel nach ungefähr einer Stunde 95 % durch beide Nieren ausgeschieden wurden.

Die durch die kontralaterale, also nicht infundierte Niere ausgeschiedene Inulinmenge mit zwei multipliziert entspricht der Inulinmenge, die über die Blutbahn gelaufen und in den Glomerula beider Nieren durch Filtration in den Urin gelangt ist. In der nebenstehenden Tabelle sind diese Werte als Prozent der insgesamt ausgeschiedenen Inulinmenge aufgeführt. Um die Methodik besser überprüfen zu können, wurden die Mikroinfusionen von zwei erfahrenen Experimentatoren durchgeführt. Experimentator B fand im ersten Experiment 4.8 %. Da in den folgenden Experimenten eine große Inulinmenge kontralateral ausgeschieden wurde, haben wir die Infusionslösung sehr stark mit Lissamingrün gefärbt, so daß die Flußrichtung der infundierten Lösung und auch Lecks gesehen werden konnten. Die genauere Analyse ergab, daß der Tubulus vor der Infusion nicht mit Öl gefüllt sein durfte — dies war in den mit Stern bezeichneten Experimenten der Fall. Hierbei sind vermutlich bei der Infusion der inulinhaltigen Lösung kleine Öltröpfchen in der Henleschen Schleife liegen geblieben und haben das Lumen etwas verengt. Die Drucke im Tubulus sind dadurch angestiegen und Infusionslösung ist nach proximal zum

25*

Glomerulum hin gelaufen. Experimentator R hat etwas später mit seiner Versuchsserie begonnen und konnte sich die Erfahrung von Experimentator B zunutze machen. Deshalb finden sich bei ihm viel niedrigere Werte. Wichtig ist, daß in einem Drittel der Versuche weniger als 2 % der insgesamt ausgeschiedenen Inulinmenge über die Blutbahn gelaufen ist. Es erscheint uns keineswegs sicher, nicht einmal wahrscheinlich, daß dieses Inulin durch die Tubuluswand rückresorbiert wurde. Vor und nach der eigentlichen Infusion muß die Infusionskapillare durch den die Niere bedeckenden Flüssigkeitsfilm geführt werden, und es ist leicht möglich,

Tabelle. *Ausscheidung von Inulin, das in ein proximales Konvolut infundiert wurde.* Die mit * bezeichneten Experimente waren technisch unzulänglich, da ein Teil der Perfusionsflüssigkeit nach proximal zum Glomerulum hin lief. A = Antidiurese, W = Wasserdiurese, M = leichte Mannitoldiurese

Nr.	Punktionsstelle im proximalen Konvolut 1., 2. bzw. 3. Drittel	ausgeschiedene Inulinmenge = 100%			Diureseart
		ipsilateral	kontralateral	kontralateral × 2	
		Experimentator B			
2	2.	84,3	15,7	31,4*	W
5	3.	87,6	12,4	24,8*	W
4	1.	89,2	10,8	21,6*	W
7	1.	92,3	7,7	15,4	M
6	3.	94,3	5,7	11,4	M
3	2.	95,2	4,8	9,6	W
9	2.	97,2	2,8	5,6	M
1	2.	97,6	2,4	4,8	A
8	1.	98,4	1,6	3,2	M
12	1.	98,8	1,2	2,4	M
11	1.	99,0	1,0	2,0	M
10	2.	99,4	0,6	1,2	A
13	2.	99,8	0,2	0,4	M
		Experimentator R			
3	2.	93,5	6,5	13,0	M
4	1.	96,15	3,85	7,7	M
7	1.	97,7	2,3	4,6	A
5	2.	98,15	1,85	3,7	M
8	3.	99,2	0,8	1,6	A
2	2.	99,3	0,7	1,4	M
1	3.	99,55	0,45	0,9	M
6	2.	99,8	0,2	0,4	A

daß dabei kleine Perfusatmengen aus der Kapillare auslaufen. Vergleichende Experimente zeigten, daß Inulin, das auf die Nierenkapsel gegeben wurde, rasch über beide Nieren gleichmäßig ausgeschieden wurde.

Die aufgeführten Befunde bestärken uns darin, daß kein Grund besteht, die Inulinclearance als Maß für die glomeruläre Filtration zu verlassen und den vielen Arbeiten, die auf der Inulinclearance basieren, nun skeptisch gegenüberzustehen.

Kuyper: Dr. Solomon, did you ever try to measure the radioactivity in the changing compartment with the photographic method ?

Solomon: This is the next thing we are going to do. We get now to the question of measuring small quantities. The amount of fluid in this small compartment is of the order of 10^{-9} litre and we are now developing methods by which we can measure, I hope simultaneously, the length of the column within the tubular lumen, the inulin radioactivity, and the sodium concentration. I might say also with respect to Dr. ULLRICH's point that we spread J^{131} labelled human serum albumin on the oil layer on top of the capsule of the rat kidney, because we were worried about the problem of contamination from that source, and under those conditions we found nothing in the blood[1].

Tosteson: Dr. MORRELL in Paris has injected C^{14} labelled inulin into the proximal tubules of rats and been able to recover all of the injected inulin in the urine from that kidney. He has suggested as a possible explanation for the difference in the experimental results obtained by Dr. SOLOMON and his group, as compared with MORRELL's laboratory and ULLRICH's laboratory, that a factor might be damage to the tubular epithelium by the injected oil, and I would like to have Dr. SOLOMON's comments on this possibility.

Solomon: I do not know whether the oil damages it in the rat; it may be. We really believe that the oil does not damage the *Necturus* tubule; at least in the presence of oil it seems to be able to absorb water and we use the same oil in the rat that we do in the *Necturus*. But it may also be that we are dealing with a species difference. I find it obviously quite disturbing to have made what I hope you will agree are a series of very careful and thoughtful experiments and to have found results that differ substantially from those of others. I suppose I am the more concerned with Dr. ULLRICH's results than anyone else's because for a long time he has been engaged in this field. I take comfort in part that he gets numbers which are not indeed totally zero, which seems to me to mean that this process also takes place in his rats as well as in ours. I feel that what we are discussing is an order of magnitude and the "cause of" rather than whether an event happens or whether it does not. The cause of the different values I think, can be any one of a variety of factors[2].

Junqueira: It seems to me that the rate in which these columns flow through the tubules might be one of the important factors and I would like to ask both Prof. SOLOMON and ULLRICH if they have measurements as to approximately what speed would these columns have in each species.

Solomon: So far actually no-one has studied the *Necturus* save us, and no-one has questioned on experimental grounds our findings in the *Necturus*, so I would like to leave that aside. But with respect to the rat, as Prof. ULLRICH showed, his velocities were 0.02 μls. per minute, and our velocities ranged from 0.01 to 0.018 μls. per minute. So we were slightly slower than he was, perhaps by a factor of 2.

Junqueira: That is the rate of injection. I was thinking the rate of speed of the column of liquid along the tubule.

Solomon: Well, this is the rate it goes in. It goes down at approximately the same rate (save for fluid absorption).

Junqueira: I wonder if oil would not decrease the dynamics of this column running down the tubule.

Solomon: I do not think so.

Niesel: Ich glaube, es gäbe eine Möglichkeit, beide Ergebnisse zu vereinigen, und zwar folgende: Wenn das Inulin resorbiert wird und wieder in derselben Niere sofort sezerniert wird, wenn also Inulin einen Kurzschluß macht von einem proximalen Tubulus zu einem proximalen Tubulus oder zu einem distalen Tubulus eines anderen Nephrons, würde Herr SOLOMON messen, daß das Inulin verschwindet, und Herr ULLRICH würde finden, daß es nicht in der kontralateralen Niere erscheint.

Solomon: That would be possible in the rat, if it happened in the rat, whereas it would not account for it in the *Necturus*.

---

[1] *Note added in proof:* In 4 subsequent experiments C^{14} inulin was layered on top of the capsule with no intervening layer of oil. In 90 minutes about 50 % of the radioactivity could be accounted for in the excreted urine and the inulin space.

[2] *Note added in proof:* We have now had the opportunity to repeat our preliminary experiments in the rat [Physiologist **7**, 254 (1964)] using improved micropuncture techniques. In 6 experiments with C^{14} labelled inulin, an average of 3.0 % (range 0.5 to 6.4 %) of the C^{14} injected into the tubule was found in the inulin space of the animal after 10 to 25 minutes. These new results are in good agreement with those obtained by Prof. ULLRICH and other workers.

Niesel: Beim *Necturus* bleibt das Inulin etwa 20 min in einem Tubulus, bei der Ratte dagegen nur 1 min. Möglicherweise werden dadurch die Größenordnungsunterschiede bedingt.

Ullrich: Ich glaube nicht, daß diese Erklärung zutrifft. Es gibt natürlich manche Umstände, bei denen man nicht mehr von einem normalen Verhalten der Tubuluszellen reden kann. Deetjen hat z. B. gezeigt, daß schon bei einer geringen Erniedrigung der Körpertemperatur die Sekretion von PAH im proximalen Tubulus fast völlig sistiert. Daß die Nierentubuli durch Öl stark beeinträchtigt werden ist unwahrscheinlich. Gertz fand nämlich eine gleich rasche Resorption von Kochsalzlösung aus dem proximalen Tubulus — ganz gleich ob dieser vorher mit Paraffin —, Rizinus —, Silikon- oder Knochenöl gefüllt war. Aber vielleicht kannHerr Thoenes einen Kommentar zu dem Problem der Schädigung des Bürstensaumes durch Öl geben.

Thoenes: Sorry, till now I have not investigated such pieces of tubule. We only have investigated tubules fixed by free flow perfusion.

Solomon: I do not think we can be clear about the effect of inulin in the rat, but on the basis of the fact that we obtain the same results with albumin both in the *Necturus* and in the rat, if inulin damages the tubule so does albumin, so it seems to me that as a possible explanation is not likely.

Hartmann: I have two questions regarding the reabsorption of albumin. The first question is, you used a foreign protein, human albumin, for the reabsorption experiment. Is the same true even for the *Necturus* albumin ? The second question is, does the *Necturus* have in the glomerular filtrate albumin or not ?

Solomon: I know that the *Necturus* does not have proteins in its filtrates, certainly only in quite small amounts. We have not tried it with *Necturus* albumin.

Hartmann: Both substances you used for the tubules were unphysiological substances, inulin and human albumin. This was the reason for my questions.

Solomon: If you can name me a molecule which is physiological —

Kuyper: Water

Solomon: — which is larger than glucose. The problem is any molecule which is metabolized in the normal course of events is a very hard molecule to use as a marker, because then it has more than one fate.

Diamond: With regard to the suggestion of Dr. Hartmann that foreign molecules such as inulin and albumin from animals other than the animal in question might have unphysiological effects on the preparation: we know from other epithelial organs such as the gall bladder, the small intestine, the stomach, the urinary bladder, the frog skin and others that these organs will continue to transport water and salt in the presence of inulin, in the presence of albumin and other foreign molecules. In the gall bladder I know that albumin on the one hand and on the other hand, concentrations of inulin several hundred times greater than those used by Dr. Solomon in the kidney, have no effect on the transport processes.

M. Hokin: I was wondering whether there was any possibility that the oil in the system stimulated or initiated pinocytosis which would not normally occur, and whether there was any morphological evidence about this at all.

Solomon: I do not know.

Thoenes: Wir müssen — das ging auch aus dem Referat von Herrn Prof. Staubesand hervor — zwischen Pinocytose und Cytopempsis unterscheiden. Das Bild von Dr. Maunsbach, das Sie zeigten, weist nur das Phänomen der Pinocytose nach, das wir für Eiweißstoffe kennen. Wenn Sie gefunden haben, daß Inulin unverändert die Zelle passiert hat, müssen wir den Vorgang der Cytopempsis fordern, d. h. das Durchschleusen ganzer Bläschen, in denen das Inulin erhalten bleibt und an der basalen Seite der Zelle ausgeschieden wird. Dieser Vorgang müßte morphologisch erst noch gezeigt werden. Bisher haben wir keine sicheren Hinweise darauf, daß eine Cytopempsis im Tubulus vor sich geht. Das wäre eine Frage an die Morphologen, die sich aus Ihren Befunden ergibt.

Solomon: I would agree, yes. I have not seen this. I am not sure what the fate of these vacuoles is. Our findings are not morphological and I was very careful to say that *a possible* morphological explanation would be, I should say, cytopempsis rather than pinocytosis. But we do not know what the fate of this absorbed material is once it has left the tubular wall, except that the inulin and the albumin appear again in the blood stream. It seems to me to matter only that the major barrier be the luminal membrane, and that as far as our findings are concerned they throw no light about what has to happen to these molecules after they get across.

Schmidt: Meine Antwort richtet sich an Herrn THOENES. Bei Eisendextranangebot erfolgt eine Aufnahme durch Pinocytose. Das Eisendextran wird an der Basis der Zelle unverändert abgegeben und läßt sich dort histochemisch nachweisen. Das spricht also dafür, daß auch in der Nierentubuluszelle eine Cytopempsis stattfindet.

Permeabilität der kortikalen Nephronabschnitte in Beziehung zu Transportvorgängen und Struktur

Von

K. J. Ullrich, Berlin

Mit 3 Abbildungen

Jedes funktionelle Geschehen läuft an Strukturen ab und ist letzten Endes mit strukturellen Änderungen identisch. Erforschung der Funktion wird dann besonders erfolgreich, wenn die Struktur ganz mit einbezogen wird. Eines der besten Beispiele ist der quergestreifte Muskel.

Huxley [13] hat jüngst gezeigt, daß offene Ausmündungen des endoplasmatischen Retikulums mit den Stellen identisch sind, wo durch lokalen Reiz eine lokale Kontraktion ausgelöst werden kann. Diesem Schritt folgte unmittelbar der nächste. Im elektronenmikroskopischen Bild sieht man, daß sich die calciumspeichernden Vesikel direkt an die nach außen offenen Röhren anlagern (Hasselbach [12]). Wie sehr wird dadurch die Hypothese gestützt, daß die Erregungen durch die offenen Röhren ,,nach innen steigen'' und über Freisetzung von Calciumionen die Muskelkontraktionen auslösen.

Leider können wir beim Transport durch Membranen Funktion und Struktur noch nicht in diesem Maße miteinander verbinden. Es liegen aber bereits Befunde vor, die sich am einfachsten und zwanglosesten durch Einbeziehung der Ultrastruktur erklären lassen [29]. Über solche an der Rattenniere erhobenen Befunde möchte ich im folgenden berichten und vor allem über die Methoden, die uns die gewünschten funktionellen und morphologischen Analysen erlauben.

Wesentlich für die Analyse des Stofftransportes durch die Nierentubuli ist, daß die Konzentration der zu untersuchenden Substanz zumindest auf der luminalen Seite der Tubuluszellschicht beliebig variiert werden kann. Das bedeutet, daß an der freigelegten Niere in dem jeweiligen Tubulus die glomeruläre Filtration durch Ölinjektion blockiert und dann eine künstliche Lösung in das Tubuluslumen injiziert wird. Sind lange Kontaktzeiten, d. h. über 20 sec erwünscht, so kann man mit einer Genauigkeit von ± 2 sec die injizierte Lösung wieder aus dem Tubulus entnehmen und analysieren [7, 14, 35]. Sind kurze Kontaktzeiten, d. h. unter einer Sekunde erwünscht, so muß man zu einem ,,fließenden'' System übergehen, d. h. den Tubulus mit einer hochpräzisen Perfusionspumpe durchströmen und in verschiedenem Abstand von der Infusionsstelle einen Teil der dort vorbeilaufenden Lösung wieder absaugen und analysieren [24]. Wie man aus den gewonnenen Daten den transtubulären Stofftransport und die transtubuläre Permeabilität berechnet, wurde ausführlich mitgeteilt [25, 29]. In analoger Weise ist es möglich, den Tubulus mit Fixierungsflüssigkeiten zu durchströmen und dann der elektronenmikroskopischen Betrachtung zuzuführen. Es ist zu hoffen, daß bei dieser Momentanfixation Strukturveränderungen, die durch den Stofftransport bedingt sind, sichtbar werden, wenn es solche in dem mit dem Elektronenmikroskop erfaßbaren Größenbereich überhaupt gibt.

A. Transtubulärer Wasserfluß

Aus der Geschwindigkeit des osmotischen Angleichs von anisotonen, in das Tubuluslumen infundierten Lösungen an den osmotischen Druck des Plasmas konnte die Wasserpermeabilität des proximalen und distalen Konvolutes errechnet werden. Die in Tab. 1 aufgeführten Werte zeigen,

Tabelle 1. *Wasserpermeabilität des proximalen und distalen Konvolutes von Ratten in Antidiurese und Diurese (Diabetes insipidus).* Zum Vergleich sind die Werte der Wasserpermeabilität von Muskel- und Glomerulumkapillaren mit angegeben [20]

Wasserpermeabilität	10^{-8} cm³/cm²/sec/cm H_2O
Proximales Konvolut (Ratten in Antidiurese) . . .	17,4
Distales Konvolut (Ratten in Antidiurese)	7,6
Proximales Konvolut (Diabetes insipidus-Ratten) . .	15,8
Distales Konvolut (Diabetes insipidus-Ratten) . . .	2,6
Glomerulumkapillarmembran	300—600
Muskelkapillarmembran	2,5

daß die Wasserpermeabilität im proximalen Konvolut in Antidiurese zweieinhalb mal größer ist als im distalen Konvolut und siebenmal größer als in den Muskelkapillaren. Die Wasserpermeabilität der Glomerulumkapillaren ist dagegen 17—34mal größer als die des proximalen Tubulus. Die größere Wasserpermeabilität des proximalen Konvolutes im Vergleich zum distalen könnte man allein der enormen Oberflächenvergrößerung durch die Fortsätze des Bürstensaumes, bei sonst gleicher Beschaffenheit der maßgeblichen Membranen zuschreiben. Es sei hier gleich erwähnt, daß bei sämtlichen Permeabilitätsberechnungen an Nierentubuli die Wandfläche eines Zylinders mit einem dem Tubuluslumen entsprechenden Durchmesser in die Rechnung eingesetzt wurde, und zwar 20 μ für das proximale Konvolut und 16 μ für das distale Konvolut [29][1]. Die Flächen der wahrscheinlich die Permeabilität begrenzenden Membranen, des Bürstensaumes und der basalen Einfältelungen sind nicht berücksichtigt.

Ob auch die Zwischenzellräume als Passagewege benutzt werden, ist fraglich. Aber auch wenn Stoffe in nennenswertem Ausmaße zwischen den Zellen hindurchlaufen sollten, würde der Befund von FARQUHAR und PALADE [6], daß die Verbindung der "terminal bars" im distalen Konvolut fester ist als im proximalen, in das allgemeine Bild passen, nämlich, daß nicht nur die Wasserpermeabilität, sondern auch, wie wir später sehen werden, die Permeabilität von Elektrolyten und Nichtelektrolyten im proximalen Konvolut größer ist als im distalen.

---

[1] Bei Wasserdiurese wird der Durchmesser im distalen Konvolut unter den Standarddurchströmungsbedingungen 20 μ.

Die viel größere Durchlässigkeit der Glomerulumkapillaren im Vergleich zu den Muskelkapillaren ist morphologisch gut erklärbar. Das Endothel der Muskelkapillaren ist in der Regel eine geschlossene zelluläre Schicht, und sämtliche Stoffe müssen, wie auch bei den Tubuluszellen, durch den Zellkörper hindurch, wenn sie nicht den schmalen Interzellularspalt benutzen. Das Endothel der Nierenkapillaren hat dagegen Poren, die z. B. im Glomerulum 20—30% der gesamten Kapillaroberfläche einnehmen und einen Durchmesser von etwa 500 Å haben [21, 27].

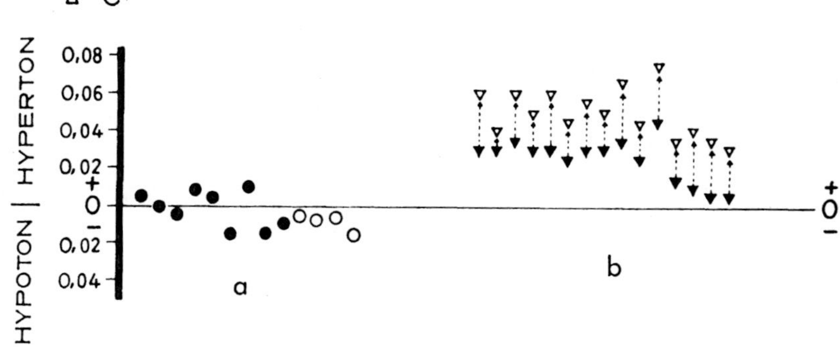

Abb. 1a u. b. a Abweichung der Gefrierpunktserniedrigung der proximalen Tubulusflüssigkeit von der des Plasmas. Jeder Punkt entspricht einer vergleichenden Messung. ● = Antidiurese; ○ = Wasserdiurese. b dgl. bei Durchströmung mit isotoner Mannitollösung. Die infundierte Lösung (▼) ist bereits leicht hyperton, die nach 150—400 msec Verweilzeit entnommene Lösung (▽) ist wegen raschem Einstrom von NaCl noch stärker hyperton [29]

Renkin und Pappenheimer [20] glauben, daß man die Permeabilität von Zellen nicht mit der von Kapillaren vergleichen darf. Wenn es aber eine Eigenschaft der submikroskopischen Organisation ist, daß durch unterschiedliche Kombination von verhältnismäßig wenigen Grundstrukturen eine Fülle spezifischer Organisationsformen mit verschiedenen Leistungen entsteht [22, 27], so muß man sich fragen, ob nicht durch Kombination molekular und damit physikalisch sehr ähnlicher Membranen die unterschiedliche „physiologische" Permeabilität zustande kommt.

Es ist seit den Untersuchungen von Walker u. Mitarb. [33] vor über 20 Jahren bekannt, daß die proximale Tubulusflüssigkeit stets isoton mit dem Blutplasma ist. Sorgfältige Nachuntersuchungen haben diesen Befund immer wieder bestätigt [10, 37]. Nachdem nun Daten über die Wasserpermeabilität vorliegen und auch die bei freiem Fluß im proximalen Tubulus rückresorbierte Wassermenge bekannt ist, kann man fragen, ob der benötigte osmotische Gradient wirklich so klein ist, daß er sich der Messung entzieht. Wenn die bei freiem Fluß aus dem proximalen Tubulus der Ratte rückresorbierte Wassermenge von $2{,}7 \cdot 10^{-4}$ µl/sec analog den Verhältnissen im distalen Konvolut durch den transtubulären osmotischen Gradienten bedingt wäre, so müßte bei der in Tab. 1 angegebenen Wasserpermeabilität die Tubulusflüssigkeit um 23 mosmol gegenüber dem Plasma hypoton sein, was einer Gefrierpunktdifferenz von 0,04° C entspricht. Diese ist sicher nicht vorhanden. Der Gefrierpunkt der

Tubulusflüssigkeit und des Plasmas sind gleich (Abb. 1a). Ähnliche Verhältnisse liegen übrigens an allen Organen vor, durch die ein isotoner Flüssigkeitstransport erfolgt: am Dünndarm [4], der Gallenblase [5], dem Ciliarkörper [1], den Tubuli von *Necturus* [34] und auch an der Krötenblase [15]. Überall ist der für den osmotischen Wasserfluß benötigte Gradient zwischen Plasma und Lumen nicht vorhanden. DIAMOND [5] wies nun darauf hin, daß der Fluß von gelösten Teilchen das Wasser durch Reibung mitnehmen könne, und zwar in den genannten Fällen in

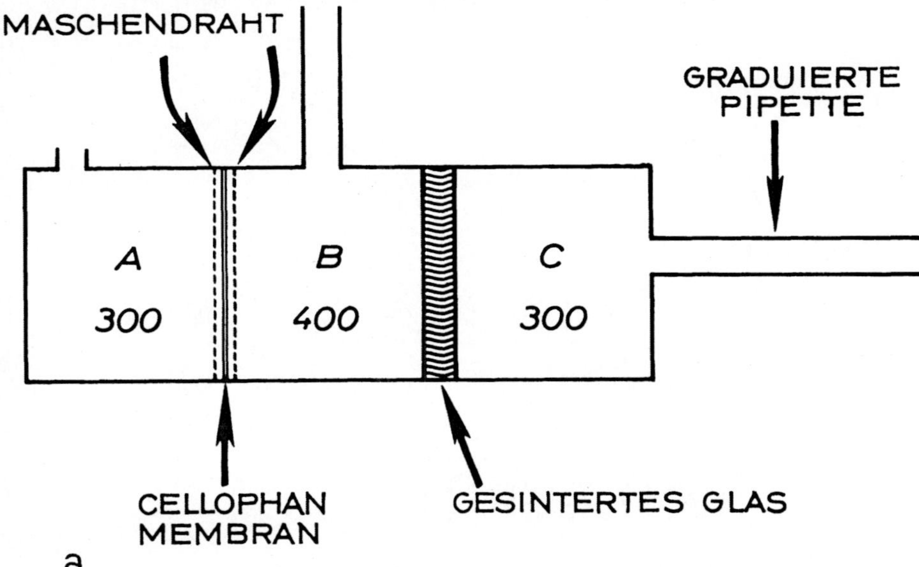

Abb. 2a Modell zur Erklärung der isotonen Reabsorption

isotonem Verhältnis. Für einen derartigen Vorgang würde der Begriff Kodiffusion zutreffen. Fände sich ein solcher Vorgang, z. B. am proximalen Tubulus, dann sollte man, wenn nicht unbedingt, so doch mit großer Wahrscheinlichkeit erwarten, daß nicht nur der transtubuläre Flüssigkeitsausstrom, sondern auch der Einstrom isoton erfolgt. Dies ist jedoch, wie die Abb. 1b zeigt, am proximalen Tubulus nicht der Fall, denn der Einwärtsfluß von gelösten Stoffen und Wasser in die das Lumen durchströmende isotone Mannitollösung ist nicht isoton. Natriumchlorid fließt zuerst in das Lumen und macht die intratubuläre Flüssigkeit hyperton. Nach der Durchströmung eines Tubulussegmentes von 150−400 μ entsprechend einer Kontaktzeit von 0,15−0,4 sec war der osmotische Druck des an sich schon leicht hypertonen Perfusates im Mittel um weitere 27 mosmol/l angestiegen. Das Wasser fließt dann wegen des entstehenden transtubulären osmotischen Gradienten in das Lumen. Wegen der unterschiedlichen Art des transtubulären Wasserflusses möchten wir einer Erklärungsmöglichkeit unter Zuhilfenahme des Modells von CURRAN und MACINTOSH [3] den Vorzug geben. In diesem Modell (Abb. 2a) sind drei starrwandige nebeneinanderliegende Kompartments (ABC) durch verschiedene Membranen getrennt. Zwischen A und B ist eine Cellophanmembran, zwischen B und C gesintertes Glas. Osmotischer Wasserfluß ist nur zwischen A und B möglich, nicht zwischen B und C, da dort die Poren zu groß sind. Erhöhung des osmotischen Druckes in B führt zu einem Wasserfluß von A nach B. In B steigt der hydrostatische Druck und führt zu einem hydrodynamischen Wasserfluß von B nach C, unabhängig vom osmotischen Druck in C. Diese

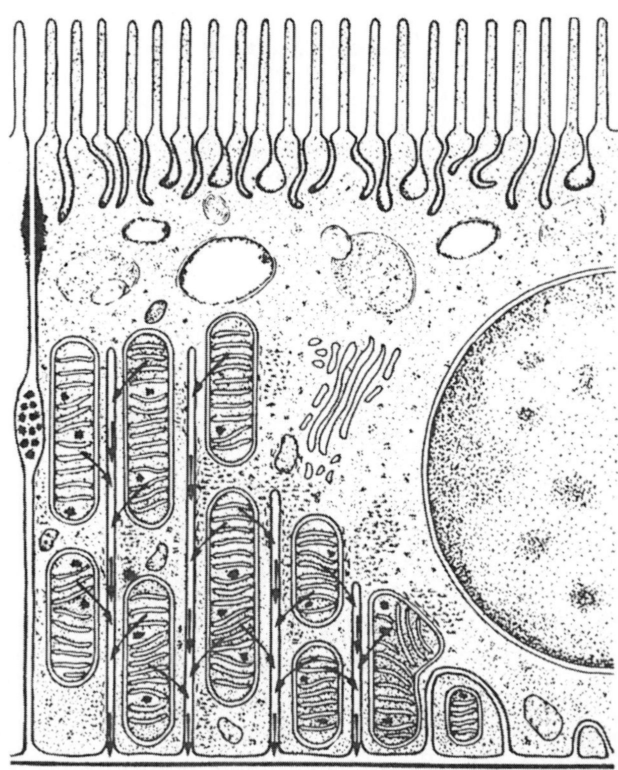

Abb. 2b. Modell zur Erklärung der isotonen Reabsorption auf die Verhältnisse im Tubulus übertragen. Nähere Erläuterung siehe im Text

Modellvorstellung könnte man folgendermaßen auf die Strukturen an der proximalen Tubuluszelle übertragen (Abb. 2b). NaCl wird aus der Zelle in die basalen Einfältelungen transportiert und macht die Flüssigkeit in diesen Kanälchen leicht hyperton. Wasser folgt: der hydrostatische Druck steigt etwas und drückt die Flüssigkeit zur Zellbasis hin. So könnte der Wasserfluß über eine relativ große Wegstrecke in der Zelle hydrodynamisch sein. Man kann auch plausibel erklären, daß ein solcher Vorgang nicht umkehrbar ist, da in einem solchen Falle der hydrostatische Druck „negativ" werden müßte, und dann die von den basalen Einfältelungen gebildeten Kanäle wegen ihrer nicht starren Wände kollabieren müßten.

Übrigens hatten Ruska [22] und später Thoenes [28] allein aufgrund morphologischer Aspekte die Funktion der basalen Einfältelungen in ähnlicher Weise gedeutet.

B. Zellvolumen und Transportrate

Verschiedene Untersucher haben gefunden [9, 11], daß bei Erhöhung der glomerulären Filtrationsrate, z. B. durch Infusion hypertoner Kochsalzlösung, der TF/P-Quotient für Inulin am Ende des proximalen Konvolutes unverändert ist. Das heißt: Die Rückresorption von NaCl und Wasser aus dem proximalen Konvolut ist proportional der Filtrationsrate angewachsen. Umgekehrt kann man aus Clearanceversuchen schließen, daß bei Verkleinerung der Filtrationsrate auch die proximale Rückresorption von NaCl und Wasser kleiner werden muß. Um die Aufklärung dieses Phänomens, der sog. glomerulo-tubulären Balance, haben sich mehrere Autoren bemüht [8, 16, 17].

GERTZ [8] konnte mit Hilfe verschiedener Methoden zeigen, daß die Natriumchlorid- und Wasserrückresorption aus dem proximalen Tubulus proportional dem Quadrat des Radius des Tubuluslumens ist. Da nun Veränderung der Filtrationsrate gleichlaufende Veränderung der Lumenweite auslöst und diese zu parallelen Veränderungen der NaCl und Wasserrückresorption führt, ist eine einfache Erklärung für die glomeruläre tubuläre Balance gegeben. Man darf demnach annehmen, daß Dehnung der Tubuluszellen und möglicherweise auch Entfaltung des Bürstensaumes oder der basalen Einfältelungen zu einem erhöhten transtubulären NaCl-Transport führt. In Experimenten an der Froschhaut fand USSING [31] etwas Ähnliches, nämlich, daß jede Zellschrumpfung, gleichzeitig ob durch Wasserausstrom allein oder durch Wasser- und Elektrolytausstrom bedingt, zu einer Verminderung des transzellulären Natriumtransportes führt, jede Zellschwellung dagegen zu einer Erhöhung. Es wäre bei diesen Befunden interessant zu wissen, ob bei diesen Veränderungen des Natriumtransportes bereits mit dem Elektronenmikroskop Veränderungen der Struktur darstellbar sind.

C. Vergleich der Permeabilität des proximalen und distalen Konvolutes

Ich möchte nun zu einem weiteren Punkt kommen, nämlich, daß nicht nur, wie bereits beschrieben, die Permeabilität von Wasser, sondern auch die von Mannitol [7], Harnstoff [2] und NaCl [30] im distalen Konvolut kleiner ist als im proximalen. Die Permeabilität von Mannitol hat GERTZ [7] aus der steady state Rückresorption einer in das Tubuluslumen injizierten Kochsalz-Mannitollösung errechnet. Wie die Tab. 2 zeigt, ist die Permeabilität des distalen Konvolutes für Mannitol nur 40% der des proximalen Konvolutes. Das gleiche Verhältnis haben wir bereits bei Vergleich der distalen und proximalen Wasserpermeabilität in Antidiurese gesehen, und etwa das gleiche Verhältnis liegt auch — worauf ich später noch eingehen werde — für die NaCl-Permeabilität vor. Diese Übereinstimmung der relativen Permeabilität verschiedener Substanzen ist überraschend und deutet wieder darauf hin, daß die verschieden große physiologische Permeabilität des proximalen und distalen Konvolutes nicht durch verschiedene Bauelemente, sondern einfach durch verschiedene räumliche Anordnung des gleichen Bauelementes bedingt ist. Allerdings paßt die aus Durchströmungsexperimenten mit radioaktiv markiertem Harnstoff errechnete Permeabilität für Harnstoff nicht so schön in dieses Schema. Die Harnstoffpermeabilität des distalen Konvolutes ist demnach 70% des proximalen. Dies ist überraschend, da man in Analogie zu Befunden, die an der Krötenblase [15] erhoben

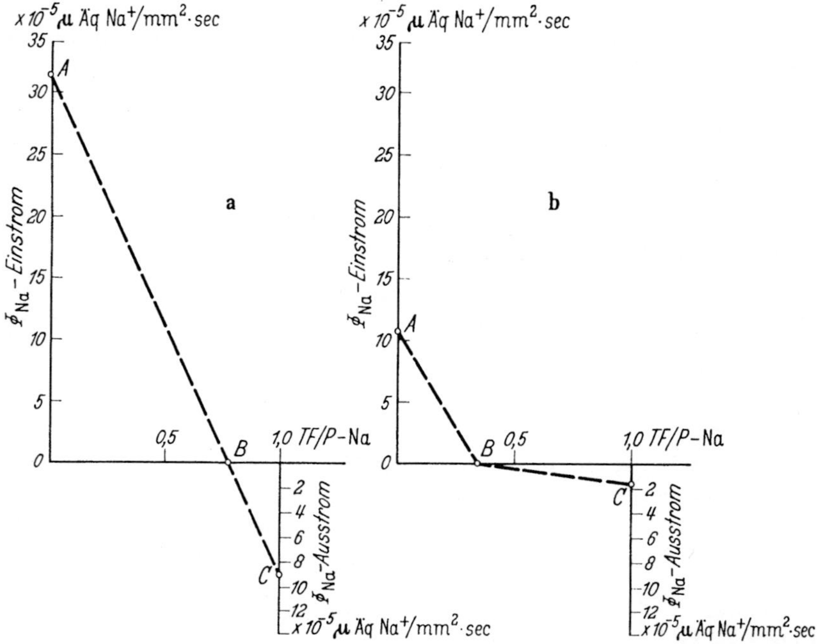

Abb. 3a u. b. Transtubuläre Natriumflüsse am proximalen a und distalen b Konvolut in Abhängigkeit von der Natriumkonzentration in der Tubulusflüssigkeit. A = Einstrom bei niedriger Na-Konzentration in der Tubulusflüssigkeit; B = Gleichgewichtskonzentration, bei der weder ein Einstrom noch ein Auswärtstransport stattfindet, also der Nettofluß Null ist. C = Auswärtstransport bei isotoner NaCl-Lösung im Tubuluslumen. TF/P = Na⁺ Konzentration in der Tubulusflüssigkeit bezogen auf die Na⁺-Konzentration im Plasma

wurden, gerade für Harnstoff ein enges Parallelgehen mit der Wasserpermeabilität erwartet hätte.

Der passive Ionentransport durch Membranen hängt von der Permeabilität und von dem elektrochemischen Potentialunterschied für die betreffende Ionenart ab. Beim aktiven Transport kommt eine fixe treibende Kraft (E_a) dazu [23, 32]. Ist nun eine diesem aktiven Transportpotential entgegengerichtete größere elektrochemische Potentialdifferenz vorhanden, so erfolgt ein Nettofluß der betreffenden Ionenart in umgekehrter Richtung. Durchströmt man z. B. den Tubulus mit einer isotonen Mannitollösung, so strömt NaCl in das Tubuluslumen hinein. Die Anfangsgeschwindigkeit des NaCl-Einstroms ist ein Maß für die NaCl-Permeabilität der Membran. Bei Durchströmung kurzer Tubulusstrecken wurde ein NaCl-Einstrom in das proximale Konvolut von $31{,}5 \cdot 10^{-5}$ μeq/mm²/sec, in das distale von nur 34% davon, nämlich von

$10,7 \cdot 10^{-5}$ μeq/mm²/sec gemessen (Punkt A in Abb. 3). Die geringere Permeabilität des distalen Tubulus hat zur Folge, daß die NaCl-Konzentration im Lumen viel tiefer herabgesetzt werden kann als im proximalen (Punkt B in Abb. 3), wahrscheinlich weil infolge der geringeren Permeabilität für NaCl der passive Rückfluß aus dem Interstitium ins Lumen geringer ist. Umgekehrt scheint auch der bei Durchströmung des Lumens mit isotoner Kochsalzlösung viel kleinere Austransport von NaCl aus dem distalen Tubulus im Vergleich zum proximalen (Punkt C in Abb. 3) mit der unterschiedlichen passiven Permeabilität für NaCl erklärbar zu sein. Schon vor mehreren Jahren hat LINDERHOLM [18] auf ähnliche

Tabelle 2. *Permeabilität (D/Δx) einzelner Tubulussegmente (· 10⁻⁴ mm/sec)*

	proximales Konvolut	distales Konvolut	distale Permeabilität in % der proximalen
Mannitol	1	0,4	40
Sulfamerazin pH 5	92		
pH 8	48		
Harnsäure pH 5	3		
pH 8	13		
Harnstoff	15	10.7	
Wasser	420	184 (Antidiurese)	44
NaCl	21,8	7,4	34

Verhältnisse an der Froschhaut aufmerksam gemacht und ein entsprechendes Modell nebst mathematischer Ableitung angegeben.

In der Tab. 2 sind die Permeabilitäten des proximalen Konvolutes für zwei weitere Substanzen angegeben [25, 26], nämlich für Sulfamerazin, einem gut lipoidlöslichen Sulfonamid, und für Harnsäure, die lipoidunlöslich ist. Sulfamerazin ist eine schwache Säure und liegt bei pH 5 zu 99% in der nichtionisierten Form vor, während bei pH 8 nur 9% nicht ionisiert sind. Die nichtionisierte Form ist in Lipoidlösungsmitteln viel stärker löslich als die ionisierte. Dies mag die Ursache dafür sein, daß Sulfamerazin bei pH 5 viel rascher aus dem Tubulus hinausdiffundiert als bei pH 8. Eine solche "non ionic diffusion" scheint es bei nicht lipoidlöslichen Stoffen nicht zu geben. Die Permeabilität des proximalen Konvolutes für Harnsäure ist an und für sich schon gering. Die nichtionisierte Form bei pH 5 diffundiert kaum aus dem Lumen heraus. Wider Erwarten ist aber die Auswärtsdiffusion der bei pH 8 überwiegenden ionisierten Form größer. Eine Interpretation dafür ist nicht leicht zu geben, doch ist es möglich, daß aktive Transportprozesse oder Veränderungen der Zellmembran bei der Durchströmung mit verschiedenem pH mitspielen. Das unterschiedliche Verhalten der lipoidlöslichen Stoffe steht am besten im Einklang mit der Vorstellung von TOSTESON (s. S. 67), wonach die Zellmembran aus 200 Å langen Plättchen von Lipoiddoppelmembranen zusammengesetzt ist. Lipoidlösliche Stoffe könnten dann durch die ganze Fläche, wasserlösliche Stoffe dagegen nur durch die dazwischenliegenden Lücken hindurchtreten.

Literatur

[1] Auricchio, G., and E. H. Bárány: Acta physiol. scand. 45, 190 (1959).
[2] Capek, K., G. Rumrich u. K. J. Ullrich: Pflügers Arch. ges. Physiol. (im Druck).
[3] Curran, P. F., and J. R. Macintosh: Nature (Lond.) 193, 347 (1962).
[4] —, and A. K. Solomon: J. gen. Physiol. 41, 143 (1957).
[5] Diamond, J. M.: J. Physiol. (Lond.) 161, 503 (1962).
[6] Farquhar, M. J., and G. E. Palade: J. Cell Biol. 17, 375 (1963).
[7] Gertz, K. H.: Pflügers Arch. ges. Physiol. 276, 336 (1963).
[8] — Habil.-Schrift. Freie Universität Berlin 1964.
[9] Giebisch, G., R. M. Klose, and E. E. Windhager: Amer. J. Physiol. 206, 687 (1964).
[10] Gottschalk, C. W., and M. Mylle: Amer. J. Physiol. 196, 927 (1959).
[11] — Harvey Lect. series 58, 99 (1963).
[12] Hasselbach, W.: persönl. Mitteilung.
[13] Huxley, A. F.: Internat. Biophysics Meeting. Paris-Orsay, June 1964, B III 2.
[14] Kashgarian, M., H. Stöckle, C. W. Gottschalk, and K. J. Ullrich: Pflügers Arch. ges. Physiol. 277, 89 (1963).
[15] Leaf, A., and R. M. Hays: J. gen. Physiol. 45, 921 (1962).
[16] Kelman, R. B.: Bull. Math. Biophys. 24, 303 (1962).
[17] Kruhoffer, P.: Handb. d. experim. Pharmakologie, Ergänzungswerk Bd. 13 p. 233. Ed. O. Eichler, and A. Farah. Berlin-Göttingen-Heidelberg: Springer-Verlag 1960.
[18] Linderholm, H.: Nierensymposion Göttingen 1959. Ed. K. Kramer u. K.J. Ullrich. S. 1. Stuttgart: Thieme-Verlag 1960.
[19] Peachey, L. D.: Internat. Biophysics Meeting. Paris-Orsay, Juni 1964, B III 1.
[20] Renkin, E. M., u. J. R. Pappenheimer: Ergebn. Physiol. 49, 59 (1957).
[21] Rhodin, J. A. G.: Glom. und tub. Nierenerkrankungen. Ed. E. Wollheim, H. E. Schäfer u. A. Heidland. p. 92. Stuttgart: Thieme-Verlag 1962.
[22] Ruska, H., D. H. Moore u. J. Weinstock: J. biophys. biochem. Cytol. 3, 249 (1957).
[23] Schlögl, R.: Fortschr. physik. Chemie 9. Darmstadt: Steinkopff-Verlag 1964.
[24] Sonnenberg, H., u. P. Deetjen: Pflügers Arch. ges. Physiol. 278, 669 (1964).
[25] — Dissertation. Freie Universität Berlin, 1964.
[26] —, H. Oelert, and K. Baumann: (in Vorbereitung).
[27] Thoenes, W.: Klin. Wschr. 39, 504 (1961).
[28] — Diskussion, S. 110. In: Glomeruläre und tubuläre Nierenerkrankungen. Ed. E. Wollheim, H. E. Schäfer u. A. Heidland. Stuttgart: Thieme-Verlag 1962.
[29] Ullrich, K. J., G. Rumrich u. G. Fuchs: Pflügers Archiv ges. Physiol. 280, 99 (1964).
[30] Eigene unveröffentlichte Befunde.
[31] Ussing, H. H.: Internat. Biophys. Meeting Paris-Orsay, Juni 1964, B V 3.
[32] — The alkali metal ions in biology. Handb. d. experim. Pharmakologie. Ergänzungswerk Bd. 13 Ed. O. Eichler, and A. Farah. p. 122. Berlin-Göttingen-Heidelberg: Springer-Verlag 1960.
[33] Walker, A. M., P. A. Bott, J. Oliver, and M. C. Macdowell: Amer. J. Physiol. 134, 580 (1941).
[34] Whittembury, G., D. E. Oken, E. E. Windhager, and A. K. Solomon: Amer. J. Physiol. 197, 1121 (1959).
[35] Windhager, E. E., G. Whittembury, D. E. Oken, H. J. Schatzmann, and A. K. Solomon: Amer. J. Physiol. 197, 313 (1959).
[36] —, and G. Giebisch: Nature (Lond.) 191, 1205 (1961).
[37] Wirz, H.: Helv. physiol. pharmacol. Acta 14, 353 (1956).

Summary

Perfusion of cortical nephron-segments of rats with various solutions, using a microperfusion pump, gave the following results:

1. In antidiuresis the water permeability of the proximal convolution is $2^1/_2$ times greater than that of the distal convolution (Table 1). Water diuresis does not alter water permeability in the proximal convolution, but greatly reduces it in the distal convolution, during maximal diuresis to as little as $^1/_{10}$ of the value during antidiuresis. NaCl reabsorption from the proximal tubule is accomplished isotonically with efflux of water; but the observed influx of water, during perfusion of the proximal convolution with isotonic mannitol solution, although accompanied by NaCl influx, is not isotonic but occurs under the influence of a readily measured osmotic pressure gradient. This discrepancy can be explained with the aid of a model supplied by CURRAN and McINTOSH (Fig. 2): NaCl is transported out of the cells into the basal infoldings rendering the fluid in these interspaces hypertonic. Water follows, the hydrostatic pressure increases somewhat and moves the fluid to the basal cell border. In this manner the movement of water could be accomplished hydrodynamically over much of the pathway through the cell. It is likely that such a mode of transport would not be reversible. A reversible mode would require a "negative" hydrostatic pressure; and then the interspaces which are formed by infoldings of the basal border, would collapse because their walls are not rigid.

2. Table 2 lists the permeability of the proximal and distal convolutions to various substances. The permeability of the distal convolution to mannitol, water and NaCl[1] is 34—44 % that of the proximal convolution. This similarity of the relative permeabilities favours the view that the cell membranes of proximal and distal convolutions contain similar constituents in different spatial arrangements, rather than different constituents.

3. The wall of the proximal tubule is particularly permeable to fat-soluble substances like Sulfamerazine (Table 2). At pH 5, 99% of this substance is in the non-ionized form, at pH 8 only 9%. That the non ionic form diffuses more readily through the wall of the tubule is readily explained by its greater solubility in the lipoid phase of the cell wall, as in lipoid solvents. This situation is reversed in the case of fat-insoluble substances like uric acid, where the cell wall is more permeable to the ionized form. It may be that the negatively charged uric acid passes more easily through positively charged pores, but it is also possible that active transport processes are implicated or that the cell membranes are affected by the varying H^+ concentrations of the perfusates. Further experiments are required to establish whether these interpretations of the findings with a limited number of compounds can be generalized.

Diskussion

Diamond: Was den Mechanismus des Wassertransports anbetrifft, so hat Herr ULLRICH zwischen zwei Theorien zu entscheiden versucht, und zwar zwischen der Kodiffusion und der Doppelmembrantheorie. In der Zwischenzeit hat es sich herausgestellt, daß beide Theorien falsch sind. In der Niere liegt das Problem vor, daß Wasser isotonisch reabsorbiert wird, ohne daß ein osmotischer Gradient vorhanden ist. Das hat Herr ULLRICH bewiesen. Das ist auch der Fall in der Gallenblase, im Dünndarm, in der Leber, im Magen und noch in einer Reihe anderer Organe.

Um den Vorgang der Wasserreabsorption ohne das Vorhandensein eines transmuralen osmotischen Gradienten zu erklären, kann man annehmen, daß dieser Gradient nicht zwischen den beiden äußeren Flüssigkeiten liegt, sondern innerhalb der Zellen. Man könnte sich bei der Niere vorstellen, daß Natriumchlorid durch die Zellmembran transportiert wird und dann in einem kleinen Gebiet einen höheren osmotischen Druck aufbaut. Durch diese lokale osmotische Differenz wird das Wasser osmotisch nachgezogen. Das ist die eine Möglichkeit.

Es gibt eine Reihe anderer Theorien auf nicht osmotischer Basis. Dazu gehören beispielsweise die Kodiffusion und die Doppelmembranlehre. In der Gallenblase habe ich versucht, zwischen diesen beiden Möglichkeiten zu unterscheiden, und zwar

---

[1] Measured as NaCl influx into an isotonic mannitol solution not corrected for electrical potential-differences.

auf folgende Weise: Nach den nichtosmotischen Theorien mußte man annehmen, daß die Struktur der Membran so gebaut ist, daß die absorbierte Lösung dieselbe Osmolarität besitzt wie das Plasma des Tieres. Ich habe eine Gallenblase mit verschiedenen Lösungen gefüllt, die nicht mit dem Plasma des Kaninchens isotonisch waren, also nicht 300 mosmol besaßen, sondern verschiedene Osmolaritäten zwischen 50 mosmol und 600 mosmol. Wenn die richtige Erklärung des Wassertransports Lokalosmose wäre, dann würde man erwarten, daß die absorbierte Lösung immer isotonisch der im Lumen der Gallenblase vorhandenen Lösung sein müßte. Wenn aber eine andere Theorie stimmte, dann würde es sich anders verhalten. Beispielsweise bei der Kodiffusion würde man erwarten, daß bei 50 mosmol im Lumen die reabsorbierte Lösung immer noch näher an 300 mosmol bleibt. Umgekehrt, wenn man 600 mosmol in die Gallenblase einspritzt, sollte man nicht 600 herausbekommen, sondern etwas näher an 300. Das gleiche gilt auch für die Doppelmembrantheorie. Das heißt, die absorbierte Lösung darf nur isotonisch sein, wenn das Organ mit einer Lösung gefüllt wird, die dem Plasma des Tieres isotonisch ist.

Das Ergebnis zeigte nun, daß die reabsorbierte Lösung immer isotonisch der Lösung in der Gallenblase war — innerhalb einer Fehlergrenze von 2%. Das heißt, bei 600 mosmol werden reabsorbiert 600 ± 2%, nicht 300. Und bei 50 mosmol im Lumen der Gallenblase werden 50 mosmol reabsorbiert. Das beweist, daß diese anderen Theorien in der Gallenblase nicht stimmen können und daß die Theorie der lokalen Osmose richtig ist. Es ist wahrscheinlich, aber es bleibt noch zu beweisen, daß lokale Osmose auch die Erklärung des Wassertransportes in anderen Organen ist, wie den Nieren, dem Dünndarm usw., wo Wasser isotonisch absorbiert wird.

Ullrich: Die Nierenphysiologen hinken immer etwas hinter den Membranphysiologen, die an besser zugänglichen Gebilden wie Gallenblase, Froschhaut, Krötenblase oder Darm arbeiten, hinterher. Sie sind froh, wenn sie mit relativ einfachen Experimenten nachweisen können, daß ein Vorgang in der Niere genau so abläuft wie an diesen Organen. So glaube ich, daß unsere Befunde über die transtubulären Wasserflüsse am proximalen Tubulus mit Ihrem Konzept der lokalen Osmose erklärbar sind.

Solomon: I just wanted to point out that we have often discussed, Dr. Diamond and I, this business of nomenclature. If you consider the double membrane model of Curran and MacIntosh and make the requirement that the membrane which was a sintered glass membrane, have a reflection coefficient of zero which is, indeed, true in that model, then this is the same as the local osmosis that Dr. Diamond has, so that we argue very little about the basic biophysics of the process, but there is still some discussion as to what the names are.

Diamond: I shall show you the difference between the double membrane theory and local osmosis. Under the term 'double membrane theory' — I understand the complete model of Patlak, Goldstein and Hoffmann which, in its most general case, pictures two membranes with different reflection coefficients. These two theories merge into each other in the borderline case where one of the membranes has a reflection coefficient of one, that is to say, it is completely impermeable to solute, and the other has a reflection coefficient of zero, that is to say it is freely permeable to solute. I should, nevertheless, like to distinguish between these two terms, local osmosis and the double membrane effect, for two reasons. First of all: physiologically, one can show in the gall bladder that if there are two membranes, they do have reflection coefficients of 1 and zero, so we do not have the more general double membrane case, and mathematically it is far more easy to handle the algebra of local osmosis than the extremely complicated mathematics which Patlak, Goldstein and Hoffmann had to go through to treat the general case. Secondly, I think we should reserve the term 'double membrane effect' for the more general case where the reflection coefficients are not 1 and zero, because this general case may still prove to be significant in biology in explaining osmotic rectification. As far as the anatomy is concerned, in the gall bladder we are certainly dealing with one membrane, not two membranes or three membranes or four membranes. Although a double membrane model would fit the results, if the second membrane in practice did not exist, that is to say, had a reflection coefficient of zero, still that is no advantage to say that we have got the 'double membrane effect'. However, I agree with Dr. Solomon completely what the physical facts are and the only question is, what is the more useful term for the situation.

Tosteson: While we are on the subject of terminology I would like to make one brief comment about the use of the term, E_{Na}. That term has been used by LINDERHOLM and USSING in one sense and in quite another sense by the neurophysiologists and muscle physiologists. In the latter case of excitable membranes the term E_{Na} is used to indicate the equilibrium potential for the sodium ion and for that reason I wondered whether we might not better choose some other term for this LINDERHOLM E_{Na}, just because it is confusing to people who are not in the field. I think that the formal difference in the two definitions is that you include the active transport potential in the definition of E_{Na}, but in the sense that the term is used in treating the physiology of excitable membranes, that is treated separately, so that they are certainly both the potential where there is no net movement of sodium, but in your case and in LINDERHOLM's case and USSING's case there is no net movement because of the existence of the active transport which balances the diffusion of the sodium, whereas in the case of the nerve and muscle membranes, this active transport term is not included.

Ullrich: Of course, you are right. E_{Na} in LINDERHOLM and USSING sense is in the equilibrium state with zero netflux $E_{Na} = \dfrac{RT}{\pm\, zF} \cdot \ln \dfrac{Na_i}{Na_a} + E$ (E = trans membran-electrical potential) and as the neurophysiologists use it:

$$E_{Na} = \frac{RT}{\pm\, zF} \cdot \ln \frac{Na_i}{Na_a}.$$

Kuyper: In der ersten Tabelle haben Sie eine Reihe von Zahlen über die Wasserpermeabilität gegeben. Wenn die Membran nun keine Widerstände hätte, wie groß sollten dann die Zahlen sein? Das heißt, wie groß ist die Wasserpermeabilität in Prozenten freier Diffusion?

Ullrich: Um einen vergleichbaren Wert angeben zu können, müßte das Konzentrationsprofil durch die Zelle bekannt sein. Da das nicht der Fall ist, können sämtliche Permeabilitätswerte, die die ganze Zelle betreffen, nicht mit freier Diffusion verglichen werden. Ihre Frage wäre nur dann zu beantworten, wenn wir die Permeabilität mit markiertem Wasser gemessen hätten und diesen Wert mit dem Selbstdiffusionskoeffizienten für Wasser vergleichen würden.

Komnick: Herr Prof. ULLRICH, wenn ich Sie richtig verstanden habe, dann stellen Sie sich die Ableitung des Resorbats aus den basalen Interzellularspalten in Form eines freien Wasserflusses vor. Wenn wir uns einmal diese Strukturen als starre Röhren oder Spalten vorstellen, möchte ich Sie fragen: Wie hoch muß der lokale osmotische Gradient bzw. letztlich der hydrostatische Druck sein, um in diesen engen Schlitzen einen freien Wasserfluß zu erzeugen?

Ullrich: Wenn die im elektronenmikroskopischen Bild gesehenen Dimensionen der basalen Spalträume mit denen in vivo vorhandenen übereinstimmen, können wir bestenfalls die Lineargeschwindigkeit des Wassers in diesen Spalträumen berechnen. Die hydrostatische Druckdifferenz könnte man nur dann mit einer modifizierten Hagen-Poiseuilleschen Gleichung berechnen, wenn die Lineargeschwindigkeit der Wassermoleküle an der Wand gleich Null ist, und di es ist sicher nicht der Fall.

Schmidt: Bisher war der Histologe der Meinung, daß in der Niere nicht alle Nephrone gleichzeitig funktionieren. Es gibt morphologische Bilder, die dafür sprechen, daß Sperrmechanismen vorhanden sind. Haben Sie so etwas je beobachtet?

Ullrich: Nein, die Tubuli sind in einer normalen Niere immer gleichmäßig durchströmt. Nur in einem schlechten Präparat und nach längerer Versuchsdauer wird die Durchströmung ungleichmäßig. Es gibt einen guten Test, um die gleichmäßige Funktion der Tubuli zu überprüfen, nämlich die Lissamingrüninjektion von STEINHAUSEN.

Kurz nachdem Lissamingrün in eine Vene injiziert wurde, sieht man die Niere ganz kurz diffus grün werden. In diesem Augenblick läuft das die Farbe enthaltende Blut durch die Niere. Kurz darauf sieht man den grünen Farbstoff in den proximalen Tubuli. Schließlich verschwindet die Farbsäule in den Henleschen Schleifen und kommt stark konzentriert für etwa 20 Sekunden im distalen Konvolut wieder. Die Steinhausensche Lissamingrünpassage wird von allen Mikropunkteuren öfter während eines Versuches durchgeführt, um den Zustand der Niere zu prüfen.

Schlußwort

G. Bruns, Jena

Meine sehr verehrten Damen und Herren!

Wir sind am Ende des wissenschaftlichen Teils unserer Konferenz. Im Namen der Gesellschaft darf ich zum Abschluß einige Worte an Sie richten.

Wenn wir uns in die Zeit des Don Carlos zurückversetzen könnten, würde ich sagen: „Die schönen Tage von Aranjuez sind nun vorüber." Wir bedauern das sehr. Unser Symposion hat etwas aufgegriffen, was man bei Ihnen "to travel and to talk" nennt. Das Miteinander-Sprechen haben wir besorgt.

Hier müssen wir uns der Gründung unserer Gesellschaft durch Lorenz Oken erinnern, der am 22.9.1822 in Leipzig *deutsche* Gelehrte zu einer Gesellschaft *Deutscher* Naturforscher und Ärzte zusammenführte mit der Aufgabe, sich in dem damals sehr zersplitterten Land zu vereinigen und bei offenen Türen miteinander zu sprechen. Der Kreis um Oken hat hier im thüringischen Raum das Pressegesetz im Großherzogtum Sachsen-Weimar durchgesetzt und die Freiheit der Meinung und Gedanken gefordert.

In der ganzen Welt wird die öffentliche Meinung von Ideologien beherrscht. Sie ist institutionalisiert. Eine geistige Koexistenz soll es nicht geben. Uns hat aber ein eminent wichtiges soziologisches Problem zusammengeführt, *das* Humanum schlechthin, dem wir uns verbunden fühlen. Dabei denke ich besonders an die Versammlung in Weimar. Sie ermöglichte eine Begegnung zwischen den Gelehrten und der Deutschen untereinander. Das wird das bedeutende geschichtliche Ereignis der 103. Versammlung bleiben.

Wollen Sie bitte ermessen, was dieses Symposion gerade für unsere jungen Leute bedeuten muß. Sie können nicht zu den Hokins oder zu Dr. Solomon fahren, wie sie gern möchten. Auch hier hat das Symposion seine Wirkung gehabt.

Meine Damen und Herren! In der ganzen Welt hängen in den Straßen wie hier Transparente. Ein Freund von uns hat gesagt: „Losungen sind keine Lösungen". Was wir brauchen, ist die Begegnung und der Austausch zwischen den Menschen, auch im Geistigen. Gestatten Sie mir, daß ich einmal Karl Marx zitiere. Er wird oft strapaziert und wohl auch mißverstanden. Er hat aber gesagt: Laßt uns das Denken von der Ideologie befreien! Ich denke, in diesem Zusammenhang können wir von unserem Symposion sagen: Quod erat demonstrandum.

Unsere guten Wünsche begleiten Sie in die Heimat. Wir danken Ihnen allen. Wir danken den deutschen Stenographen wie auch den Damen Tennyson und Ward für ihre große Hilfe. Ich möchte einen Wunsch aussprechen: Wenn wir Sie einmal mehr rufen sollten, dann kommen Sie bitte wieder, eingedenk meiner Abschiedsworte.

Nunquam retrorsum.